Selected Titles in This Series

For a complete list of titles in this series, visit the
AMS Bookstore at **www.ams.org/bookstore/**.

Discrete Mathematical Chemistry

DIMACS
Series in Discrete Mathematics
and Theoretical Computer Science

Volume 51

Discrete Mathematical Chemistry

DIMACS Workshop
Discrete Mathematical Chemistry
March 23–24, 1998
Rutgers University

Pierre Hansen
Patrick Fowler
Maolin Zheng
Editors

NSF Science and Technology Center
in Discrete Mathematics and Theoretical Computer Science
A consortium of Rutgers University, Princeton University,
AT&T Labs–Research, Bell Labs (Lucent Technologies),
Telcordia Technologies, and NEC Research Institute

American Mathematical Society

This DIMACS volume contains papers from a DIMACS workshop on Discrete Mathematical Chemistry, held March 23–24, 1998.

2000 *Mathematics Subject Classification.* Primary 05C50; Secondary 92E10.

Library of Congress Cataloging-in-Publication Data

Discrete mathematical chemistry : DIMACS workshop, discrete mathematical chemistry, March 23–24, 1998, Rutgers University / Pierre Hansen, Patrick Fowler, Maolin Zheng, editors.
p. cm. — (DIMACS series in discrete mathematics and theoretical computer science, ISSN 1052-1798 ; v. 51)
Includes bibliographical references.
ISBN 0-8218-0987-3 (alk. paper)
1. Chemistry—Mathematics—Congresses. I. Hansen, P. (Pierre) II. Fowler, P. W. III. Zheng, Maolin, 1955– IV. Series.
QD39.3.M3 D57 2000
540′.1′51—dc21 99-059470

∞ The paper used in this book is acid-free and falls within the guidelines
established to ensure permanence and durability.
Visit the AMS home page at URL: http://www.ams.org/

10 9 8 7 6 5 4 3 2 1   05 04 03 02 01 00

Contents

Photo courtesy of Professor Milan Randić

Participants in the Workshop

1. M. Zheng
2. R.E. Valdez-Perez
3. P. Vismara
4. K. Rogers
5. R. Aringhieri
6. A. Lawrenz
7. B.T. Fan
8. S.C. Basak
9. L. Quintas
10. A.V. Zeigarnik
11. H. Terrones
12. S. Fajtlowicz
13. H. Sachs
14. Unidentified
15. P.W. Fowler
16. G. Caporossi
17. J.M. Wills
18. P. Hansen
19. P.G. Mezey
20. G. Brinkmann
21. K. Cameron
22. O. Delgado-Friedrichs
23. T. Pisanski
24. M. Boisen
25. L.L. Beverly
26. V. Janota
27. P. Pancoska
28. S. Klavzar
29. J.L.R. Chandler
30. R.B. King
31. J. Malkevitch
32. A. Dietz
33. S. Elk
34. C.M. Quinn
35. F. Zhang
36. D. Huson
37. X. Guo
38. P. Rowlinson
39. A. Kerber
40. M. Randić
41. K. Mislow
42. S. Hyde
43. D. Cvetković
44. B. Gruber
45. T. Harmuth
46. E.K. Lloyd

Not on the picture:

H. Abeledo
G.W. Atkinson
K. Collins
E.C. Constable
S. Divakaran
A. Dress
J.-E. Dubois
M.L. Gargano
R. Hefferlin
H.M. Ohlenbusch
D. Rockmore
T. Shinbrot
W.L. Smith
K.K. Tysdal

Contributing Authors

H. Abeledo
The George Washington University
abeledo@seas.gwu.edu

T. Aste
LDFC, Universite Louis Pasteur
aste@ldfc.u-strasbg.fr

G. Atkinson
Lucent Technologies
atkinson@bell-labs.com

S.C. Basak
Natural Resources Research Institute
sbasak@wyle.nrri.umn.edu

S. Bobst
Drake University
smb009@drake.edu

G. Brinkmann
Universität Bielefeld
gunnar@mathematik.uni-bielefeld.de

L. Bytautas
Texas A&M University
bytautal@arctic.tamug.tamu.edu

G. Caporossi
École Polytechnique de Montréal
gillesc@crt.umontreal.ca

D. Cvetković
University of Belgrade
ECVETKOD@etf.bg.ac.yu

O. Delgado Friedrichs
Universität Bielefeld
delgado@Mathematik.Uni-Bielefeld.de

M. Deza
DMI-LIENS, Ecole Normale Superieure
deza@dmi.ens.fr

A. Dietz
Université Montpellier II
dietz@lirmm.fr

J.P. Doucet
Université Paris 7
doucet@paris7.jussieu.fr

B. Dubertret
LDFC, Universite Louis Pasteur
benoit@ldfc.u-strasbg.fr

B.T. Fan
Université Paris 7
fan@paris7.jussieu.fr

C. Fiorio
Université Montpellier II
fiorio@lirmm.fr

P.W. Fowler
University of Exeter
P.W.Fowler@exeter.ac.uk

A. Graovac
Rugjer Bošković Institute
Graovac@rudjer.irb.hr

X. Guo
Xinjiang University
xfguo@xju.edu.cn

B.D. Gute
Natural Resources Research Institute
bgute@wyle.nrri.umn.edu

M. Habib
Université Montpellier II
habib@lirmm.fr

P. Hansen
École des Hautes Études Commerciales
pierreh@crt.umontreal.ca

R. Hefferlin
Southern Adventist University U.
hefferln@southern.edu

S. Hyde
Australian National University
stephen.hyde@anu.edu.au

P.E. John
Technical University of Ilmenau
Peter.John@Mathematik.TU-Ilmenau.DE

A. Kerber
University of Bayreuth
kerber@uni-bayreuth.de

R.B. King
University of Georgia
rbking@sunchem.chem.uga.edu

S. Klavžar
University of Maribor
sandi.klavzar@uni-lj.si

D.J. Klein
Texas A&M University
kleind@tamug.tamu.edu

R. Laue
University of Bayreuth
laue@uni-bayreuth.de

C. Laurenço
UPR 9023 CNRS — CCIPE
cl@lirmm.fr

H. Li
Zhengzhou University
huaienli1@263.net

J. Malkevitch
York College (CUNY)
joeyc@cunyvm.cuny.edu

P.G. Mezey
University of Saskatchewan
mezey@sask.usask.ca

H.M. Ohlenbusch
Worcester Polytechnic Institute
helgo@cs.wpi.edu

E. Osawa
Toyohashi University of Technology
osawa@cochem.tutkie.tut.ac.jp

A. Panaye
Université Paris 7
panaye@paris7.jussieu.fr

T. Pisanski
University of Ljubljana
Tomaz.Pisanski@fmf.uni-lj.si

C.M. Quinn
The National University of Ireland Maynooth
cm.quinn@may.ie

S. Ramsden
Australian National University
stuart.ramsden@anu.edu.au

M. Randić
Drake University
milan.randic@drake.edu

D.B. Redmond
The National University of Ireland Maynooth
dredmond@maths.may.ie

N. Rivier
LDFC, Universite Louis Pasteur
nick@ldfc.u-strasbg.fr

P. Rowlinson
University of Stirling
p.rowlinson@stir.ac.uk

H. Sachs
Technical University of Ilmenau
Horst.Sachs@Mathematik.TU-Ilmenau.DE

H. Terrones
Instituto de Física, UNAM
terrones@fenix.ifisicacu.unam.mx

M. Terrones
University of Sussex
kapa4@sussex.ac.uk

P. Vismara
ENSA-M
vismara@ensam.inra.fr

T. Wieland
Deutsches Zentrum für Luft- und Raumfahrt
thomas.wieland@dlr.de

J.M. Wills
Universität Siegen
wills@mathematik.uni-siegen.de

J.H. Yao
Université Paris 7
yao@paris7.jussieu.fr

M. Yoshida
Exploratory Chemistry Research Laboratories
mitsuh@shina.sankyo.co.jp

S.G. Yuan
Institute of Organic Chemistry of Shanghai
yuansg@pub.sioc.ac.cn

A.V. Zeigarnik
Lomonosov Academy of Fine Chemical Technology
azeigarn@mitht.rssi.ru

J. Žerovnik
University of Maribor
Janez.Zerovnik@fmf.uni-lj.si

F. Zhang
Xiamen University
fjzhang@jingxian.xmu.edu.cn

M. Zheng
Lexis-Nexis
Maolin.Zheng@lexis-nexis.com

Foreword

The DIMACS Workshop on Discrete Mathematical Chemistry was held on March 23-25, 1998 at Rutgers University. We would like to express our appreciation to Pierre Hansen, Patrick Fowler, and Maolin Zheng for their efforts to organize and plan this successful workshop.

The aim of the workshop was to bring together chemists working in the mathematical chemistry field; mathematicians and computer scientists working in that field; mathematicians and computer scientists who do not yet work on chemical problems but whose research topic and/or recent results hold much potential for solving problems in chemistry; and chemists, mathematicians and computer scientists working in the chemical industry. The workshop participants discussed theoretical problems of chemistry expressed by discrete mathematics; chemical graph algorithms; coding theory applied to chemistry; applications of discrete mathematics in the chemical industry; open problems and directions for future research in discrete mathematical chemistry; and software for discrete mathematical chemistry.

DIMACS gratefully acknowledges the generous support that makes these programs possible. The National Science Foundation, through its Science and Technology Centers program, the New Jersey Commission on Science and Technology, and DIMACS's partners at Rutgers, Princeton, AT&T Labs-Research, Bell Labs, NEC Research Institute, and Telcordia Technologies generously supported this workshop.

Fred S. Roberts
Director

Robert Sedgewick
Co-Director for Princeton

Preface

It can be argued that chemistry, of all the sciences, has the strongest link with discrete mathematics and the most to gain from closer contact with it. There are natural mappings at many levels from the abstract objects of mathematical theory to the materials and molecules in the test tube. In the early days of graph theory (itself a term derived from chemistry), isomerism supplied the motivation for the development of enumeration techniques, and chemists have been working ever since on problems that demand discrete mathematical formulation. As quantum mechanics began to be applied in chemistry, graph theory gave simple pictures of electronic structure and group theory became the basic language of spectroscopy. With known compounds numbered in millions, and the rise of combinatorial chemistry, synthetic chemists look to mathematics to supply complete schemes of nomenclature and flexible coding to handle reaction systems and structure-activity relationships involving thousands of components. Recently, mathematicians have gained something in return from chemistry, in the shape of the fullerenes, which gave a fillip to investigations on polyhedra, graphs, symmetry and optimisation theory, in addition to their huge impact in chemistry, physics and materials science. Many of the papers in this volume reflect the combined mathematical and physical interest of these new molecules.

Interest in the interface between chemistry and discrete mathematics is now perhaps greater than ever before, and judge from the number of meetings, new journals and research collaborations, has a healthy future. The present volume is a record of the first DIMACS meeting on 'Discrete Mathematical Chemistry' held at Rutgers University in March 1998. The meeting was made possible by financial support from DIMACS, GERAD (Montréal), École des Hautes Études Commerciales (Montréal) and the EU TMR contract FMRX-CT97-0126 on Usable Fullerene Derivatives, for which the organisers are profoundly grateful. 59 delegates attended. The contributed peer-reviewed papers amplify the lecture presentations and reflect the breadth of current work, from the Benzenoid Clar problem to the Wulff-shape of sphere packings. We thank the referees for their careful analysis of all papers in this volume. We are also grateful to GERAD's secretarial team for help in processing papers and getting them into final form.

Our aim in convening the meeting was to highlight the range of opportunities for fruitful and informed collaboration across the subject boundary, and to show that 'real' chemistry and 'real' mathematics have something to say to each other. Completion of this volume of interdisciplinary papers is an encouraging start, but we hope that the many collaborations initiated in the DIMACS tradition of lively discussion before, during and after the lecture sessions will continue in the years to come.

In the week before the conference we were shocked to hear of the sudden and premature passing of Dr. Mitsutaka Fujita, an invited speaker and a friend of many of the delgates. The proceedings volume is dedicated to his memory. Professor Osawa and Dr. Yoshida agreed to provide a fitting tribute to him in the form of a note on their colleague's contributions to fullerene and nanotube science, which is printed at the outset of this volume.

Pierre Hansen (Montréal), Patrick Fowler (Exeter), Maolin Zheng (Dayton)
September, 1999.

Professor Mitsutaka Fujita

Mitsutaka Fujita, 1959 – 1998

It is our sad task to report the sudden passing of Professor M. Fujita from a sub-arachnoid haemorrhage in Tsukuba Medical Center on March 18, 1998, at the early age of 38. It emerged later that he had suffered from chronic high blood pressure, but this was known to few as this active and cheerful man was not a person to complain about his health problems. His death occurred only a few days before the start of the DIMACS Conference on Discrete Mathematics, at which he was to be an invited speaker. We wish to fill his place in the Proceedings volume with this brief record of his short but brilliant scientific career.

Mitsutaka Fujita was educated in Kyoto University as a physicist. After obtaining his Dr. Sci. degree by a thesis on theoretical physics in 1986, he won an open competition for an assistant position in the Department of Material Science, University of Tsukuba. He quickly rose to lecturer in 1989, and to assistant professor in 1995. His early research interests were directed to the electronic states of magnetic materials including superconductors [**1-25**]. His interest soon turned to the emerging field of fullerene science when he joined the noted research group of physics professor Mildred Dresselhaus at the Massachusetts Institute of Technology during 1991-1992 [**26**]. Together with M. and G. Dresselhaus and R. Saito, he wrote an early paper on the application of honeycomb net diagrams to the analysis of topological properties of fullerenes [**27**]. Clearly at MIT he was in the right place at the right time to participate in the theoretical analysis of the once-in-a-lifetime discovery of the new forms of carbon [**28, 29**]. Furthermore, he could see how to apply the powerful technique of 2D projection to correlate electronic states and 3D topology of carbon networks.

By coincidence, we had been working independently on a net algorithm for automatic generation and drawing of fullerene structures [M. Yoshida, E. Osawa, Bull. Chem. Soc. Jpn. 1995, 68, 2073], and were starting to apply the algorithm to structure enumeration, reactivity problems, and growth mechanism of fullerenes when Dr. Fujita returned from the US [**30-34**]. Henceforth we worked closely together. This time it was our good fortune to meet the right man at the right time. The list of our joint publications includes enumeration of fullerene isomers, drawing net diagrams of fullerenes containing heptagonal and octagonal rings, and extension of the concept of the soliton to fullerenes (in the form of the phason) [**35-37, 39-40**]. The last mentioned was Dr. Fujita's own invention [**39**] and at first as chemists we hesitated to propose a term that sounded like a new elementary particle, but the phason line proved to be a novel measure for estimating the relative stabilities for fullerene isomers. Its theoretical significance was analyzed and extended in later work [**43**]. Our joint work prompted him to pursue new models for correlating interesting molecular orbital properties with the shape and other topological characteristics of larger electronic systems. Most recently he was working on the description of edge- and bulk-localized states in variously shaped microscopic pieces of graphite which he referred to 'nano-graphites' [**41-42, 44-50**].

Nano-graphite research was developing into an exciting area of carbon physics under Professor Fujita's leadership and his ideas will continue to influence the field. We greatly regret the loss of a good friend and colleague who impressed and stimulated his many collaborators and discussion partners in different disciplines and departments all around the world with his inexhaustible curiosity, sense of humour and enthusiasm for the 'shape and fantasy' of carbon materials [**IV-6**]. He is survived by his wife Naoko and two small daughters of 2 and 8 years old, his parents and a brother.

We thank the conference organizers for arranging for this tribute to appear in the Proceedings.

Complete Publication List of Professor Mitsutaka Fujita

I. Original Papers.

1. "Theory of Magnetic Polarizable Superconductors II", M. Fujita, K. Machida and T. Matsubara, *Journal of Low Temperature Physics*, **1984**, Vol. 54, pp. 535–546.
2. "Soliton Lattice Structure of Incommensurate Spin-Density-Waves: Application to Cr and Cr-rich Cr-Mn and Cr-V Alloys", K. Machida and M. Fujita, *Physical Review B*, **1984**, Vol. 30, pp. 5284–5299.
3. "Spin-Peierls Transition in Magnetic Fields: Thermodynamic Properties of a Soliton Lattice State", M. Fujita and K. Machida, *Journal of the Physical Society of Japan*, **1984**, Vol. 53, pp. 4395–4404.
4. "Charge Density Waves under Magnetic Fields", M. Fujita, K. Machida and H. Nakanishi, *Journal of the Physical Society of Japan*, **1985**, Vol. 54, pp. 3820–3832.
5. "Electrons on One-Dimensional Quasi-Lattices", M. Fujita and K.Machida, *Solid State Communications*, **1986**, Vol. 59, pp. 61–65.
6. "Phonons in One-Dimensional Quasi-Lattices", K. Machida and M. Fujita, *Journal of the Physical Society of Japan*, **1986**, Vol. 55, pp. 1799–1801.
7. "Quantum Energy Spectra and One-Dimensional Quasi-Periodic Systems", K. Machida and M. Fujita, *Physical Review B*, **1986**, Vol. 34, pp. 7367–7370.
8. "Electrons on One-Dimensional Quasi-Periodic Lattices", M. Fujita and K. Machida, *Synthetic Metals*, **1987**, Vol. 19, pp. 39–44.
9. "Spectral Properties of One-Dimensional Quasi-Crystalline and Incommensurate Systems", M. Fujita and K. Machida, *Journal of the Physical Society of Japan*, **1987**, Vol. 56, pp. 1470–1477.
10. "Spin Density Waves in Cr and Cr-rich Cr-Mn and Cr-V Alloys", M. Fujita and K. Machida, *Journal of Magnetism and Magnetic Materials*, **1987**, Vol. 70, pp. 442–444.
11. "Soliton Lattice Structure in the Spin-Peierls Systems", M. Fujita, K. Machida, K. H. Korving, H. B. Brom and L.J. de Jongh, *Japanese Journal of Applied Physics*, **1987**, Vol. 26, pp. 591–592.
12. "The Intermediate State of the Spin-Peierls System TTF-AuBDT: A Magnetic and Specific Heat Study", K.H. Korving, G.J. Kramer, R.A. Steeman, H.B. Brom, L.J. de Jongh, M. Fujita and K. Machida, *Physica*, **1987**, Vol. 145B, pp. 299–310.
13. "Magnetic Structure in a Soliton Lattice State of a Spin-Peierls System under Applied Fields", M. Fujita and K. Machida, *Journal of Physics C: Solid State Physics*, **1988**, Vol. 21, pp. 5813–5823.
14. "Magnetic Structure in Spin-Peierls Systems", M. Fujita and K. Machida, *Journal de Physique*, **1988**, Vol. C8, pp. 1405–1406.
15. "Charge Density Wave with Imperfect Nesting and Superconductivity I", M. Ichimura, M. Fujita and K. Nakao, *Physical Review B*, **1990**, Vol. 41, pp. 6387–6393.
16. "Soliton Lattice Modulation of Incommensurate Spin Density Wave in Two Dimensional Hubbard Model: A Mean Field Study", M. Kato, K. Machida, H. Nakanishi and M. Fujita, *Journal of the Physical Society of Japan*, **1990**, Vol. 59, pp. 1047–1058.
17. "A Novel Magnetic Structure of Vortex Lattice in Adsorbed ^{3}He Films", K. Machida and M. Fujita, *Physical Review B*, **1990**, Vol. 42, pp. 2673–2676 (Rapid Communication).
18. "Magnetic Structure of Vortex Lattice State in the Adsorbed Second Layer of ^{3}He Films", M. Fujita and K. Machida, *Physica*, **1990**, Vol. 165–166, pp. 711–712.
19. "Effect of Charge Density Wave on Superconducting Critical Temperature", M. Ichimura, M. Fujita and K. Nakao, *Physica*, **1990**, Vol. 165–166, pp. 1035–1036.

20. "Charge Density Wave with Imperfect Nesting and Superconductivity II", M. Ichimura, M. Fujita and K. Nakao, *Physical Review B*, **1991**, Vol. 43, pp. 175–180.
21. "Helical States vs. Soliton and Vortex Lattice States in the Two-dimensional Hubbard Model", M. Fujita, M. Ichimura and K. Nakao, *Journal of the Physical Society of Japan*, **1991**, Vol. 60, pp. 2831–2834.
22. "Non-colinear Spin Density Wave in a Two-dimensional Hubbard Model", M. Ichimura, M. Fujita and K. Nakao, *Physica C*, **1991**, Vol. 185–189, pp. 1665–1666.
23. "Spin Density Wave States in Two-dimensional Hubbard Model", M. Ichimura, M. Fujita and K. Nakao, *Journal of the Physical Society of Japan*, **1992**, Vol. 61, pp. 2027–2039.
24. "Triple-Q Vortex Lattice Magnetic Structures of Itinerant Systems on a Triangular Lattice", M. Fujita, T. Nakanishi and K. Machida, *Physical Review B*, **1992**, Vol. 45, pp. 2190–2195.
25. "Field-Induced Instability of Dimer Structures near Single-Layered Steps on Si(001) Surfaces", T. Miyazaki and M. Fujita, *Journal of Physics: Condensed Matter*, **1992**, Vol. 4, pp. 6539–6548.
26. "Electronic Structure of Chiral Graphene Tubules", R. Saito, M. Fujita, G. Dresselhaus and M.S. Dresselhaus, *Applied Physics Letters*, **1992**, Vol. 60, pp. 2204–2206.
27. "Formation of General Fullerenes by Their Projection on a Honeycomb Lattice", M. Fujita, R. Saito, G. Dresselhaus and M.S. Dresselhaus, *Physical Review B*, **1992**, Vol. 45, pp. 13834–13836 (Rapid Communication).
28. "Electronic Structure of Graphene Tubules Based on C_{60}", R. Saito, M. Fujita, G. Dresselhaus and M.S. Dresselhaus, *Physical Review B*, **1992**, Vol. 46, pp. 1804–1811.
29. "Electronic Structure and Growth Mechanism of Carbon Tubules", R. Saito, M. Fujita, G. Dresselhaus and M.S. Dresselhaus, *Materials Science and Engineering*, **1993**, Vol. B19, pp. 185–191.
30. "Dimerization and Energy-Level Structures in Fullerene Tubules Investigated with an Electron-Phonon Model", K. Harigaya and M. Fujita, *Synthetic Metals*, **1993**, Vol. 55–57, pp. 3196–3201.
31. "Dimerization Structures on the Metallic and Semiconducting Fullerene Tubules with Half-filled Electrons", K. Harigaya and M. Fujita, *Physical Review B*, **1993**, Vol. 47, pp. 16563–16569.
32. "Geometrical Understanding of Fullerene Structures and Electronic States", M. Fujita, *Fullerene Science and Technology*, **1993**, Vol. 1, pp. 365–377.
33. "Ab-initio Molecular Orbital Calculation for C_{70} and Seven Isomers of C_{80}", K. Nakao, N. Kurita and M. Fujita, *Physical Review B*, **1994**, Vol. 49, pp. 11415–11420.
34. "Stable Structures and Electronic Properties of C_{70} and Seven Isomers of C_{80}", M. Fujita, N. Kurita and K. Nakao, *Synthetic Metals*, **1995**, Vol. 70, pp. 1509–1510.
35. "Morphology and Electronic Structure of Graphitic Sponges", M. Fujita, M. Yoshida, T. Umeda and E. Osawa, *SyntheticMetals*, **1995**, Vol. 70, pp. 1897–1898.
36. "Morphology of New Fullerene Families with Negative Curvature", M. Fujita, M. Yoshida and E. Osawa, *Fullerene Science and Technology*, **1995**, Vol. 3, pp. 93–105.
37. "Polymorphism of Carbon Forms: Polyhedral Morphology and Electronic Structures", M. Fujita, T. Umeda and M. Yoshida, *Physical Reveiw B*, **1995**, Vol. 51, pp. 13778–13780.
38. "Role of Hydrogen in C and Si(001) Homoepitaxy", T. Ogitsu, T. Miyazaki, M. Fujita and M. Okazaki, *Phys. Rev. Lett.*, **1995**, Vol. 75, pp. 4226–4229.
39. "Phason Line: A New Criterion for the Stability of Fullerenes", M. Yoshida, M. Fujita, Hitoshi Goto and E. Osawa, *Electron J. Theor. Chem.*, **1996**, Vol. 1, pp. 151–162.
40. "Polymorphism of Extended Fullerene Networks: Geometrical Parameters and Electronic Structures", M. Fujita, M. Yoshida and K. Nakada, *Fullerene Science and Technology*, **1996**, Vol. 4, pp. 565–582.

41. "Peculiar Localized State at Zigzag Graphite Edge", M. Fujita, K. Wakabayashi, K. Nakada and K. Kusakabe, *Journal of Physical Society of Japan*, **1996**, Vol. 65, pp. 1920–1923.
42. "Edge State in Graphene Ribbons: Nanometer Size Effect and Edge Shape Dependence", K. Nakada, M. Fujita, G. Dresselhaus and M.S. Dresselhaus, *Phys. Rev. B*, **1996**, Vol. 54, pp. 17954–17963.
43. "Leapfrog, Fries, Clar and Phason-line-free Fullerenes", P.W. Fowler, M. Fujita and M. Yoshida, *J. Chem. Soc., Faraday Trans.*, **1996**, Vol. 92, pp. 3673–3675.
44. "Lattice Distortion in Nanographite Ribbons", M. Fujita, M. Igami and K. Nakada, *J. Phys. Soc. Jpn.*, **1997**, Vol. 66, pp. 1864–1867.
45. "Magnetism of Nanometer-Scale Graphite with Edge or Topological Defects", K. Kusakabe, K. Wakabayashi, M. Igami, K. Nakada and M. Fujita, *Molecular Crystals and Liquid Crystals*, **1997**, Vol. 305, pp. 445–454.
46. "Non-bonding orbitals in graphite, carbon tubules, toroids and fullerenes", M. Yoshida, M. Fujita, P. W. Fowler and E. C. Kirby, *J. Chem. Soc. Faraday Trans*, **1997**, Vol. 93, pp. 1037–43.
47. "Phonon Dispersion of Nano-graphite Ribbons", M. Igami, M. Fujita and S. Mizuno, *Appl. Surf. Sci.*, **1998**, Vol. 130-132, pp. 870–875.
48. "Spin Wave Mode of Edge-Localized Magnetic States in Nanographite Zigzag Ribbons", K. Wakabayashi, M. Sigrist and M. Fujita, *J. Phys. Soc. Jpn.*, **1998**, Vol. 67, pp. 2089–2093.
49. "Magnetic Field Effect on Graphite Ribbons", K. Wakabayashi, M. Fujita, K. Kusakabe and H. Ajiki, *J. Mag. Mat.*, **1998**, Vol. 177–181, pp. 1484-1485.
50. "Electron-Phonon and Electron-Electron Interactions in Nanographite Ribbons", M. Fujita, M. Igami, K. Wakabayashi and K. Nakada, *Molecular Crystals and Liquid Crystals*, **1998**, in press.
51. "Localized π Electronic Edge State in Nanographite", EK. Nakada, M. Igami and M. Fujita, *J. Phys. Soc. Jpn.*, **1998**, in press.

II. Conference Proceedings.

1. "Electronic Properties of Carbon Fiber Based on C_{60}", R. Saito, M. Fujita, G. Dresselhaus and M.S. Dresselhaus, in *Electrical, Optical and Magnetic Properties of Organic Solid State Materials, MRS Symposia Proceedings, Boston*, **1992**, Vol. 247, pp. 333–336.
2. "Polyhedra and Morphology of General Fullerene", M. Fujita, in *Advanced Materials '93, I/B: Magnetic, Fullerene, Dielectric, Ferroelectric, Diamond and Related Materials*, edited by M. Homma *et al.*, *Trans. Mat. Res. Soc. Jpn.*, **1994**, Vol. 14B, pp. 1157–1160.
3. "Stability of Boron and Nitrogen Substituted Buckminsterfullerene", D. Inomata, N. Kurita, M. Fujita, S. Suzuki and K. Nakao, *Advanced Materials '93, I/B: Magnetic, Fullerene, Dielectric, Ferroelectric, Diamond and Related Materials*, edited by M. Homma *et al.*, *Trans. Mat. Res. Soc. Jpn.*, **1994**, Vol. 14B, pp. 1231–1234.
4. "Ab-initio Study of Hydrogenated (100)Surfaces of Diamond and Silicon", T. Ogitsu, T. Miyazaki, M. Fujita and M. Okazaki, in *Advanced Materials '93, III/A: Computations, Glassy Materials, Microgravity and Non-Destructive Testing*, edited by T. Masumoto *et al.*, *Trans. Mat. Res. Soc. Jpn.*, **1994**, Vol. 16A, pp. 263–266.
5. "Graphitic Sponges: Polymorphic Forms and Electronic Structures", M. Fujita, T. Umeda, M. Yoshida and E. Osawa, in *Proc. 22nd International Conf. On the Physics of Semiconductors, Vancouver 1994*, edited by D.J. Lockwood, (World Scientific, Singapore, **1994**), pp. 2069–2072.
6. "Phason Lines in Higher Fullerenes", M. Fujita, M. Yoshida, E. Osawa and K. Harigaya, in *Proc. 22nd International Conf. Phys. Semiconductors, Vancouver 1994*, edited by D.J. Lockwood, (World Scientific, Singapore, **1994**), pp. 2077–2080.

7. "Ab-initio Study of the Elementary Process of Epitaxial Diamond Growth", T. Ogitsu, T. Miyazaki, M. Fujita and M. Okazaki, in *Proc. 22nd International Conf. Phys. Semiconductors, Vancouver 1994*, edited by D.J. Lockwood, (World Scientific, Singapore, **1994**), pp. 537–540.
8. "Electronic States and Phason Lines in Higher Fullerenes", M. Yoshida, M. Fujita, and E. Osawa, *Synthetic Metals*, **1995**, Vol. 70, pp. 1487–1488. (Synthetic Metals, Proc. of the International Conf. on Science and Technology of Synthetic Metals, Seoul, 1994).
9. "Electronic States of Graphitic Heterocompounds of Carbon, Boron and Nitrogen", S. Mizuno, M. Fujita and K. Nakao, *Synthetic Metals*, **1995**, Vol. 71, pp. 1869–1870. (Proc. of the International Conf. on Science and Technology of Synthetic Metals, Seoul, 24–29 July, 1994.)
10. "Localized Electronic States on Graphite Edge", K. Nakada, M. Fujita, K. Wakabayashi and K. Kusakabe, *Czech Journal of Physics*, **1996**, Vol. 46, pp. 2429–2430 (Proc. of 21st International Conf. of Low Temperature Physics, Prague, Aug. 8–14, 1996).
11. "Magnetic Structure of Graphite Ribbon", K. Wakabayashi, M. Fujita, K. Kusakabe and K. Nakada, *Czech Journal of Physics*, **1996**, Vol. 46, pp. 1865–1866 (Proc. of 21st International Conf. of Low Temperature Physics, Prague, Aug. 8–14, 1996).
12. "Effect of Topological Defects in Graphite", M. Igami, K. Nakada, M. Fujita and K. Kusakabe, *Czech Journal of Physics*, **1996**, Vol. 46, pp. 2715–2716 (Proc. of 21st International Conf. of Low Temperature Physics, Prague, Aug. 8–14, 1996).
13. "Electronic and Magnetic Structure of Micro-Graphite", K. Wakabayashi, K. Nakada, M. Fujita and K. Kusakabe, *Proceedings of J-MRS*, **1996**, Vol. 20, p. 490.

III. Book Chapters.

1. "Texture of Spin Density Waves for Various Electron Fillings on Two-dimensional Systems", M. Fujita, M. Ichimura, K. Nakao and K. Machida, in *Electronic Properties and Mechanisms of High TC Superconductors*, edited by T. Oguchi, K. Kadowaki and T.Sasaki (Elsevier, Amsterdam, **1992**), pp. 261–265.
2. "Theoretical Study of Bond Exchange Mechanism during CVD Growth of Diamond", T. Ogitsu, T. Miyazaki, M. Fujita and M. Okazaki, in *Advances in New Diamond Science and Technology*, edited by S. Saito et al. (MYU, Tokyo, **1994**), pp. 473–476.
3. "Spin Density Waves in the Itinerant-Electron Systems on a Two-dimensional Lattice", M. Ichimura, M. Fujita and K. Nakao, in *Elementary Excitations in Solids*, edited by J. L. Birman, C. Sebenne and R. F. Wallis (Elsevier, Amsterdam, **1992**), pp. 51–79.

IV. Review Articles (in Japanese apart from item 6).

1. "Energy Gap in Incommensurate Systems", K. Machida and M. Fujita, *Kotai Butsuri (Solids State Physics)*, **1986**, Vol. 21, pp. 315–325.
2. "How to Understand the Geometrical Structure of General Fullerenes", M. Fujita, *Kotai Butsuri (Solids State Physics)*, **1993**, Vol. 28, pp. 199–205.
3. "Geometrical Study for Fullerene Structures and Construction of New Carbon Forms", M. Fujita, M. Yoshida and E. Osawa, *Tanso (Carbon)*, **1994**, Vol. 162, pp. 100–109.
4. "Geometries of Fullerene Families and Polyhedra", M. Fujita, M. Yoshida and E. Osawa, *HyperSpace*, **1994**, Vol. 3, pp. 25–35.
5. "An Ab Initio Study of the Elementary Process of Hydrogen-Mediated Diamond Growth", T. Ogitsu, M. Fujita, T. Miyazaki and M. Okazaki, *Hyomen Kagaku (J. Surf. Sci. Soc. Jpn.)*, **1994**, Vol. 15, pp. 604–609.
6. "Shape and Fantasy of Fullerenes", E. Osawa, M. Yoshida and M. Fujita, *MRS Bulletin*, **1994**, Vol. 19, No.11, pp. 33–36.
7. "Polymorphism of Fullerene Networks", M. Fujita and M. Yoshida, *Kotai Butsuri (Solid State Physics)*, **1995**, Vol. 30, pp. 567–574.

8. "Polymorphism of Fullerene Network and the Electronic States", M. Fujita and M. Yoshida, *Butsuri (Physics)*, **1995**, Vol. 50, 945–952.
9. "Surface Electronic States and Phason Line of Higher Fullerenes", M. Yoshida, M. Fujita and E. Osawa, *Hyoumen Kagaku (J. Surf. Sci. Soc. Jpn.)*, **1996**, Vol. 17, No. 2, pp. 62–67.
10. "Electronic States of Fullerene Networks with Edges", K. Nakada, M. Fujita and H. Hosoya, *Fullerene Quarterly*, **1996**, Vol. 4, No. 2, pp. 101–111.
11. "Microscale Graphite Fragment - Effect of Edge on π Electronic States", K. Nakada and M. Fujita, *Hyoumen (Surface)*, **1996**, Vol. 34, No. 4, pp. 237–244.
12. "Electronic Properties of Nanographites - Nanometer Effects in the π Electronic System with Edges", M. Fujita, K. Wakabayashi, M. Igami, K. Nakada and K. Kusakabe, *Kotai Butsuri (Solid State Physics)*, **1997**, Vol. 32, pp. 179–188.
13. "Defects and Edges in Nanographites - Nanometer Effects in the π Electronic System", M. Fujita, M. Igami, K. Nakada, K. Wakabayashi and K. Kusakabe, *Hyoumen Kagaku (J. Surf. Sci. Soc. Jpn.)*, **1998**, Vol. 19, pp. 35–42.

by

Eiji Osawa
Department of Knowledge-based Information Engineering, Toyohashi University of Technology, Toyohashi, Aichi 441-8580
E-mail address: osawa@cochem.tutkie.tut.ac.jp

and

Mitsuho Yoshida
Exploratory Chemistry Research Laboratories, Sankyo Co., Ltd., 2-58, Hiromachi, 1-Chome, Shinagawa-ku, Tokyo 140-8710, Japan
E-mail address: mitsuh@shina.sankyo.co.jp

DIMACS Series in Discrete Mathematics
and Theoretical Computer Science
Volume 51, 2000

The Clar and Fries Problems for Benzenoid Hydrocarbons are Linear Programs

Hernán Abeledo and Gary Atkinson

ABSTRACT. Many chemical properties of benzenoid hydrocarbons can be understood in terms of the maximum number of mutually resonant hexagons, or Clar number, of the molecules. Hansen and Zheng (1994) formulated this problem as an integer program and, based on computational evidence, conjectured that solving the linear programming relaxation always yields integral solutions. We establish their conjecture by showing that the constraint matrices of these problems are unimodular. We also show that a similar result holds for the Fries number, another index associated with benzenoids. Previously, Hansen and Zheng (1992) showed that a certain minimum weight cut cover problem defined for benzenoids yields an upper bound for the Clar number and conjectured that equality always holds. We use a network flow formulation of the cut cover problem to prove this conjecture and, in the process, obtain an alternative means of computing the Clar number. All results hold for generalizations of the Clar, Fries, and cut cover problems defined for plane graphs that are bipartite and 2-connected.

1. Introduction

Benzenoid hydrocarbons are organic molecules composed of carbon and hydrogen atoms organized into connected (hexagonal) benzene rings. The structure of such a molecule is usually represented by a *benzenoid system*, a bipartite and two-connected plane graph whose interior faces are all regular hexagons. Each node of a benzenoid system represents a carbon atom, and each edge corresponds to a single or double bond between a pair of carbon atoms. The hydrogen atoms are not explicitly represented since their location can be easily deduced from the configuration of the carbon atoms. One interesting property of benzenoid molecules is that, as empirical evidence shows ([**4**],[**6**]), their graphs always have a perfect matching, i.e., a disjoint set of edges covering all nodes.

The focus of this work is on the optimal values of two combinatorial optimization problems over the set of perfect matchings of a benzenoid, namely, the Clar and Fries numbers ([**8**],[**9**]). The Clar number is a key index in the aromatic sextet

1991 *Mathematics Subject Classification.* Primary: 90C35, Secondary: 15A36, 92E10, 90C05, 05C70.

This is an extended abstract of a paper that will be submitted for publication elsewhere.

theory developed by E. Clar [3] to explain benzenoid phenomenology. For example, it has been observed that as the Clar number increases within an isomeric series of benzenoid hydrocarbons, (1) isomeric stability increases, so chemical reactivity decreases, and (2) the absorption bands shift towards shorter wavelengths, so the isomer colors change from dark blue-green to red, yellow or white [6]. It has also been demonstrated that the Clar number provides a rough estimate of the Dewar-type resonance energy [1]. Along the same lines, the Fries number is an index proposed by K. Fries [5] in an early attempt to understand properties of benzenoids.

Hansen and Zheng [8, 9] modeled the Clar and Fries numbers as integer programming problems. In particular, their computational results led them to conjecture that solving the linear programming relaxation of the Clar number integer program yields integer solutions for all benzenoids. We prove this conjecture is true by showing that the linear relaxation polytope is integral, i.e., all its extreme points are integer vectors. We also show that similar properties hold for integer programming formulations of the Fries number. These results imply that the Clar and Fries numbers of a benzenoid can be efficiently computed using linear programming algorithms.

Central to our developments is proving that the constraint matrices of these integer programs are always unimodular. Interestingly, these unimodular constraint matrices constitute an unusual case since they are not, in general, totally unimodular as often occurs with optimization problems on graphs that give rise to integral polyhedra. However, for the Clar problem, we remark that a subset of them are (also) totally unimodular, specifically, those constraint matrices corresponding to a natural subset of benzenoids called catacondensed benzenoid hydrocarbons.

In another paper, Hansen and Zheng [7] considered a minimization problem for benzenoid systems that we call here the minimum weight cut cover problem. They showed that the optimal value of the cover problem is an upper bound for the Clar number and conjectured that equality always holds. We prove this is indeed the case by formulating the minimum weight cut cover problem as a network flow problem and using its dual to construct a solution for the Clar number problem. As a consequence, the Clar number and the minimum weight cut cover problems can be solved in strongly polynomial time using a minimum cost network flow algorithm.

In the course of our research, we found it informative to experiment computationally with various optimization models for benzenoids. To facilitate this task, we developed an interactive software program with a graphical interface that might be of interest to our readers. So, we briefly describe it here.

Finally, we mention that our results on the Clar and Fries numbers extend beyond benzenoid systems to plane graphs that are bipartite and 2-connected. The presentation given here will cover the more general case. Proofs are omitted or at most sketched since a full version will be published elsewhere.

2. Definitions and Preliminaries

In this paper we consider combinatorial optimization problems defined for plane graphs that are bipartite and 2-connected. The undefined graph theory terminology we use is standard and may be found in [2] or [10]. Here (V, E, F) will always denote the plane map with set of finite faces F that results from embedding a bipartite and 2-connected planar graph $G = (V, E)$. We will further assume that the nodes of V are colored black or white so that all edges connect nodes of different color.

Since G is bipartite and 2-connected, the boundary of each face $f \in F$ is an even cycle which can be perfectly matched in two different ways. Each of these two possible matchings is called a *frame* of the face. A frame of a face f is *clockwise oriented* if each frame edge is drawn from a black node to white node in a clockwise direction along the boundary of f. Otherwise, the frame is *counterclockwise oriented.* A *framework* for (V, E, F) is a mapping $\varphi : F \mapsto 2^E$ such that $\varphi(f)$ is a frame for each $f \in F$. A framework φ is *simple* if $\varphi(f) \cap \varphi(f') = \emptyset$ whenever $f \neq f'$. An *oriented* framework has all frames with the same orientation. It follows that oriented frameworks are simple.

A set of faces $F' \subseteq F$ is *framed* in a perfect matching for G if the matching contains a frame for each face in F'. We denote by $fra(V, E, F)$ the maximum cardinality of a framed set of faces for (V, E, F). A framed set of faces that are also node-disjoint is called *resonant.* The maximum cardinality of a resonant set of faces for (V, E, F) is denoted by $res(V, E, F)$. When (V, E, F) is a benzenoid system, $fra(V, E, F)$ is called its *Fries number* and $res(V, E, F)$ is its *Clar number* ([**8**, **9**]). Given a plane map (V, E, F), the *maximum framed set problem* and the *maximum resonant set problem* seek to determine $fra(V, E, F)$ and $res(V, E, F)$, respectively.

A *basis* of a matrix is a maximal linearly independent column submatrix. An integer matrix is called *unimodular* if, for each basis, the determinants of the largest submatrices are relatively prime. A matrix is *totally unimodular* if the determinant of every square submatrix is 0, 1, or -1. Totally unimodular matrices are unimodular. The fundamental relationship between unimodular matrices and integral polyhedra is stated in a theorem established by Veinott and Dantzig [**14**] and extended by Truemper [**13**] to matrices that do not have full row rank.

THEOREM 2.1 ([**14**], [**13**]). *An integer matrix A is unimodular if and only if the polyhedron $\{x : Ax = b,\ x \geq 0\}$ is integral for any integer vector b.*

3. Integer Programs for Resonant Sets and Framed Sets

To compute the Clar number of a benzenoid, Hansen and Zheng [**8**] proposed the following integer program formulation (IP1) of the maximum resonant set problem:

$$\max \{1^T y : Kx + Ry = 1,\ x, y \geq 0,\ x \in \mathrm{Z}^E,\ y \in \mathrm{Z}^F\}$$

where K is the $V \times E$ node-edge incidence matrix of $G = (V, E)$, and R is the $V \times F$ node-face incidence matrix of (V, E, F). All entries in a feasible solution to IP1 must be 0 or 1. Thus, the vector y is the incidence vector of the resonant faces and the vector x identifies the edges in the underlying perfect matching that are not in the frames of resonant faces.

Computational evidence led Hansen and Zheng [**8**] to conjecture that, for every benzenoid system, the solution of the linear relaxation of IP1 is an integer vector. They also noted that, in general, the constraint matrix $[K\ R]$ of IP1 is not totally unimodular - if it was, it would immediately imply their conjecture. However, since the constraints involving matrix $[K\ R]$ are equations, it suffices to establish that $[K\ R]$ is unimodular in order to prove that the linear relaxation polytope is integral.

With this goal in mind, it is convenient to reformulate problem IP1. Given a framework φ for (V, E, F), we define its edge-frame incidence matrix U to be an $E \times F$ matrix such that, for each face $f \in F$, its corresponding column of U is the edge incidence vector of the frame $\varphi(f)$. The incidence matrix U of a framework is related to the matrices K and R defined above via the following useful factorization.

LEMMA 3.1. *Let U be the incidence matrix of a framework for (V, E, F). Then $R = KU$.*

Introducing the vector of variables $z = x + Uy$ yields an alternative formulation of the maximum resonant set problem (IP2).

$$\max \{1^T y : Kz = 1,\ x + Uy - z = 0,\ x, y, z \geq 0,\ y \in \mathbb{Z}^F,\ x \in \mathbb{Z}^E, z \in \mathbb{Z}^E\}$$

with constraint matrix

$$A_2 = \begin{bmatrix} 0 & 0 & K \\ I & U & -I \end{bmatrix}.$$

All feasible solutions to problem IP2 are necessarily binary vectors. Hence, constraints $Kz = 1$ express that z is the incidence vector of a perfect matching for G. Constraints $x+Uy-z = 0$ partition the edges in the perfect matching represented by z between those in the frames of faces considered resonant (represented by Uy), and the remaining edges identified by x. As in problem IP1, the constraint matrix A_2 is generally not totally unimodular. In this case, however, it can be shown directly to be unimodular for any plane map (V, E, F) of a bipartite and 2-connected planar graph.

THEOREM 3.2. *Let U be the incidence matrix of a framework for (V, E, F). Then the matrix A_2 is unimodular.*

To establish Theorem 3.2, we show that conditions given by Truemper in [**13**] are satisfied, namely that every column Eulerian column submatrix of A_2 is linearly dependent. We remark that it is easier to apply Truemper's conditions to matrix A_2, instead of directly to $[K\ R]$, since the rows of A_2 have fewer nonzero entries. In this sense, the proof of Theorem 3.2 is even easier if one assumes that U is the incidence matrix of a simple framework.

This theorem can now be used to establish the unimodularity of the constraint matrix $[K\ R]$ of problem IP1. For any given framework incidence matrix U, let $P_2(b)$ denote the polyhedron defined by the linear system

$$\begin{bmatrix} 0 & 0 & K \\ I & U & -I \end{bmatrix} \begin{bmatrix} x \\ y \\ z \end{bmatrix} = \begin{bmatrix} b \\ 0 \end{bmatrix}$$

where $x, y, z \geq 0$. Similarly, let $P_1(b)$ be the solution set of $Kx + Ry = b,\ x, y \geq 0$. The following lemma is straightforward.

LEMMA 3.3. *Let b be a vector such that $P_1(b)$ is nonempty. Then (x, y) is an extreme point of $P_1(b)$ if and only if $(x, y, x + Uy)$ is an extreme point of $P_2(b)$.*

Combining Theorem 3.2, Lemma 3.3, and Theorem 2.1 establishes the unimodularity of the constraint matrix of problem IP1.

COROLLARY 3.4. *For all (V, E, F), the matrix $[K\ R]$ is unimodular.*

Benzenoid hydrocarbons are called *catacondensed* if their graphical representation has no internal nodes, i.e., all nodes lie on the exterior face boundary cycle. Otherwise, if there are internal nodes, they are called *pericondensed.* Interestingly, there is an alignment of these two classes of benzenoids with unimodularity properties of the matrix $[K\ R]$.

THEOREM 3.5. *Let (V, E, F) be a benzenoid system. Then the matrix $[K\ R]$ is totally unimodular if and only if the benzenoid is catacondensed.*

We now turn to integer programming models of the framed set problem. Let φ and $\overline{\varphi}$ denote the clockwise and counterclockwise frameworks of (V, E, F), respectively, and let U and $\overline{U}$ be their corresponding edge-face incidence matrices. The following formulation of the maximum framed set problem (IP3) was proposed in [**9**] to compute the Fries number of a benzenoid.

$$\max\ \{1^T y + 1^T \overline{y} : Kz = 1,\ Uy \leq z,\ \overline{U}\overline{y} \leq z,\ y, \overline{y}, z \geq 0,\ y \in Z^F,\ \overline{y} \in Z^F,\ z \in Z^E\}$$

Introducing slack variables x and $\overline{x}$, the constraints of IP3 are as follows

$$\begin{bmatrix} 0 & 0 & 0 & 0 & K \\ I & U & 0 & 0 & -I \\ 0 & 0 & I & \overline{U} & -I \end{bmatrix} \begin{bmatrix} x \\ y \\ \overline{x} \\ \overline{y} \\ z \end{bmatrix} = \begin{bmatrix} 1 \\ 0 \\ 0 \end{bmatrix},$$

where all variables are nonnegative and integer.

As in the resonant set problem, Lemma 3.1 yields an alternative formulation of the maximum framed set problem (IP4) without the vector of variables z. The constraints of problem IP4 are

$$\begin{bmatrix} K & R & 0 & 0 \\ I & U & -I & -\overline{U} \end{bmatrix} \begin{bmatrix} x \\ y \\ \overline{x} \\ \overline{y} \end{bmatrix} = \begin{bmatrix} 1 \\ 0 \end{bmatrix}$$

where all variables are nonnegative and integer. Arguing substantially as before, we can show that the constraint matrices of the two formulations of the maximum framed set problem are unimodular.

THEOREM 3.6. *For all (V, E, F), the constraint matrices of problems IP3 and IP4 are unimodular.*

In conclusion, the constraint matrices of both formulations of the maximum resonant set problem and of the maximum framed set problem are unimodular. A key consequence of this is that linear programming can be used to compute $res(V, E, F)$ and $fra(V, E, F)$.

4. Minimum Weight Cut Cover Problem

Hansen and Zheng [**7**] proposed a class of cuts for benzenoids systems and defined a combinatorial optimization problem based on these cuts. In this section we apply their definitions to a larger class of plane graphs. The proofs of results established by Hansen and Zheng [**7**] for benzenoid systems extend directly to the more general version considered here. As before, (V, E, F) continues to denote the plane map of a bipartite and 2-connected planar graph $G = (V, E)$, where nodes are colored white or black and every edge connects nodes of different color. We also assume in this section that G has a perfect matching.

DEFINITION 4.1. A cut of (V, E, F) is a simple (possibly closed) plane curve c such that

(1) c intersects G only at some edge subset E_c of E, (i.e., c does not go through any nodes);
(2) the subgraph $G_c = (V, E \setminus E_c)$ is disconnected such that the nodes of each edge in E_c belong to different components of G_c; and
(3) all nodes of edges in E_c in the same component of G_c are of the same color.

DEFINITION 4.2. A cut cover of (V, E, F) is a set of cuts C such that each face in F is intersected by at least one cut in C.

DEFINITION 4.3. Let $M \subseteq E$ be a perfect matching for (V, E, F). The weight m_c with respect to M of a cut c is $m_c = |E_c \cap M|$, and the weight $m(C)$ of a cut cover C is $m(C) = \sum_{c \in C} m_c$.

Hansen and Zheng [7] proved the following two results.

THEOREM 4.4 ([7]). *The weight of a cut is the same for all perfect matchings.*

Thus, for any cut cover C of (V, E, F), its weight $m(C)$ is also independent of the perfect matching. Let $cov(V, E, F)$ denote the value of a minimum weight cut cover for (V, E, F). Then

THEOREM 4.5 ([7]). *For all* (V, E, F), $res(V, E, F) \leq cov(V, E, F)$.

Hansen and Zheng [7] conjectured that the above inequality is satisfied as an equality by all benzenoid systems. Using a network flow formulation of the cut cover problem and duality arguments, we establish their conjecture and extend it to the larger class of graphs considered here.

THEOREM 4.6. *For all* (V, E, F), $res(V, E, F) = cov(V, E, F)$.

We briefly describe next the network problem used to prove Theorem 4.6 and which can also be used to compute $res(V, E, F)$ in strongly polynomial time using a network flow algorithm. The directed graph of this flow problem is based on the geometric dual (multi)graph of a plane map as defined, for example, in Harary [10].

Let $G^* = (F \cup \{t\}, E^*)$ denote the geometric dual of (V, E, F), where t denotes the external face of G. The construction of G^* defines a one-to-one correspondence between E and E^* where each edge in E is crossed by an edge in E^*. Since G is 2-connected, G^* has no loops but may have multiple edges connecting the same pair of nodes.

We now transform G^* into a directed graph. Let φ denote the clockwise oriented framework of (V, E, F). Let $e \in E$ and let f and i denote the two faces that contain e in their boundary (one of them could be the external face). Without loss of generality, we will assume $f \in F$. Let $\{f, i\}$ denote the edge in E^* that crosses e. If $e \in \varphi(f)$, then $\{f, i\}$ becomes arc (f, i). Otherwise, $\{f, i\}$ becomes arc (i, f). Next, for each node $f \in F$, we perform a node splitting transformation to obtain nodes $f_1 \in F_1$ and $f_2 \in F_2$ connected by the arc $(f_1, f_2) \in A_F$, where F_1 and F_2 are two copies of F and A_F is the set of resulting arcs connecting nodes in these two sets. Outgoing (incoming) arcs of node f become outgoing (incoming) arcs of node f_2 (f_1). Let $D = (F_1 \cup F_2 \cup \{t\}, A_E \cup A_F)$ denote the resulting directed graph, where A_E is the set of arcs connecting nodes in different faces. Each directed cycle in $D = (F_1 \cup F_2 \cup \{t\}, A_E \cup A_F)$ corresponds to a cut of (V, E, F).

We now define a minimum cost circulation problem on the directed graph D. A lower bound of 1 is imposed on the flow value of each arc in A_F (this enforces that all faces in F are covered by cuts). Flow on all remaining arcs is required to be nonnegative. The cost assigned to each arc in A_F is zero. Finally, let $p \in \{0,1\}^E$ be the incidence vector of a perfect matching for $G = (V, E)$. The cost of arcs in A_E is given by the corresponding entry of the vector p. Let γ denote the optimal value of this network flow problem. Since the right hand side data for this network flow problem is integer, there exists an optimal solution to the problem that is integer. As integer solutions can be interpreted as cut covers, it follows that $cov(V, E, F) \leq \gamma$.

To prove that $\gamma \leq res(V, E, F)$, we examine the dual of the network flow problem and show that any feasible dual solution contains, as a subvector, a feasible solution to the linear relaxation of problem IP1 with same objective function value.

5. A Computational Tool

In the course of this work, we found it useful to experiment computationally with various optimization models for benzenoid systems. However this could not be done efficiently for benzenoid systems of meaningful size because it was too time-consuming to input the associated constraint data into available optimization packages. To address this problem, we developed an interactive software program written in MATLAB that we call *Benzenoid Builder*.

This program has a graphical user interface that permits the user to easily custom design a benzenoid system and solve various linear programs for it, including all the ones mentioned earlier in this article. Based on the designed graph, the software automatically generates the mathematical formulations of several optimization models, computes statistics such as the number of nodes, edges, and perfect matchings. It also constructs and displays the adjacency and incidence matrices of the graph, as well as the various matrices that appear in the optimization models. The user interacts with the program via three screens.

1. **Grid Screen**: The user defines a hexagonal grid on which to work by choosing its number of rows and columns. From this screen, the user can also recall and reload previously defined benzenoid systems.
2. **Design Screen**: Using a mouse to point and click on a hexagonal cell, the user designs or modifies a benzenoid. This screen also displays statistics and determines if the benzenoid has a perfect matching.
3. **Model Screen**: The user can choose from a menu different linear programming problems for the current benzenoid, view their solution as well as various statistics, and also display different matrices associated with the benzenoid and the linear programs.

The linear programming optimization is performed with a straightforward implementation of the simplex method. One interesting feature is that the user can choose to either observe only the optimal solution, or actually view the optimization in progress. In the latter case, each simplex pivot is displayed after it is performed in the form of graphical structures on the benzenoid such as color-coded matching edges, circled (resonant or framed) hexagons, or cuts. The user can also select to step through the pivots one at a time. The program runs on any platform where

MATLAB (version 4.2c.1) runs on. Not designed for speed, this software running on a Pentium PC requires about a minute to find the Clar number of a benzenoid with 100 hexagons.

References

[1] J. Aihara, On the Number of Aromatic Sextets in a Benzenoid Hydrocarbon, *Bulletin of the Chemical Society of Japan*, 49 (1976) 1429-1430.

[2] B. Bollobás, *Graph Theory: An Introductory Course*, Springer-Verlag, New York, 1979.

[3] E. Clar, *The Aromatic Sextet*, John Wiley & Sons, London, 1972.

[4] S.J. Cyvin and I. Gutman, *Kekulé Structures in Benzenoid Hydrocarbons*, Springer-Verlag, Berlin, 1988.

[5] K. Fries, Über Byclische Verbindungen und ihren Vergleich mit dem Naphtalin, *Annals of Chemistry*, 454 (1927) 121- 324.

[6] I. Gutman and S.J. Cyvin, *Introduction to the Theory of Benzenoid Hydrocarbons*, Springer-Verlag, Berlin, 1989.

[7] P. Hansen and M. Zheng, Upper Bounds for the Clar Number of Benzenoid Hydrocarbons, *Journal of the Chemical Society, Faraday Transactions*, 88 (1992) 1621-1625.

[8] P. Hansen and M. Zheng, The Clar Number of a Benzenoid Hydrocarbon and Linear Programming, *Journal of Mathematical Chemistry*, 15 (1994) 93-107.

[9] P. Hansen and M. Zheng, Numerical Bounds for the Perfect Matching Vectors of a Polyhex, *Journal of Chemical Information and Computer Sciences* 34 (1994) 305-308.

[10] F. Harary, *Graph Theory*, Addison-Wesley, Boston, 1969.

[11] G.L. Nemhauser and L.A. Wolsey, *Integer and Combinatorial Optimization*, John Wiley & Sons, New York, 1988.

[12] A. Schrijver, *Theory of Integer and Linear Programming*, John Wiley & Sons, New York, 1986.

[13] K. Truemper, Algebraic Characterizations of Unimodular Matrices, *SIAM Journal of Applied Mathematics*, 35 No. 2 (1978) 328-332.

[14] A.F. Veinott and G.B. Dantzig, Integral Extreme Points, *SIAM Review* 10 No. 3 (1968) 371-372.

DEPARTMENT OF OPERATIONS RESEARCH, THE GEORGE WASHINGTON UNIVERSITY, WASHINGTON, DC 20052

E-mail address: abeledo@seas.gwu.edu

BELL LABORATORIES, LUCENT TECHNOLOGIES, HOLMDEL, NJ 07733

E-mail address: atkinson@bell-labs.com

DIMACS Series in Discrete Mathematics
and Theoretical Computer Science
Volume 51, 2000

Use of Graph Invariants in QMSA and Predictive Toxicology

S.C. Basak and B.D. Gute

ABSTRACT. Mathematical invariants are frequently used for the characterization of molecular graphs. Such invariants quantify structural features of chemicals like size, shape, symmetry, cyclicity, complexity, branching, etc. Numerical graph invariants or topological indices (TIs) have been used in developing quantitative structure-property/ activity/ toxicity relationship models and in defining intermolecular similarity. In this paper, we have used a set of TIs and a class of substructures called atom pairs (APs) in selecting analogs of probe chemicals from a set of mutagens. The result shows that both of the similarity methods select analogs which have reasonable structural similarity with the query chemicals. Such analogs, selected computationally, can be useful in the hazard assessment of chemicals for which very little or no toxicity data are available.

1. Introduction

A contemporary interest in mathematical chemistry is the characterization of molecular structure using graph theoretic formalism [1]–[11]. A graph $G = [V, E]$ consists of an ordered pair of two sets V and E, representing the vertices and edges, respectively. G becomes a molecular graph when the set V represents the set of atoms in a molecule and the set E symbolizes chemical bonds between adjacent atoms [8].

Mathematical characterization of molecular graphs (structures) may be accomplished using graph invariants. An invariant may be a polynomial, a sequence of numbers, or a real number. A real number characterizing a molecular graph is called a topological index (TI). TIs quantify different aspects of molecular architecture, viz., size, shape, cyclicity, branching, symmetry, etc [8].

TIs have been used extensively in quantitative structure-property/activity relationships (QSPR and QSAR respectively) and the quantification of intermolecular similarity/dissimilarity of chemicals [10]–[24]. In quantitative molecular similarity analysis (QMSA) studies, TIs have been used to derive high dimensional structure spaces where the Euclidean distance D_{ij} between a pair of molecules i and j is used

1991 *Mathematics Subject Classification.* Primary 54C40, 14E20; Secondary 46E25, 20C20.

This is contribution number 239 from the Center for Water and the Environment of the Natural Resources Research Institute. Research reported in this paper was supported by grants F49620-94-1-0401 and F49620-96-1-0330 from the United States Air Force. The authors would like to extend their thanks to Greg Grunwald for technical support.

to quantify the similarity between them. Similarity measures can be used either for the selection of analogs of chemicals or in the prediction of the property/activity of a molecule from the property of its selected neighbor(s).

In some of our recent QSAR/QMSA studies we have used different similarity measures derived from TIs in the selection of analogs and prediction of properties/activities for diverse sets of chemicals. We have also used orthogonal descriptors derived from a set of over 100 graph invariants to estimate bioactivity/toxicity of different graphs of molecules. In this paper we have used similarity measures derived from TIs in: a) selecting analogs of an isospectral graph from a diverse set of 221 compounds, and b) predicting the mutagenicity of a set of 113 mutagens and non-mutagens using QMSA methods.

2. Methods

2.1. Databases. A set of 19 pairs of isospectral graphs from the work of Balasubramanian and Basak [**25**] were added to a set of 107 benzamidines [**26**] and a composite set of 76 diverse compounds used in an earlier study by Basak and Grunwald [**23**] to create a varied library of 221 compounds. This composite library was created to provide a large set containing both congeneric and non-congeneric sets to test analog selection methods. The chemical structures for the 19 pairs of isospectral graphs have been presented previously [**25**].

A second data set, representing a subset of the set of 277 chemicals presented by Yamaguchi *et al.* [**27**] was also used in the current study. This subset consisted of all the chemicals in the set of 277 chemicals that had reported results for mutagenicity in the Ames test, mutagenicity in the medium term liver carcinogenesis bioassay, and carcinogenicity in the two-year rodent bioassay in rat and/or mouse. This subseting resulted in a set of 113 chemicals, 68 of which are classified as non-mutagens and 45 of which are classified as mutagens in the Ames test. This set of chemicals and their observed mutagenicity are reported in Table 1.

TABLE 1: Mutagenicity in the Ames test for 113 chemicals

No.[a]	Compound Name	Obs. Ames Mutagenicity
1.5	butylated hydroxyanisole (BHA)	0
1.6	caffeic acid	0
1.7	catechol	0
1.8	clofibrate	0
1.9	di(2-ethylhexyl)phthalate (DEHP)	0
1.10	hydroquinone	0
1.11	p-methoxyphenol	0
1.12	sesamol	0
1.13	tamoxifen	0
1.14	acetaminophen	0
1.15	benzoin	0
1.16	EPN	0
1.17	gallic acid	0
1.18	a-tocopherol	0
2.2	2-acethylaminofluorene (AAF)	1

TABLE 1: Mutagenicity in the Ames test for 113 chemicals

No.[a]	Compound Name	Obs. Ames Mutagenicity
2.3	adriamycin	1
2.4	aflatoxin B1	1
2.5	benzo[a]pyrene	1
2.7	captafol	1
2.8	captan	1
2.9	carbazole	1
2.10	dibutylnitrosamine (DBN)	1
2.11	diethylnitrosamine (DEN)	1
2.12	3,2'-dimethyl-4-aminobiphenyl (DMAB)	1
2.14	dimethylnitrosamine (DMN)	1
2.15	N-ethyl-N-hydroxyethylnitrosamine (EHEN)	1
2.16	N-ethyl-N-nitrosourea (ENU)	1
2.20	hydrazobenzene	1
2.22	laciocarpine	1
2.26	3'-methyl-4-dimethylaminoazobenzene (3'-Me-DAB)	1
2.27	3-amino-9-ethylcarbazole	1
2.28	N-nitrosooxazolidine	1
2.29	N-nitrosodi-n-propylamine (NDPA)	1
2.30	N-nitrosomorpholine	1
2.31	N-nitrosopiperidine	1
2.32	N-nitrosopyrrolidine	1
2.33	quinoline	1
2.34	sterigmatocystin	1
2.35	4,4'-thiodianiline	1
2.42	alachlor	0
2.43	aldrin	0
2.44	auramine O	0
2.45	barbital	0
2.46	chlordane	0
2.47	chlorendic acid	0
2.48	chlorobenzilate	0
2.49	DDT	0
2.50	dieldrin	0
2.51	diethylstilbestrol	0
2.53	ethenzamide	0
2.54	17α-ethinyl estradiol	0
2.55	DL-ethionine	0
2.56	hexachlorobenzene (HCB)	0
2.57	a-hexachlorocyclohexane (a-HCH)	0
2.58	d-limonene	0
2.59	monoclotaline	0
2.60	N-nitrosodiethanolamine	0
2.61	phenobarbital	0
2.64	safrole	0

TABLE 1: Mutagenicity in the Ames test for 113 chemicals

No.[a]	Compound Name	Obs. Ames Mutagenicity
2.66	thioacetamide	0
2.67	triadimefon	0
2.68	trifluralin	0
2.69	urethane	0
2.70	polychlorinated biphenyl (PCB)	0
2.71	malathion	0
2.72	vinclozolin	0
3.1	acetophenetidine (phenacetin)	1
3.2	azathioprine	1
3.3	N-butyl-N-(4-hydroxybutyl)nitrosamine (BBN)	1
3.4	chrysazin (danthron)	1
3.5	4,4'-diaminodiphenylmethane (DDPM)	1
3.6	7,12-dimethylbenz[a]anthracene (DMBA)	1
3.7	N-ethyl-N-(4-hydroxybutyl)nitrosamine (EHBN)	1
3.8	folpet	1
3.9	hydrogen peroxide	1
3.11	3-methylcholanthrene (3-MC)	1
3.12	N-methyl-N'-nitro-N-nitrosoguanidine (MNNG)	1
3.13	N-methyl-N-nitrosourea (MNU)	1
3.14	8-nitroquinoline	1
3.17	streptozotocin	1
3.18	o-toluidine	1
3.20	6-methylquinoline	1
3.21	8-methylquinoline	1
3.22	nitrofrantoln	1
3.23	6-nitroquinoline	1
3.24	quercetin	1
3.32	acetaldehyde	0
3.33	atrazine	0
3.34	di(2-ethylhexyl)adipate (DEHA)	0
3.35	1,1-dimethylhydrazine	0
3.39	trichloroacetic acid	0
3.42	4-acethylaminofluorene (AAF)	0
3.43	aspirin	0
3.44	butylated hydroxytoluene (BHT)	0
3.45	caffeine	0
3.46	caprolactam	0
3.47	chenodeoxicholic acid	0
3.49	cypermethrin	0
3.50	deltamethrin	0
3.51	diltiazem	0
3.52	dimethylsulfoxide (DMSO)	0
3.53	diazinon	0
3.54	fenvalerate	0

TABLE 1: Mutagenicity in the Ames test for 113 chemicals

No.[a]	Compound Name	Obs. Ames Mutagenicity
3.55	glutathione	0
3.56	4-o-hexyl-2,3,6-trimethylhydroquinone (HTHQ)	0
3.58	lithocolic acid	0
3.59	d-mannitol	0
3.61	phenol	0
3.64	propyl galiate	0
3.65	propylparaben	0
3.66	pyrene	0
3.67	resorcinol	0
3.71	trimorphamide	0

[a] The numbering scheme refers to the enumeration of the chemicals in the presentation by Yamaguchi *et al.* [**27**] where the numeral before the decimal place refers to the table in which the compound was listed (see below) and the numerals after the decimal refer to the compounds location within the table.
Table 1 - Association between inhibitory results in the medium-term liver bioassay (Ito test) and reported mutagenicity and carcinogenicity.
Table 2 - Association between positive results in the medium-term liver bioassay (Ito test) and reported mutagenicity and carcinogenicity.
Table 3 - Association between negative results in the medium-term liver bioassay (Ito test) and reported mutagenicity and carcinogenicity.

2.2. Calculation of Topological Indices. The TIs calculated for this study are listed in Table 2 and include Wiener number [**28**], molecular connectivity indices as calculated by Randić [**29**] and Kier and Hall [**4**], frequency of path lengths of varying size, information theoretic indices defined on distance matrices of graphs using the methods of Bonchev and Trinajstić [**30**] as well as those of Raychaudhury *et al.* [**31**], parameters defined on the neighborhood complexity of vertices in hydrogen-filled molecular graphs [**32**]–[**34**], and Balaban's J indices [**35**]–[**37**]. The majority of the TIs were calculated using POLLY 2.3 [**38**]. The J indices were calculated using software developed by the authors.

The Wiener index (W) [**28**], the first topological index reported in the chemical literature, may be calculated from the distance matrix $D(G)$ of a hydrogen-suppressed chemical graph G as the sum of the entries in the upper triangular distance submatrix. The distance matrix $D(G)$ of a nondirected graph G with n vertices is a symmetric $n \times n$ matrix (d_{ij}), where d_{ij} is equal to the distance between vertices v_i and v_j in G. Each diagonal element d_{ii} of D(G) is zero. We give below the distance matrix $D(G_1)$ of the unlabeled hydrogen-suppressed graph G_1 of thioacetamide (Fig. 1):

$$D(G_1) = \begin{array}{c|cccc} & 1 & 2 & 3 & 4 \\ \hline 1 & 0 & 1 & 2 & 2 \\ 2 & 1 & 0 & 1 & 1 \\ 3 & 2 & 1 & 0 & 2 \\ 4 & 2 & 1 & 2 & 0 \end{array}$$

W is calculated as:

$$W = 1/2 \sum_{ij} d_{ij} = \sum_h h \cdot g_h \tag{2.1}$$

where g_h is the number of unordered pairs of vertices whose distance is h. Thus for $D(G_1)$, W has a value of nine.

FIGURE 1. Unlabeled, hydrogen-suppressed graph of thioacetamide (G_1)

Randić's connectivity index [**29**], and higher-order connectivity path, cluster, path-cluster and chain types of simple, bond and valence connectivity parameters were calculated using the method of Kier and Hall [**4**]. The generalized form of the simple path connectivity index is as follows:

$$^h\chi = \sum (v_i v_j \dots v_{h+1})^{-1/2} \tag{2.2}$$

where $v_i, v_j, \dots, v_{h+1}$ are the degrees of the vertices in the path of length h. The path length parameters (P_h), number of paths of length h ($h = 0, 1, \dots, 10$) in the hydrogen-suppressed graph, are calculated using standard algorithms.

Information-theoretic topological indices are calculated by the application of information theory on chemical graphs. An appropriate set A of n elements is derived from a molecular graph G depending upon certain structural characteristics. On the basis of an equivalence relation defined on A, the set A is partitioned into h disjoint subsets A_i of order $n_i (i = 1, 2, \dots, h; \sum_{i=1}^{h} n_i = n)$. A probability distribution is then assigned to the set of equivalence classes:

$$\begin{array}{c} A_1, A_2, \dots, A_h \\ p_1, p_2, \dots, p_h \end{array}$$

where $p_i = n_i/n$ is the probability that a randomly selected element of A will occur in the ith subset.

TABLE 2: Symbols and brief definitions for 101 topological indices

I_D^W	Information index for the magnitudes of distances between all possible pairs of vertices of a graph
$\bar{I}_D^W$	Mean information index for the magnitude of distance
W	Wiener index = half-sum of the off-diagonal elements of the distance matrix of a graph
I^D	Degree complexity
H^V	Graph vertex complexity
H^D	Graph distance complexity
$\overline{IC}$	Information content of the distance matrix partitioned by frequency of occurrences of distance h
I_{ORB}	Information content or complexity of the hydrogen-suppressed graph at its maximum neighborhood of vertices
O	Order of neighborhood when IC_r reaches its maximum value for the hydrogen-filled graph
M_1	A Zagreb group parameter = sum of square of degree over all vertices
M_2	A Zagreb group parameter = sum of cross-product of degrees over all neighboring (connected) vertices
IC_r	Mean information content or complexity of a graph based on the rth ($r = 0-6$) order neighborhood of vertices in a hydrogen-filled graph
SIC_r	Structural information content for r^{th} ($r = 0-6$) order neighborhood of vertices in a hydrogen-filled graph
CIC_r	Complementary information content for r^{th} ($r = 0-6$) order neighborhood of vertices in a hydrogen-filled graph
${}^h\chi$	Path connectivity index of order $h = 0-6$
${}^h\chi_C$	Cluster connectivity index of order $h = 3-6$
${}^h\chi_{Ch}$	Chain connectivity index of order $h = 3-6$
${}^h\chi_{PC}$	Path-cluster connectivity index of order $h = 4-6$
${}^h\chi^b$	Bond path connectivity index of order $h = 0-6$
${}^h\chi_C^b$	Bond cluster connectivity index of order $h = 3-6$
${}^h\chi_{Ch}^b$	Bond chain connectivity index of order $h = 3-6$
${}^h\chi_{PC}^b$	Bond path-cluster connectivity index of order $h = 4-6$
${}^h\chi^v$	Valence path connectivity index of order $h = 0-6$
${}^h\chi_C^v$	Valence cluster connectivity index of order $h = 3-6$
${}^h\chi_{Ch}^v$	Valence chain connectivity index of order $h = 3-6$
${}^h\chi_{PC}^v$	Valence path-cluster connectivity index of order $h = 4-6$
P_h	Number of paths of length $h = 0-10$
J	Balaban's J index based on distance
J^B	Balaban's J index based on bond types
J^X	Balaban's J index based on relative electronegativities
J^Y	Balaban's J index based on relative covalent radii

The mean information content of an element of A is defined by Shannon's relation [**39**]:

$$IC = -\sum_{i=1}^{h} p_i \log_2 p_i \tag{2.3}$$

The logarithm is taken at base 2 for measuring the information content in bits. The total information content of the set A is then $n \times IC$. Figure 2 provides a sample calculation for IC_1.

It is to be noted that the information content of a graph G is not uniquely defined. It depends on how the set A is derived from G as well as on the equivalence relation which partitions A into disjoint subsets A_i. For example, when A constitutes the vertex set of a chemical graph G, two methods of partitioning have been widely used: a) chromatic-number coloring of G where two vertices of the same color are considered equivalent, and b) determination of the orbits of the automorphism group of G thereafter vertices belonging to the same orbit are considered equivalent.

Rashevsky was the first to calculate the information content of graphs where "topologically equivalent" vertices were placed in the same equivalence class [**40**]. In Rashevsky's approach, two vertices u and v of a graph are said to be topologically equivalent if and only if for each neighboring vertex $u_i (i = 1, 2, \ldots, k)$ of the vertex u, there is a distinct neighboring vertex v_i of the same degree for the vertex v. While Rashevsky used simple linear graphs with indistinguishable vertices to symbolize molecular structure, weighted linear graphs or multigraphs are better models for conjugated or aromatic molecules because they more properly reflect the actual bonding patterns, i.e., electron distribution.

To account for the chemical nature of vertices as well as their bonding pattern, Sarkar *et al.* [**41**] calculated information content of chemical graphs on the basis of an equivalence relation where two atoms of the same element are considered equivalent if they possess an identical first-order topological neighborhood. Since properties of atoms or reaction centers are often modulated by stereo-electronic characteristics of distant neighbors, i.e., neighbors of neighbors, it was deemed essential to extend this approach to account for higher-order neighbors of vertices. This can be accomplished by defining open spheres for all vertices of a chemical graph. If r is any non-negative real number and v is a vertex of the graph G, then the open sphere $S(v, r)$ is defined as the set consisting of all vertices v_i in G such that $d(v, v_i) < r$. Therefore, $S(v, 0) = \emptyset$, $S(v, r) = v$ for $0 < r < 1$, and $S(v, r)$ is the set consisting of v and all vertices v_i of G situated at unit distance from v, if $1 < r < 2$.

One can construct such open spheres for higher integral values of r. For a particular value of r, the collection of all such open spheres $S(v, r)$, where v runs over the whole vertex set V, forms a neighborhood system of the vertices of G. A suitably defined equivalence relation can then partition V into disjoint subsets consisting of vertices which are topologically equivalent for r^{th} order neighborhood. Such an approach has been developed and the information-theoretic indices calculated based on this idea are called indices of neighborhood symmetry [**34**].

In this method, chemicals are symbolized by weighted linear graphs. Two vertices u_o and v_o of a molecular graph are said to be equivalent with respect to r^{th} order neighborhood if and only if corresponding to each path $u_o, u_1, \ldots, u_r$ of

G_2: thioacetamide

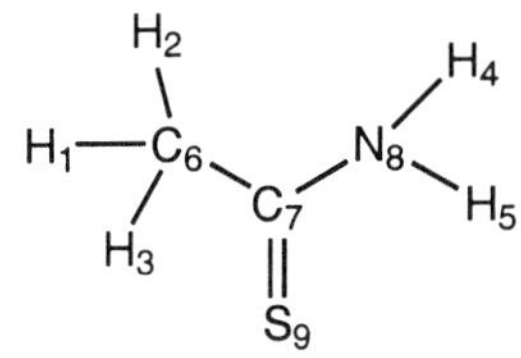

First order neighbors:

I	II	III	IV	V	VI

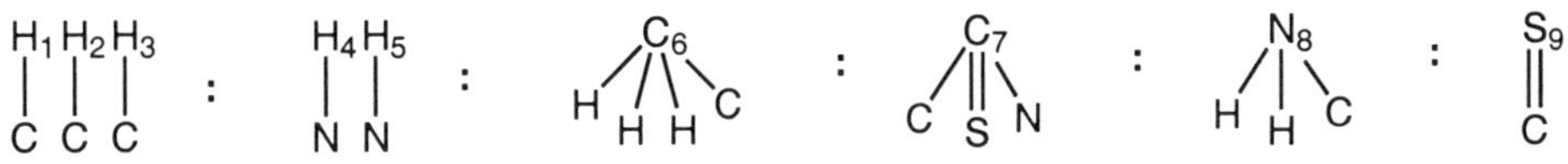

Subsets:

I	II	III	IV	V	VI
(H_1-H_3)	(H_4-H_5)	C_6	C_7	N_8	S_9

Probability:

I	II	III	IV	V	VI
3/9	2/9	1/9	1/9	1/9	1/9

$$IC_1 = 4 * 1/9 * \mathrm{Log}_2\, 9 + 2/9 * \mathrm{Log}_2\, 9/2 + 3/9 * \mathrm{Log}_2\, 9/3 \qquad = 2.419 \text{ bits}$$

$$SIC_1 = IC_1/\mathrm{Log}_2\, 9 \qquad = 0.763 \text{ bits}$$

$$CIC_1 = \mathrm{Log}_2\, 12 - IC_2 \qquad = 0.751 \text{ bits}$$

FIGURE 2. Labeled, hydrogen-filled graph of thioacetamide (G_2) and sample calculations for IC_1, SIC_1 and CIC_1

length r, there is a distinct path $v_o, v1, \ldots, v_r$ of the same length such that the paths have similar edge weights, and both u_o and v_o are connected to the same number and type of atoms up to the rth order bonded neighbors. The detailed equivalence relation has been described in earlier studies [**34, 42**].

Once partitioning of the vertex set for a particular order of neighborhood is completed, IC_r is calculated by Eq. (2.2). Basak *et al.* [**32**] defined another information-theoretic measure, structural information content (SIC_r), which is calculated as:

$$SIC_r = IC_r / \log_2 n \tag{2.4}$$

where IC_r is calculated from Eq.(2.2) and n is the total number of vertices of the graph.

Another information-theoretic invariant, complementary information content (CIC_r) [**43**], is defined as:

$$CIC_r = \log_2 n - IC_r \tag{2.5}$$

CIC_r represents the difference between maximum possible complexity of a graph (where each vertex belongs to a separate equivalence class) and the realized topological information of a chemical species as defined by IC_r. Sample calculations for SIC_1 and CIC_1 have been included in Figure 2.

The information-theoretic index on graph distance, I_D^W is calculated from the distance matrix $D(G)$ of a chemical graph G as follows [**30**]:

$$I_D^W = W \log_2 W - \sum_h g_h \cdot h \log_2 h \tag{2.6}$$

The mean information index, $\bar{I}_D^W$, is found by dividing the information index I_D^W by W. The information theoretic parameters defined on the distance matrix, H^D and H^V, were calculated by the method of Raychaudhury *et al.* [**31**].

Balaban defined a series of indices based upon distance sums within the distance matrix for a chemical graph that he designated as J indices [**35**]–[**37**]. These indices are highly discriminating with low degeneracy. Unlike W, the J indices range of values are independent of molecular size. The general form of the J index calculation is as follows:

$$J = q(\mu + 1)^{-1} \sum_{i,j,edges} (s_i s_j)^{-1/2} \tag{2.7}$$

where the cyclomatic number μ (or number of rings in the graph) is $\mu = q - n + 1$, with q edges and n vertices and s_i is the sum of the distances of atom i to all other atoms and s_j is the sum of the distances of atom j to all other atoms [**35**]. Variants were proposed by Balaban for incorporating information on bond type, relative electronegativities, and relative covalent radii [**36, 37**].

2.3. Calculation of Atom Pairs. Atom pairs (APs) were calculated using the method of Carhart *et al.* [**3**]. An atom pair is defined as a substructure consisting of two non-hydrogen atoms i and j and their interatomic separation:

$$< \text{atom descriptor}_i > - < \text{separation} > - < \text{atom descriptor}_j >$$

where $<$ atom descriptor $>$ contains information about the atomic type, number of non-hydrogen neighbors and the number of π electrons. The interatomic separation of two atoms is the number of atoms traversed in the shortest bond-by-bond path containing both atoms. APs used in this study were calculated by the APProbe software [**43**].

2.4. Statistical Methods and Computation of Intermolecular Similarity.

2.4.1. *Data Reduction.* Initially, all TIs were transformed by the natural logarithm of the index plus one. This was done since the scale of some TIs may be several orders of magnitude greater than other TIs.

A principal component analysis (PCA) was used on the transformed indices to minimize the intercorrelation of indices. The PCA was conducted using the SAS procedure PRINCOMP [**44**]. The PCA produces linear combinations of the TIs, called principal components (PCs) which are derived from the correlation matrix. The first PC has the largest variance, or eigenvalue, of the linear combination of TIs. Each subsequent PC explains the maximal index variance orthogonal to the previous PCs, eliminating any redundancies that could occur within the set of TIs. The maximum number of PCs generated is equal to the number of TIs available. For the purposes of this study, only PCs with eigenvalues greater than one were retained. A more detailed explanation of this approach has been provided in a previous study by Basak *et al.*[**13**]. These PCs were subsequently used to determine similarity scores as described below.

2.4.2. *Similarity Measures.* Intermolecular similarity was measured using two distinct methods. The AP method uses an associative measure described by Carhart *et al.* [**3**] and is based on atom pair descriptors. The measurement is the ratio of the number of shared atom pairs between two molecules over the total number of atom pairs present in the two molecules. Similarity (S) between molecules i and j is defined as:

$$S_{ij} = 2C/(T_i + T_j) \tag{2.8}$$

where C is the number of atom pairs common to molecule i and j. T_i and T_j are the total number of atom pairs in molecule i and j, respectively. The numerator is multiplied by a factor of 2 to reflect the presence of shared atom pairs in both compounds.

The second similarity method, Euclidean distance (ED) within an n-dimensional PC space derived from TIs was used. ED between molecules i and j is defined as:

$$ED_{ij} = \left[\sum_{k=1}^{n}(D_{ik} - D_{jk})^2\right]^{1/2} \tag{2.9}$$

where n equals the number of dimensions or PCs retained from the PCA. D_{ik} and D_{jk} are the data values of the kth dimension for molecules i and j, respectively.

2.4.3. *Analog / K-Nearest Neighbor Selection.* Following the quantification of intermolecular similarity of the molecules, analogs or nearest neighbors are determined on the basis of both S and ED. In the case of the AP method, two molecules are considered identical if $S = 1$, while they have no atom pairs in common if $S = 0$. The ED method measures a distance between molecules, thus the lower the value of ED the greater the similarity between two molecules.

2.4.4. *Property Estimation.* Since the data presented in the work of Yamaguchi *et al.* [**27**] represented mutagenicity as non-mutagen (−) or mutagen (+) this data was treated as a zero-one relationship, where non-mutagens have a value of zero and mutagens have a value of one. In estimating the mutagenicity of the probe compound, the mean of the observed mutagenicity of the K-nearest neighbors was used as the estimate. Thus, if the mean resulted in a value greater than 0.5, the compound was classified as a mutagen. However, if the mean was equal to 0.5, the compound was not classified as the results were inconclusive.

3. Results

3.1. Principal Component Analysis. From the PCA of the 102 TIs, eight PCs with eigenvalues greater than one were retained. These eight PCs explained, cumulatively, 95.2% of the total variance within the TI data. Table 3 lists the eigenvalues of the eight PCs, the proportion of variance explained by each PC, the cumulative variance explained, and the two TIs most correlated with each individual PC.

TABLE 3. Eigenvalues, variance explained and two TIs most correlated with the eight principal components

PC	Eigenvalue	Percent variance explained	Cumulative variance explained	First most correlated TI		Second most correlated TI	
PC_1	55.52	54.97	54.97	${}^4\chi^b$	(96.5%)	${}^3\chi$	(96.4%)
PC_2	12.38	12.26	67.23	SIC_3	(86.4%)	SIC_4	(85.5%)
PC_3	11.73	11.61	78.84	${}^5\chi^b_{Ch}$	(77.3%)	${}^5\chi^v_{Ch}$	(76.1%)
PC_4	6.78	6.71	85.55	IC_0	(55.0%)	${}^4\chi^v_{Ch}$	(49.7%)
PC_5	4.60	4.55	90.10	J	(68.9%)	J^Y	(62.4%)
PC_6	2.35	2.32	92.43	IC_0	(−47.2%)	SIC_0	(−36.4%)
PC_7	1.65	1.63	94.06	${}^4\chi^b_C$	(44.4%)	${}^4\chi^v_C$	(43.5%)
PC_8	1.16	1.14	95.21	${}^4\chi^v_C$	(−34.6%)	${}^6\chi^b_C$	(23.0%)

3.2. Analog Selection. Figure 3 shows the results of the analog selection for isospectral graph 10.1.1 using atom pairs to derive a similarity space and PCs to derive a Euclidean distance space. The first five analogs (neighbors) for the probe compound, 10.1.1, are presented for each of the similarity methods.

3.3. *K*-Nearest Neighbor Estimation. Table 4 presents the results for the prediction of mutagenicity for the 113 molecules over a range of K values ($K = 1-5$) for both the AP and ED methods. The results are presented as percent correctly classified and over-all percent correct prediction rates are provided as a means of comparing the efficacy of the individual models. The variability between the K levels is easily explained by the problematic nature of using a binary relationship such as this one in estimation. When the number of neighbors was even, the potential for unclassified compounds led to lower prediction rates than in the case of an odd number of neighbors.

4. Discussion

The major objective of this paper was to study the effectiveness of mathematical invariants in the characterization of molecular structure and the estimation of the toxicity of chemicals. An invariant maps a chemical structure into the set R of real numbers. A specific invariant may be used for the ordering or partial ordering of sets of molecules or in structure-activity relationship studies [**45**]. A particular

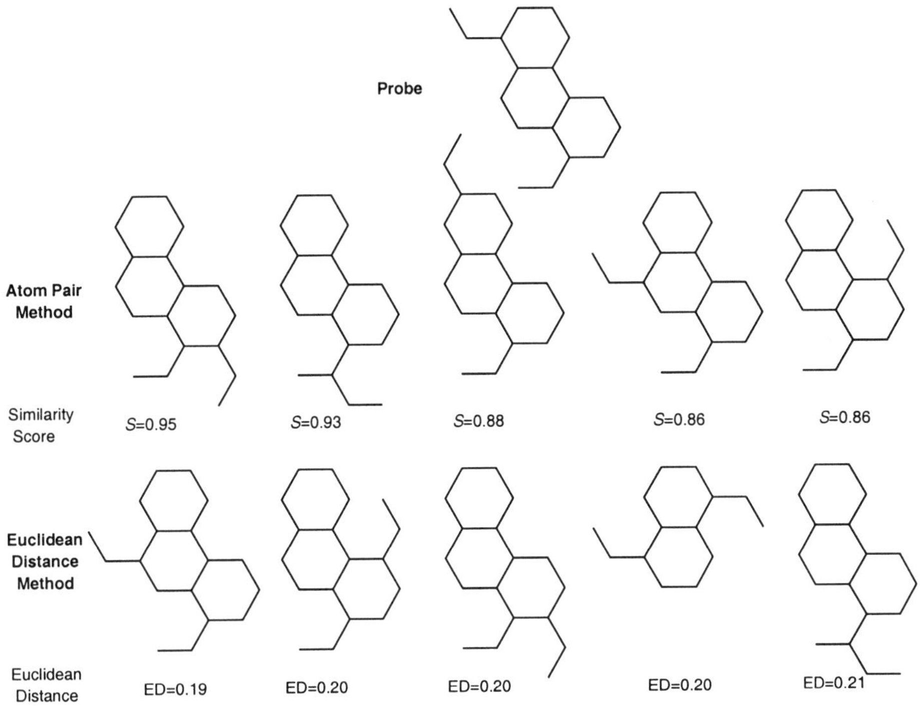

FIGURE 3. Analogs selected for isospectral graph 10.1.1

TABLE 4. KNN results for the prediction of mutagenicity for 113 chemicals

	Percent Negative Correct		Percent Positive Correct		Total Percent Correct	
K	*AP*	*ED*	*AP*	*ED*	*AP*	*ED*
1	73.5	75.0	84.1	66.7	77.7	71.7
2	66.2	64.7	72.7	33.3	68.8	52.2
3	77.9	80.9	88.6	53.3	82.1	69.9
4	70.6	69.1	77.3	42.2	73.2	58.4
5	79.4	77.9	86.4	53.3	82.1	68.1

structural invariant quantifies distinct aspects of molecular structure. Therefore, a combination of such indices might be more powerful in the mathematical characterization of molecular structure as compared to the use of one specific invariant. The problem arises out of the fact that often the various graph theoretic indices of molecular structures are strongly correlated. We have attempted to resolve this problem through the implementation of a PCA to derive orthogonal variables from a large set of calculated TIs, and using the orthogonal parameters in the characterization of structure [**10, 12, 15, 17, 18, 22, 23**].

In the present study we have used calculated atom pairs and principal components derived from TIs to select structural analogs for a probe compound from a diverse set containing closely related structures. The result of this analog selection, depicted in Figure 3, shows that the five neighbors selected by each of the methods exhibit sufficient power to reject dissimilar structures. In other words, we may conclude that both the atom pair and Euclidean distance methods are capable of choosing similar molecules from a collection of structurally diverse structures. This is in line with our earlier studies with various diverse sets of molecules [**10, 12, 15, 17, 18, 22, 23**].

The central paradigm of QSAR holds that similar structures usually have similar properties. To test this idea, we selected K-nearest neighbors ($K = 1 - 5$) for each molecule from a set of 113 mutagens and non-mutagens using the ED and AP methods and used the selected nearest neighbors in estimating mutagenicity. The results in Table 4 show that both methods lead to reasonably good estimates, although the AP method was superior to the ED method.

In conclusion, both the ED and AP methods, based on calculated graph theoretic structural invariants, did reasonably well in the selection of structural analogs and in the estimation of chemical properties based on nearest neighbors.

References

[1] H. Narumi, H. Hosoya, Topological Index and Thermodynamic Properties. II. Analysis of the Topological Factors on the Absolute Entropy of Acyclic Saturated Hydrocarbons, *Bull. Chem. Soc. Jpn.* **53** (1980), 1228–1237.

[2] M. Randić, Nonempirical Approaches to Structure-Activity Studies, *Int. J. Quantum Chem: Quant. Biol. Symp.***11** (1984), 137–153.

[3] R.E. Carhart, D.H. Smith, and R. Venkataraghavan, Atom Pairs as Molecular Features in Structure-Activity Studies: Definition and Applications, *J. Chem. Inf. Comput. Sci.* **25** (1985), 64–73.

[4] L.B. Kier, and L.H. Hall, Molecular Connectivity in Structure-Activity Analysis. Research Studies Press: Letchworth, Hertfordshire, U.K, 1986.

[5] D.H. Rouvray, and R.B. Pandey, The Fractal Nature, Graph Invariants and Physicochemical Properties of Normal Alkanes. *J. Chem. Phys.* **85** (1986), 2286–2290.

[6] S.C. Basak, Use of Molecular Complexity Indices in Predictive Pharmacology and Toxicology: A QSAR Approach. *Med. Sci. Res.* **15** (1987), 605–609.

[7] S.C. Basak, G.J. Niemi, and G.D. Veith, In Computational Chemical Graph Theory, D.H. Rouvray, Ed.; NOVA: New York, 1990, pp. 235–277.

[8] N. Trinajstić, Chemical Graph Theory, Klein, D. J., and Randić, M., Eds.; CRC Press: Boca Raton, 1992.

[9] A.T. Balaban, S. Bertelsen, and S.C. Basak, New Centric Topological Indexes for Acyclic Molecules (Trees) and Substituents (Rooted Trees) and Coding of Rooted Trees. *Math. Chem.* **30** (1994), 55–72.

[10] S.C. Basak, S. Bertelsen, and G.D. Grunwald, Application of Graph Theoretical Parameters in Quantifying Molecular Similarity and Structure-Activity Studies. *J. Chem. Inf. Comput. Sci.* **34** (1994), 270–276.

[11] S.C. Basak, G.D. Grunwald, and G.J. Niemi, Use of Graph-Theoretic and Geometrical Molecular Descriptors in Structure-Activity Relationships. In From Chemical Topology to Three Dimensional Molecular Geometry, Balaban, A. T., Ed.; Plenum Press: New York, 1997; pp 73–116.

[12] M. Johnson, S.C. Basak, and G. Maggiora, A Characterization of Molecular Similarity Methods for Property Prediction. *Mathematical and Computer Modelling* **II** (1988), 630–635.

[13] S.C. Basak, V.R. Magnuson, G.J. Niemi, and R.R. Regal, Determining Structural Similarity of Chemicals using Graph-Theoretic Indices. *Discrete Appl. Math.* **19** (1988), 17–44.

[14] S.C. Basak, G.J. Niemi, and G.D. Veith, Predicting Properties of Molecules Using Graph Invariants. *J. Math. Chem.* **7** (1991), 243–272.

[15] S.C. Basak, and G.D. Grunwald, Estimation of Lipophilicity from Structural Similarity. *New J. Chem.* **19** (1995), 231–237.

[16] S.C. Basak, and G.D. Grunwald, Predicting Mutagenicity of Chemicals Using Topological and Quantum Chemical Parameters: A Similarity Based Study. *Chemosphere* **31** (1995), 2529–2546.

[17] S.C. Basak, B.D. Gute, and G.D. Grunwald, Estimation of Normal Boiling Points of Haloalkanes Using Molecular Similarity. *Croat. Chim. Acta* **69** (1996), 1159–1173.

[18] S.C. Basak, and B.D. Gute, Use of Graph Theoretic Parameters in Predicting Inhibition of Microsomal p-Hydroxylation of Anilines by Alcohol: A Molecular Similarity Approach. In Proceedings of the International Congress on Hazardous Waste: Impact on Human and Ecological Health; Johnson, B. L., Xintaras, C., Andrews, J. S., Jr., Eds.; Princeton Scientific Publishing Co., Inc.: Princeton, NJ, 1997; pp 492–504.

[19] S.C. Basak, B.D. Gute, and G.D. Grunwald, Relative Effectiveness of Topological, Geometrical, and Quantum Chemical Parameters in Estimating Mutagenicity of Chemicals, Quantitative Structure-Activity Relationships. In Quantitative Structure-Activity Relationships in Environmental Sciences; Chen, F., Schuurman, G., Eds.; SETAC Press: Pensacola, FL, 1997; Vol. 7, Chapter 17, pp 245.

[20] S.C. Basak, B.D. Gute, and G.D. Grunwald, Use of Topostructural, Topochemical and Geometric Parameters in the Prediction of Vapor Pressure: A Hierarchical QSAR Approach, *J. Chem. Inf. Comput. Sci.* **37** (1997), 651–655.

[21] B.D. Gute, and S.C. Basak, Predicting Acute Toxicity (LC50) of Benzene Derivatives Using Theoretical Molecular Descriptors: A Hierarchical QSAR Approach, *SAR QSAR Environ. Res.* **7** (1997), 117–131.

[22] S.C. Basak, B.D. Gute, and G.D. Grunwald, Development and Applications of Molecular Similarity Methods using Nonempirical Parameters. *Mathl. Modelling Sci. Computing*, in press, 1999.

[23] S.C. Basak, and G.D. Grunwald, Use of Topological Space and Property Space in Selecting Structural Analogs. *Mathl. Modelling Sci. Computing*, in press, 1999.

[24] B.D. Gute, G.D. Grunwald, and S.C. Basak, Prediction of the Dermal Penetration of Polycyclic Aromatic Hydrocarbons (PAHs): A Hierarchical QSAR Approach, *SAR QSAR Environ. Res.* **10** (1999), 1–15.

[25] K. Balasubramanian, and S.C. Basak, Characterization of Isospectral Graphs Using Graph Invariants and Derived Orthogonal Parameters. *J. Chem. Inf. Comput. Sci.* **38** (1998), 367–373.

[26] S.C. Basak, B.D. Gute, and G.D. Grunwald, Prediction of Complement-Inhibitory Activity of Benzamidines Using Topological and Geometric Parameters. *J. Chem. Inf. Comput. Sci.* **39** (1999) 255–260.

[27] T. Yamaguchi, R. Hasegawa, A. Hagiwara, M. Hirose, K. Imaida, N. Ito, and T. Shirai, Results for 277 Chemicals in the Medium Term Liver Carcinogenesis Bioassay of Rats.

[28] H. Wiener, Structural Determination of Paraffin Boiling Points. *J. Am. Chem. Soc.* **69** (1947), 17–20.

[29] M. Randić, On Characterization of Molecular Branching. *J. Am. Chem. Soc.* **97** (1975), 6609–6615.

[30] D. Bonchev, and N. Trinajstić, Information Theory, Distance Matrix and Molecular Branching. *J. Chem. Phys.* **67** (1977), 4517–4533.

[31] C. Raychaudhury, S.K. Ray, J.J. Ghosh, A.B. Roy, S.C. Basak, Discrimination of Isomeric Structures Using Information Theoretic Topological Indices. *J. Comput. Chem.* **5** (1984), 581–588.

[32] S.C. Basak, A.B. Roy, and J.J. Ghosh, Study of the Structure-Function Relationship of Pharmacological and Toxicological Agents Using Information Theory. In Proceedings of the 2nd International Conference on Mathematical Modelling, Avula, X. J. R., Bellman, R., Luke, Y. L., and Rigler, A. K., Eds.; University of Missouri-Rolla: Rolla, Missouri, 1980; Vol. II, pp. 851–856.

[33] S.C. Basak, and V.R. Magnuson, Molecular Topology and Narcosis: A Quantitative Structure-Activity Relationship (QSAR) Study of Alcohols Using Complementary Information Content (CIC). *Arzneim. Forsch.* **33** (1983), 501–503.

[34] A.B. Roy, S.C. Basak, D.K. Harriss, and V.R. Magnuson, Neighborhood Complexities and Symmetry of Chemical Graphs and Their Biological Applications. In Mathematical Modelling in Science and Technology, Avula, X. J. R., Kalman, R. E., Lipais, A. I., and Rodin, E. Y., Eds.; Pergamon Press: New York, 1984, pp. 745–750.

[35] A.T. Balaban, Highly Discriminating Distance-Based Topological Index. *Chem. Phys. Lett.* **89** (1982), 399–404.
[36] A.T. Balaban, Topological Indices Based on Topological Distances in Molecular Graphs. *Pure & Appl. Chem.* **55** (1983), 199–206.
[37] A.T. Balaban, Chemical Graphs. Part 48. Topological Index J for Heteroatom-Containing Molecules Taking into account Periodicities of Element Properties. *Math. Chem. (MATCH)* **21** (1986), 115–122.
[38] S.C. Basak, D.K. Harriss, and V.R. Magnuson, POLLY 2.3: Copyright of the University of Minnesota, 1988.
[39] C.E. Shannon, A Mathematical Theory of Communication. *Bell Syst. Tech. J.* **27** (1948), 379–423.
[40] N. Rashevsky, Life, Information Theory and Topology. *Bull. Math. Biophys.* **17** (1955), 229–235.
[41] R. Sarkar, A.B. Roy, and R.K. Sarkar, Topological Information Content of Genetic Molecules — I. *Math. Biosci.* **39** (1978), 299–312.
[42] V.R. Magnuson, D.K. Harriss, and S.C. Basak, Topological Indices Based on Neighborhood Symmetry: Chemical and Biological Applications. In Studies in Physical and Theoretical Chemistry, King, R. B., Ed.; Elsevier: Amsterdam, 1983, pp. 178–191.
[43] S.C. Basak, and G.D. Grunwald, APProbe: Copyright of the University of Minnesota, 1993.
[44] SAS Institute Inc, in: SAS/STAT User's Guide, Release 6.03 Edition (SAS Institute Inc., Cary, NC, 1988 p. 751.
[45] C.L. Wilkins, and M. Randić, A Graph Theoretical Approach to Structure-Property and Structure-Activity Correlations. *Theor. Chim. Acta* **58** (1980), 45–68.

NATURAL RESOURCES RESEARCH INSTITUTE, 5013 MILLER TRUNK HWY., DULUTH, MN, 55811 USA
E-mail address: `sbasak@wyle.nrri.umn.edu`

NATURAL RESOURCES RESEARCH INSTITUTE, 5013 MILLER TRUNK HWY., DULUTH, MN, 55811 USA
E-mail address: `bgute@wyle.nrri.umn.edu`

DIMACS Series in Discrete Mathematics
and Theoretical Computer Science
Volume 51, 2000

Isomorphism Rejection in Structure Generation Programs

Gunnar Brinkmann

ABSTRACT. Efficient methods for structure generation are applicable in mathematics as well as in chemistry and even in industry. One of the problems when generating lists of structures is to avoid the generation of isomorphic copies. For this task in the last 20 years mathematicians have developed some techniques which will be communicated in this article.

1. Introduction

While in mathematics lists of interesting structures are used to test conjectures, search for counterexamples or solve finite problems, in chemistry these lists can be used to detect energetically suitable isomers for a given gross formula or as an approach to determine the structure automatically from a mass spectrum. In industry, sometimes properties of a product (e.g. a tire) can be modelled by a combinatorial structure. In this case structure generation programs can be applied for determining the combinatorial structure that optimizes this property.

One of the most important classes of graphs from the mathematical point of view is the class of *cubic* or *3-regular* graphs, that is: graphs where every vertex is adjacent with exactly 3 other vertices. While the first complete lists of cubic graphs on up to 10 vertices were generated by mathematicians already in the last century (see [**11**],[**12**]), the first computer approach for this important class was done by a chemist: In 1966/67 (see [**1**]) A.T. Balaban was interested in isomers of $C_{2n}H_{2n}$ without double bonds and every C-atom adjacent to an H-atom. He realized that this problem is equivalent to generating non-isomorphic cubic graphs and solved the problem for up to 12 vertices. Unfortunately this work was not well known in the mathematical community, so the lists were much later independently generated by mathematicians (see [**18**], [**25**]). After that various articles were published on this topic and for the time being the fastest program for this task is the one described in [**5**]. A more general algorithm to generate *k-regular graphs* for every k and not only $k = 3$ is described in [**24**].

The first approach to generate all isomers for an arbitrary chemical formula was started in 1965 in the well known *DENDRAL* project [**19**]. Since then various programs have been developed for this and similar tasks, some of them described in a special issue (number 27, 1992) of the journal *MATCH*. Nowadays the fastest program for this task is *MOLGEN* (see [**17**][**3**]), a program capable of generating all

1991 *Mathematics Subject Classification.* Primary 05C85; Secondary 05C90,68Q20,68R10.

isomers for a given chemical formula and accepting also restrictions such as forbidden or forced substructures which play an important role in practical applications e.g. automatic structure elucidation. The program described in [**24**] and a very efficient program to generate all graphs with a given set of vertex valencies (see [**16**]) are an essential part of *MOLGEN*.

2. Isomorphism Rejection

None of the programs mentioned can – a priori – guarantee to construct only non-isomorphic structures. So after the structures have been generated, isomorphic copies have to be sorted out. The first idea how this could be done is possibly to store all the structures generated so far in a file and whenever a new structure is generated to look for an isomorphic copy already in the file. Unfortunately this is only possible if a very small number of graphs shall be generated – and even then the time for searching the list will decrease the generation rate – that is the number of non-isomorphic structures generated per second – dramatically.

So it must be the aim to decide for every structure whether it is necessary for the list without comparing it to the others. The aim of this article is to give an informal and easy-to-read introduction into the most important techniques that achieve this. Due to this aim, at some places the text is not as formal as it should be for a deeper treatment. In these cases the reader is also referred to other articles with an explicit and formal treatment of the method. For the same reason we will restrict the exposition to graphs. It is not difficult to generalize the methods to deal with other combinatorial structures if necessary.

The essential concept in all the methods is the concept of *canonicity* of a certain labelling or numbering of the graph. Because of the importance of this concept, we will first give – as an example – a formal definition of a *canonical vertex labelling* before using it more informally (and sometimes also more generally, e.g. also for *canonical labellings of faces*) in the rest of this text.

Let $\mathcal{G}$ denote the set of all labelled graphs. A canonical representative function (short: *canonical form*) is a function $c : \mathcal{G} \to \mathcal{G}$ with the properties that for all $G \in \mathcal{G}$ the graph $c(G)$ is isomorphic to G and that $c(G) = c(G')$ whenever G and G' are isomorphic. The graph $c(G)$ is also called the canonical representative of G. Note that in every isomorphism class c fixes exactly one element as the canonical representative of the class. An isomorphism $\phi : G \to c(G)$ is called a canonical labelling of the vertices of G. While canonical representatives are unique, canonical labellings are unique only up to automorphisms of G.

The easiest and simultaneously most important canonical form *max* can be explained as follows: Given a (labelled) graph on n vertices. Assign the new labels $1, \ldots, n$ to the vertices in every possible way and compute the adjacency matrix – that is the $n \times n$ matrix with the entry a_{ij} in line i and row j equal to 1 if vertex i and j are adjacent and 0 otherwise. The binary number obtained when writing the rows of the adjacency matrix after each other is called the *number* of the labelled graph. Take that labelled graph as the canonical representative that gives the largest number. This canonical form was used e.g. in [**13**],[**27**],[**24**],[**5**] and many other algorithms in the mathematical literature and was independently invented and used by Randić in [**26**] and some follow-up papers.

An algorithm to compute another canonical form very efficiently is described in [**22**]. In [**20**] Klein and Liu also invent a method for the canonical ordering of vertices based on the computation of eigenvalues.

Of course in a similar way we can define canonical labellings of edges and in case we talk about graphs embedded on a surface even canonical labellings of faces.

3. Generating Canonical Representatives

The method of *generating canonical representatives* can be described in 4 steps:

a: Define a way to code the structures that is unique for every labelled structure occurring in the generation.

b: Among all possible codes for a certain structure that can be obtained by relabelling, choose one as *the canonical code* (e.g. the lexicographically smallest or largest one)

c: Generate structures together with their associated codes and make sure that for every isomorphism class one structure is generated together with the canonical code.

d: Accept a generated structure if the associated code is canonical – reject it otherwise.

This method will be illustrated for the example of fullerene generation. Though the fastest algorithm for generating fullerenes (see [**6**]) also uses this technique, we will not choose this algorithm, since the canonical form used there is very technical and needs a lot of constructions that would have to be explained first. We will instead use the spiral algorithm proposed by Fowler and Manolopoulos (see [**14**]) which is not complete (for a structure that cannot be generated by this algorithm see [**14**]), but because of its simple structure well suited for demonstrating the technique.

Step a: Define a way to code the structures:

We want to generate all fullerenes that can be encoded in the way depicted in Figure 1 – that is: Starting at one face and deciding on the next (neighbouring) face and an orientation (clockwise or counterclockwise), all faces of the fullerene can be visited by recursively choosing that face as the next one to be visited, which has not been visited before, is adjacent to the last face visited so far and is the first new neighbour running around the last face in clockwise (counterclockwise) direction from the intersection with the next to last face. The code is simply the sequence of face sizes in the order they were visited.

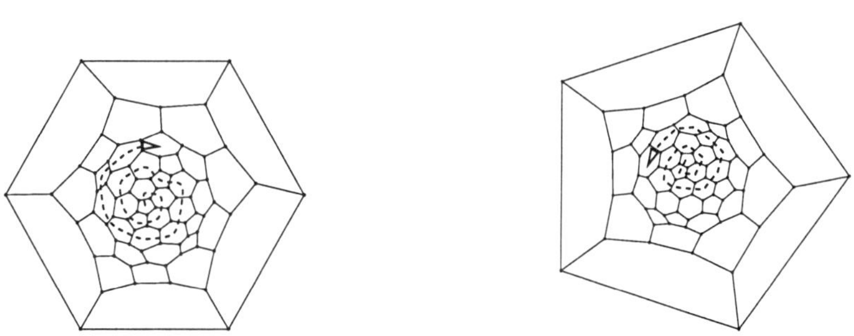

A spiral corresponding to the code
566656665666666566566556666566666556655666

The lexicographically minimal code
556666666665666666556656656655666665666655

FIGURE 1.

Step b: Among all possible codes for a certain structure choose one as *the canonical code*

In general a fullerene will allow spirals for many different starts and orientations. Some few fullerenes even have spirals starting at every face pairs and for every orientation. These fullerenes have been called *omnispiral fullerenes* by Patrick Fowler. Among all possible codes we choose the lexicographically smallest one to be *the canonical code*. So to determine the canonical code of a fullerene, we must try to construct a spiral for every possible start and orientation and in case we succeed compare it to the best code found so far. Of course in those cases where the beginning spiral already is lexicographically larger than the best spiral found so far, we do not need to proceed. The time consumption for this test is quadratic in the number of vertices. In the case of fullerenes that also have spirals starting at a pentagon (which is by far most of them), the time consumption is linear in the number of vertices if the possible spirals starting at a pentagon are checked first.

Step c: Generate structures together with their associated codes and make sure that every structure is once generated together with the canonical code:

To this end we generate the fullerenes in a spiral fashion – that is: We generate all sequences with the pregiven number of "6's" and twelve "5's" and interpret them as corresponding to a spiral patch formed in e.g. counterclockwise direction. Since every such sequence is generated exactly once, we can be sure that every such fullerene is generated exactly once together with its canonical code.

If we generate the sequences recursively by constructing the sequences of length n out of those of length $n-1$ and enlarge our spiral patch together with the code, we can sometimes realize that a sequence will not lead to a fullerene at all. So starting with e.g. eleven "5's" in our sequence, we can see at once that this will not close to give a fullerene unless we have no hexagons, so we would not have to proceed generating sequences of this type.

Step d: Accept a generated structure if the associated code is canonical – reject it otherwise.

To this end we try to construct the fullerene corresponding to the sequence generated. If the sequence gives a spiral that closes to a fullerene, we compute the canonical code. If the canonical code is lexicographically smaller than the sequence used to construct the fullerene, the structure is rejected – otherwise it is accepted. Of course we can reject the structure as soon as **any** smaller code is found, since already this tells us that the canonical code will be smaller than the original sequence. So the structure in Figure 1 would be rejected when generated together with the first code and accepted when generated together with the second one.

The main advantages of this approach are:

- One never has to look into lists or compute invariants other than the canonical form.
- If one wants to test properties of the graphs, they can be piped directly into the testing program – one does not need disk space or even main memory for storing them.

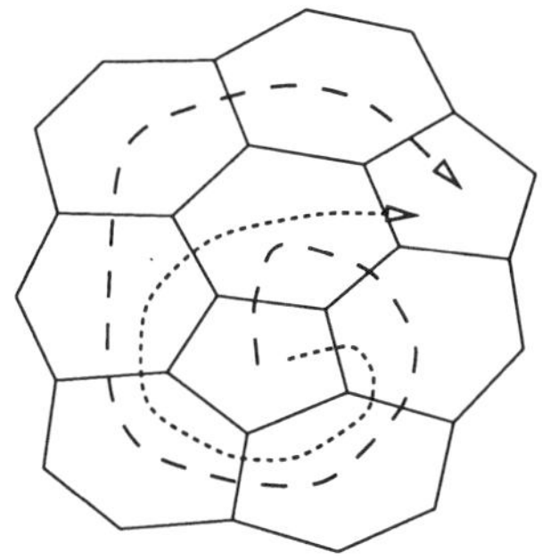

FIGURE 2. The partial spiral with code 566666665 (dashed line) cannot lead to a canonical omnispiral fullerene, since the dotted line will give a smaller code.

3.1. Read/Faradžev-Type Orderly Algorithms. Sometimes, we can even improve on this approach: **Assume** that only omnispiral fullerenes shall be generated. So we can be sure that every possible start of a spiral will close to give a spiral.

In this case, some early bounding criterion can be inserted, similar to the early check whether the structure can possibly close to give a fullerene:

We can check beginning spirals in the partial structure as depicted in Figure 2. In case a partial structure already contains a beginning spiral with a code smaller than the one used to construct it, we can backtrack already at this early point, since in every structure being constructed out of this partial structure the smaller beginning spiral will lead to a code that is smaller than the original one.

Algorithms with such an early bounding criterion with respect to the canonicity criterion are called Read/Faradžev-type orderly algorithms. Some applications of this technique, sometimes with very special and interesting canonical forms, can be found e.g. in [**13**],[**27**],[**9**], [**10**],[**6**],[**24**],[**7**],[**5**].

It should be mentioned that the exact definition of orderly algorithms in [**27**] is slightly different, while in [**13**] it is closer to the way we introduced it. Nevertheless the differences are mainly of a technical nature and there is no danger of misunderstanding.

But there are also some disadvantages of the method of generating canonical representatives:

- The canonical code and the construction must be "compatible" in the sense that one must make sure that a structure corresponding to the canonical code is constructed. So we cannot necessarily use the same canonical code for different ways to construct our structures.
- In some cases the corresponding canonical form may be hard to compute.
- Often an early bounding criterion is not possible and so the approach is not as efficient.

4. Application of this Approach to the Homomorphism Principle

The homomorphism principle is an essential tool e.g. in the *MOLGEN* generator. See e.g. [**17**] for a formally exact treatment of this topic. It is not only a method aiming at isomorphism rejection, but also a method for construction, nevertheless it allows very efficient isomorphism rejection combined e.g. with the previous approach. If a set of structures is to be generated, the main strategy is as follows:

a: Find an underlying coarser structure so that any isomorphism between two structures generated induces an isomorphism of the underlying coarser structures. In most cases the number of coarser structures is much smaller and they are easier to handle.
b: Generate all (isomorphism classes of) coarser structures.
c: Generate all finer structures out of these coarser structures.

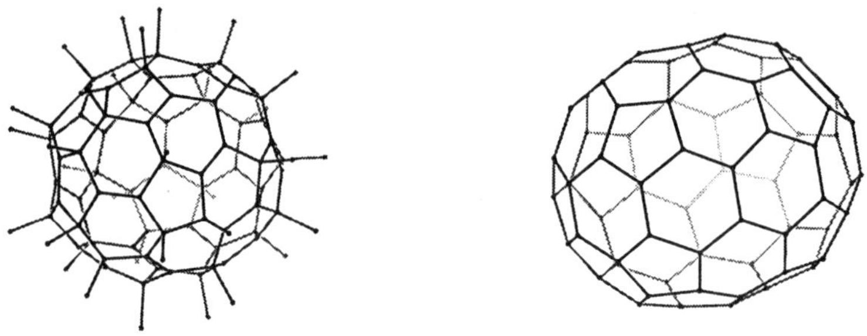

FIGURE 3. A hydrogenated fullerene and its underlying structure

The two generation processes in this strategy can again be organized using the method of *generating canonical representatives* . We will make this explicit as an example: Suppose we want to generate hydrogenated fullerenes, that is chemical isomers of $C_x H_y$ with $x \geq y$, so that the C-atoms form a fullerene. For an approach that just counts such structures without constructing them, Polya's theorem can be applied (see e.g. [**2**]).

It is obvious that an isomorphism from one hydrogenated fullerene onto another will map C-atoms onto C-atoms, since even if we do not explicitly label the vertices with H and C, the H atoms can be identified by being all the vertices with valency one. So a natural choice in step (a) for the underlying coarser structure is the graph spanned by all vertices with valency more than 1 – the underlying fullerene.

In this case an algorithm for step b was already given in the first part (if we restrict ourselves to fullerenes which can be generated with the spiral algorithm), but of course we could also use the much faster algorithm in [**6**]. Note that in fact in the first part we not only get the fullerene in canonical form, but we also get the automorphism group of the fullerene without extra computations: There is a one-to-one correspondence between elements of the automorphism group and spirals leading to a canonical code. The fact that at least a generating set for the automorphism group is generated is typical for the computation of canonical forms.

For step c we again apply the *generation of canonical representatives*: Given an underlying fullerene we define the following code: For a canonical labelling of the vertices of the fullerene, we **additionally** label the vertices with 0 (not bonded to a Hydrogen atom) and 1 (bonded to a Hydrogen atom) and write these additional labels into one line according to the canonical numbering of the underlying fullerene.

The canonical code of the hydrogenated fullerene (with respect to the underlying fullerene) now is defined as the lexicographically smallest one among all those codes obtained from the different canonical labellings of the fullerene. Note that the canonical labelling of the faces coming from the spiral code can easily be converted into a canonical labelling of the vertices.

To check canonicity, we just have to check whether any permutation given by the automorphism group of the fullerene (this way we get all possible canonical labellings of the fullerene) gives a smaller code.

The restrictions imposed by the underlying coarser structure (in this case the underlying fullerene) makes the problem much easier: For C_{60} there is a ratio of 83 percent of isomers with trivial automorphism group, for C_{80} the ratio is 94 percent and for C_{100} it is already more than 98 percent. For these structures no check at all must be performed – there is only one canonical labelling, so we know that every additional labelling with 0 and 1 is canonical. And even in the remaining cases we mostly have to deal with comparatively small groups. Computing e.g. isomers of $C_{100}H_{20}$ for each of the 98 percent of the fullerenes with trivial group, $\binom{100}{20}$ non-isomorphic isomers can be constructed without any isomorphism checking. So the homomorphism principle is indeed a very powerful method and leads to extraordinarily fast programs in most cases where it is applicable.

5. McKay-Type Orderly Algorithms

Now we will present an alternative approach, which has some advantages compared with Read/Faradžev-type orderly algorithms. For details about this approach see [**23**]. In fact there is some common basis for both kinds of orderly algorithms and formally one could even define the previous examples as an application of *McKay-type orderly algorithms.* Nevertheless from the intuitive point of view it is easier to handle them as different approaches.

The aim is to construct larger structures out of smaller ones and to avoid the construction of isomorphic copies during such a construction step.

If a recursive construction for the class of structures in question is given – that is: Larger structures are generated by some well defined operations from smaller ones of the same kind starting from some finite set of initial elements – the method can be applied recursively to assure that on every level we have only non-isomorphic structures. In some cases even an infinite set of initial elements can serve as a start if one already has a method for constructing these efficiently.

Again the strategy can be described in some easy steps:

a: For every structure assign a unique ancestor from which it must be generated and a way how it must be generated from it.

b: Accept a structure if and only if it is generated this way.

c: Make sure that from one ancestor the same larger structure is never generated twice in the same way.

We will make this method explicit for the example of planar polycyclic hydrocarbons with a fixed number of pentagons and hexagons. This method has been applied for similar structures in [8] and [**15**]. To be exact: by *planar polycyclic hydrocarbon* we mean a graph that can be embedded into the plane, all faces except one (the outer face) are hexagons or pentagons, all vertices have valency two or three and all vertices with valency two lie in the boundary of the outer face. Furthermore we do not allow edges with the same face on both sides (bridges).

Note that we do not require the graph to be embeddable into the plane by regular hexagons – our definition allows topological deformations of the faces.

Without pentagons, these structures are sometimes called *fusenes* in the chemical literature and must be distinguished from *benzenoids* which might be defined as those helicenes that are subgraphs of the graphite lattice.

There is an easy recursive structure for planar polycyclic hydrocarbons. They can be generated by adding one face (hexagon or pentagon) at a time, starting at an initial pentagon or hexagon as depicted in Figure 4.

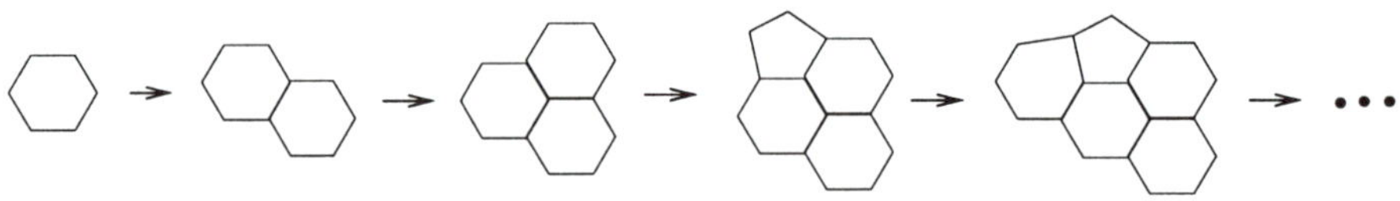

FIGURE 4. The recursive construction of a planar polycyclic hydrocarbon

So how can we carry out the three steps ?

Step a: For every structure assign a unique ancestor from which it must be generated and a means by which to generate it.

Since the construction operation adds a face, we must define what is the last face to be added – removing this again gives the ancestor. One possibility to fix such a face is the following:

Again the concept of canonicity plays a central role. Suppose we are given an **arbitrary** canonical form that assigns numbers to the faces (like the spiral code), edges or vertices. We want a face to be the last one to be added, that is in the boundary and can be deleted without disconnecting the structure. If the numbers are assigned to the faces, among all such faces we can choose that face as the last one to be added, that has the smallest canonical number. If the numbers are assigned to the edges, we could e.g. choose that face which contains the edge with the smallest canonical number in the common boundary with the outer face (to make sure it is just contained in one possible face). If the numbers are assigned to the vertices, we could e.g. choose that face which contains the edge with the lexicographically smallest vertex pair in the boundary. The fact that we are very free in our choice of the canonical labelling routine gives us the possibility to choose a very efficient algorithm for this purpose, without any respect to the compatibility with our construction method. Nevertheless for the simplicity of the presentation let us assume that the canonical labelling routine assigns numbers to the faces.

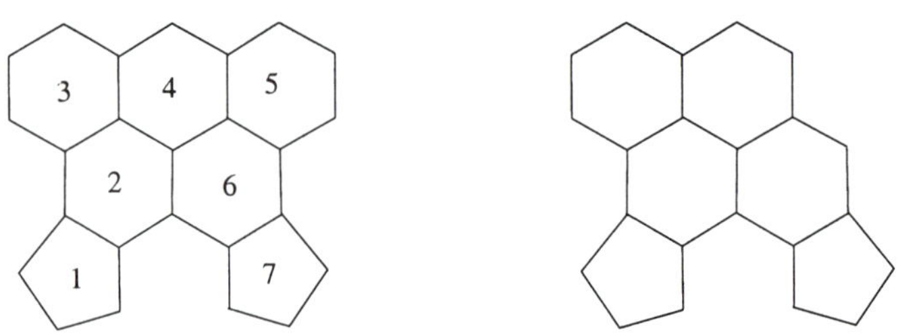

FIGURE 5. A canonically labelled planar polycyclic hydrocarbon and its ancestor

Step b: Accept a structure if and only if it is generated this way.

To this end, we must check whether the last face added is one that **can** yield the smallest canonical label among all those that can be removed. In this case it is constructed from the correct ancestor by adding the correct face. Note that there is some ambiguity in canonical labelling: it is unique only up to isomorphisms of the graph. And in fact we get the same result if we remove the canonical face or any other face in the orbit of the canonical one. So if we compute a canonical labelling of the new structure, we must not check whether the last face added is the canonical one, but whether it is in the same orbit. As noted before, in general the automorphism group is a byproduct of the canonical numbering, so it is easy to compute the orbit of the canonical face. In Figure 5 the faces numbered 3,4 and 5 are the only faces which can be removed without disconnecting the structure, so the face numbered 3 is the canonical one. So we must accept the structure if and only if the last face added was the one labelled 3 or the one labelled 5 – which is equivalent to the one labelled 3.

Step c: Make sure that from a small structure the same larger structure is never generated twice in the same way.

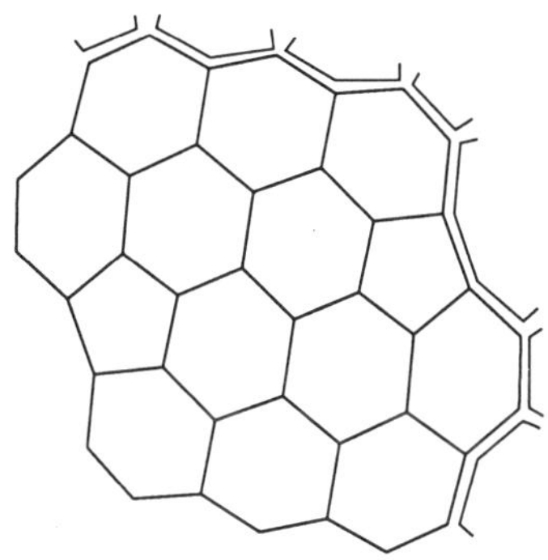

FIGURE 6. The places where a face must be added.

Of course the vague term *in the same way* must be made precise. This means that – up to isomorphism – the two operations used to construct the larger structure are indistinguishable. So we could not distinguish between two operations adding e.g. a hexagon to two places that can be mapped onto each other by an automorphism of the smaller structure, while we could distinguish between two operations adding hexagons to two places that cannot be mapped onto each other. So what we have to do is to compute the orbits of possible places where new faces can be added and choose one representative out of every orbit to add the next face.

It can be easily seen that this indeed leads to our aim that the same graph is never accepted twice:

Assume we construct two isomorphic structures and both are accepted. Then there is an isomorphism mapping the last faces that were added onto each other, since both were canonical. But this means an isomorphism of the smaller structures is induced, mapping the positions where the faces have been added onto each other. So on one hand both copies were generated from the same ancestor (since there is an isomorphism of the smaller structures) – and on the other hand the positions where the faces are added can be mapped onto each other by an automorphism of the graph – so they have been in the same orbit. But this is a contradiction to the fact that we use only one position in every orbit.

The main advantage of the generation of canonical representatives – that is essentially that no lists of graphs have to be used – is also an attribute of this approach – but there are also other advantages:

- The canonical numbering routine must not fulfill any compatibility requirements, so fast canonical numbering programs and even existing routines like e.g. nauty [**21**] can be used.
- Since only an inverse operation must be fixed, in many cases it can even be checked without canonical numbering whether the structure is generated from the correct ancestor. In the example one could e.g. require a pentagon to be deleted if possible – or some other fast heuristics that might determine a face uniquely.

6. McKay-Type Orderly Algorithms and the Double Coset Method

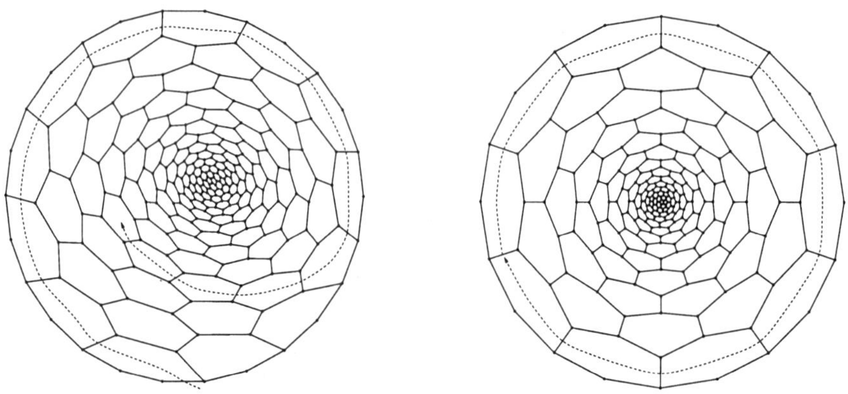

FIGURE 7. Cuts through two tube-like fullerenes: The first is through a twisted, the second through an untwisted tube.

While in the last example, the requirement in step c could easily be fulfilled by computing some orbits, sometimes the situation looks more complicated. If the construction step requires to glue two nontrivial structures to each other, the double coset method (see [**17**][**4**]) is some convenient means. The double coset method reduces the problem of constructing only non-isomorphic ancestors – which may look different for every kind of generation program we are aiming for – to a standard problem in group theory: the generation of double cosets.

This will again be made explicit at an example. Assume we want to generate all tube-like fullerenes with a fixed circumference h and an untwisted tube body – that is: traversing hexagons in a way that one enters through the midpoint of one edge and leaves them through the midpoint of the opposite one, one can run around the tube body by traversing h hexagons and ends up in the same hexagon. In the twisted case, one does not end up in the same hexagon, but runs around the tube body like a helix. This is shown in Figure 7. It can be shown that an untwisted tube can be cut along a zigzag path (e.g. the boundary path of the second tube in Figure 7) while a twisted one can be cut along a path that is zigzag with two exceptions – one point with two right turns after each other and one with two left turns after each other. Two parts of a twisted tube-like fullerene cut along such a path can be glued to each other or to the end of a tube body only in one way, so in this case there is no problem. In case of an untwisted tube, the caps can be glued to the body in various positions that might or might not give different results and this is the problem we want to discuss here. An efficient method to generate all combinatorially different caps can easily be developed by slight modifications of the constructions described in [**6**].

Assume we want to glue the two structures in Figure 8 to the end of a long tube body in a way that we get every descendant structure exactly once. If there is an isomorphism between two descendant structures, it must clearly map the first cap onto the first cap, the second cap onto itself and the tube also. So this isomorphism must induce automorphisms of the caps. It is easy to see that in fact the length of the tube does not play any role as long as we can be sure that the caps must be fixed by any isomorphism, so we can as well look at the problem of gluing the caps to each other in every possible way, so that no isomorphism of descendant structures exists that fixes the boundary cycles of the caps. In general, of course both caps can be isomorphic, but here we will restrict to the case where they are different, so that any isomorphism fixes the caps as a whole.

Gluing the caps to each other along the boundary can be formally expressed by an isomorphism ϕ mapping the boundary of cap 1 onto that of cap 2, if we define the resulting graph as the one obtained by identifying the boundary vertices and edges of cap 1 with their images. For the sake of simplicity we still address these vertices by the names they have as parts of cap 2. So we are interested in all those isomorphisms of the boundaries that give non-isomorphic structures when identifying the boundary vertices and edges of cap 1 with their images.

Note that the automorphisms of the caps induce automorphisms of the boundary cycles. We denote these induced automorphisms by $\alpha_i \in A_i$ if α is the automorphism in the automorphism group A of the cap.

For a fixed isomorphism ϕ it will of course produce an isomorphic structure, if we first apply an automorphism α_i with α in the automorphism group A_1 of cap 1 and then apply ϕ (e.g. first do a rotation by 180 degrees – that is 6 positions – in the case depicted in Figure 8). This can be easily seen by the isomorphism acting like the identity on cap 2 and like α^{-1} on the interior of cap 1. The same is obviously true, if we first apply ϕ and then an automorphism β_i with β in the automorphism group A_2 of cap 2 (e.g. a rotation by 4 positions in the case depicted in Figure 8). If on the other hand an isomorphism ϕ' of the boundary cycles produces a structure isomorphic to the one obtained by ϕ', then the isomorphism of the structures must induce automorphisms α of cap 1 and β of cap 2, since they are fixed as a whole. But first applying α_i^{-1}, then ϕ and then β_i^{-1} we get a structure that is isomorphic to the one obtained by ϕ by the same automorphism as the one obtained by ϕ'.

If we define two isomorphisms ϕ, ϕ' of the boundaries to be equivalent if and only if there are automorphisms α of cap 1 and β of cap 2, so that $\phi' = \beta_i \circ \phi\alpha_i$ we obtain a one-to-one correspondence between non-isomorphic structures obtained by identifying the boundaries and equivalence classes of isomorphisms.

If ϕ and ϕ' are such isomorphisms of the boundary, then $\phi' \circ \phi^{-1}$ is an automorphism in the automorphism group B_2 of the boundary of cap 2, so $\phi' = \phi' \circ \phi^{-1} \circ \phi$ can be written in the form $\gamma \circ \phi$ with γ an automorphism of the boundary of cap 2. On the other hand for every such automorphism γ the function $\gamma \circ \phi$ is an isomorphism of the boundaries.

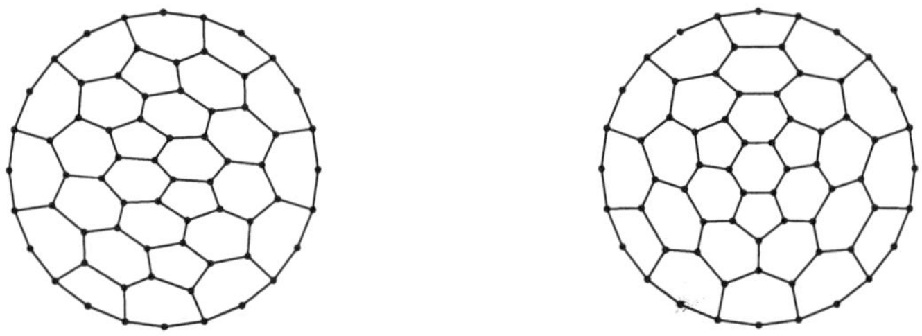

FIGURE 8. Two caps of a tube-like fullerene that must be glued to the ends of the tube body in every possible way.

Following the strategy above, for a fixed ϕ we have a one-to-one correspondence between isomorphism classes of structures obtained by identifying the boundaries and equivalence classes of isomorphisms of the form $\gamma \circ \phi$ with $\gamma \in B_2$.

Furthermore any fixed isomorphism ϕ of the boundaries induces an isomorphism between the automorphism groups A_1 of cap 1 and a subgroup A_1' of B_2 by mapping $\alpha \in A_1$ onto $\phi \circ \alpha_i \circ \phi^{-1}$.

By definition $\gamma \circ \phi$ is equivalent to $\gamma' \circ \phi$ if and only if $\gamma \circ \phi = \beta_i \circ \gamma' \circ \phi \circ \alpha_i$. Letting ϕ^{-1} operate from the right, we get $\gamma = \beta_i \circ \gamma' \circ \phi \circ \alpha_i \circ \phi^{-1}$.

So we get a one-to-one correspondence between isomorphism classes of structures obtained by identifying the boundaries and equivalence classes of automorphisms of the boundary of cap 2 if we define two such automorphisms γ, γ' to be equivalent if and only if there is an automorphism $\beta_i \in (A_2)_i$ and an automorphism $\alpha' \in A_1'$, so that $\gamma = \beta_i \circ \gamma' \circ \alpha'$. These equivalence classes are called double cosets and are written as $(A_2)_i \gamma A_1'$. That is: we reduced the combinatorial problem to the group theoretic problem of generating double cosets. Having solved this problem, we can take one representative out of every double coset and compute the corresponding isomorphism to identify the boundaries.

In case of trivial automorphisms of the caps we get that every isomorphism leads to a new structure, which will again make the algorithm very efficient. For nontrivial groups double cosets have to be computed, but this is a well studied standard problem in group theory and very efficient algorithms (see e.g. **[28][29]**) exist. In some cases it is even possible to use existing programs like those implemented in GAP (see `http://www-gap.dcs.st-and.ac.uk/~gap`).

7. Conclusions

A conclusion of this article could be:

In General isomorphism rejection isn't the problem in Structure Generation.

Of course this is not a mathematical theorem and there are some examples, where none of these techniques is easily applicable or leads to an efficient program. Furthermore when aiming for **very** fast programs, every part must be optimized to the limit of course and in this case again the isomorphism rejection might need some efforts.

Nevertheless the real problems in structure generation are in the construction method itself. This can be easily seen by revisiting the example of planar polycyclic hydrocarbons. This time we do not want to give the number of hexagons as the input, but – more realistically – the constitutional formula C_xH_y and the number

p of pentagons. The Euler formula gives us that the number of bounded faces (hexagons or pentagons) is exactly $f = 1 + (x - y)/2$, so we know exactly the number h of hexagons and the number p of pentagons. Unfortunately in some cases only a very small ratio of the structures with this number of hexagons and pentagons correspond to this constitution formula – and sometimes even none of them. There are for example already more than $4.5 \cdot 10^{11}$ isomers with 20 hexagons. So asking for all isomers of $C_{384}H_{48}$ and no pentagons – they have 169 hexagons by the Euler formula – at first sight might seem to be asking for an enormous number of isomers. But there is only one such isomer depicted in Figure 9. This example shows at once that the generation strategy used for generating just the structures with a prescribed number of hexagons and pentagons cannot be applied here. Even if some bounding criteria would be inserted that at some point of the generation tree can recognize that none of the ancestors will correspond to the required formula, this will surely be at a relatively high level in the generation tree, so it will be much too late for an efficient generation.

This means that an alternative generation strategy is necessary. Developing generation strategies for restricted classes of structures that are able to construct efficiently exactly the structures inside that class is normally the main problem with which one is confronted. Once the construction method has been found, in most cases one of the above mentioned methods can be applied for isomorphism rejection.

A method for generating planar polycyclic hydrocarbons with a given number of pentagons and a given chemical formula has been developed and will be described in another article.

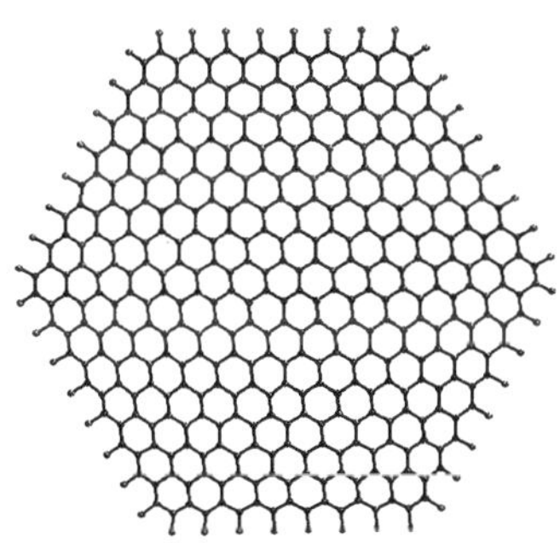

FIGURE 9. The only possible isomer of $C_{384}H_{48}$ with only hexagons (169).

References

[1] A.T. Balaban. Valence-isomerism of cyclopolyenes. *Revue Roumaine de chimie 11*, pages 1097–1116, 1966. No. 12 (1967) p.103 (erratum).

[2] K. Balasubramanian. Enumeration of chiral and positional isomers of substituted fullerene cages (c_{20}–c_{70}). *J. Phys. Chem.*, 97:6990–6998, 1993.

[3] C. Benecke, T.Grüner, A.Kerber, R.Laue, and T.Wieland. MOLecular structure GENeration with MOLGEN, new features and future developments. *Fresenius Journal of Anal. Chem.*, 359:23–32, 1997.

[4] A. Betten, A. Kerber, A. Kohnert, R. Laue, and A. Wassermann. The discovery of simple 7-designs with automorphism group $P\Gamma L(2,32)$. In *Applied algebra, algebraic algorithms and error-correcting codes (Paris, 1995)*, volume 948 of *Lecture Notes in Comput. Sci.*, pages 131–145. Springer, Berlin, 1995.
[5] G. Brinkmann. Fast generation of cubic graphs. *Journal of Graph Theory*, 23(2):139–149, 1996.
[6] G. Brinkmann and A.W.M. Dress. A constructive enumeration of fullerenes. *Journal of Algorithms*, 23:345–358, 1997.
[7] G. Brinkmann, B.D. McKay, and C. Saager. The smallest cubic graphs of girth 9. *Combinatorics, Probability and Computing*, 5:1–13, 1995.
[8] G. Caporossi and P. Hansen. Enumeration of polyhex hydrocarbons to $h = 21$. *Journal of Chemical Information and Computer Sciences*, 38:610–619, 1998.
[9] C.J. Colburn and R.C. Read. Orderly algorithms for generating restricted classes of graphs. *Journal of Graph Theory, Vol.3*, pages 187–195, 1979.
[10] C.J. Colburn and R.C. Read. Orderly algorithms for graph generation. *Intern. J. Computer Math.1979, Sec.A.7*, pages 167–172, 1979.
[11] J. de Vries. Over vlakke configuraties waarin elk punt met twee lijnen incident is. *Verslagen en Mededeelingen der Koninklijke Akademie voor Wetenschappen, Afdeeling Natuurkunde (3) 6*, pages 382–407, 1889.
[12] J. de Vries. Sur les configurations planes dont chaque point supporte deux droites. *Rendiconti Circolo Mat. Palermo 5*, pages 221–226, 1891.
[13] I.A. Faradžev. Constructive enumeration of combinatorial objects. *Colloques internationaux C.N.R.S. No260 - Problèmes Combinatoires et Théorie des Graphes, Orsay 1976*, pages 131–135, 1976.
[14] P.W. Fowler and D.E. Manolopoulos. *An Atlas of Fullerenes.* Oxford University Press, 1995.
[15] P.W. Fowler, D. Mitchell, and G. Brinkmann. Electronic and steric factors in the stability of proto-fullerene hydrocarbons. In W. Andreoni, editor, *The chemical physics of the fullerenes 10 (and 5) years later.* Kluwer, 1995. to appear.
[16] T. Grüner. Ein neuer Ansatz zur rekursiven Erzeugung von schlichten Graphen. Master's thesis, Universität Bayreuth, 1995.
[17] R. Grund, A. Kerber, and R. Laue. MOLGEN – ein Computeralgebrasystem für die Konstruktion molekularer Graphen. *match 27*, pages 87–131, 1992.
[18] W. Imrich. Zehnpunktige kubische Graphen. *Aequationes Math.6*, pages 6–10, 1971.
[19] R.K. Lindsay, B.G. Buchanan, E.A. Feigenbaum, and J. Lederberg. *Artificial Intelligence for Organic Chemistry: The Dendral Project.* McGraw-Hill, New York, 1980.
[20] X. Liu and D.J. Klein. The graph isomorphism problem. *Journal of Computational Chemistry*, 12, No 10:1243–1251, 1991.
[21] B.D. McKay. nauty user's guide (version 1.5). Technical Report TR-CS-90-02, Australian National University, Department of Computer Science, 1990.
http://cs.anu.edu.au/ bdm/index.html.
[22] B.D. McKay. Practical graph isomorphism. *Congressus Numerantium Vol. 30*, pages 45–87, 1981.
[23] B.D. McKay. Isomorph-free exhaustive generation. *Journal of Algorithms*, 26:306–324, 1998.
[24] M. Meringer. Fast generation of regular graphs and construction of cages. to appear in Journal of Graph Theory.
[25] A.N. Petrenjuk and L.P. Petrenjuk. On constructive enumeration of 12 vertex cubic graphs (russian). *Combinatorial analysis, no.3 Moscow*, 1974.
[26] M. Randić. On discerning symmetry properties of graphs. *Chem. Phys. Letters*, 42(2):283–287, 1976.
[27] R.C. Read. Every one a winner. *Annals of Discrete Mathematics 2*, pages 107–120, 1978.
[28] B. Schmalz. Verwendung von Untergruppenleitern zur Bestimmung von Doppelnebenklassen. *Bayreuth. Math. Schr.*, 31:109–143, 1990.
[29] B. Schmalz. t-Designs zu vorgegebener Automorphismengruppe. *Bayreuth. Math. Schr.*, 41:1–164, 1992. Dissertation, Universität Bayreuth, Bayreuth, 1992.

Fakultät für Mathematik, Universität Bielefeld, D33501 Bielefeld
E-mail address: gunnar@mathematik.uni-bielefeld.de

DIMACS Series in Discrete Mathematics
and Theoretical Computer Science
Volume 51, 2000

Foldedness in Linear Polymers: A Difference Between Graphical and Euclidean Distances

L. Bytautas, D.J. Klein, M. Randić, and T. Pisanski

ABSTRACT. A formal measure of the difference between two distance functions for a molecule is made, and for the case of through-bond and through-space distances this distance-function difference is formulated as a useful characterization of molecular "shape". In particular for linear-polymer conformations such a difference is proposed as a measure of the degree of a polymer's "foldedness". The measure is computed for a few sequences of self-similar conformations of linear-chain paths on the square-planar lattice. An analysis is made of the asymptotic long-chain behavior to obtain accurately extrapolated estimates for this limit, which then are compared with the fractal dimensions for these limiting structures. A few comments on other distance-function distances are made.

1. Introduction

Molecular shape is recognized (as in [**1**]) to be important in governing physical, chemical, and biological properties. In turn one key characteristic of the overall shape of a linear-chain polymer is its "degree of foldedness" as it appears embedded in Euclidean space. In fact in dealing with protein chains the detailed manner of folding is now realized [**2**]–[**6**] to be crucial to understanding their bioactivities, as they vary with shape which in turn is governed by amino-acid sequence and physico-chemical environment. For regular polymers the extent to which the polymer chains are folded is recognized to be qualitatively different depending on chemical conditions or type of model used:

- random-walk models appropriate for "poor" solvents (where there is a low solvation energy for the polymer), and
- self-avoiding-walk models appropriate for "good" solvents (which solvate strongly with the polymer).

In particular these different circumstances lead to [**7, 8**] different exponential dependences of extensional size on chain length, and this difference is sometimes identified [**9, 10**] with different so-called "fractal" dimensions [**11**] for the infinite-chain limit in the two circumstances. Indeed one might plausibly imagine that as

---

1991 *Mathematics Subject Classification.* Primary 05C12; Secondary 92E10.

Two of the authors (LB & DJK) acknowledge the support of the Welch Foundation of Houston, Texas.

the fractal dimension increases so also does the "degree of foldedness". That is, if the N monomer units of a typical conformation are all confined within a Euclidean distance of L of one another, it is typically argued (as by Flory [**7**] and deGennes [**8**]) that on the average there is a power-law relation with L scaling as $N^{1/d}$, in approaching the limit $N \to \infty$, so that as d increases the polymer may be anticipated to be more "folded". But with this scaling relation $L \sim N^{1/d}$ inverted to $N \sim L^d$ the exponent d may be recognized [**9, 10**] as a *fractal* dimension (such as discussed in Section 3 here). And now even with reference to protein folding the fractal view is sometimes emphasized [**12**]–[**16**]. Different properties varying with fractal dimension might equally be imagined to vary with a "degree of foldedness", which we define without restriction to the $N \to \infty$ limit.

Here we consider a fundamental approach to "degree of foldedness" (or more briefly just "foldedness") independent of the $N \to \infty$ limit. This approach entails the comparison of two distance functions, one measuring distances through space and the other measuring distances through the bonding network. The first distance function here is the usual Euclidean distance function, and the second might be identified as the *graphical* (or *shortest-path* or *topological* or *through-bond*) distance measuring lengths of shortest paths along bonds of the molecule. The Euclidean distance function is typically used in specifying "molecular geometry", and the graphical distance function has been widely studied in mathematical graph theory, as extensively reviewed in [**18**], and it has also been studied in "chemical graph theory", as reviewed in [**19**]. Comparisons between the two distances have been suggested [**20**] to be of chemical relevance, and inter-related types of comparisons to measure "foldedness" have been proposed [**21, 22**], with the first of them illustrated [**21, 23**] for several smaller graphs. Here such foldedness measures are to be studied developing an efficient computational scheme, which is applied for a range of linear-chain conformations. Special attention is paid to "self-similar" chain structures, the variations as a function of chain length, and extrapolation to the infinite-chain limit, where comparison to the fractal dimensions associated to the different types of infinite-length chain conformations is relevant. The foldedness measure used here is however not necessarily restricted to fractal limits or to self-similar structures.

Further the generality of some of the ideas beyond that of foldedness may be noted. That is, the measures considered apply to graphs other than chains. And also since the measures are for the difference between two distance functions (or metrics), neither of the distance functions under comparison need be the Euclidean metric, or alternatively both could be Euclidean metrics, say for different conformations of the same molecule. Thus we generally presume a discrete set V on which there are defined two (or more) metrics. The set V is in our applications associated to a molecular graph G, which in turn is specified: first by the set V whose elements are *vertices* or *sites*, and second by a set E of *edges* or *bonds* each consisting of a pair of distinct (near-neighbor) elements of V. A *distance* function or *metric* ρ on V is standardly defined as a function mapping pairs of vertices of V to real numbers such that

$$
(1.1) \qquad
\begin{array}{lll}
\text{(a)} & \rho(i,i) = 0 & \\
\text{(b)} & \rho(i,j) = \rho(j,i) > 0, & \text{for } i \neq j \\
\text{(c)} & \rho(i,j) + \rho(j,k) \geq \rho(i,k) &
\end{array}
$$

If the elements of V are embedded in a Euclidean space $\mathcal{E}$, then one natural choice for ρ is that induced by the space $\mathcal{E}$. If V is the vertex-set of a connected graph G, then one (graph-intrinsic) choice for $\rho(i,j)$ is as the number of steps in the shortest path (of bonds) between vertices i and j. Now given one metric ρ_1, another metric ρ_s may be defined for any positive real s by the *rescaling* relation

$$\rho_s(i,j) \equiv s \cdot \rho_1(i,j) \tag{1.2}$$

and the two metrics may be described as *scale equivalent.* The factor s may be viewed to reflect a change in numerical values under a change in the "units" (or scale) used to measure distances. Or even with fixed units one can view s to represent a magnification (or dilatation) of a fixed "shape", so that scale equivalence of two objects may also be described by saying the two objects have the "same shape" (but different sizes perhaps). The mathematical ideas involved in a measure of the difference between two metrics is detailed in Section 2, both with respect to measuring an absolute difference between two metrics and with respect to measuring the difference between their associated scale equivalence classes. For the choice of the shortest-path graphical distance and the Euclidean distance of an embedding in Euclidean space, the measure of such a difference seems a natural measure of "foldedness", as emphasized [**21**] previously. That is, one naturally expects that a polymer conformation should be more highly folded the greater the difference there is between "through-space" and "through-bond" distances. In Section 3 the different types of self-similar sequences of linear-chain conformations chosen for study are described, the fractal dimensions for their limiting forms are identified, and an efficient computational scheme for our foldedness measure is described. This method is used to obtain the foldedness for rather long chains (of up to about 10^5 sites), and the ensuant numerical results are reported. In Section 4 arguments for asymptotic forms for the long-chain limit are developed, and the numerical data is fit to these forms, to obtain accurate infinite-chain "degrees of foldedness" (as distance-differences). In Section 5 comparisons of these extrapolated distance-differences to conventional "fractal" dimensions are made, to reveal some degree of correspondence, and finally some discussion is made including comments on different uses of distance-difference measures.

2. Distance-Difference Measures

It is wished to compare the shortest-path graphical metric $\rho = \rho_g$ and a Euclidean metric $\sigma = \rho_{\mathcal{E}}$, but we initially phrase our discussion more generally. The comparison between two metrics ρ and σ on a common set V is conveniently expressible in terms of ratios of their values at different arguments. Thence two *distance-ratio* matrices $\mathbb{M}(\eta,+)$ and $\mathbb{M}(\eta,-)$ may be defined with rows & columns labeled by elements of V and $(i,j)^{\text{th}}$ elements giving the ratios of distances between these two sites

$$\mathbb{M}(\eta,+)_{ij} \equiv \begin{cases} \eta, & i=j \\ \rho(i,j)/\sigma(i,j), & i \neq j \end{cases} \tag{2.1}$$

and

$$\mathbb{M}(\eta,-)_{ij} \equiv \begin{cases} 1/\eta, & i=j \\ \sigma(i,j)/\rho(i,j), & i \neq j \end{cases} \tag{2.2}$$

The corresponding elements of the two matrices may be viewed as magnification (or dilatation) factors needed to bring the corresponding distances into coincidence. Thence, corresponding elements are inverses of one another, and the diagonal elements $\eta > 0$ and $1/\eta$ are indeterminate in terms of ratios between $\rho(i,i)$ and $\sigma(i,i)$ (since these "diagonal" ρ- and σ-values are 0). The value of η presumably is chosen in some "natural" way, as is taken up at the end of this section.

The $\mathbb{M}$-matrices encoding a comparison between the two distance functions may now be manipulated to present this comparison more compactly. The respective matrices of (2.1) and (2.2) have maximum eigenvalues denoted $\lambda(\eta,+)$ and $\lambda(\eta,-)$, and the corresponding eigenvectors are denoted $\boldsymbol{C}(\eta,+)$ and $\boldsymbol{C}(\eta,-)$. That is

$$\mathbb{M}(\eta,s)\boldsymbol{C}(\eta,s) \quad = \quad \lambda(\eta,s)\boldsymbol{C}(\eta,s), \qquad s = + \text{ or } - \tag{2.3}$$

Since these matrices have all elements positive, a theorem 16 of Frobenius & Perron applies to say that the maximum-eigenvalue eigenvectors are *nodeless* in the sense that their phase may be chosen so that all their components $\boldsymbol{C}(\eta,s)_i$ are positive. Then the Rayleigh-Ritz quotient

$$\frac{\sum_{ij}\boldsymbol{C}(\eta,s)_i\mathbb{M}(\eta,s)_{ij}\boldsymbol{C}(\eta,s)_j}{\sum_i \boldsymbol{C}(\eta,s)_i\boldsymbol{C}(n,s)_i} \quad = \quad \lambda(\eta,s) \tag{2.4}$$

appears much like an average over the elements of the matrix $\mathbb{M}(\eta,s)$, except that the non-negative coefficients of these elements do not necessarily add up to 1. But this lack of normalization is readily rectified with a multiplication by a suitable "normalization" factor, which is conveniently expressed in terms of the eigenvector $\boldsymbol{C}(\eta,s)$ and the vector $\mathbf{1}$ with all components 1, as well as their transposes, $\boldsymbol{C}^\dagger(\eta,s)$ and $\mathbf{1}^\dagger$. That is, the expression

$$\frac{\lambda(\eta,s)\left(\boldsymbol{C}^\dagger(\eta,s)\boldsymbol{C}(\eta,s)\right)}{\left(\mathbf{1}^\dagger\boldsymbol{C}(\eta,s)\right)\left(\boldsymbol{C}^\dagger(\eta,s)\mathbf{1}\right)} \quad \equiv \quad c(\eta,s) \tag{2.5}$$

is a proper probabilistic average over all the elements of the matrix $\mathbb{M}(\eta,s)$. Then $c(\eta,+)$ is an average factor by which the σ-distances would need to be dilatated in order to match the ρ-distances (with the average weighting the larger magnitudes more strongly), and $c(\eta,-)$ is an average factor by which the ρ-distances would need to be dilatated to match the σ-distances. The (magnitudes of the) logarithms of these $c(\eta,s)$ then are interpretable as average "magnification powers" (needed to bring one set of distances in closer coincidence with another). Indeed, the quantity

$$|\log c(\eta,+)| + |\log c(\eta,-)| \quad \equiv \quad \Delta(\rho,\sigma) \tag{2.6}$$

is a measure of the *difference* between ρ and σ. More rigorously stated: Δ is **[22]** a *semi-metric* on the set of distance functions on the vertices V of G. That is, Δ is symmetric in its arguments, all its values are non-negative, and it takes the value 0 iff ρ and σ are the same distance function.

But there are other possible measures of the difference between two metrics. One measure based on the maximum and minimum off-diagonal values in $\mathbb{M}(\eta,+)$ is **[22]** possible, but the result ends up being very sensitive to what may be atypical extreme distance ratios in $\mathbb{M}(\eta,+)$. That is, in working with averages as in the $c(\eta,s)$ going into Δ, there is an avoidance of sensitivity to atypical extreme values in distance ratios for the distance functions under comparison. In the case of a chain embedded in Euclidean space such that all bond lengths are the same, there has also been proposed **[21]** an alternative related measure

$$\lambda(0,+)/N \tag{2.7}$$

where N is the order of V, the diagonal element η has been taken to be 0, ρ is the Euclidean distance, σ is the shortest-path graphical distance, and the units of length are chosen to match the bond lengths. Here the smallness of the excess over 1 was considered to be a measure of similarity of the two distances, but though usually no less than 1, this quantity turns out to be (just barely) less than 1 for a geometrically straight chain. A better alternative would be to use a distance-ratio matrix $\mathbb{M}(1,+)$ and take a logarithm. [The logarithm of course is one way to obtain an excess over 1, but especially it has another fundamental feature in removing dependence on scale: e.g., with the logarithmic choice, if a metric ρ_1 is scaled up and down to rescaled metrics ρ_s and $\rho_{1/s}$ each to be a common difference δ from ρ_1, then the difference from $\rho_{1/s}$ to ρ_s is 2δ.] But now in fact this logarithmic measure can be modified even further to make it independent of the presumptions of the uniformity of bond lengths, and even independent of the particular nature of the distances ρ and σ being compared: one takes the difference between these two measures to be

$$\left|\log\left(\lambda(\eta,+)/N\right)\right| + \left|\log\left(\lambda(\eta,-)/N\right)\right| \quad \equiv \quad \Delta'(\rho,\sigma) \tag{2.8}$$

This possible definition is fairly close to our definition of Δ, and indeed the proof from ref. [**22**] that Δ is a semi-metric applies also to Δ'.

Finally it is possible to measure the difference between whole classes of scale equivalent distance functions. Here two distance functions ρ and σ are *scale equivalent* iff $\rho(i,j) = k \cdot \sigma(i,j)$ for all i,j. The quantity

$$\log\left(c(\eta,+)c(\eta,-)\right) \quad \equiv \quad D(\rho,\sigma) \tag{2.9}$$

is [**22**] a measure of the difference between the two scale equivalence classes of ρ and σ. That is, this measures (again in a semi-metric manner) the difference between ρ and σ modulo scale. And there are other possible alternative measures, including

$$\log\left(\lambda(\eta,+)\lambda(\eta,-)/N^2\right) \quad \equiv \quad D'(\rho,\sigma) \tag{2.10}$$

Here we will be more interested in these differences modulo scale, thereby avoiding any particular choice of scale.

There still remains the choice of the parameter η (as appears in the diagonal elements of the distance-ratio matrices). One fairly natural choice would be to take η to be the ratio of smallest non-zero distances for the compared distance functions ρ and σ, with the ratio taken in inverse manners for the ρ/σ & σ/ρ matrices. Another natural choice would be that obtained by varying η so that the difference (Δ or D) becomes as small as possible. And indeed this variation may be done analytically (especially for D), making use of the fact that the eigenvalues $\lambda(\eta,+)$ and $\lambda(\eta,-)$ depend especially simply on η while the eigenvectors are independent of η (since a change of η corresponds to just adding a multiple of the identity matrix to the $\mathbb{M}$-matrices). That is, noting that $\lambda(0,+)$ and $\lambda(\infty,-)$ represent the eigenvalues when there are only 0-elements along the diagonals of the corresponding $\mathbb{M}$-matrices, one sees that

$$\lambda(\eta,+) = \lambda(0,+) + \eta \qquad \& \qquad \lambda(\eta,-) = \lambda(\infty,-) + 1/\eta \tag{2.11}$$

Then upon substitution of this first into (2.5) and then (2.9) there results an expression whose derivative with respect to η may be set to 0 to obtain the optimal value

$$\eta = \{\lambda(0,+)/\lambda(\infty,-)\}^{1/2} \tag{2.12}$$

In the particular cases described in the next section we find little difference between the two indicated natural choices for η. Our "optimal" choice seems to us especially "nice", e.g., in avoiding a choice for the association of the minimum Euclidean distance with a particular value of bond-length of which there will likely be slightly different values for a real polymer (with bond lengths near the end of a polymer chain being different than in the interior).

Overall there seems to be especially nice features for the distance-difference of (2.9) with the η-choice of (2.12). This measure involves proper averages of distance ratios combined so as to be independent both of scale and other details of a conformation which one might otherwise imagine in a specification of η. The formulation indeed is sufficiently general to measure the difference (modulo scale) of a general pair of distance functions, though in the following we focus on a particular pair, and then as applied to a few particular lattice embeddings of linear-chain path graphs.

3. Generation of Self-Similar Conformations & Eigen-Solutions

The linear-chain conformations of polymers are generally viewed [**8, 9**] to be self-similar on the average, so that it is reasonable to consider a few such conformations, here chosen to be built in a deterministic way with a range of so-called "fractal" dimensions, presumably correlating with a range of different experimental circumstances. Further to aid in making extrapolations to infinite chain lengths the conformations are confined to a two-dimensional square-planar lattice. These deterministically self-similar paths are generated in stages: blocks with walks at one stage are combined in a fixed pattern to obtain walks at the next stage, as indicated in a general way in Figure 1, and as illustrated for a particular pattern through the first 3 stages in Figure 2. Indeed this pattern of Figure 2 leads (upon iteration) toward the famous [**25**] "area-filling" Hilbert curve, and being "area-filling" it might reasonably be imagined to have some sort of effective dimension of 2. Indeed such ideas of effective dimension can be given a precise meaning in terms of so-called "fractals". With a number of "self-similar" structures of N sites and characteristic length one finds a large-N scaling relation $N \sim L^d$, which (via comparison with "classical" structures) might suggest that d be identified as a dimension. More formally a *fractal* dimension can be defined [**11**] to be a limit

$$d_f \equiv \lim_{N\to\infty} \{(\log N)/(\log L)\} \tag{3.1}$$

where N is a measure of the substance (here the number of sites) of the structure (here a path or polymer) and L is the characteristic length of the region of space in which the structure is confined. Sometimes d_f is instead called the "Besicovitch-Hausdorff" dimension. For the Hilbert curve the number of sites quadruples at each stage while the length scale doubles, so that at the nth stage $N = 4^n$ and (with a suitable length unit) $L = 2^n$, whence the formal definition of the fractal dimension agrees with our intuition of the 2-dimensionality. Though we continue with the Hilbert curve sequence as our proto-typical example of self-similar structures, it is desirable that the relevancy for self-similar structures in a wider context be addressed.

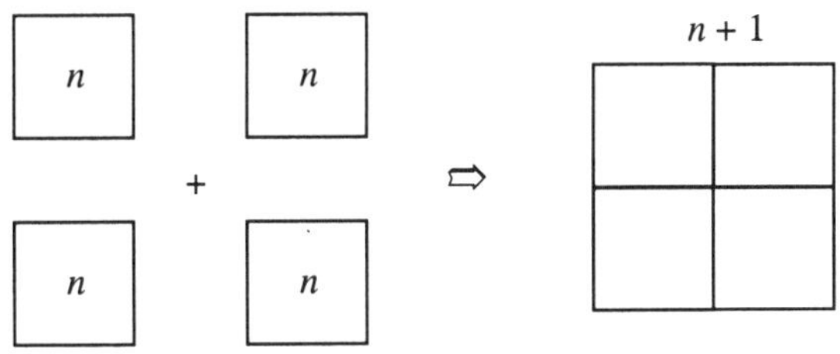

FIGURE 1. The manner in which four n^{th} stage blocks may be combined together to form an $(n+1)^{\text{th}}$ stage block.

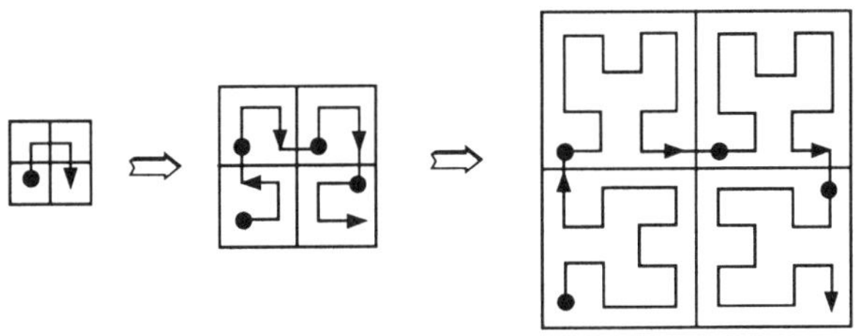

FIGURE 2. The first three stages for the construction which in the infinite-stage limit approaches the "area-filling" Hilbert curve.

Some further examples of sequences of self-similar conformations may be developed in a comparable manner. That is, $L_1 \times L_1$-sized first-stage blocks each with a number N_1 of sites in a path are to be built up in a self-similar pattern to yield second-stage blocks, and then the whole process iterated. As an example consider $L_1 = 3$ and $N_1 = 7$ with first-stage 3×3 blocks of the form indicated in Figure 3. In contrast to Figure 1 one now takes 7 first-stage blocks (as well as 2 empty blocks) to obtain each second-stage $L_2 \times L_2$ block, as also indicated in Figure 3. Here it is seen that this second stage contains $N_2 = N_1^2$ sites and the length of the block is $L_2 = L_1^2$. And at the nth stage $N = N_n = 7^n$ and $L = L_n = 3^n$, so that the fractal dimension approached is

$$\begin{aligned} d_f &= \lim_{n\to\infty} (\log N_n)/(\log L_n) = (\log N_1)/(\log L_1) \\ &= (\log 7)/(\log 3) \approx 1.77124375 \end{aligned} \tag{3.2}$$

which notably is non-integer.

In addition to the two self-similar sequences of Figures 2 & 3, several others are to be treated. One such now based on $L_1 \times L_1 = 5 \times 5$ sets of blocks has the first two stages in Figure 4, and with n^{th} stage values $N_n = N_1 N_{n-1} = N_1^n$ for $N_1 = 13$ and $L_n = L_1 L_{n-1} = L_1^n$, the associated fractal dimension is $d_f = (\log 13)/(\log 5) \approx 1.59369264$. As yet further cases the four additional patterns with first stages shown in Figure 5(a), (b), (c), & (d) are considered, covering a range of respective fractal dimensions of $d_f = (\log 8)/(\log 4) = 3/2$, $d_f = (\log 9)/(\log 5) \approx 1.3652124$, $d_f = (\log 11)/(\log 7) \approx 1.2322744$, & $d_f = (\log 3)/(\log 3) = 1$. And also to check variations which may occur at the same fractal dimension, the patterns in Figure 5(e) & (f) are considered at the (repeated) dimensions of $d_f = 2$ & $d_f = 3/2$.

But now it is desirable to have an efficient scheme to compute the distance-ratio matrices $\mathbb{M}(\eta,+)$ and $\mathbb{M}(\eta,-)$, with the distances (ρ & σ) involved being the graphical shortest-path and Euclidean distances. To this end, e.g., for $\mathbb{M}(0,+)$, one

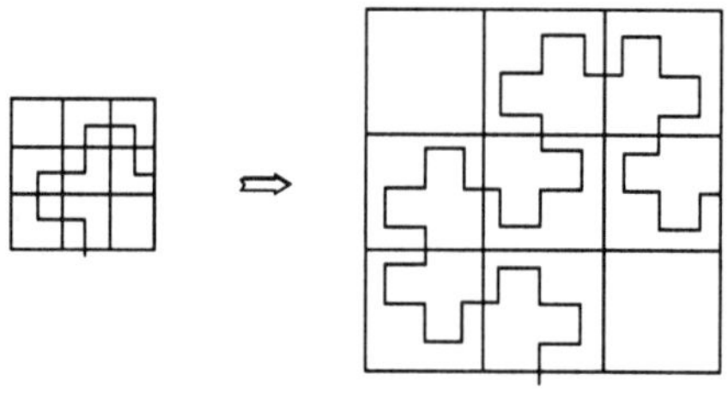

FIGURE 3. The first two stages for the construction leading toward a curve of fractal dimension $d_f \approx 1.771$. Here instead of combining four n^{th} stage blocks to form one $(n+1)^{\text{th}}$ stage block (as for the Hilbert curve), one now combines nine nth stage blocks (though some of them are empty).

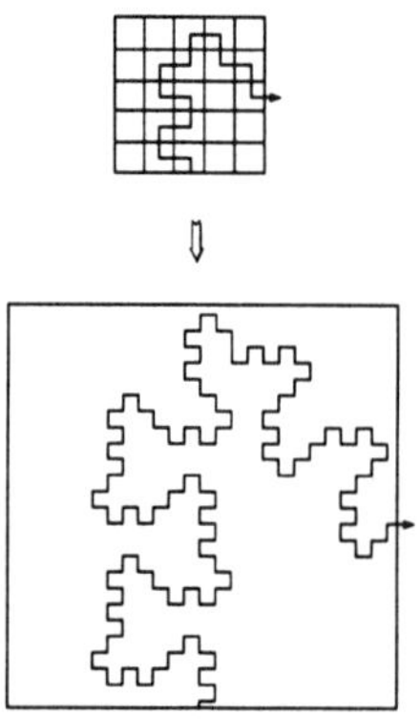

FIGURE 4. The first two stages for the construction leading toward a curve of fractal dimension $d_f \approx 1.594$.

may avoid storing such matrices, and instead generate their elements

$$\mathbb{M}(0,+)_{ij} \quad = \quad |i-j|\left\{(x_i-x_j)^2+(y_i-y_j)^2\right\}^{-1/2} \tag{3.3}$$

one row at a time whenever needed from the stored values of x- & y-position vectors x & y. Thus for an N-site chain (which in our present work may reach $N = 10^5$) storage requirements are reduced from $\sim N^2$ to $\sim N$. Further these x- & y-position vectors may be conveniently generated in close analogy to the pattern recursion, as of Figures 1, 2, or 3, especially if x- & y-position vectors for the different orientations of a block are introduced. That is, let $x(n;O)$ & $y(n;O)$ denote the position vectors at the nth stage for blocks in orientation O, with say $O = E, N, W, S$ (in correspondence with the compass direction of the side of the block near to the in- & out-going parts of the path through that block). For the Hilbert curve the four orientations at stage 1 are indicated in Figure 6 (and one could also be led to suggest orientation names ⊏, ⊔, ⊐, ⊓). Still for the Hilbert curve the position vectors of size 4^{n+1} at stage $n+1$ may be seen to be obtained recursively from the block vectors at the previous stage and an nth stage "size" vector $s(n)$ of 4^n components all equal. When the coordinate system for the positions in a block has origin at the center of the block, and the distance between two neighbor sites is taken as 1 so that the elements of $s(n)$ are all 2^{n-1}, the recursions are

$$
(3.4)\quad
\begin{aligned}
x(n+1;E) &= \begin{pmatrix} +s(n)+x(n;S) \\ -s(n)+x(n;E) \\ -s(n)+x(n;E) \\ +s(n)+x(n;N) \end{pmatrix} \quad \& \quad y(n+1;E) = \begin{pmatrix} -s(n)+y(n;S) \\ -s(n)+y(n;E) \\ +s(n)+y(n;E) \\ +s(n)+y(n;N) \end{pmatrix} \\
x(n+1;N) &= \begin{pmatrix} -s(n)+x(n;W) \\ -s(n)+x(n;N) \\ +s(n)+x(n;N) \\ +s(n)+x(n;E) \end{pmatrix} \quad \& \quad y(n+1;N) = \begin{pmatrix} +s(n)+y(n;W) \\ -s(n)+y(n;N) \\ -s(n)+y(n;N) \\ +s(n)+y(n;E) \end{pmatrix} \\
x(n+1;W) &= \begin{pmatrix} -s(n)+x(n;N) \\ +s(n)+x(n;W) \\ +s(n)+x(n;W) \\ -s(n)+x(n;S) \end{pmatrix} \quad \& \quad y(n+1;W) = \begin{pmatrix} +s(n)+y(n;N) \\ +s(n)+y(n;W) \\ -s(n)+y(n;W) \\ -s(n)+y(n;S) \end{pmatrix} \\
x(n+1;S) &= \begin{pmatrix} -s(n)+x(n;W) \\ -s(n)+x(n;S) \\ +s(n)+x(n;S) \\ +s(n)+x(n;E) \end{pmatrix} \quad \& \quad y(n+1;S) = \begin{pmatrix} -s(n)+y(n;W) \\ +s(n)+y(n;S) \\ +s(n)+y(n;S) \\ -s(n)+y(n;E) \end{pmatrix}
\end{aligned}
$$

For the other self-similar structures of Figures 3, 4, & 5 there are similar sorts of recursions, though the sizes of the nth stage vectors are N_1^n with N_1 the number of steps in the basic first-stage pattern. Further the special "sizing" vectors (as the $s(n)$ above) with elements all L_1^n occur in the recursions with other small integer coefficients other than ± 1 when $L_1 \geq 3$.

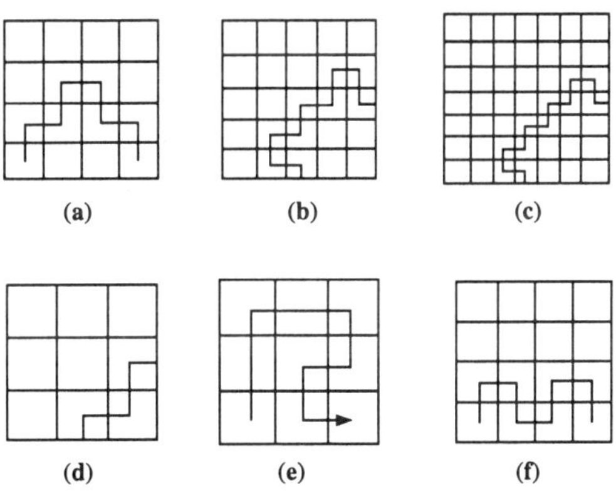

FIGURE 5. The first stage for five more fractal curves, with fractal dimensions: in (a) $d_f = 3/2$; in (b) $d_f \approx 1.365$; in (c) $d_f \approx 1.232$; in (d) $d_f \approx 1$; in (e) $d_f = 2$; & in (f) $d_f = 3/2$.

Next granted an efficient storage/generation scheme for the $\mathbb{M}$-matrices (with 0 on the diagonal) the eigen-problem for the maximum eigenvalue is to be efficiently handled. At a given stage a rather straight-forward approach is to apply the power method or some simple modification of it, such as that of Lanczos [**26**]. In these schemes one starts from an initial vector $v_{[0]}$ here conveniently chosen as a vector with all elements 1, and therefrom iteratively computes a sequence $v_{[1]}, v_{[2]}, v_{[3]}, \ldots$ converging to the desired maximum-eigenvalue eigenvector. A slight extension of the power method (or alternatively a simplification of the Lanczos algorithm) may be used, where having completed the kth stage, one first applies the considered

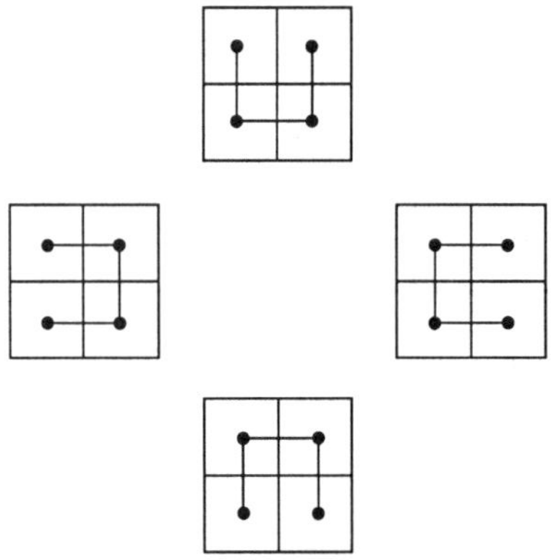

FIGURE 6. The four orientations (E, N, W, & S) for the first stage of the Hilbert curve. The second and third stages of Figure 2 both occur in the W direction.

matrix $\mathbb{M}$ to the kth vector $v_{[k]}$ and second combines this result ($\mathbb{M}v_{[k]}$) optimally with the kth vector to obtain the $(k+1)^{\text{th}}$ vector $v_{[k+1]}$. This optimal combination is conveniently determined from the (maximum-eigenvalue) eigen-solution for $\mathbb{M}$ on the subspace spanned by $v_{[k]}$ and $\mathbb{M}v_{[k]}$. That is,

$$v_{[k+1]} = \alpha v_{[k]} + \beta \mathbb{M} v_{[k]} \tag{3.5}$$

where α and β are obtained as the eigenvectorial solution in

$$\begin{pmatrix} \mu_1 & \mu_2 \\ \mu_2 & \mu_3 \end{pmatrix} \begin{pmatrix} \alpha \\ \beta \end{pmatrix} = \lambda_{[k+1]} \begin{pmatrix} \mu_0 & \mu_1 \\ \mu_1 & \mu_2 \end{pmatrix} \begin{pmatrix} \alpha \\ \beta \end{pmatrix} \tag{3.6}$$

with $\mu_j \equiv v_{[k]}^{\dagger} \mathbb{M}^j v_{[k]}$. Of the two solutions of (3.6) the one with the larger eigenvalue $\lambda_{[k+1]}$ to this 2×2 problem is taken as an estimate to the maximum eigenvalue λ of $\mathbb{M}$. The Rayleigh-Ritz principle then leads to $\lambda_{[k]} \leq \lambda_{[k+1]} \leq \lambda$, and one simply iterates the procedure till successive $\lambda_{[k]}$ are the same to within some tolerance. In application to our present distance-ratio matrices we choose this tolerance to be 10^{-10}, and typically this scheme seems to give convergence of the eigenvalue to within 9 decimal places in about a dozen iterations.

The different linear-chain conformations at different stages of size have been treated, to obtain the maximum eigenvalues λ to the (diagonally zeroed) distance-ratio matrices and thereafter to obtain the associated foldedness measures. These eigenvalues $\lambda(+,0)$ & $\lambda(-,\infty)$ are reported in Table 1, where the distances ρ & σ of Section 3 are respectively identified as the shortest-path-graphical & Euclidean metrics ρ_g & $\rho_{\mathcal{E}}$. For these results the bond lengths are taken to be 1. That $\lambda(+,0)$ becomes so ever much bigger with stage number n occurs because of the ever greater excesses of some of the shortest-path distances over the corresponding Euclidean distances, at least for the more folded structures with $d_f > 1$. In Table 2 the degree of foldedness D_n at each stage n is reported, with our optimal choice of η, as in (2.12). Also for each nth stage structure there is reported the *density* taken to be given as N_n/L_n^2. With the recollection that N_n scales as $L_n^{d_f}$ it is seen that this density is related to the fractal dimension d_f, via

$$N_n/L_n^2 \sim L_n^{d_f-2} \tag{3.7}$$

TABLE 1. Eigenvalue data for finite self-similar conformations at different stages n

Variables*		$d_f = 1.0$	$d_f \approx 1.23$	$d_f \approx 1.36$	$d_f = 1.5^{(5a)}$	$d_f = 1.5^{(5f)}$	$d_f \approx 1.59$	$d_f \approx 1.77$	$d_f = 2^{(2)}$	$d_f = 2^{(5e)}$
L_1		3	7	5	4	4	5	3	2	3
N_1		3	11	9	8	8	13	7	4	9
$n = 1$	$\lambda(+)$	2.28824	15.4527	12.9225	10.8706	11.9879	24.3335	10.8094	4.6131	17.6125
	$\lambda(-)$	1.81129	7.1238	5.6719	5.0151	4.7157	7.1564	4.2272	2.4060	5.1965
$n = 2$	$\lambda(+)$	10.4526	286.4283	222.2857	179.3726	206.1098	848.2524	188.8319	43.4779	521.8849
	$\lambda(-)$	6.2465	54.1290	33.1461	26.0689	23.5007	44.9378	18.0451	7.9231	22.6857
$n = 3$	$\lambda(+)$	35.7477	4972.8773	3614.5735	2866.0239	3371.2973	28724.0467	3128.6296	364.6790	14349.7791
	$\lambda(-)$	19.0656	382.3388	170.1240	110.2516	98.0508	234.1399	59.8181	19.3649	75.5821
$n = 4$	$\lambda(+)$	112.0429	85981.7419	58569.0042	45836.7888	54282.852	970964.97	51226.57	2969.0586	388721.778
	$\lambda(-)$	57.2896	2679.6963	854.9446	446.8399	395.873	1180.1748	185.2168	42.3767	234.3532
$n = 5$	$\lambda(+)$	341.1151	1486277.694	948830.0405	733341.2104	869840.740		837025.8767	23901.8286	10500957.1
	$\lambda(-)$	171.8576	18761.189	4279.0412	1793.1073	1586.974		561.4264	88.4480	710.67
$n = 6$	$\lambda(+)$	1028.4106						13672206.33	191615.19	
	$\lambda(-)$	515.5181						1690.055	180.6055	
$n = 7$	$\lambda(+)$	3090.3292							1533935.15	
	$\lambda(-)$	1546.4824							364.9238	
$n = 8$	$\lambda(+)$	9276.0975							12273940.78	
	$\lambda(-)$	4639.3684							733.5603	

* the superscript on d_f value indicates the Figure picturing the corresponding fractal (in cases when there are several different fractals having the same d_f value).

TABLE 2. Foldedness and density for finite self-similar conformations at different stages n

Variables*		$d_f = 1.0$	$d_f \approx 1.23$	$d_f \approx 1.36$	$d_f = 1.5^{(5a)}$	$d_f = 1.5^{(5f)}$	$d_f \approx 1.59$	$d_f \approx 1.77$	$d_f = 2^{(2)}$	$d_f = 2^{(5e)}$
L_1		3	7	5	4	4	5	3	2	3
N_1		3	11	9	8	8	13	7	4	9
$n = 1$	$D_n(opt)$	0.02984	0.11208	0.15615	0.11018	0.14252	0.20920	0.27156	0.20737	0.39197
	density	0.33333	0.22448	0.36000	0.50000	0.50000	0.52000	0.77778	1.00000	1.00000
$n = 2$	$D_n(opt)$	0.01834	0.08724	0.16461	0.17894	0.21655	0.33996	0.47011	0.50026	0.72367
	density	0.11111	0.05040	0.12960	0.25000	0.25000	0.27040	0.60494	1.00000	1.00000
$n = 3$	$D_n(opt)$	0.00791	0.08561	0.17340	0.20760	0.26397	0.37422	0.56239	0.69343	0.84145
	density	0.03704	0.01131	0.04666	0.12500	0.12500	0.14061	0.47051	1.00000	1.00000
$n = 4$	$D_n(opt)$	0.00291	0.08578	0.17639	0.21715	0.27993	0.38169	0.59391	0.79451	0.87827
	density	0.01235	0.00254	0.01680	0.06250	0.06250	0.07312	0.36595	1.00000	1.00000
$n = 5$	$D_n(opt)$	0.00101	0.08585	0.17717	0.21991	0.28424		0.60433	0.84372	0.88972
	density	0.00411	0.00057	0.00605	0.03125	0.03125		0.28463	1.00000	1.00000
$n = 6$	$D_n(opt)$	0.00034						0.60777	0.86712	
	density	0.00137						0.22138	1.00000	
$n = 7$	$D_n(opt)$	0.00011							0.87827	
	density	0.00046							1.00000	
$n = 8$	$D_n(opt)$	0.00004								
	density	0.00015								

* the superscript on d_f value indicates the Figure picturing the corresponding fractal (in cases when there are several different fractals having the same d_f value).

That is, the higher the fractal dimension, the less rapidly should the density fall with increasing size. Upon examination of the D_n values it seems that there is agreement with intuition as to how the different conformations of different types are ordered in degree of foldedness - and the foldedness is seen to vary only slowly with stage number. An important limit is that of infinite size, so that from the reported numerical data at various stages extrapolation to the infinite-stage limit may be attempted, as done in the next section after an argument for the expected functional form of the asymptotic behavior is developed.

4. Asymptotic Behavior

The large-system behavior of the maximum-magnitude eigenvalues of the distance-ratio matrices, and the associated distance-difference measure are of interest if one is to address the fractal limit. For the present self-similar structures, one has at the nth stage $N_n = N_1 N_{n-1} = N_1^n$ sites and a block with length scale $L_n = L_1 L_{n-1} = L_1^n$. Now first note that the sequence number of a site along a chain is conveniently presented as a number in base-N_1, with the jth digit in this number identifying the stage-j block in which the considered site occurs. That is, the n-digit number

$$i(n) \equiv (i_n, i_{n-1}, \ldots, i_2, i_1) \tag{4.1}$$

locates a site in position i_1 of a first-stage block, which in turn is in position i_2 of a second-stage block, ..., and finally the occupied $n-1$-stage block is in position in of an n-stage block. For instance, the site in the position in the 3rd row & 5th column of the third-stage block for the Hilbert curve is identified by $(i_3, i_2, i_1) = (2, 0, 3)$. With this notation in hand, the graphical distance between two sites is then presentable as

$$\begin{aligned} \rho_g\big(i(n), j(n)\big) &= \big|\{i_1 + i_2 N_1 + i_3 N_1^2 + \cdots + i_n N_1^{n-1}\} \\ &\quad -\{j_1 + j_2 N_1 + j_3 N_1^2 + \cdots + j_n N_1^{n-1}\}\big| \end{aligned} \tag{4.2}$$

which may be rearranged to

$$\begin{aligned} \rho_g\big(i(n), j(n)\big) &= N_1^n \cdot \big|(i_n - j_n) \cdot N_1^{-1} + (i_{n-1} - j_{n-1}) \cdot N_1^{-2} + \ldots \\ &\quad +(i_2 - j_2) \cdot N_1^{1-n} + (i_1 - j_1) \cdot N_1^{-n}\big| \end{aligned} \tag{4.3}$$

and the final magnitude here represents just the fraction that the difference between the two numbers $i(n)$ and $j(n)$ is of the maximum difference $N(= N_1^n)$. That is, if one lets $f = f_{i(n)}$ denote the fraction that $i(n)$ (viewed as an n-digit base-N_1 number) is of $N = N_n$, and f' the similar fraction that $j(n)$ is of N, then

$$\rho_g\big(i(n), j(n)\big) = N_1^n \cdot |f - f'| \tag{4.4}$$

so that the essential dependence on stage number n is seen to be through the factor N_1^n with the remaining factor ranging uniformly and presumably rather smoothly over the interval from 0 to 1. There is a similar manner of expression for the

Euclidean distance between two sites

$$\begin{aligned}\rho_{\mathcal{E}}\big(i(n),j(n)\big) \quad = \quad & \Big\{\Big[\big(x(i_1)+x(i_2)L_1+x(i_3)L_1^2+\cdots+x(i_n)L_1^{n-1}\big) \\ & \quad -\big(x(j_1)+x(j_2)L_1+x(j_3)L_1^2+\cdots+x(j_n)L_1^{n-1}\big)\Big]^2 \\ & +\Big[\big(y(i_1)+y(i_2)L_1+y(i_3)L_1^2+\cdots+y(i_n)L_1^{n-1}\big) \\ & \quad -\big(y(j_1)+y(j_2)L_1+y(j_3)L_1^2+\cdots+y(j_n)L_1^{n-1}\big)\Big]^2\Big\}^{1/2} \end{aligned} \tag{4.5}$$

where L_1 is the length of a side of the first-stage block, and $x(i)$ & $y(i)$ are x- & y-positions of the ith site in a stage-1 block. Again the essential stage dependence may be factored out

$$\begin{aligned}\rho_{\mathcal{E}}\big(i(n),j(n)\big) \quad = \quad & L_1^n\cdot\Big\{\Big[\big(x(i_n)-x(j_n)\big)\cdot L_1^{-1}+\cdots+\big(x(i_2)-x(j_2)\big)\cdot L_1^{1-n} \\ & \quad +\big(x(i_1)-x(j_1)\big)\cdot L_1^{-n}\Big]^2 \\ & +\Big[\big(y(i_n)-y(j_n)\big)\cdot L_1+\cdots+\big(y(i_2)-y(j_2)\big)\cdot L_1^{1-n} \\ & \quad +\big(y(i_1)-y(j_1)\big)\cdot L_1^{-n}\Big]^2\Big\}^{1/2} \end{aligned} \tag{4.6}$$

and the remaining factor (expressed as a square root) presumably varies rather smoothly between 0 & 1. Here as in (4.4) one might imagine a smooth function, perhaps more naturally in terms of the coordinates $x(i_k)$ & $y(j_k)$ viewed in terms of numbers now in base-L_1. That is, one lets $x = x_{i(n)}$ denote the fraction that $x(i_n)$, $x(i_{n-1})$, ..., $x(i_1)$ (viewed as an n-digit base-L_1 number) is of L_1^n, and similarly $x' = x_{j(n)}$, $y = y_{i(n)}$, & $y' = y_{j(n)}$ are defined as fractions in terms of $x(j_a)$, of $y(i_a)$, & of $y(j_a)$. Then

$$\rho_{\mathcal{E}}\big(i(n),j(n)\big) \quad = \quad L_1^n\cdot\Big\{(x-x')^2+(y-y')^2\Big\}^{1/2} \tag{4.7}$$

with the square-root again presumably varying smoothly in the same range (0 to 1) as x, y, x', & y'. Now in turn we would like to think of the fractions x & y as functions of f while x' & y' are functions of f'. Of course base-L_1 numbers are involved in x, y, x', & y' while f & f' involve base-N_1 numbers, so that one might view the smoothness of the resultant functions of f & f' to be more problematic, with some further qualifications mentioned in Appendix A.

Granted the identification of the dominant stage-number (n) dependence of the distance functions in equations (4.4) and (4.7), one next turns to the matrix elements and eigenvalues of the distance-ratio matrices. Indeed since these matrices have elements which are just ratios of corresponding shortest-path and Euclidean distances, one expects to have

$$\begin{aligned}\mathbb{M}(0,+)_{i(n),j(n)} \quad &\approx \quad (N_1/L_1)^n\xi(f,f') \\ \mathbb{M}(\infty,-)_{i(n),j(n)} \quad &\approx \quad (L_1/N_1)^n\zeta(f,f')\end{aligned} \tag{4.8}$$

where $\xi(f,f')$ & $\zeta(f,f')$ are smooth functions of their arguments f & f' which range over the interval from 0 to 1 with f & f' representing the fractional part that the N_1-base number $i(n)$ and $j(n)$ are of $N = N_n$. Next the (maximum-magnitude) eigenvalues to the $\mathbb{M}$-matrices may be given by Rayleigh-Ritz expressions as

$$\lambda = \frac{C^\dagger \mathbb{M} C}{C^\dagger C} \tag{4.9}$$

where C is the maximum-eigenvalue eigenvector. Here when the vector inner products are expanded out in terms of their components, there arise $N^2 = N_1^{2n}$ terms in the numerator and $N = N_1^n$ in the denominator, so that one might guess that the asymptotic stage dependence for the eigenvalues of interest involves a factor of $N_1^{2n}/N_1^n = N_1^n$ times a typical matrix element (as in (4.8)). In fact the matrix element $C^\dagger \mathbb{M} C$ may be viewed as an un-normalized matrix-element average, which aside from the identified exponential n-dependences, differs but little from one stage to the next. Indeed the numerators and denominators in the matrix elements might be imagined to deviate by only exponentially small amounts in a manner dependent on the matrix or block sizes N_1^n or L_1^n. That is, we might guess

$$\begin{aligned} \lambda_n(0,+) &\approx (N_1/L_1)^n N_1^n \{A_+ + B_+ \cdot L_1^{-n} + C_+ \cdot N_1^{-n}\} \\ \lambda_n(\infty,-) &\approx (L_1/N_1)^n N_1^n \{A_- + B_- \cdot L_1^{-n} + C_- \cdot N_1^{-n}\} \end{aligned} \tag{4.10}$$

with the A_s, B_s, & C_s for $s = \pm$ being constants (for the particular fractal). There is some further support for this form in the argument of Appendix A. Then the distance-difference measures D_n at stage n involve comparators which are just (weighted) averages over the elements of the $\mathbb{M}$-matrices, so that one might argue that they should exhibit related asymptotic forms,

$$D_n(\rho_{\mathcal{E}}, \rho_g) \approx D_\infty + B \cdot L_1^{-n} + C \cdot N_1^{-n} \tag{4.11}$$

with B & C being suitable constants.

The asymptotic forms suggested in equations (4.10) and (4.11) might be tested. These asymptotic formulas indicate that ratios of corresponding eigenvalues at successive stages should approach simple rational numbers (often integers): N_1^2/L_1 for $\lambda(0,+)$ and L_1 for $\lambda(\infty,-)$ - and leading corrections to this should be of order L_1^{-n} or N_1^{-n}. To test this expectation a plot of these ratios *vs.* L_1^{-n} or N_1^{-n} for the case of the Hilbert curve, yields the result of Figure 7. That is, letting $\lambda_n(0,+)$ and $\lambda_n(\infty,-)$ denote n^{th}-stage eigenvalues, there are shown plots of

$$\lambda_{n+1}(0,+)/\lambda_n(0,+) \ vs.\ L_1^{-n} \quad \& \quad \lambda_{n+1}(\infty,-)/\lambda_n(\infty,-) \ vs.\ L_1^{-n} \tag{4.12}$$

which we have argued should respectively approach N_1^2/L_1 and L_1 (at the $L_1^{-n} = 0$ axis). For the Hilbert curve these expected intercepts are 8 & 2, as from Figure 7 are seen to be closely realized. Similarly the result for the fractal-curve case of Figure 3 results in the plot of Figure 8, and again the expected intercepts seem to be realized, now at 49/3 and 3. Somewhat similar results are obtained with the other self-similar patterns of Figure 5, so that evidently the argued leading stage-dependence is correct. Whether such plots are made *vs.* L_1^{-n} or *vs.* N_1^{-n} depends on what one thinks the next non-vanishing higher-order term is, but the same intercept of $\approx N_1^2/L_1$ and of $\approx L_1$ is obtained in all these cases independently of this choice or horizontal axis. That the L_1^{-n} term seems to be absent in the eigenvalue of the first expansions of (4.10) is suggested from a second type of plot, involving the rescaled eigenvalues

$$\pi_{n,+} \equiv (L_1/N_1^2)^n \lambda_n(0,+) \quad \& \quad \pi_{n,-} \equiv L_1^{-n} \lambda_n(\infty,-) \tag{4.13}$$

Then plots of

$$(4.14)\qquad (\pi_{n,\star}-\pi_{n-1,\star})/(\pi_{n+1,\star}-\pi_{n,\star})\quad vs.\quad L_1^{-n}(\text{or } vs.\, N_1^{-n}),\qquad \star=+,\,-$$

should approach either an intercept of L_1 if $B_\star \neq 0$ (in (4.10)) or an intercept of N_1 if $B_\star = 0$ while $C_\star \neq 0$. From Figure 9 for the Hilbert curve this second circumstance with $B_\star = 0$ seems to be the case for $\lambda_n(0,+)$ but not for $\lambda_n(\infty,-)$. The same is true for the $d_f \approx 1.77$-dimensional fractal and the $d_f = 1$-dimensional case. Generally the approach of the ratio involving the + eigenvalues toward the asymptote of N_1 seems rather slow. For the other fractal structures there are fewer stages available, so that the extrapolations are not so definitive – but these plots (which we do not explicitly show) are consistent with this same circumstance. On the other hand for the distance-difference quantities this next leading term seems to be with $B \neq 0$ – that is, it seems to involve a non-zero term $\sim L_1^{-n}$.

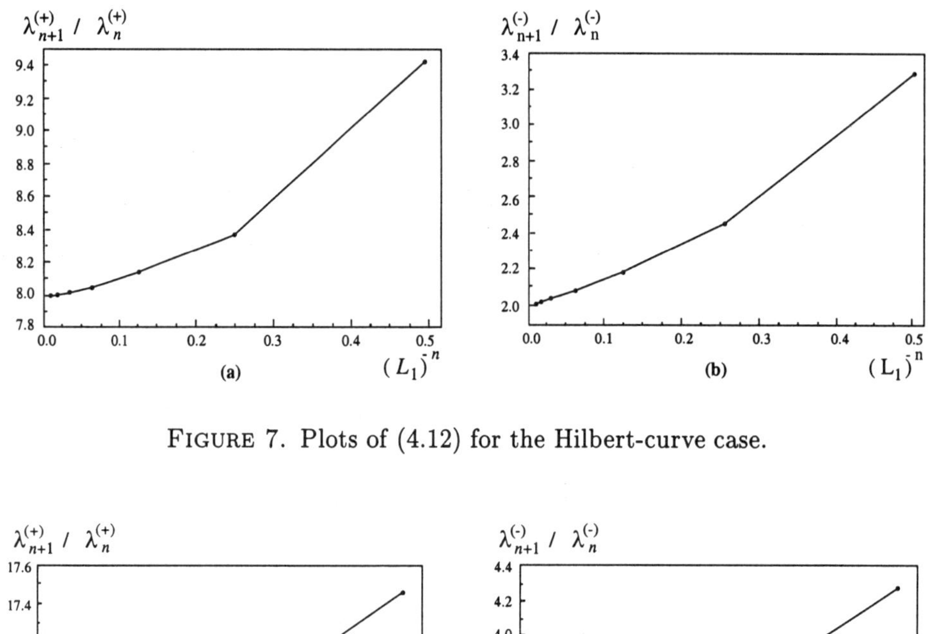

FIGURE 7. Plots of (4.12) for the Hilbert-curve case.

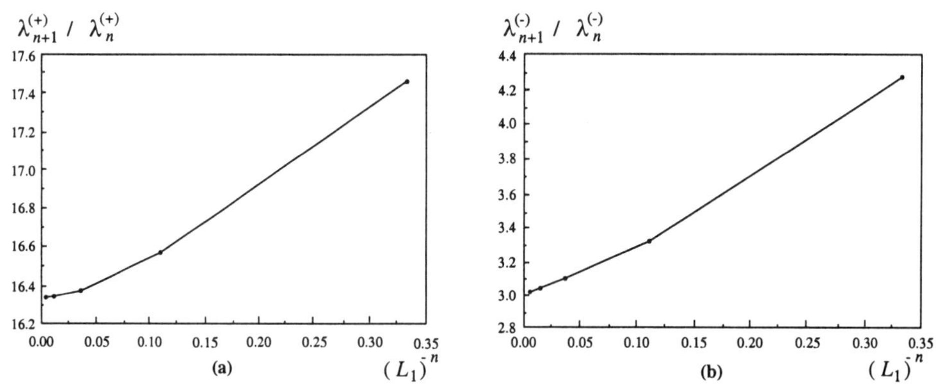

FIGURE 8. Plots of (4.12) for the $d_f \approx 1.771$ fractal case.

Consequent fitted values for the various A, B, C parameters as well as the infinite limits for D are thence made, and reported in Table 3. The numerical extrapolations for the eigenvalues are made with $B_+ = 0$, $C_+ \neq 0$, $B_- \neq 0$, $C_- = 0$ in (4.10), and higher-order terms neglected, except that a weighting is chosen for the least-squares fitting to give a proportionately higher weight to points with lesser error: that is, an nth stage point is given a weight L_1^{+4n}, the inverse of the square of the next highest order term neglected. For the comparators of (4.11) we take $B \neq 0$, neglect the CN_1^{-n} term, and choose a weight of N_1^{2n}. For the special case of the conformational structures associated to $d_f \approx 1.23$ the approach of D_n toward the limit is not so monotonic, seeming to go through a minimum at a fairly high order n, so that we use just the last three data points in extrapolating to the limit. From the table there is evident a progression of foldedness values D_∞ for the $n \to \infty$ fractal structures as a function of their fractal dimension d_f. The particular coefficients $A_+ = 2^{1/2}$ & $A_- = 2^{-1/2}$ for the $d_f = 1$ structure can in fact be readily understood, with these values reflecting average sorts of ratios $\rho_g(i,j)/\rho_\mathcal{E}(i,j)$ & $\rho_\mathcal{E}(i,j)/\rho_g(i,j)$, respectively. And it is clear from Figure 5(d) that this $d_f = 1$ structure is naught but a long wiggly structure (like that with all-*trans* polyethylene, except with 90° bond angles along the chain). Then for this case each step away from a given site i increases the graphical distance from i by 1, while the along-the-chain coordinate ultimately dominating the Euclidean distance increases by just $2^{-1/2}$. (The perpendicular-to-chain coordinate does not really increase but simply fluctuates between 0 & $2^{-1/2}$.) Thence the longer distances of which there are many more than the shorter distances ultimately dominate the distance-ratio averages and since they give ratios ever closer to $2^{1/2}$ & $2^{-1/2}$ end up giving (exactly) $A_+ = 2^{1/2}$ & $A_- = 2^{-1/2}$.

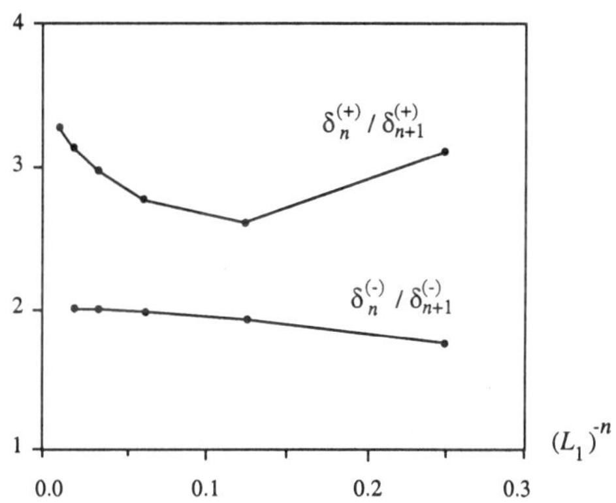

FIGURE 9. Plots of (4.14) for the case of the Hilbert curve.

5. Discussion & Further Prospects

The foldedness measure D proposed here seems to be "reasonable". It is seen to be efficiently computable for a range of linear-chain conformations. For the infinite-length limit of self-similar conformations some asymptotic aspects of D have been found (in equations (4.10) & (4.11)), and this limiting measure D_∞ may be compared with fractal dimensions, which as we have noted also may be argued (in the infinite-length limit) to correlate with "foldedness". Such a comparison is made in Figure 10, whence it is seen that there is some fair degree of correlation. The measure D_∞ does distinguish the different fractal conformations of the same fractal dimension, though only very slightly for the two structures with $d_f = 2$, so that

TABLE 3. Asymptotic results for self-similar (fractal) conformations

const*	$d_f = 1$	$d_f \approx 1.23$	$d_f \approx 1.36$	$d_f = 1.5^{(5a)}$	$d_f = 1.5^{(5f)}$	$d_f \approx 1.59$	$d_f \approx 1.77$	$d_f = 2^{(2)}$	$d_f = 2^{(5e)}$
L_1	3	7	5	4	4	5	3	2	3
N_1	3	11	9	8	8	13	7	4	9
D_∞	0.0	0.0859	0.1773	0.2208	0.2856	0.3835	0.6095	0.8897	0.8955
B	0.1996	-0.1981	-0.5351	-0.8850	-1.4038	-1.1447	-1.2536	-1.4591	-1.4075
A_+	1.4142	0.9631	0.8504	0.6994	0.8297	0.7439	0.7201	0.7316	0.7319
C_+	-2.5536	-0.2282	-0.0938	0.1916	-4.9880	-0.1944	-0.8091	-1.6733	-1.6531
A_-	0.7071	1.1163	1.3696	1.7530	1.5509	1.8921	2.3223	2.8800	2.9402
B_-	0.0387	-0.5524	-1.0794	-1.9184	-1.1667	-2.3683	-2.8877	-3.7114	-3.8056

* the superscript on d_f value indicates the Figure picturing the corresponding fractal (in cases when there are several different fractals having the same d_f value).

their points lie nearly atop one another in the figure. As earlier noted D applies regardless of whether one is near the infinite-length limit or not, though proper fractals (and the associated dimension) are only properly defined in the infinite limit. It can be expected that this infinite limit sometimes leads to problems, in that structures described as "fractal" often in practice seem to exhibit [**27**] self-similarity only over a rather limited range of scales of perhaps two decades or so. Thence, for instance, even the determination of the value for a fractal dimension might be expected to lead to ambiguities. Indeed such ambiguity is rather dramatically manifested [**16, 17**] when two estimates for the fractal dimensions for a collection of protein structures are found to give a scatter seemingly greater than that we find between foldedness and fractal dimension in Figure 10. For the general case of linear-chain polymers there are physical quantities (e.g., hydrodynamic sizes, light-scattering behavior, and extents of intra-polymer interaction) which have been proposed to depend upon a fractal dimension, so that one might comparably surmise dependences on foldedness (now definable without reference to the infinite-chain limit). We reiterate, D is smaller the closer the two types of distances (Euclidean and "topological") agree in a scale-independent manner with one another in an average fashion for the conformation as a whole.

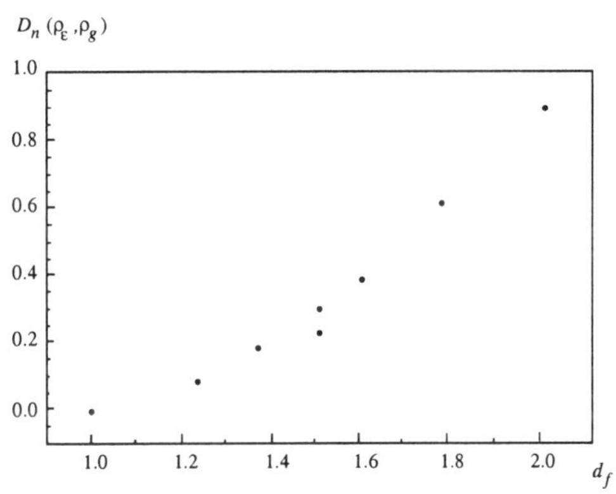

FIGURE 10. Plot of distance-difference $D_n(\rho_\mathcal{E}, \rho_g)$ *vs.* fractal dimension d_f, for the 9 different infinite-stage limit self-similar structures considered.

It may be noted that the type of computation of the foldedness measure proposed here is rather efficiently done. The type of algorithm used for finding the critical maximum eigenvalue to an $N \times N$ distance-ratio matrix requires times proportional to N^2 which is the same scale of time that it takes to generate the matrix (which in turn minimally records all the distance ratios). Moreover, the proportionality to N^2 seems to be rather small since the distance-ratio matrices are of a Frobenius-Perron type, which further seem in our examples to have the maximum-magnitude eigenvalue well separated from the next.

The type of distance difference measure considered here could be useful in other manners of characterization of molecular shape. Even when the molecular graph is not a linear-chain polymer, the present type of measure could be used. For such general graphs the distance differences can be imagined to be a measure of "convolutedness".

Instead of comparing two different distance functions for the same conformation, the same Euclidean distance function for two different conformations could be compared [**22, 23**]. Letting $\rho_A(i,j)$ & $\rho_B(i,j)$ denote the two sets of distances for conformations A & B, the distance difference $D(\rho_A, \rho_B)$ measures the difference in shape between the two different conformations. This averaged distance difference $D(\rho_A, \rho_B)$ should in general be rather distinct from the extremal distance difference

$$D_{\max}(\rho_A, \rho_B) \equiv \max_{i \neq j} \left| \log \frac{\rho_A(i,j)}{\rho_B(i,j)} \right| + \max_{k \neq l} \left| \log \frac{\rho_B(k,l)}{\rho_A(k,l)} \right| \tag{5.1}$$

measure recently considered elsewhere [**22, 27**]. Further for a given conformation A lacking some point-group symmetry one might consider other conformations S exhibiting this symmetry, and take the smallest distance difference $D(\rho_A, \rho_S)$ as minimized over all symmetric conformations S as a measure of the degree of dissymmetry of conformation A. Alternative dissymmetry measures of an "extremal" type (depending on maximum distance deviations rather than averages) have already been a recent topic of consideration [**29, 30**]. And the related circumstance of "extremal" achirality measures have been rather similarly considered [**31**], so that again there is an "averaged" distance difference to a closest achiral conformation.

Another type of possibility would be to compare the same shortest-path graphical distance for different structural isomers, e.g., of the different alkanes, whence the distance difference to the normal (linear-chain) alkane could be a measure of the degree of "branching", also a topic of some past interest [**32**]–[**37**]. [Here the distance ratios for corresponding pairs of sites would depend on the correspondence made between the sites of the linear and branched structures, whence a "natural" choice for the correspondence would be such that the distance-difference turns out to be a minimum.]

Besides the shortest-path and Euclidean distances there are [**38**]–[**40**] several other (even intrinsic) graphical distances which could be incorporated in various comparisons. Particularly the so-called "resistance" distance and the "quasi-Euclidean" distance pay attention to multiple pathways of connection between pairs of vertices, and have been suggested to correlate more closely with Euclidean distances for ordinary stable molecular structures. At any rate in paying attention to multiple pathways the distance-difference between the resistance-distance & shortest-path distance functions on a given (connected) graph might reasonably be taken to be a measure of the *degree of cyclicity* of the graph.

It seems that there are several avenues for possible utilization of distance differences in chemistry.

Appendix A.

There are some qualifications that should be made to the development leading up to equation (4.6), e.g., concerning the orientation dependence of the blocks. Indeed the coordinate functions $x(i)$ & $y(i)$ should depend on the orientation of the block, and this could be indicated by adding a subscript O for the different orientations, $x_O(i)$ & $y_O(i)$. Then in equations (4.5) & (4.6) in place of any $x(i_a)$ one should have $x_O(i_a)$ for some suitable O which in fact depends on i_a and the orientation of the stage-$(a+1)$ block in which the stage-a blocks are being embedded. All this makes a complete identification of these different dependences rather a mess, so that only the summary notation most relevant to the presented arguments is made in (4.5) & (4.6). Thence the arguments are suggestive, and the ensuant asymptotic forms are then to be subjected to empirical testing, as done toward the end of Section 4.

Granted (4.4) and (4.7), with smooth functions ξ & ζ, an elaborated argument for the asymptotic forms of the eigenvalues applies. E.g., for the case of $\mathbb{M}(0,+)$ we use the rescaled matrix ξ of equation (4.11) and introduce a relatedly rescaled companion eigenvalue

$$(L_1/N_1)^n \cdot \lambda(0,+)/N \equiv \Lambda \tag{A.1}$$

Then the eigen-problem becomes

$$\sum_z \xi(z,z')C_{z'} = N\Lambda C_z \tag{A.2}$$

where the variables z & z' are ratios of the ordinary integer indexes to their full range N. But because of the nice continuous behavior of the function (and thence presumably too of C_z) one may use the Euler-Maclaurin expansion

$$\begin{aligned}\sum_z \xi(z,z')C_{z'} &= N\int_0^1 \xi(z,z')C_{z'}dz' \\ &\quad +\frac{1}{2}\Big[-\xi(z,0)C_0 + \xi(z,1)C_1\Big] + \mathcal{O}(N^{-1})\end{aligned} \tag{A.3}$$

Thus there is evidently a 0-order eigen-problem

$$\int_0^1 \xi(z,z')C_{z'}^{(0)}dz' = \Lambda^{(0)}C_z^{(0)} \tag{A.4}$$

which gives an infinite-stage limiting eigenvalue $\Lambda^{(0)}$. If the convolution-product notation

$$\int_0^1 \xi(z,z')C_{z'}^{(0)}dz' \equiv (\xi \star C^{(0)})_z \tag{A.5}$$

is used, then the 0-order eigen-problem can be written in a more suggestive form

$$\xi \star C^{(0)} = \Lambda^{(0)}C^{(0)} \tag{A.6}$$

The higher-order terms in the Euler-Maclaurin expansion can be viewed as perturbative corrections to the 0-order eigen-problem, and perturbation formulas developed along rather standard lines [**32**]–[**37**]. The overall result is an expansion in powers of N_1^{-n}, for both the eigenvalues and eigenvectors. Terms $\sim L_1^{-n}$ might arise if there are "corrections" of such type inherent in the presumption of the functions ξ & ζ.

References

[1] P.G. Mezey, *Shape in Chemistry*, VCH Press, New York, 1993.

[2] K.A. Dill, Dominant Forces in Protein Folding, *Biochem.*, **29** (1990) 7133–7155.

[3] M. Vasquez, G. Nemethy, and H.A. Scheraga, Conformational Energy Calculations on Polypeptides and Proteins, *Chem. Rev.*, **94** (1994) 2183–2239.

[4] H. Frauenfelder, S.G. Sligar, and P.G. Wolynes, The Energy Landscapes and Motions of Proteins, *Science*, **254** (1991) 1598–1603.

[5] C. Mumenthaler, and W. Braun, Folding of Globular Proteins by Energy Minimization and Monte Carlo Simulations with Hydrophobic Surface Area Potentials, *J. Mol. Mod.*, **1** (1995) 1–10.

[6] B.A. Shoemaker, J. Wang, and P.G. Wolynes, Structural Correlations in Protein Folding Funnels, *Proc. Natl. Acad. Sci. USA 94*, (1997) 777–782.

[7] P.J. Flory, *Principles of Polymer Chemistry*, Cornell University Press, Ithaca, New York, 1953.
[8] P.G. deGennes, *Scaling Concepts in Polymer Physics*, Cornell University Press, Ithaca, New York, 1979.
[9] H.E. Stanley, P.J. Reynolds, S. Redner, and F. Family, pages 169–206 in *Real-Space Renormalization*, T.W. Burkhardt, and J.M.J. van Leeuwen, (eds.) Springer-Verlag, Berlin, 1982.
[10] D.J. Klein, and W.A. Seitz, pages 430–453 in *Topology & Graph Theory in Chemistry*, R.B. King, (ed.), Elsevier, Amsterdam, 1983.
[11] B. Mandelbrot, *The Fractal Geometry of Nature*, W.H. Freeman, San Francisco, 1983.
[12] J.S. Helman, A. Coniglio, and C. Tsallis, Fractons & Fractal Structure of Proteins, *Phys. Rev. Lett.*, **53** (1984) 1195–1197.
[13] G.C. Vagner, J.T. Colvin, J.P. Allen, and H.J. Stapleton, Fractal Models of Protein Structure, Dynamics, and Magnetic Relaxation, *J. Am. Chem. Soc.*, **107** (1985) 5589–5594.
[14] J.T. Colvin, and H.J. Stapleton, Fractal & Spectral Dimensions of Biopolymer Chains: Solvent Studies of Electron Spin Relaxation Rates in Myoglobin Azide, *J. Chem. Phys.*, **82** (1985) 4699–4706.
[15] B.A. Fedorov, B.B. Fedorov, and P.W. Schmidt, An Analysis of the Fractal Properties of the Surfaces of Globular Proteins, *J. Chem. Phys.*, **99** (1993) 4076–4083.
[16] C.X. Wang, Y.Y. Shi, and F.H. Huang, Fractal Structure of Proteins, *Phys. Rev.*, **A 41** (1990) 7043–7048.
[17] Y. Xiao, Comment on 'Fractal Structure of Proteins', *Phys. Rev.*, **E 49** (1994) 5903–5905.
[18] F. Buckley, and R. Harary, *Distance in Graphs*, Addison-Wesley, Reading, Massachusetts, 1989.
[19] N. Trinjastić, *Chemical Graph Theory*, CRC Press, Boca Raton, Florida, 1992.
[20] D.J. Klein, Chemical Graph-Theoretic Cluster Expansions, *Intl. J. Quantum Chem.*, **S 20** (1986) 153–171.
[21] M. Randić, A.F. Kleiner, and L.M. De Alba, Distance/Distance Matrices, *J. Chem. Inf. & Comp. Sci.*, **34** (1994) 277–286.
[22] D.J. Klein, Similarity & Dissimilarity in Posets, *J. Math. Chem.*, **18** (1995) 321–348.
[23] M. Randić, and M. Razinger, Sec. 6.7 of "On the Characterization of 3D Molecular Structure", pages 159–236 of *From Chemical Topology to Three-Dimensional Geometry*, Balaban, A.T. (ed.), Plenum Press, New York, 1997.
[24] F. Gantmacher, Chap. 13 of *Theory of Matrices*, vol. II, Chelsea Pub., New York, 1959.
[25] D. Hilbert, *Math. Ann.*, **38** (1891) 157.
[26] C. Lanczos, Solution of Systems of Linear Equations by Minimized Iterations, *J. Res. Nat. Bur. Standards*, **B49** (1952) 33–53.
[27] D. Avnir, O. Biham, D. Lidar, and O. Malcai, Is the Geometry of Nature Fractal?, *Science*, **279** (1998) 39–40.
[28] P.G. Mezey, A Proof of the Metric Properties of the Symmetric Scaling-Nesting Dissimilarity Measure and Related Symmetry Measures, *Intl. J. Quantum Chem.*, **63** (1997) 105–109.
[29] H. Zabrodsky, and D. Avnir, Measuring Symmetry in Structural Chemistry, *Adv. Molec. Struct. Res.*, **1** (1995) 1–31.
[30] H. Zabrodsky, S. Peleg, and D. Avnir, Continuous Symmetry Measures, *J. Am. Chem. Soc.*, **114** (1992) 7843–7851.
[31] N. Weinberg, and K. Mislow, Distance Functions as Generators of Chirality Measures, *J. Math. Chem.*, **14** (1993) 427–450.
[32] M. Randić, On Characterization of Molecular Branching, *J. Am. Chem. Soc.*, **97** (1975) 6609–6615.
[33] D. Bonchev, and N. Trinajstić, Information Theory, Distance Matrix, and Molecular Branching, *J. Chem. Phys.*, **67** (1977) 4517–4533.
[34] D. Bonchev, J.V. Knop, and N. Trinajstić, Mathematical Models of Branching, *MATCH*, **6** (1979) 21–47.
[35] D. Bonchev, O. Mekenyan, and N. Trinajstić, Topological Characterization of Cyclic Structures, *Intl. J. Quantum Chem.*, **17** (1980) 845–893.
[36] B. Barysz, J.V. Knop, S. Pejakovi, and N. Trinajstić, Characterization of Branching, *Polish J. Chem.*, **59** (1985) 405–432.
[37] S.H. Bertz, Branching in Graphs & Molecules, *Disc. Appl. Math.*, **19** (1988) 65–83.

[38] H.Y. Zhu, and D.J. Klein, Graph-Geometric Invariants for Molecular Structures, *J. Chem. Inf. & Comp. Sci.*, **36** (1997) 1067–1075.
[39] D.J. Klein, Graph Geometry, Graph Metrics, & Wiener, *Match*, **35** (1997) 7–27.
[40] D.J. Klein, and H.Y. Zhu, Distances & volumina for graphs, *J. Math. Chem.*, **23** (1998) 179–195.

TEXAS A&M UNIVERSITY – GALVESTON, GALVESTON, TEXAS 77553-1675, USA AND INSTITUTE OF THEORETICAL PHYSICS & ASTRONOMY, GOSTAUTO 12, 2600 VILNIUS, LITHUANIA
E-mail address: bytautal@arctic.tamug.tamu.edu

TEXAS A&M UNIVERSITY – GALVESTON, GALVESTON, TEXAS 77553-1675, USA
E-mail address: kleind@tamug.tamu.edu

DEPT. OF MATHEMATICS & COMPUTER SCIENCE, DRAKE UNIVERSITY, DES MOINES, IOWA 50311, USA
E-mail address: milan.randic@drake.edu

INSTITUTE OF MATHEMATICS, PHYSICS, & MECHANICS, AND MATHEMATICS DEPT., UNIVERSITY OF LJUBLJANA, SI-1000 LJUBLJANA, SLOVENIA
E-mail address: Tomaz.Pisanski@fmf.uni-lj.si

DIMACS Series in Discrete Mathematics
and Theoretical Computer Science
Volume 51, 2000

Enumeration of Fusenes to $h = 20$

Gilles Caporossi, Pierre Hansen, and Maolin Zheng

ABSTRACT. Fusenes are simply connected, geometrically planar or non-planar polyhexes. Enumeration of such systems with up to 20 hexagons, *i.e.*, a set 4 000 times larger than the largest previously enumerated is reported. Results are broken down by isomers and symmetry classes.

1. Introduction

Polyhexes are systems of congruent regular hexagons which are pairwise disjoint or have a single edge in common [**1**] [**2**]. We only consider here simply connected polyhexes, which are called *fusenes* [**3**]. Fusenes may be geometrically non-planar and look like a helix, in which case they are called *helicenes* [**4**]. In other words, helicenes cannot be projected on the infinite hexagonal lattice without some edges being superposed. Polyhexes may possess or not interior vertices, which belong to 3 hexagons and hence are not on the boundary (the *boundary* is formed by the edges which belong to only one hexagon). In the former case these polyhexes are said to be *pericondensed* and in the latter, *catacondensed*. For short, pericondensed and catacondensed fusenes will be referred to as *perifusenes* and *catafusenes* [**3**]. Similarly, one will speak of *perihelicenes* and *catahelicenes*.

Polyhexes correspond to known or unknown polycyclic hydrocarbons with only six-membered rings. Much work has been devoted to enumeration of all polyhexes with a given number of hexagons and of particular subsets thereof.

Indeed, counting and enumeration of molecules of various families have been important activities in mathematical chemistry for a long time. One may consult Balaban [**5**], Bababan *et al.*[**6**], Balasubramanian [**7**], Bonchev and Rouvray [**8**] [**9**], Brunvoll *et al.* [**4**], Dias [**10**], Gutman and Cyvin [**11**], Hosoya [**12**]), Knop *et al.* [**13**], Mercier *et al.* [**14**], Trinajstic *et al.* [**15**] [**16**] and Brunvoll *et al* [**17**] for surveys and history.

Recently, enumeration of benzenoids was extended to $h = 21$ [**18**]. This result was achieved by using the *boundary-edges code* (or BE code for short) [**19**] (which is equivalent in this case to the PC-2 code [**20**]), and *orderly generation* [**21**] [**22**] [**23**] (in fact, the algorithm was devised by specializing Avis and Fukuda's *reverse search* [**24**] which, in this case is close to the cited procedures, particularly McKay's orderly generation [**23**]). Efficiency of the enumeration is due in large part to a

1991 *Mathematics Subject Classification.* 19A15, 05C30.

new father-son relationship: the father is the polyhex obtained when removing the hexagon corresponding to the first digit of the son's code. This entails a technical problem: with this definition, not all polyhexes have fathers. Those who do not are called *orphans*. However, the smallest benzenoid orphan has 29 hexagons. This approach is extended in the present paper to enumeration of fusenes. Another technical problem arises: it must be shown that any given sufficiently small BE code corresponds to a unique polyhex. While this is obvious for benzenoids, it turns not to be the case for helicenes: indeed there are some pairs of helicenes with the same BE code. Examples are given in a companion paper [**25**]. However, we will show that here that helicenes with $h \leq 21$ hexagons have a unique BE code, which will suffice to justify the enumeration made. Moreover, one must determine the size of the smallest fusene orphans and, if this size is not larger than the maximum number h of hexagons considered, enumerate them.

This paper is organized as follows: the algorithm is described in the next section. The two main theoretical results on uniqueness of small fusenes with a given BE code and on the size of the smallest fusene orphans, are given in Section 3. Computational results are presented in Section 4 and conclusions drawn in Section 5.

2. Algorithm

We first describe the BE code [**19**] which is basic to the algorithm. For simply connected polyhexes, it is equivalent to the PC-2 code [**20**] and obtained as follows: beginning at any external vertex of degree 3, which thus belongs to 2 hexagons, follow the boundary of the polyhex noting by a digit the number of edges on this boundary for each successive hexagon encountered. Observe that the same hexagon may appear up to three times on the boundary, and hence may correspond to 1,2 or 3 digits in the code. Then apply, if needed, a circular shift and/or an order reversal to the code to make it lexicographically maximum. Construction of the BE code of a polyhex is illustrated in Figure 1. Note that it may be obtained in several ways in case of symmetry.

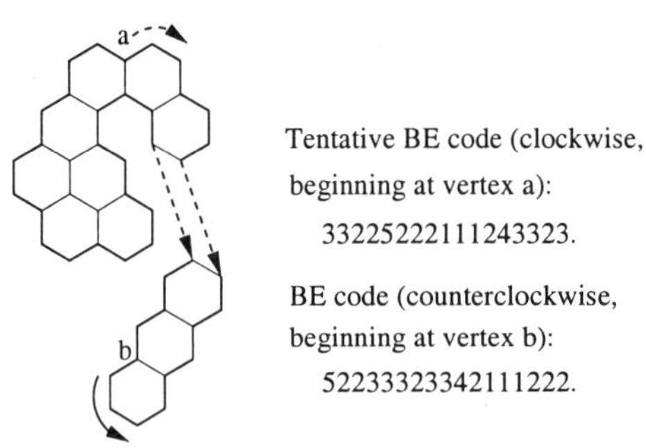

FIGURE 1. A helicene and its BE code

For benzenoids, it is clear that any given boundary code defines a unique polyhex (*i.e.*, that one inside the boundary when tracing it on the infinite hexagonal lattice). At first sight, it appears plausible that this remains the case for helicenes (and hence for fusenes). This is true at least for $h \leq 21$ as shown in the next section. But, as mentioned above, this property does not hold in general [**25**].

As in most polyhex enumeration methods, we proceed by successive additions of hexagons. This can be done in a depth-first way, adding hexagons until the desired number h_{max} is attained, then backtracking and progressing again to the point where there are h_{max} hexagons, and so forth. There might be duplications, as the same polyhex with $h+1$ hexagons may be obtained from several ones with h hexagons. To avoid that, each polyhex with $h+1$ hexagons is legitimately generated from one and only one polyhex with h hexagons, using McKay [**23**] orderly generation. Validation, *i.e.*, checking if a polyhex has been generated from its father, a specific polyhex with $h-1$ hexagons, is easy and quick, due to a very efficient choice of the father-son relationship [**18**].

Rule 1: *A polyhex is viewed as legitimate if and only if the first digit of its BE code corresponds to the last added hexagon.*

In other words, *Rule 1* states that the father of a polyhex is the polyhex obtained by removing the hexagon which corresponds to the first digit of its BE code. The use of the so defined enumeration tree (see Figure 2 for an illustration) reduces significantly the complexity of the program. Instead of scanning the code of the polyhex and finding all its potential fathers as described above, we just consider one of them.

An ideal enumeration scheme would allow direct identification of all the positions where hexagons may be added to generate a legitimate polyhex instead of any latter validation. This appears to be out of reach. However, *a priori* identification of some illegitimate additions is often possible, even before computation, due to the definition of the enumeration tree (for example: adding an hexagon appearing as a 3 in the code while another hexagon will be represented as a 4 or a 5 in the same code). When generation has not been shown to be illegitimate, canonization is easy as the starting point of the code of a legitimate polyhex is known before such an operation. Moreover, validation becomes a part of canonization, as the identification of a better starting point of the code brings this last operation to an end.

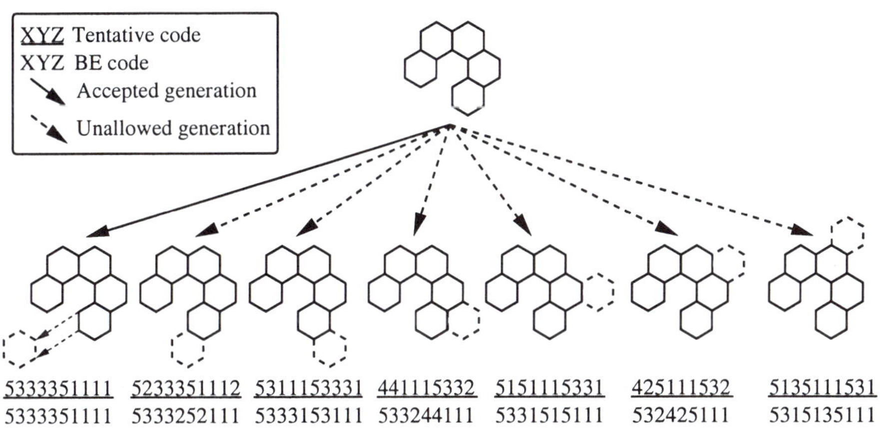

FIGURE 2. Enumeration tree

3. Theoretical Results

In this section we address the two problems mentioned in the previous section and on the solutions of which depends the correctness of the algorithm of this paper: uniqueness of small fusenes with a given BE code and existence and enumeration of orphans. While for benzenoids it is straightforward to see that a given BE code defines a single well-defined polyhex, the question is much more difficult for helicenes, and hence for fusenes. Indeed, the boundary of a helicene can be such that several edges are superposed when projecting it on the infinite hexagonal lattice. Moreover, the hexagons may be on either side of its edges. As mentioned above, to obtain the desired result, we limit the size of polyhexes considered. This suffices for our purpose, i.e., enumeration of fusenes with up to 20 hexagons.

For a helicene H let $B(H)$ denote its boundary. The *characteristic graph* $C(H)$ of H is defined as follows: the vertex set of $C(H)$ is the set of hexagons of H and two vertices are adjacent in $C(H)$ if and only if they have an edge in common in H. By convention, H is drawn such that one of it edges is vertical. A helicene H and its characteristic graph $C(H)$ are presented in Figure 3

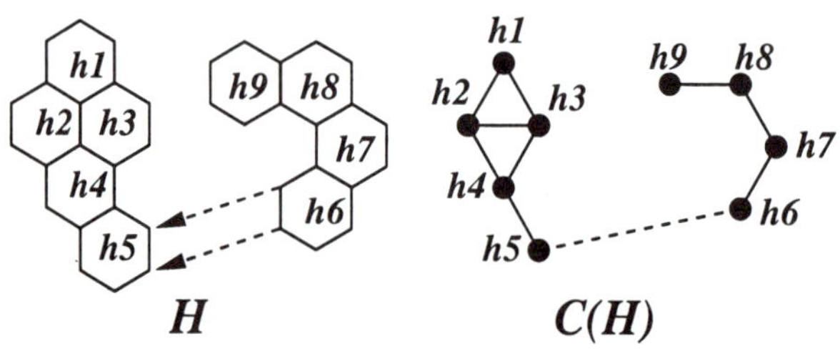

FIGURE 3. A helicene H and its characteristic graph $C(H)$.

THEOREM 3.1. *All fusenes with up to 21 hexagons have different boundary-edges codes.*

PROOF. Let H be a fusene. A boundary hexagon h is called *removable* if after removing from H the edges of h on the boundary and then some isolated vertices a fusene is obtained. Otherwise it is called *non-removable* (see Figure 4 for an illustration). One can check that any boundary hexagon which corresponds to a digit 4 or 5 in the BE code is removable.

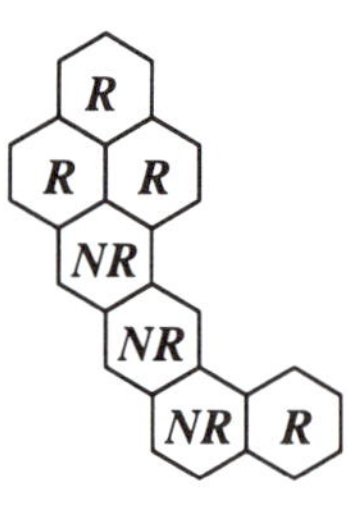

FIGURE 4. Removable (R) and non-removable (NR) hexagons.

Clearly, a boundary hexagon h is removable if and only if the intersection of h and $B(H)$ is a path. It is easily shown that any fusene H with at least two hexagons has at least two removable hexagons.

Let H_1 and H_2 be two fusenes with the same BE code (or in other words, with the same boundary). If they are not equal, let us choose H_1 and H_2 in such a way that the sum of their numbers of hexagons is minimum. Since H_1 and H_2 have the same BE code, $B(H_1)$ can be mapped onto $B(H_2)$ such that they are geometrically identical. Then for every removable hexagon h of H_1, the hexagon h' of H_2 to which h is mapped into by the mapping between boundaries is not removable in H_2 and vice versa. So in the BE code, there are no 4 or 5.

Let $C(H_1)$ and $C(H_2)$ be the characteristic graphs of H_1 and H_2 Then both of $C(H_1)$ and $C(H_2)$ have cut vertices (each non-removable hexagon corresponds to a cut vertex). So $C(H_1)$ ($C(H_2)$) has two maximal 2-connected subgraphs each of which contains only one cut vertex of $C(H_1)$ ($C(H_2)$). Let D_1 and D_2 be two such subgraphs of $C(H_1)$. Let $G(D_1)$ and $G(D_2)$ be the subgraphs of H_1 consisting of the hexagons which correspond to the vertices in D_1 and D_2 respectively. Then $G(D_1)$ (as well as $G(D_2)$) contains only one non-removable hexagon of H_1. Since the BE code of H_1 has no 4 or 5, $G(D_1)$ ($G(D_2)$) contains at least 7 hexagons. Let u be the hexagon of $G(D_1)$ which is a cut vertex in $C(H_1)$. Without loss of generality, we assume that $G(D_1)$ is drawn such that u is one of the lowest hexagons. Also u has degree at least 3 in $C(H_1)$. Then consider the leftmost and the rightmost hexagons (there are at least 2). Similarly, consider the leftmost and the rightmost hexagons on the highest level. There are 2 such hexagons too (since we assume u is on the lowest level). Each of these four hexagons corresponds to a digit 3 in the BE code of H_1. So the hexagons corresponding to these four in H_2 are distinct, non-removable and have degree 2 in $C(H_2)$. By the choice of D_1 and D_2, there are at least 8 hexagons (let M be the set of them) in H_2 such that each of them has degree 2 in $C(H_2)$.

In the same way as selecting D_1 and D_2, we can select D'_1 and D'_2 for $C(H_2)$. Then each hexagon of M does not belong to D'_1 or D'_2 for they have degree 2 in $C(H_2)$. By the same reasoning as above, D'_1 and D'_2 have at least 14 hexagons together. Thus H_2 has at least 22 hexagons. The same is true for H_1. The proof is completed. □

We next turn to the study of orphans. We first recall a few definitions from [**18**]. The nth *hexagon* is the hexagon represented by the nth digit in the code of the orphan. The *isthmus* is the maximal set of consecutive hexagons including the first hexagon, which each appear twice in the code and are adjacent to exactly two hexagons which also appear twice in the code. A *peninsula* is one of the two polyhexes obtained by removing the isthmus of the orphan (the *first peninsula* appears first in the orphan's code and the *second* after). These concepts are illustrated in Figure 5.

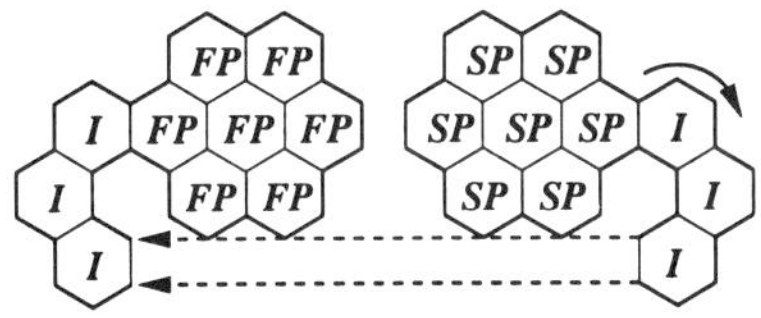

FIGURE 5. An orphan (with BE Code 333333133333111111111333331), its isthmus I, first FP and second SP peninsulas.

LEMMA 3.2. *An orphan's code begins by a 3 and cannot contain a 4 or a 5.*

PROOF. The first hexagon of an orphan must appear twice in the code, hence it cannot be a 4 or a 5. If the BE code only contains 1's and 2's, the boundary can only turn anticlockwise (beginning with an hexagon below or to the right of its edge(s) on the boundary) and hence cannot close around a finite polyhex. So this code must begin with a 3. Then from the lexicographical rule, it cannot contain a 4 or a 5. □

The second main result we need is the following:

THEOREM 3.3. *The smallest fusene orphans have $h = 20$ hexagons. Moreover, there are exactly 4 of them.*

PROOF. The proof is by case enumeration, combining peninsulas and isthmuses.

1. As by Lemma 3.2, an orphan's code cannot contain a 4 or a 5, it is easily checked that any peninsula must contain at least 7 hexagons (see Figure 6a). Moreover, there is only one peninsula with 7 hexagons, one with 8 and two with 9 (see Figure 6a, b and c). The part of the BE Code corresponding to a 7 hexagon (resp. 8 or 9 hexagon) peninsula contains 5 (resp. 4) successive 3's. Hence the corresponding isthmus in an orphan must contain at least 5 (resp. 4) hexagons.

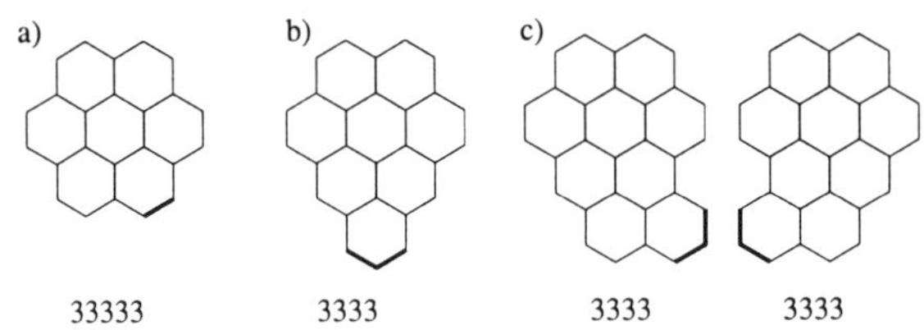

FIGURE 6. Peninsulas with 7, 8 and 9 hexagons, and their largest sequence in the BE Code, with possibly connections to the isthmus (in bold).

   So the smallest fusene orphans must have at least 19 hexagons. There is a unique way to combine two 7 hexagon peninsulas and a 5 hexagon isthmus. The tentative BE code, beginning with the first hexagon of the isthmus, is then 3333313333311111111333331. However, this code is not lexicographically maximum.

   Shifting it to begin with the second hexagon of the second peninsula yields the code 3333313333313333311111111 which is lexicographically larger. Hence the smallest fusene orphans must have at least 20 hexagons.
2. Consider now two 7 hexagon peninsulas and a 6 hexagon isthmus. There are several cases. If the isthmus code contains the sequence 333333, there is a unique way to connect it to the peninsulas.

   The tentative code 33333313333311111111333331 is lexicographically maximal. A first smallest orphan with 20 hexagons is found (see Figure 8a). If the isthmus code does not contain the sequence 333333, as it must contain the sequence 33333, one of the peninsulas must be connected to an hexagon of the isthmus at an edge (noted (a) in Figure 7) such that 3 edges of that hexagon are on the first side of the boundary. Moreover, this peninsula must be the second one. Indeed, if not, the first hexagon would be the second one of the isthmus (should the fusene be an orphan) and the code would begin with 33333133333111... but beginning at the last 3 of the first peninsula's code and reversing the code yields a code beginning with 33333133333213... or 33333133333113... which is lexicographically larger. Then connecting the first peninsula at the edge of the last hexagon of the isthmus which leaves 2 edges on each side (noted (b) in Figure 7) yields the tentative code 3333321333331211111133331 which is lexicographically maximum. A second smallest fusene orphan is found (see Figure 8 b). Connecting the first peninsula to the isthmus at an edge having 1 edge of the last hexagon on the first side of the father (noted (c) in Figure 7) yields a tentative code 3333311333331311111333331 which is not lexicographically maximum.

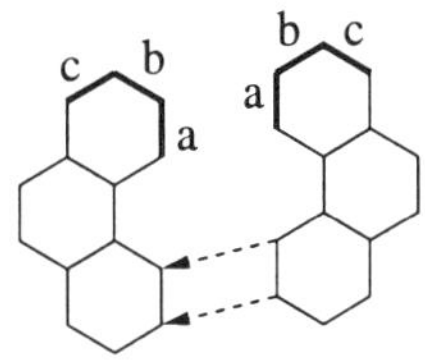

FIGURE 7. An isthmus with possible connections to the peninsulas (in bold).

3. Consider next a 7 hexagon and a 8 hexagon peninsula. The latter has a unique geometric form, but may be connected to the isthmus on two different edges (bold edges on Figure 6 b), which are not equivalent as the code is computed clockwise. Again if one peninsula has 7 hexagons the isthmus must have at least 5. If the isthmus has 5 hexagons, the first peninsula has 7 hexagons and the second 8, there are two cases:

   (i) The second peninsula shares edge *(a)* with the isthmus. The tentative code is 3333313333311111122333321 which is not lexicographically maximum.

   (ii) The second peninsula shares edge *(b)* with the isthmus. The tentative code is 3333313333311111112333322 which is again not lexicographically maximum.

   If the isthmus has 5 hexagons, the first peninsula has 8 hexagons and the second 7 there are again two cases:

   (i) The first peninsula shares edge *(a)* with the isthmus . The tentative code is 3333322333332111111333331 which is lexicographically maximum. A third smallest fusene orphan has been found (see Figure 8 c)

   (ii) The first peninsula shares edge *(b)* with the isthmus. The tentative code is 3333312333332211111133331 which is not lexicographically maximum.
4. Finally, consider two 8 hexagon peninsulas and a 4 hexagon isthmus. As the isthmus part of the BE Code must contain the sequence 3333 for the fusene to be an orphan, the edges of the isthmus on which the peninsulas are connected are fixed and both *(a)*. Then connecting both peninsulas at edges *(a)* yields the tentative code 333322333321111122333321 which is lexicographically maximum. A fourth smallest fusene orphan has been found (see Figure 8d). Connecting the first (resp. second or both) peninsula(s) yields the tentative

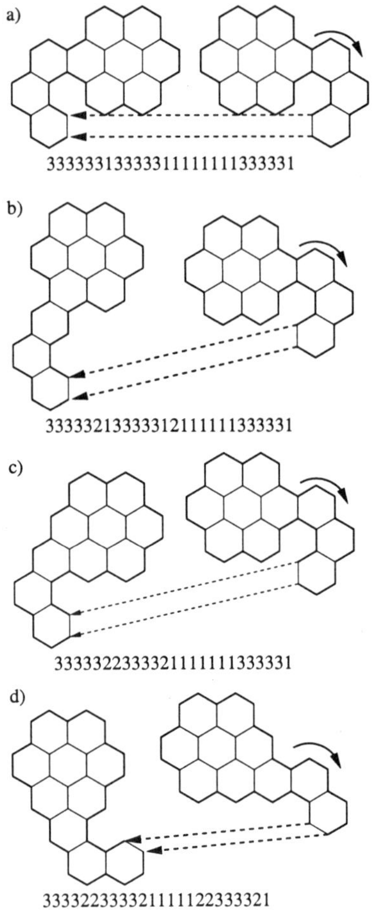

FIGURE 8. Fusene orphans with 20 hexagons

codes 333312333322111122333321, 33332233332111112333322, and 333312333322111112333322 none of which is lexicographically maximum.

□

It follows from Theorem 3.3 that fusenes with up to $h = 20$ hexagons can be enumerated with *Algorithm 1*, adding the four orphans described there for the set found for $h = 20$

4. Computational Results

The algorithm of Section 2 has been coded in C and tested on a (Linux) PC and on various Sparc Sun 4,5,10,20.

The number of catafusenes, perifusenes and fusenes with $h = 1$ to 20 hexagons are given in Table 1.

The number of perifusenes and fusenes for $h = 16$ to 20 are, to the best of our knowledge, given here for the first time. The set of fusenes with 20 hexagons is more than 4 000 times larger than the set of fusenes with 15 hexagons, the largest previously enumerated [17]

Computing times for $h \geq 17$ (using a variety of machines) are also reported in Table 1. They appear to grow linearly with the number of fusenes enumerated and with the number of hexagons in these fusenes. About 21500 fusenes are enumerated per second when $h = 20$. Using the numbers of catacondensed and pericondensed benzenoids with up to $h = 20$ hexagons obtained in [**18**], one can deduce from Table 1 the number of catahelicenes, perihelicenes and helicenes with up to $h = 20$ hexagons. These results are given in Table 1. Moreover, numbers of fusenes with $h = 10$ to $h = 20$ broken down by symmetry classes are given in Tables 2 to 12.

5. Conclusion

Enumeration of various types of chemical graphs and similar representations of molecules is an important research topic. In particular, polyhexes have attracted much attention as evidenced in the extensive survey of Cyvin, Brunvoll and Cyvin [**4**]. In a previous paper, it was shown how combining the BE code [**20**] [**19**] with orderly generation [**21**] [**22**] [**23**] [**24**] yields an effective tool for enumeration of benzenoids. In the present paper, this approach is extended to the enumeration of fusenes. There are a couple of difficulties, *i.e.*, one must prove uniqueness of small fusenes with a given BE code and be more careful in treatment of orphans than in the case of benzenoids. But then, the approach works well again: a set of over 500 000 000 000 fusenes has been enumerated. It is expected that the proposed approach will be adapted and extended to enumerate various other classes of polyhexes and other molecular graphs. The next steps might be to enumerate coronoids, multiple coronoids, circulenes and all polyhexes.

TABLE 1. Number of catafusenes and perifusenes according to h

h	Catafusenes	Perifusenes	total	
1	1	0	1	
2	1	0	1	
3	2	1	3	
4	5	2	7	
5	12	10	22	
6	37	45	82	
7	123	216	339	
8	446	1059	1505	
9	1689	5347	7036	
10	6693	27143	33836	
11	27034	139212	166246	
12	111630	718357	829987	
13	467262	3730011	4197273	
14	1981353	19475091	21456444	
15	8487400	102229185	110716585	
16	36695369	539332368	576027737	
17	159918120	2859067920	3018986040	2 days 4h 31'
18	701957539	15225372566	15927330105	11 days 18h 22'
19	3101072051	81429798404	84530870455	46 days 16h 31'
20	13779935438	437289403625	451069339063	242 days 4h 40'

TABLE 2. Fusene isomers for $h = 10$ according to symmetry

Isomer	Number	C_{2h}	D_{2h}	C_{3h}	D_{3h}	C_{6h}	D_{6h}	C_{2v}
$C_{32}H_{14}$	1	0	1	0	0	0	0	0
$C_{33}H_{15}$	9	0	0	0	1	0	0	2
$C_{34}H_{16}$	42	5	0	0	0	0	0	7
$C_{35}H_{17}$	123	0	0	0	0	0	0	3
$C_{36}H_{18}$	416	13	1	0	1	0	0	21
$C_{37}H_{19}$	1032	0	0	0	0	0	0	5
$C_{38}H_{20}$	2440	28	3	0	0	0	0	49
$C_{39}H_{21}$	4871	0	0	1	0	0	0	11
$C_{40}H_{22}$	8562	59	2	0	0	0	0	102
$C_{41}H_{23}$	9647	0	0	0	0	0	0	0
$C_{42}H_{24}$	6693	66	5	4	2	0	0	130
Total	33836	171	12	5	4	0	0	330

TABLE 3. Fusene isomers for $h = 11$ according to symmetry

Isomer	Number	C_{2h}	D_{2h}	C_{3h}	D_{3h}	C_{6h}	D_{6h}	C_{2v}
$C_{35}H_{15}$	2	0	0	0	0	0	0	0
$C_{36}H_{16}$	26	1	1	0	0	0	0	8
$C_{37}H_{17}$	100	0	0	0	0	0	0	6
$C_{38}H_{18}$	333	3	2	0	0	0	0	15
$C_{39}H_{19}$	1010	0	0	0	0	0	0	12
$C_{40}H_{20}$	2875	9	3	0	0	0	0	42
$C_{41}H_{21}$	6743	0	0	0	0	0	0	38
$C_{42}H_{22}$	14598	30	3	0	0	0	0	118
$C_{43}H_{23}$	27283	0	0	0	0	0	0	86
$C_{44}H_{24}$	42621	71	0	0	0	0	0	284
$C_{45}H_{25}$	43621	0	0	0	0	0	0	72
$C_{46}H_{26}$	27034	66	5	0	0	0	0	269
Total	166246	180	14	0	0	0	0	950

TABLE 4. Fusene isomers for $h = 12$ according to symmetry

Isomer	Number	C_{2h}	D_{2h}	C_{3h}	D_{3h}	C_{6h}	D_{6h}	C_{2v}
$C_{37}H_{15}$	1	0	0	0	1	0	0	0
$C_{38}H_{16}$	13	2	1	0	0	0	0	3
$C_{39}H_{17}$	61	0	0	0	0	0	0	5
$C_{40}H_{18}$	279	11	1	1	0	0	0	17
$C_{41}H_{19}$	878	0	0	0	0	0	0	10
$C_{42}H_{20}$	2733	32	1	0	0	0	0	50
$C_{43}H_{21}$	7601	0	0	3	2	0	0	14
$C_{44}H_{22}$	19510	79	4	0	0	0	0	134
$C_{45}H_{23}$	42674	0	0	0	0	0	0	27
$C_{46}H_{24}$	86286	165	8	7	0	0	0	290
$C_{47}H_{25}$	149483	0	0	0	0	0	0	42
$C_{48}H_{26}$	211071	283	2	0	0	0	0	501
$C_{49}H_{27}$	197767	0	0	10	0	0	0	0
$C_{50}H_{28}$	111630	269	5	0	0	0	0	538
Total	829987	841	22	21	3	0	0	1631

TABLE 5. Fusene isomers for $h = 13$ according to symmetry

Isomer	Number	C_{2h}	D_{2h}	C_{3h}	D_{3h}	C_{6h}	D_{6h}	C_{2v}
$C_{40}H_{16}$	4	0	1	0	0	0	0	2
$C_{41}H_{17}$	35	0	0	0	0	0	0	1
$C_{42}H_{18}$	187	2	1	1	0	0	1	13
$C_{43}H_{19}$	708	0	0	0	0	0	0	15
$C_{44}H_{20}$	2462	12	1	0	0	0	0	56
$C_{45}H_{21}$	7370	0	0	1	0	0	0	32
$C_{46}H_{22}$	20938	27	3	0	0	0	0	115
$C_{47}H_{23}$	54356	0	0	0	0	0	0	84
$C_{48}H_{24}$	128060	71	3	2	2	0	1	293
$C_{49}H_{25}$	264867	0	0	0	0	0	0	221
$C_{50}H_{26}$	501160	190	5	0	0	0	0	742
$C_{51}H_{27}$	807433	0	0	7	0	0	0	386
$C_{52}H_{28}$	1040995	328	6	0	0	0	0	1318
$C_{53}H_{29}$	901436	0	0	0	0	0	0	274
$C_{54}H_{30}$	467262	269	5	17	2	0	0	1079
Total	4197273	899	25	28	4	0	2	4631

TABLE 6. Fusene isomers for $h = 14$ according to symmetry

Isomer	Number	C_{2h}	D_{2h}	C_{3h}	D_{3h}	C_{6h}	D_{6h}	C_{2v}
$C_{42}H_{16}$	1	0	1	0	0	0	0	0
$C_{43}H_{17}$	16	0	0	0	0	0	0	4
$C_{44}H_{18}$	120	7	2	0	0	0	0	10
$C_{45}H_{19}$	533	0	0	0	0	0	0	10
$C_{46}H_{20}$	2001	29	1	0	0	0	0	46
$C_{47}H_{21}$	6628	0	0	0	0	0	0	26
$C_{48}H_{22}$	20826	80	3	0	0	0	0	127
$C_{49}H_{23}$	58100	0	0	0	0	0	0	50
$C_{50}H_{24}$	154318	218	2	0	0	0	0	355
$C_{51}H_{25}$	374024	0	0	0	0	0	0	90
$C_{52}H_{26}$	823379	482	4	0	0	0	0	825
$C_{53}H_{27}$	1613616	0	0	0	0	0	0	137
$C_{54}H_{28}$	2869349	930	15	0	0	0	0	1648
$C_{55}H_{29}$	4306614	0	0	0	0	0	0	166
$C_{56}H_{30}$	5119930	1338	5	0	0	0	0	2407
$C_{57}H_{31}$	4125636	0	0	0	0	0	0	0
$C_{58}H_{32}$	1981353	1102	15	0	0	0	0	2204
Total	21456444	4186	48	0	0	0	0	8105

TABLE 7. Fusene isomers for $h = 15$ according to symmetry

Isomer	Number	C_{2h}	D_{2h}	C_{3h}	D_{3h}	C_{6h}	D_{6h}	C_{2v}
$C_{45}H_{17}$	4	0	0	0	0	0	0	1
$C_{46}H_{18}$	70	2	1	1	1	0	0	12
$C_{47}H_{19}$	347	0	0	0	0	0	0	5
$C_{48}H_{20}$	1570	5	3	0	0	0	0	31
$C_{49}H_{21}$	5618	0	0	3	0	0	0	32
$C_{50}H_{22}$	18625	27	4	0	0	0	0	127
$C_{51}H_{23}$	57812	0	0	0	0	0	0	91
$C_{52}H_{24}$	166348	86	4	5	1	0	0	377
$C_{53}H_{25}$	439147	0	0	0	0	0	0	225
$C_{54}H_{26}$	1096491	197	5	0	0	0	0	806
$C_{55}H_{27}$	2503001	0	0	21	0	0	0	543
$C_{56}H_{28}$	5195186	478	9	0	0	0	0	1937
$C_{57}H_{29}$	9681616	0	0	0	0	0	0	1220
$C_{58}H_{30}$	16210990	1086	14	31	2	0	0	4253
$C_{59}H_{31}$	22755295	0	0	0	0	0	0	1739
$C_{60}H_{32}$	25132946	1541	0	0	0	0	0	6164
$C_{61}H_{33}$	18964119	0	0	36	0	0	0	1086
$C_{62}H_{34}$	8487400	1102	15	0	0	0	0	4423
Total	110716585	4524	55	97	4	0	0	23072

TABLE 8. Fusene isomers for $h = 16$ according to symmetry

Isomer	Number	C_{2h}	D_{2h}	C_{3h}	D_{3h}	C_{6h}	D_{6h}	C_{2v}
$C_{47}H_{17}$	1	0	0	0	0	0	0	1
$C_{48}H_{18}$	30	3	2	0	1	0	0	4
$C_{49}H_{19}$	223	0	0	0	0	0	0	9
$C_{50}H_{20}$	1121	20	2	0	0	0	0	31
$C_{51}H_{21}$	4501	0	0	1	1	0	0	28
$C_{52}H_{22}$	16310	80	0	0	0	0	0	121
$C_{53}H_{23}$	52795	0	0	0	0	0	0	64
$C_{54}H_{24}$	163881	215	5	4	1	0	0	348
$C_{55}H_{25}$	473555	0	0	0	0	0	0	147
$C_{56}H_{26}$	1278169	586	7	0	0	0	0	945
$C_{57}H_{27}$	3207845	0	0	7	0	0	0	280
$C_{58}H_{28}$	7579108	1404	9	0	0	0	0	2341
$C_{59}H_{29}$	16381817	0	0	0	0	0	0	497
$C_{60}H_{30}$	32280641	2864	14	17	5	0	0	4934
$C_{61}H_{31}$	57289337	0	0	0	0	0	0	676
$C_{62}H_{32}$	90549291	5057	31	0	0	0	0	9031
$C_{63}H_{33}$	119314280	0	0	39	0	0	0	675
$C_{64}H_{34}$	123231368	6335	5	0	0	0	0	11537
$C_{65}H_{35}$	87508095	0	0	0	0	0	0	0
$C_{66}H_{36}$	36695369	4635	15	66	5	0	0	9265
Total	576027737	21199	90	134	13	0	0	40934

TABLE 9. Fusene isomers for $h = 17$ according to symmetry

Isomer	Number	C_{2h}	D_{2h}	C_{3h}	D_{3h}	C_{6h}	D_{6h}	C_{2v}
$C_{50}H_{18}$	9	0	1	0	0	0	0	2
$C_{51}H_{19}$	117	0	0	0	0	0	0	6
$C_{52}H_{20}$	763	6	3	0	0	0	0	36
$C_{53}H_{21}$	3414	0	0	0	0	0	0	17
$C_{54}H_{22}$	13301	20	2	0	0	0	0	104
$C_{55}H_{23}$	46688	0	0	0	0	0	0	73
$C_{56}H_{24}$	153158	68	3	0	0	0	0	311
$C_{57}H_{25}$	467966	0	0	0	0	0	0	235
$C_{58}H_{26}$	1363062	230	11	0	0	0	0	1009
$C_{59}H_{27}$	3718409	0	0	0	0	0	0	652
$C_{60}H_{28}$	9515597	606	7	0	0	0	0	2554
$C_{61}H_{29}$	22797671	0	0	0	0	0	0	1495
$C_{62}H_{30}$	51188565	1349	10	0	0	0	0	5437
$C_{63}H_{31}$	105316193	0	0	0	0	0	0	3369
$C_{64}H_{32}$	197855777	3052	18	0	0	0	0	12209
$C_{65}H_{33}$	334891034	0	0	0	0	0	0	6509
$C_{66}H_{34}$	500794983	5941	21	0	0	0	0	23327
$C_{67}H_{35}$	621832355	0	0	0	0	0	0	7897
$C_{68}H_{36}$	603819104	7189	20	0	0	0	0	28776
$C_{69}H_{37}$	405289754	0	0	0	0	0	0	4438
$C_{70}H_{38}$	159918120	4635	15	0	0	0	0	18555
Total	3018986040	23096	111	0	0	0	0	117011

TABLE 10. Fusene isomers for $h = 18$ according to symmetry

Isomer	Number	C_{2h}	D_{2h}	C_{3h}	D_{3h}	C_{6h}	D_{6h}	C_{2v}
$C_{52}H_{18}$	3	1	0	0	1	0	0	1
$C_{53}H_{19}$	53	0	0	0	0	0	0	5
$C_{54}H_{20}$	471	14	2	0	0	0	0	23
$C_{55}H_{21}$	2437	0	0	3	0	0	0	24
$C_{56}H_{22}$	10588	59	5	0	0	0	0	91
$C_{57}H_{23}$	39334	0	0	0	0	0	0	74
$C_{58}H_{24}$	136344	206	6	8	0	0	0	315
$C_{59}H_{25}$	443380	0	0	0	0	0	0	178
$C_{60}H_{26}$	1362022	601	6	0	0	0	0	946
$C_{61}H_{27}$	3943217	0	0	15	2	0	0	410
$C_{62}H_{28}$	10891716	1641	6	0	0	0	0	2683
$C_{63}H_{29}$	28258301	0	0	0	0	0	0	876
$C_{64}H_{30}$	69045923	4025	12	39	1	0	0	6597
$C_{65}H_{31}$	158325287	0	0	0	0	0	0	1604
$C_{66}H_{32}$	339201122	8842	19	0	0	0	0	14970
$C_{67}H_{33}$	666871615	0	0	107	3	0	0	2696
$C_{68}H_{34}$	1198097947	16793	17	0	0	0	0	29202
$C_{69}H_{35}$	1936345760	0	0	0	0	0	0	3290
$C_{70}H_{36}$	2746277704	26936	65	146	0	0	0	48447
$C_{71}H_{37}$	3224895584	0	0	0	0	0	0	2809
$C_{72}H_{38}$	2957778430	30050	15	0	0	0	0	55253
$C_{73}H_{39}$	1883445328	0	0	137	0	0	0	0
$C_{74}H_{40}$	701957539	19768	51	0	0	0	0	39536
Total	15927330105	108936	204	455	7	0	0	210030

TABLE 11. Fusene isomers for $h = 19$ according to symmetry

Isomer	Number	C_{2h}	D_{2h}	C_{3h}	D_{3h}	C_{6h}	D_{6h}	C_{2v}
$C_{54}H_{18}$	1	0	0	0	0	0	1	0
$C_{55}H_{19}$	18	0	0	0	0	0	0	3
$C_{56}H_{20}$	256	3	1	0	0	0	0	17
$C_{57}H_{21}$	1647	0	0	1	0	0	0	14
$C_{58}H_{22}$	7885	17	2	0	0	0	0	92
$C_{59}H_{23}$	32086	0	0	0	0	0	0	61
$C_{60}H_{24}$	117978	66	6	4	1	0	0	328
$C_{61}H_{25}$	399935	0	0	0	0	0	0	193
$C_{62}H_{26}$	1296573	189	7	0	0	0	0	857
$C_{63}H_{27}$	3975642	0	0	6	2	0	0	585
$C_{64}H_{28}$	11566647	597	11	0	0	0	0	2625
$C_{65}H_{29}$	31990818	0	0	0	0	0	0	1824
$C_{66}H_{30}$	84280474	1758	19	26	3	1	0	7423
$C_{67}H_{31}$	209206245	0	0	0	0	0	0	4422
$C_{68}H_{32}$	490133377	4134	19	0	0	0	0	17053
$C_{69}H_{33}$	1078246412	0	0	44	0	0	0	9705
$C_{70}H_{34}$	2211468431	8933	25	0	0	0	0	35729
$C_{71}H_{35}$	4167613253	0	0	0	0	0	0	20174
$C_{72}H_{36}$	7175697743	18633	30	110	10	1	1	74104
$C_{73}H_{37}$	11087497850	0	0	0	0	0	0	33994
$C_{74}H_{38}$	14950114520	31582	60	0	0	0	0	124350
$C_{75}H_{39}$	16657928358	0	0	201	0	0	0	36177
$C_{76}H_{40}$	14487876140	33795	0	0	0	0	0	135180
$C_{77}H_{41}$	8780346115	0	0	0	0	0	0	18570
$C_{78}H_{42}$	3101072051	19768	51	269	5	0	0	79118
Total	84530870455	119475	231	661	21	2	2	602598

References

[1] J.V. Knop, W.R. Müller, K. Szymanski, and N. Trinajstić, Use of Small Computers for Large Computations: Enumeration of Polyhex Hydrocarbons. *J. Chem. Inf. Comput. Sci.*, 30:159–160, 1990.

[2] R. Tošić, D. Mašulović, I. Stojmenović, J. Brunvoll, and S.J. Cyvin, and B.N. Cyvin, Enumeration of Polyhex Hydrocarbons to $h = 17$. *J. Chem. Inf. Comput. Sci.*, 35:181–187, 1995.

[3] B.N. Cyvin, G. Xiaofeng, S.J. Cyvin, and F. Zhang, Enumeration of Helicenes. *Chemical Physics Letters*, 188:537–542, 1992.

[4] J. Brunvoll, B.N. Cyvin, and S.J. Cyvin, Enumeration of Benzenoid Systems and other Polyhexes. *Top. in Curr. Chem.*, 162:65–180, 1992.

[5] A.T. Balaban, Enumeration of Isomers. In *Chemical Graph theory*, pages 177–234. Abacus Press, Gordon and Breach, New York, 1990.

[6] A.T. Balaban, J. Brunvoll, J. Ciolowski, B.N. Cyvin, S.J. Cyvin, I. Gutman, W.J. He, J.V. Knop, M. Kovačević, W.R. Müller, K. Szymanski, R. Tošić, and N. Trinajstić, Enumeration of Benzenoid and Coronoid Hydrocarbons. *Z. Naturforsch.*, 42a:863–870, 1987.

[7] K. Balasubramanian, Recent Chemical Applications of Computational Combinatorics and Graph Theory. In D.H. Rouvray, editor, *Computational Chemical Graph Theory*. Nova Science, New York, 1990.

[8] D. Bonchev, and D.H. Rouvray, *Chemical Graph Theory, Introduction and Fundamentals*. Abacus Press, Gordon and Breach, New York, 1991.

[9] D. Bonchev and D.H. Rouvray, *Chemical Graph Theory, Reactivity and Kinetics*. Abacus Press, Gordon and Breach, New York, 1992.

TABLE 12. Fusene isomers for $h = 20$ according to symmetry

Isomer	Number	C_{2h}	D_{2h}	C_{3h}	D_{3h}	C_{6h}	D_{6h}	C_{2v}
$C_{57}H_{19}$	4	0	0	0	0	0	0	1
$C_{58}H_{20}$	129	8	1	0	0	0	0	15
$C_{59}H_{21}$	1009	0	0	0	0	0	0	17
$C_{60}H_{22}$	5726	42	5	0	0	0	0	66
$C_{61}H_{23}$	25057	0	0	0	0	0	0	69
$C_{62}H_{24}$	98126	178	5	0	0	0	0	274
$C_{63}H_{25}$	352663	0	0	0	0	0	0	195
$C_{64}H_{26}$	1189968	576	5	0	0	0	0	889
$C_{65}H_{27}$	3810058	0	0	0	0	0	0	477
$C_{66}H_{28}$	11696535	1666	11	0	0	0	0	2634
$C_{67}H_{29}$	34099127	0	0	0	0	0	0	1203
$C_{68}H_{30}$	94866661	4653	8	0	0	0	0	7453
$C_{69}H_{31}$	251646989	0	0	0	0	0	0	2647
$C_{70}H_{32}$	635388995	11750	26	0	0	0	0	19364
$C_{71}H_{33}$	1515224373	0	0	0	0	0	0	5239
$C_{72}H_{34}$	3413973431	26713	33	0	0	0	0	44397
$C_{73}H_{35}$	7220165711	0	0	0	0	0	0	9232
$C_{74}H_{36}$	14216730749	54644	60	0	0	0	0	93539
$C_{75}H_{37}$	25743534475	0	0	0	0	0	0	14436
$C_{76}H_{38}$	42549878007	96802	51	0	0	0	0	169680
$C_{77}H_{39}$	62936891732	0	0	0	0	0	0	15911
$C_{78}H_{40}$	80871221611	141596	120	0	0	0	0	256311
$C_{79}H_{41}$	85761644654	0	0	0	0	0	0	11913
$C_{80}H_{42}$	70974548080	143059	15	0	0	0	0	265101
$C_{81}H_{43}$	41052409755	0	0	0	0	0	0	0
$C_{82}H_{44}$	13779935438	85659	51	0	0	0	0	171318
Total	451069339063	567346	391	0	0	0	0	1092381

[10] J.R. Dias, Benzenoid Hydrocarbons. In *Handbook of Polycyclic Hydrocarbons*. Elsevier, Amsterdam, 1987.

[11] I. Gutman, and S.J. Cyvin, *Introduction to the Theory of Benzenoid Hydrocarbons*. Springer-Verlag, 1989.

[12] H. Hosoya, Some Recent Advances in Counting Polynomials in Chemical Graph Theory. In H.H. Rouvray, editor, *Computational Chemical Graph Theory*. Nova Science, New York, 1990.

[13] J.V. Knop, W.R. Müller, and K. Szymanski, Computer-Oriented Molecular Codes. In H.H. Rouvray, editor, *Computational Chemical Graph Theory*. Nova Science, New York, 1990.

[14] C. Mercier, Y. Sobel, and J.-E. Dubois, Darc/pelco Method: A Topological Tool for Qsar Search and its Reliable Predictive Capability. In *Chemical Graph theory*, pages 199–258. Abacus Press, Gordon and Breach, New York, 1990.

[15] N. Trinajstić, S. Nikolić, J.V. Knop, W.R. Müller, and K. Szymanski, *Computational Chemical Graph Theory*. Ellis Horwood, 1991.

[16] N. Trinajstić, *Chemical Graph Theory*. CRC Press, Boca Raton, 1992.

[17] J. Brunvoll, S.J. Cyvin, B.N. Cyvin, Z. Fuji, and X.F. Guo, Theory of Helicenic Hydrocarbons: 4. Further Enumeration. *Structural Chemistry*, 7:119–130, 1996.

[18] G. Caporossi, and P. Hansen, Enumeration of Polyhex Hydrocarbons to $h = 21$. *J. Chem. Inf. Comput. Sci.*, 38:610–619, 1998.

[19] P. Hansen, C. Lebatteux, and M. Zheng, The Boundary-Edges Code for Polyhexes. *Theochem, Journal of Molecular Structures*, 363:237–247, 1996.

[20] W. Herndon, and A.J. Bruce, Perimeter Code for Benzenoid Aromatic Hydrocarbons. *Stud. Phys. Theor. Chem.*, 51:491–513, 1987.

[21] R.C. Read, Every One a Winner. *Annals Discrete Math.*, 2:107–120, 1978.

[22] I.A. Faradzef, Constructive Enumeration of Combinatorial Objects. In *Problèmes Combinatoires et Théorie des Graphes Colloque Internat. CNRS 260.* CNRS Paris, 1978.
[23] B. McKay, Isomorph-Free Exhaustive Generation. *J. Algorithms*, 26:306–324, 1998.
[24] D. Avis, and K. Fukuda, Reverse Search for Enumeration. *Discrete Applied Mathematics*, 65:21–46, 1996.
[25] X. Guo, P. Hansen and M. Zheng, Boundary Uniqueness of Helicenes. Les Cahiers du GERAD, G99–37.

DÉPARTMENT DE MATHÉMATIQUES ET GÉNIE INDUSTRIEL, ÉCOLE POLYTECHNIQUE DE MONTRÉAL, CANADA
E-mail address: gillesc@crt.umontreal.ca

GERAD AND DÉPARTEMENT DE MÉTHODES QUANTITATIVES EN GESTION, ÉCOLE DES HAUTES ÉTUDES COMMERCIALES, 3000, CHEMIN DE LA CÔTE-SAINTE-CATHERINE, MONTRÉAL, QUÉBEC, CANADA H3T 2A7
E-mail address: pierreh@crt.umontreal.ca

LEXIS-NEXIS, DIVISION OF REED ELSEVIER INC., MIANMISBURG, OH 45342, USA
E-mail address: Maolin.Zheng@lexis-nexis.com

DIMACS Series in Discrete Mathematics
and Theoretical Computer Science
Volume 51, 2000

Characterizing Properties of some Graph Invariants Related to Electron Charges in the Hückel Molecular Orbital Theory

Dragoš Cvetković

ABSTRACT. Let G be a graph on n vertices with the adjacency matrix A. We consider cosines of angles between vectors of a standard (orthonormal) basis of $\mathbf{R}^n$ (n-dimensional real vector space) and eigenspaces of A, and call them graph angles. It turns out that electron charges from the Hückel molecular orbital theory are sums of squares of angles belonging to corresponding eigenvalues. Much of the work in both mathematical and chemical literature has been devoted to the study of isospectral graphs, i.e. isospectral molecules. We present some theoretical and computational results giving an insight as to what extent graphs are characterized when both the eigenvalues and angles are known.

Introduction

Eigenvalues, the spectrum, eigenvectors and eigenspaces of the adjacency matrix A of a graph G are called eigenvalues, the spectrum, eigenvectors and eigenspaces of G, respectively.

It is well known that graph spectra play an important role in Chemistry in the Hückel molecular orbital theory of unsaturated hydrocarbons (see, for example, [**3**], Chapter 8). In this theory carbon-atom skeletons of compounds are represented by graphs while eigenvalues represent energy levels of electrons.

Eigenvalues do not in general suffice to determine a graph up to isomorphism. Indeed, many examples of non-isomorphic graphs with the same spectrum (isospectral or cospectral graphs) have been described in the literature, both mathematical and chemical.

Since eigenvectors are not graph invariants it is reasonable to extend eigenvalue based techniques by some invariants of the eigenspaces called *graph angles.*

Let G be a graph on n vertices with distinct eigenvalues $\mu_1, \mu_2, \ldots, \mu_m$ ($\mu_1 > \mu_2 > \cdots > \mu_m$) and let $S_1, S_2, \ldots, S_m$ be the corresponding eigenspaces. Let $\{e_1, e_2, \ldots, e_n\}$ be the standard (orthonormal) basis of R^n. The numbers $\alpha_{pq} = \cos\beta_{pq} (p = 1, 2, \ldots, m; q = 1, 2, \ldots, n)$, where β_{pq} is the angle between S_p and e_q, are called *graph angles.* The sequence α_{pq} ($q = 1, 2, \ldots, n$) is called the *eigenvalue angle sequence* corresponding to the eigenvalue μ_p ($p = 1, 2, \ldots, m$).

1991 *Mathematics Subject Classification.* 05C50.

The author is grateful to I. Gutman for some useful comments and to M. Lepović for computing the data related to the set of 21 cospectral graphs.

Angles between the vector $(1, 1, \dots, 1) \in \mathbf{R}^n$ and eigenspaces $S_1, S_2, \dots, S_m$ are called *main angles* of the graph.

An overview of results on graph angles is given in [**6**] including graph properties determined by eigenvalues and angles.

In Section 1 we show that electron charges from the Hückel molecular orbital theory are related to graph angles. For a class of bipartite graphs we introduce the so called *partial electron charges*, knowledge of which is equivalent to the knowledge of angles. Section 2 contains some computational results concerning cospectral graphs with the same angles while in Section 3 we show that some graphs of interest in the Hückel theory are characterized by eigenvalues and angles.

1. Partial Electron Charges

Let G be a graph on n vertices with distinct eigenvalues $\mu_1 > \mu_2 > \cdots > \mu_m$. Let $x_i = (x_{i1}, x_{i2}, \dots, x_{in})$ $(i = 1, 2, \dots, n)$ be orthonormal eigenvectors of G. Define $M_p = \{j \mid Ax_j = \mu_p x_j\}$.

We have $\alpha_{pq}^2 = \sum_{j \in M_p} x_{jq}^2$ for squares of angles of G. This formula holds for any choice of orthonormal eigenvectors of G (cf. [**6**], p. 76).

Electron charges are $c_q^2 = \sum_{j=1}^{n} g_j x_{jq}^2$ $(q = 1, 2, \dots, n)$, where g_j is the number of electrons in orbit j [**1**], [**3**]. We can write

$$c_q^2 = \sum_{p=1}^{m} \sum_{j \in M_p} g_j x_{jq}^2 = \sum_{p=1}^{m} c_{pq}^2$$

where $c_{pq}^2 = \sum_{j \in M_p} g_j x_{jq}^2$. The quantities c_{pq}^2 could be called *partial* (*electron*) *charges*.

Consider bipartite graphs G, i.e. alternant hydrocarbons, in which zero is not in the spectrum, n and m being even. It is known that graphs without zero in the spectrum correspond to stable molecules. The well known pairing theorem says: *if λ is an eigenvalue of G then $-\lambda$ is also an eigenvalue of G, with the same multiplicity and with the same eigenvalue angle sequences.*

It is known that $g_j = 2$ for $j \leq \frac{1}{2}n$ and $g_j = 0$ for $j > \frac{1}{2}n$ (cf. [**6**], Section 9.3). Hence, we have

$$c_q^2 = \sum_{j=1}^{n/2} g_j x_{jq}^2 = 2\sum_{j=1}^{n/2} x_{jq}^2 = 2\sum_{p=1}^{m/2} \sum_{j \in M_p} x_{jq}^2 =$$

$$= 2\sum_{p=1}^{m/2} \alpha_{pq}^2 = \sum_{p=1}^{m/2} c_{pq}^2 = \sum_{p=1}^{m} \alpha_{pq}^2.$$

Further we have $c_q^2 = \sum_{p=1}^{m} \alpha_{pq}^2 = \sum_{j=1}^{n} x_{jq}^2 = 1$, a result of Coulson-Rushbrooke [**1**] from 1940. Although this result is of great importance in Chemistry, the electron charges c_q^2 in bipartite graphs are not mathematicaly interesting since they do not contain any information about the graph structure.

The connection between partial charges and angles is expressed by

$$c_{pq}^2 = \sum_{j \in M_p} 2x_{jq}^2 = 2\alpha_{pq}^2; \; p = 1, 2, \dots, m/2; \; q = 1, 2, \dots, n.$$

The knowledge of partial charges in bipartite graphs is very useful since a lot of information on the graph structure can be obtained. In fact, in bipartite graphs having no zeros in the spectrum the knowledge of partial charges is equivalent to the knowledge of graph angles. Indeed, we can derive the following formulas.

If the powers of the adjacency matrix A are represented in the following form

$$A^s = \|a_{ij}^{(s)}\|_1^n,$$

then (cf. [**6**], p. 81)

$$a_{qq}^{(s)} = \sum_{p=1}^{m} \alpha_{pq}^2 \mu_p^s.$$

In bipartite graphs this is reduced to

$$a_{qq}^{(s)} = 2\sum_{p=1}^{m/2} \alpha_{pq}^2 \mu_p^s = \sum_{p=1}^{m/2} c_{pq}^2 \mu_p^s.$$

In particular, the degree of vertex q is given by

$$d_q = \sum_{p=1}^{m/2} c_{pq}^2 \mu_p^2.$$

In chemical terms, partial π-electron charges at an atom and electron energy levels determine the number of neighbouring carbon atoms.

We can also determine the number of quadrangles, the number of closed walks at a vertex, characteristic polynomials of vertex deleted subgraphs and other graph invariants.

However, beyond the described class of bipartite graphs the importance of partial charges seems to be smaller from the mathematical point of view.

2. Some Computational Results

It is known that graphs are not in general characterized by eigenvalues and angles (see, for example, [**6**], pp. 77-79). Cospectral strongly regular graphs (more generally, distance regular graphs and walk regular graphs) have the same angles. A general tool for producing cospectral regular graphs with the same angles is Seidel switching (see, for example, [**3**], p. 184). Seidel switching explains the existance of cospectral graphs with the same angles on 10 vertices which were quoted in [**6**] as smallest known example of this kind.

It was established subsequently in [**5**] by an exhaustive computer search that there are 58 pairs of cospectral graphs with the same angles and with 10 vertices. There are no such pairs of graphs on less than 10 vertices. There is only one pair of cospectral trees with the same angles and with no more than 20 vertices; these trees have 19 vertices. In the smallest previously known example of cospectral trees with the same angles [**2**], the number of vertices was 35.

An exhaustive search for cospectral graphs on 10 vertices [**7**] shows that there exists a set of 21 cospectral graphs on 10 vertices with 20 edges. The complements of these graphs are also cospectral and have 25 edges. Computations also show that the graphs in both cases are distinguished by their angles, i.e. no two cospectral graphs from the mentioned sets have the same angles.

We reproduce here some data on the 21 graphs with 20 edges.

First we give the rows of the lower triangle of an adjacency matrix for each graph:

01	1	10	011	0111	01011	001010	0001100	00001011	000001110
02	1	10	101	1001	01011	001100	0011000	00011110	000010110
03	1	11	111	1010	01010	010001	0011001	00100100	000110010
04	1	10	011	0101	01010	001111	0010101	00011000	000011001
05	1	11	111	1010	01010	010000	0011001	00100100	000110110
06	1	11	111	1100	10010	100010	0100010	00110010	000110100
07	1	10	011	0111	01011	001100	0001001	00001111	000010010
08	1	11	110	1011	01010	010100	0100111	00100010	000101000
09	1	10	011	0111	01011	001010	0001011	00001011	000001100
10	1	11	110	1011	01100	010100	0100111	00100110	000100000
11	1	11	110	1000	01101	011001	0100101	00110100	001000001
12	1	11	110	1000	01110	011000	0101000	00101110	001001010
13	1	11	110	1000	01110	011000	0101000	00100101	000111010
14	1	11	110	1000	01101	011001	0100101	00100000	000111010
15	1	11	110	1010	10010	100101	0100101	00011101	000001000
16	1	11	111	1010	10001	010101	0100101	01000101	001000000
17	1	11	100	0111	01001	010001	0011111	00100010	000110000
18	1	01	011	0111	01010	001001	0010001	00011011	000011100
19	1	11	110	1011	10100	010101	0100101	00100000	000101100
20	1	01	011	0110	01011	001110	0010010	00001010	000001111
21	1	11	100	0110	01010	010011	0011111	00110100	000010000

Eigenvalues:

4.3803 1.6861 1.1620 0.5423 0.0000 0.0000 -1.2950 -1.5260 -2.2864 -2.6631

Coefficients of the characteristic polynomial:

1 0 -20 -18 84 76 -119 -72 56 0 0

Main angles:

0.9563 0.0248 0.0659 0.1505 0.2070 0.0436 0.1086 0.0185 0.0323

Finally, for each graph we give the positive normalized eigenvector belonging to the largest eigenvalue (i.e. the eigenvalue angle sequence corresponding to the largest eigenvalue):

01	0.4971	0.3973	0.3228	0.3098	0.3091	0.3016	0.2872	0.2496	0.2098	0.1396
02	0.4971	0.3789	0.3413	0.3098	0.3091	0.3057	0.3016	0.2312	0.2098	0.1396
03	0.4487	0.4273	0.3925	0.3413	0.3016	0.2796	0.2264	0.2230	0.2223	0.1614
04	0.4487	0.4273	0.3925	0.3126	0.3016	0.2796	0.2517	0.2510	0.2264	0.1327
05	0.4487	0.4273	0.3795	0.3413	0.3146	0.2796	0.2393	0.2230	0.2223	0.1485
06	0.4487	0.4273	0.3795	0.3413	0.3057	0.2796	0.2393	0.2318	0.2312	0.1396
07	0.4487	0.4273	0.3413	0.3098	0.3091	0.3057	0.3016	0.2796	0.1614	0.1396
08	0.4487	0.4273	0.3344	0.3126	0.3098	0.3091	0.3016	0.2796	0.1683	0.1327
09	0.4487	0.3973	0.3713	0.3357	0.3098	0.3091	0.3016	0.2496	0.1614	0.1396
10	0.4487	0.3713	0.3357	0.3344	0.3126	0.3098	0.3091	0.3016	0.2243	0.0766
11	0.4407	0.4377	0.3901	0.3413	0.3016	0.2796	0.2333	0.2264	0.2096	0.1638
12	0.4407	0.4377	0.3901	0.3102	0.3016	0.2796	0.2644	0.2407	0.2264	0.1327
13	0.4407	0.3929	0.3901	0.3549	0.3091	0.3016	0.2796	0.2407	0.1816	0.1327
14	0.4407	0.3901	0.3549	0.3413	0.3091	0.3016	0.2923	0.2796	0.2333	0.0810
15	0.4386	0.3973	0.3549	0.3362	0.3091	0.3016	0.2872	0.2817	0.2363	0.0810
16	0.4386	0.3922	0.3549	0.3413	0.3091	0.3016	0.2923	0.2817	0.2312	0.0810
17	0.4377	0.3901	0.3795	0.3628	0.3413	0.2875	0.2796	0.2333	0.1638	0.1485
18	0.3973	0.3901	0.3846	0.3549	0.3357	0.3091	0.3016	0.2363	0.2333	0.0810
19	0.3929	0.3901	0.3846	0.3549	0.3357	0.3091	0.3016	0.2407	0.2377	0.0766
20	0.3901	0.3846	0.3702	0.3628	0.3549	0.3362	0.2745	0.2363	0.2333	0.0810
21	0.3901	0.3795	0.3702	0.3628	0.3549	0.3413	0.2796	0.2333	0.2312	0.0810

In each graph the vertices are ordered in such a way that the given angle sequences are non-increasing. Graphs are ordered lexicographically by these angle sequences.

These huge sets of cospectral graphs should perhaps be exploited in experiments with ordering graphs by their angles. Experience shows that graphs are very well ordered by their eigenvalues or spectral moments. The question of how to order cospectral graphs is naturally answered by referring to angles. But there is no experience with such a recommendation. One should perhaps point out that vertex degrees are determined by eigenvalues and angles so that the ordering of graphs by angles is somehow related to ordering by vertex degrees.

3. Some Characterization Theorems

Graphs which appear in the Hückel molecular orbital theory have vertex degrees at most 3. We shall exclude regular graphs of degree 3 (cubic graphs) which correspond to fullerenes. In this section we shall show that many of the above mentioned, chemically interesting graphs, are characterized by eigenvalues and angles.

One of the basic tools in proving these characterization results is an algorithm for constructing trees with given eigenvalues and angles which is described in [**2**] (see also [**6**], pp. 111–114). The essence of this algorithm (let us call it Algorithm EA) is that we can reconstruct an edge of a tree (a bridge in general) if we know the structure of the tree (of the graph) on one side of the edge.

We start with a result related to trees obtained from three paths, each rooted at one of its endpoints, by identifying their roots.

PROPOSITION 3.1. *1 A forest in which each component has exactly one vertex of degree* 3, *others being of smaller degrees, can be reconstructed uniquely from eigenvalues and angles.*

PROOF. Applying Algorithm EA gradually, starting from a vertex of degree 1 and going along the path connecting this vertex to the vertex of degree 3, we reconstruct uniquely this path, in particular its length and the ending vertex (vertex of degree 3). We treat in this way all vertices of degree 1. Of course, vertices included in paths already constructed will not be considered as candidates for inclusion in the path currently under construction. In this way the whole forest can be constructed.

This completes the proof. □

Next we consider the set of graphs $\mathcal{S}$ in which the largest eigenvalue does not exceed 2. The connected graphs from $\mathcal{S}$ are represented in Fig. 1.

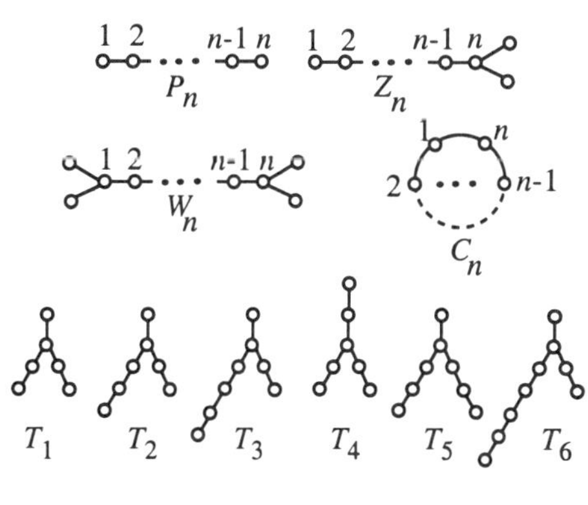

FIGURE 1

Here W_n and C_n have the largest eigenvalue equal to 2. The graph $W_1 = K_{1,4}$ is the only graph having a vertex degree equal to 4, while W_n $(n > 1)$ has two vertices of degree 3. The graphs Z_n and $T_1 - T_6$ have exactly one vertex of degree 3.

Cospectral graphs in the set $\mathcal{S}$ are very frequent. An algorithm to find all graphs cospectral to a given graph from $\mathcal{S}$ is described in [**4**].

THEOREM 3.2. *1 The graphs whose largest eigenvalue does not exceed* 2 *are characterized by eigenvalues and angles.*

PROOF. Suppose the index of the graph is equal to 2. Otherwise we proceed as in the next paragraph. Non-zero angles in the eigenvalue 2 angle sequence indicate the vertices belonging to components whose index is equal to 2. By Lemma 4.4.1 from [**6**] we can calculate the characteristic polynomial of the subgraph consisting of these components. Odd circuits can be easily extracted since the length and the number of shortest odd circuits in a graph can be determined form the characteristic polynomial of the graph. Possible components equal to $K_{1,4}$ can be recognized by vertices of degree 4 and then Algorithm EA shows which vertices of degree 1 are adjacent to vertices of degree 4. In a similar way we can treat components with two vertices of degree 3. We first determine which vertices of degree 1 are neighbours of the considered vertex of degree 3. Simple calculations show that the angle of a vertex of degree 3 of W_k belonging to eigenvalue 2 is equal to $\frac{1}{\sqrt{k+1}}$ and thus contains the information about the length of the path connecting two such vertices of degree 3. The connecting path is reconstructed by Algorithm EA. The remaining vertices are of the degree 2 and they belong to some even circuits. The angle of eigenvalue 2 for a vertex in a circuit C_s is equal to $\frac{1}{\sqrt{s}}$.

If components with the index 2 are extracted or do not exist, the components are trees from Proposition 1, paths and isolated vertices. Isolated vertices are directly recognized while Algorithm EA easily finds paths. For the rest of the graph we apply Proposition 1.

This completes the proof. □

It is possible to prove that graphs from some other classes of graphs, which are of interest in the Hückel theory, are characterized up to isomorphism by eigenvalues and angles.

References

[1] C.A. Coulson, and G.S. Rushbrooke, Note on the Method of Molecular Orbital, *Proc. Cambridge Phil. Soc.* **36** (1940) 193–200.

[2] D. Cvetković, Constructing Trees with Given Eigenvalues and Angles, *Linear Algebra and Appl.* **105** (1988) 1–8.

[3] D. Cvetković, M. Doob, and H. Sachs, *Spectra of graphs - Theory and application*, Third revised and enlarged edition, Johann Ambrosius Barth Verlag, Heidelberg-Leipzig, 1995.

[4] D. Cvetković, and I. Gutman, On the Spectral Structure of Graphs having the Maximal Eigenvalue not Greater than Two, *Publ. Inst. Math. (Beograd)* **18**(32) (1975) 39–45.

[5] D. Cvetković, and M. Lepović, Cospectral Graphs with the Same Angles and with a Minimal Number of Vertices, *Univ. Beograd, Publ. Elektrotehn. Fak., Ser. Mat.* **8** (1997) 88–102.

[6] D. Cvetković, P. Rowlinson, and S. Simić, *Eigenspaces of graphs*, Cambridge University Press, Cambridge, 1997.

[7] M. Lepović, Some Statistical Data on Graphs with 10 Vertices, *Univ. Beograd, Publ. Elektrotehn. Fak., Ser. Mat.* **9** (1998), to appear.

FACULTY OF ELECTRICAL ENGINEERING, UNIVERSITY OF BELGRADE, P.O.BOX 35–54, 11120 BELGRADE, YUGOSLAVIA

E-mail address: ECVETKOD@etf.bg.ac.yu

DIMACS Series in Discrete Mathematics
and Theoretical Computer Science
Volume **51**, 2000

Fast Embeddings for Planar Molecular Graphs

Olaf Delgado Friedrichs

ABSTRACT. We describe fast and rather simple algorithms which produce embeddings for certain classes of planar molecular graphs. This is achieved by a multi-phase iterative process based on force-models and taking into account special properties of the given graphs.

1. Introduction

The software package CAGE [**1**] contains a collection of fast generators for certain types of graphs. The emphasis is on graphs which are of interest as abstract models for molecular structures. A graphical user interface is provided which simplifies the setting of parameters and lets the user browse through the results. Since the best way to understand a graph is to look at a picture, two types of embedding algorithms are included: one for Schlegel diagrams, i.e., plane embeddings with a convex outline (c.f. Section 3), and one for "realistic" embeddings which mimic plausible molecular shapes. In this paper, we will describe and discuss these algorithms.

By the term *embedding*, we mean a function which assigns a coordinate vector to each node or vertex of a given graph. To *embed* a graph means to construct such a function. More specifically, let $G = (V, E)$ be a simple undirected graph. Each edge $e \in E$ is considered as an unordered pair $e = \{v, w\}$ of vertices. Let $f\colon V \to \mathbf{R}^d$ be a function which assigns a d-dimensional position vector to each vertex of G. For an edge $e = \{v, w\} \in E$, we define $f(e)$ to be the open line segment between $f(v)$ and $f(w)$, i.e.

$$f(e) := \{\, \lambda f(v) + (1-\lambda) f(w) \mid 0 < \lambda < 1 \,\}.$$

Now, f is called a (d-dimensional) *embedding* of G, if for any pair of different edges $e, e' \in E$, the intersection of $f(e)$ and $f(e')$ is empty.

Our goal is not to obtain an optimal Schlegel diagram layout according to some formal criterion. Neither do we try to obtain an accurate minimum energy shape for the spatial embedding. We are simply interested in a fast way to produce a plausible visual impression, one which is good enough for the naked eye. On the other hand, we need a very fast program to avoid painful delays for the user browsing through

1991 *Mathematics Subject Classification.* Primary: 05C10 Secondary: 05C85, 73K40.

the structures. To be precise, the embedding of a graph with about thousand vertices should not take more than a couple of seconds on a standard modern PC.

As of yet, our approach is rather experimental. The emphasis is on methods that will work well in our applications. We will describe our experiences and propose some general heuristics which turned out to be successful in the special cases considered. Research in this field is ongoing (see Section 6) and will eventually need and produce results in the form of lemmata and theorems. This paper does not contain such formal results.

We use the following strategies to achieve good embeddings quickly:

- Special properties of the given graphs are taken into account.
- The embedding is refined iteratively.
- Modified spring embedders, i.e. discretized force models, are used.
- Repulsion between vertices is avoided wherever possible.

For spatial embeddings, two additional strategies proved successful:

- Refinement starts from a carefully constructed initial embedding.
- Several iteration phases with different force models are used.

Currently, all the graphs generated in CaGe are planar with vertex degrees not larger than 3. Moreover, for each graph, the facial cycles of a particular planar embedding are given. This abstract embedding information is used in different phases of the algorithms to speed up calculations or to obtain better layouts. In the spatial embedding algorithm, we assume that the optimal embedding for a given graph will be vaguely ball-shaped. This is true for structures like, e.g., fullerenes, but not necessarily so for polycyclic hydrocarbons. As a consequence, our algorithm is very successful for fullerenes, where it even produces good starting points for quantum-mechanical energy calculations, whereas hydrocarbon embeddings are not always as satisfactory.

We will discuss general strategies for the embedding problem in Section 2 with a major focus on iterative methods using simulated forces, also known as *spring embedders* (c.f. [**5**, **6**]). In Section 3, we will describe and discuss several algorithms for two-dimensional Schlegel diagrams including Tutte's original method [**7**]. Spatial embeddings will be dealt with in Section 4. Section 5 contains some statistics on the very basic behaviour of our algorithms.

2. Spring Embedders

There is a vast literature on graph drawing algorithms. Various strategies have been used for different applications and different classes of graphs. Some of these are *direct* or *constructive* methods, in that they essentially construct a drawing in a single step, often using involved techniques. Others take a more dynamic or evolutionary approach by iteratively modifying an initial drawing according to certain rules. A general scheme often applied is the assignment of energy (or penalty) values to certain—undesirable—features found in a drawing. The objective is then to minimize energies and thereby to obtain an optimal layout.

Depending on the character of the total energy function, it may be sufficient to use a steepest descent method, which needs relatively few evaluations of the energy function, but easily gets stuck in local minima. In other cases, to avoid local minima, more cost-intensive strategies like, e.g., simulated annealing or genetic algorithms, may be needed. Such methods are usually much too slow if interactive speed is desired.

For steepest descent methods, it is convenient to use a model based on forces instead of energies. A model of this kind for the drawing of general graphs has been introduced by Eades [**5**] and motivated by the idea of replacing vertices by steel rings and edges by springs. There are attractive and repulsive forces between pairs of vertices, depending on their combinatorial distances. The dynamics of this "virtual physical system" are simulated discretely by repeatedly displacing each vertex by a vector proportional to the respective sum of forces acting on that vertex. Graph drawing methods based on Eades' idea are thus commonly named *spring embedders*, although not even his original implementation reflects Hooke's law, as Fruchterman and Reingold [**6**] point out.

Fruchterman and Reingold present a variant of Eades' model which is very simple and especially designed for interactive display. They introduce an attractive force between each pair of adjacent vertices which is proportional to the square of their respective distance. Additionally, there is a repulsive force between each pair of vertices which is proportional to the inverse of their distance. In effect, if $\mathrm{adj}(v) := \{ w \in V \mid \{v, w\} \in E \}$ denotes the set of vertices adjacent to v, then the total force, or, more accurately, the displacement vector applied to each vertex can be written as

(2.1)

$$\delta_{\mathrm{FR}}(f, v) = \sum_{w \in \mathrm{adj}(v)} ||f(w) - f(v)||^2 (f(w) - f(v)) + \sum_{w \in V - \{v\}} \frac{f(v) - f(w)}{||f(v) - f(w)||}.$$

In the following, we will characterize a spring embedder mainly by its displacement function

$$\delta(f, v) \colon (\mathbf{R}^d)^V \times V \to \mathbf{R}^d.$$

To speed up their algorithm, Fruchterman and Reingold propose a grid method which avoids the calculation of repulsive forces for distant vertex pairs. They also address the issues of "temperature" and "cooling" which are direct analogues to concepts from simulated annealing. In spring embedders, the maximum displacement at each step is limited by a certain "temperature", which is decreased gradually throughout the whole iteration process. A well-chosen temperature function must avoid heavy oscillation while at the same time allow the configuration to quickly approach a stable state.

3. Schlegel Diagrams

In this section, we will describe and compare three methods to construct Schlegel diagrams, namely a classical method by Tutte, a slight variation of Tutte's method and the strategy currently implemented in CaGe. All these methods have been simulated by spring embedders with identical parameter settings and with the same initial configuration. The initial configuration that we use is very simple: we choose as the outer face one which maximizes the maximal (combinatorial) distance to any other vertex in the graph. This face is then embedded as a regular polygon. Each vertex not on the outer face is positioned at the center of gravity of this outer polygon.

Originally, the term *Schlegel diagram* refers to a $(d-1)$-dimensional image of a d-dimensional polytope which is obtained by projecting its boundary into

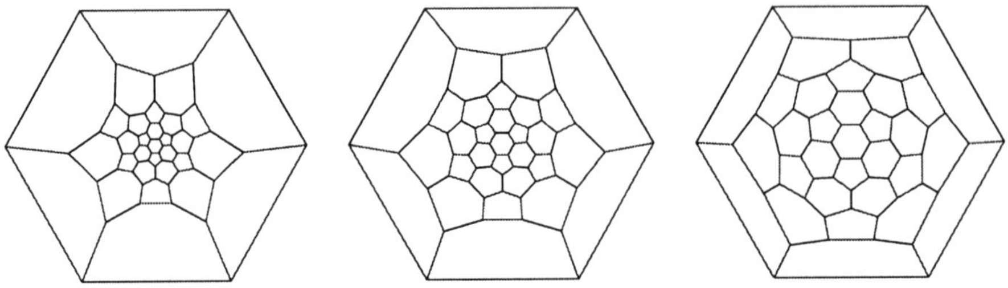

(a) Tutte (b) Tutte modified (c) CaGe

FIGURE 1. Embeddings for the buckminsterfullerene C_{60}.

one if its facets. By Steinitz's Theorem, any planar embedding of a finite, planar, 3-connected[1] graph is topologically equivalent to a Schlegel diagram of some 3-dimensional polyhedron. Even more general, the term Schlegel diagram is sometimes used for any two-dimensional drawing of a graph with some cycle chosen as the outer frame. We adopt this broader interpretation, although in the following, we will restrict our attention to planar, but not necessarily 3-connected graphs.

Tutte in his classical work [**7**] describes a very elegant method of assigning coordinates to the vertices of a planar graph, given a fixed embedding for the chosen outer face as a (strictly) convex polygon. Namely, each vertex not on the outer face is to be positioned at the center of gravity of its neighbors. This requirement can be expressed in terms of a uniquely solvable system of linear equations. Tutte also shows that for 3-connected graphs this construction always leads to embeddings with strictly convex faces.

Tutte's solution can be efficiently simulated by a spring embedder. For v a vertex not on the outer face, we can write

$$f(v) = \frac{1}{\#\operatorname{adj}(v)} \sum_{w \in \operatorname{adj}(v)} f(w)$$

or

$$\delta_{\mathrm{Tu}}(f, v) := \frac{1}{\#\operatorname{adj}(v)} \sum_{w \in \operatorname{adj}(v)} f(w) - f(v) = 0. \tag{3.1}$$

Of course, we have to set $\delta_{\mathrm{Tu}}(f, v) := 0$ for all outer vertices.

A serious drawback, though, becomes obvious in Figures 1(a) and 2(a), where we show Schlegel diagrams obtained by Tutte's method for the buckminsterfullerene C_{60} and for a 916 vertex fullerene graph[2] which we will denote by C_{916}. In the C_{60} example already, faces tend to become very small with distances from the outer polygon growing. In the drawing of the much larger graph C_{916}, a major part of the combinatorial structure is unrecognizable due to this effect.

A slightly better result is obtained by reducing the variation in edge lengths. To this end, we weigh the summand corresponding to each edge with the length of

[1] We use the term 3-*connected* in the graph-theoretical sense: a graph is called 3-connected if there is no single vertex or pair of vertices whose removal would make the graph disconnected.

[2] This graph was communicated to us by P. Fowler.

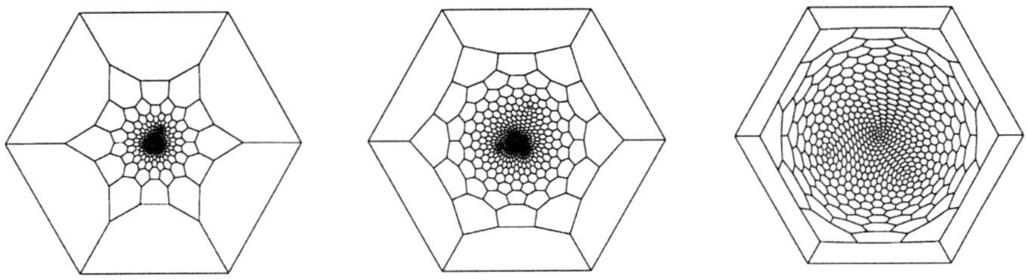

(a) Tutte (b) Tutte modified (c) CaGe

FIGURE 2. Embeddings for the 916 vertex fullerene C_{916}.

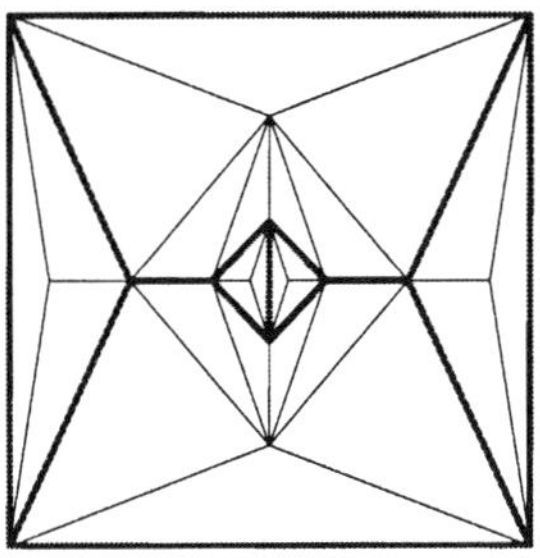

FIGURE 3. A 2-connected graph and its face-centered triangulation.

that edge, giving

$$\delta_{\mathrm{Tw}}(f,v) := \frac{\sum\limits_{w\in \mathrm{adj}(v)} \|f(w)-f(v)\|\cdot(f(w)-f(v))}{\sum\limits_{w\in \mathrm{adj}(v)} \|f(w)-f(v)\|}. \tag{3.2}$$

Results for the graphs C_{60} and C_{916} are shown in Figures 1(b) and 2(b).

Though esthetically pleasing, Figure 2(b) still only reveals a small portion of the graph's combinatorial structure. The reason, apparently, is that the outermost faces are still quite large because lengths of radial edges are coupled with lengths of concentric edges. To avoid this, it is necessary to change focus from edge lengths to face areas.

Another drawback of the original as well as the modified Tutte method is that it only works for 3-connected or "nearly 3-connected" graphs. More precisely, if a pair p, q of vertices separates the given graph G, then p and q must be nonadjacent and must both lie on the outer face. Otherwise, each component of the graph obtained by removing p and q from G will be mapped onto the straight line segment $f(p)f(q)$ except for those which intersect the outer face. A straightforward workaround would be to introduce additional edges (but no additional vertices) to make the graph 3-connected. This, however, might destroy symmetries and thus lead to very unpleasant layouts. An alternative triangulation which retains all symmetries is obtained by placing a new vertex at the center of each inner face and connecting it to every vertex of that face, as shown in Figure 3.

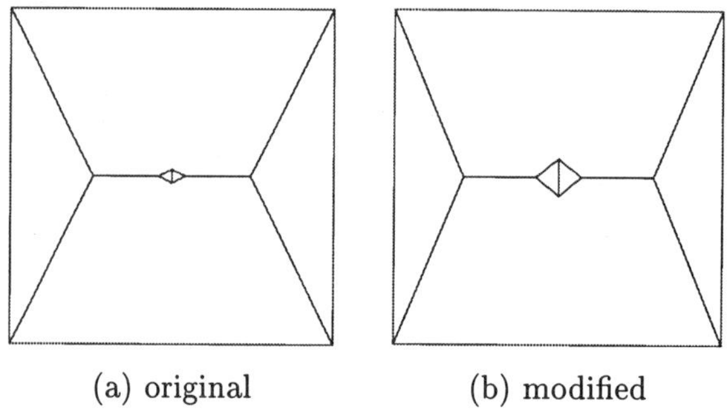

FIGURE 4. Tutte style embeddings for the graph of Figure 3 based on a face-centered triangulation (not shown).

The results of applying the Tutte style embeddings to this subdivision are shown in Figure 4. All faces are clearly visible in both cases, but even for the modified Tutte method the distribution of face sizes is extremely bad.

Yet now, we are in a situation where it is easy to apply a strategy based on face areas. Let $G' = (V', E')$ be the graph obtained from G by the subdivision described above. The inner faces of G' can be associated with triplets of vertices. Let F' be the set of all inner faces. The area of a face $Q := \{u, v, w\}$ w.r.t. the embedding $f : V' \to \mathbf{R}^2$ is

$$A(f, Q) := \frac{1}{2}(f(v) - f(u)) \times (f(w) - f(u))$$

and its center is

$$c(f, Q) := \frac{1}{3}(f(u) + f(v) + f(w)).$$

We define

$$\delta_{\mathrm{cg}}(f, v) := \frac{\sum_{v \in Q \in F'} A(f, Q)^2 \cdot (c(f, Q) - f(v))}{\sum_{v \in Q \in F'} A(f, Q)^2}. \tag{3.3}$$

This method, which we shall call the CAGE-*method*, was used to obtain Figure 3. Although this method gives much better results than the Tutte style methods, it still leaves a lot of room for improvement. On the other hand, it turns out that the CAGE-method gives very convincing results for 3-connected graphs, as shown for the graphs C_{60} and C_{916} in Figures 1(c) and 2(c). More examples can be found in [**3**]. These latter examples also show that, as opposed to Tutte style methods, the CAGE method does not guarantee convex embeddings, not even if the given graphs are 3-connected.

4. Spatial Embeddings

In spatial embeddings, our goal is to obtain layouts which resemble molecular structures. In particular, the range of edge lengths should be narrow and non-adjacent vertices should never be too close. Embeddings of this type can be produced by simple spring embedders of the Fruchterman-Reingold type, now working in three instead of two dimensions, or by energy minimizers based on analogous models. When starting from a random initial position, such programs are typically quite slow. Potentially they can deal with any kind of graph and are faster still than programs typically used in molecular modelling, which use more 'realistic' energy functions. In contrast, the algorithm which we will present here is extremely fast, but more specialized.

Our proposed new algorithm works in several phases. First, an appropriate initial configuration is constructed to avoid unnecessary iteration steps. Then, two iteration phases based on different force models are performed. In the first phase, the global shape of the embedded graph is adjusted, while the second phase has a more local effect, in particular on the distribution of edge lengths. These two phases use very simple models, which turn out to be sufficiently good for fullerene graphs. For certain other classes of graphs, more elaborate models are necessary. Sometimes these can be implemented in a third iteration phase.

Figures 5 to 8 show the results of the different phases on some typical graphs. Please note that due to space limitions, these illustrations appear in different scales, so relative sizes and lengths are not preserved.

A simple initial configuration is constructed by placing vertices on the surface of a sphere. The radius of this sphere is chosen proportional to $\sqrt{\#V}$. The layout is organized in layers depending on distances from a start vertex $v_0 \in V$. Vertices on the same layer are evenly spaced and their ordering is determined by the facial cycles. To avoid torsion, especially in structures like tube-type fullerenes, simple local heuristics are applied to determine the rotation of each new layer w.r.t. the previous one. It seems that this method works reasonably well for our examples. Improved algorithms for the initial embedding are under development.

For some graphs as e.g., tube-type fullerenes or pericondensed polycyclic hydrocarbons, whose shapes will be rather ellipsoidal than spherical, the choice of a start vertex is crucial. At present, if all vertices have degree at least 3, we choose a vertex $v_0 \in V$ for which the maximum distance to any other vertex is maximal. If a "boundary" exists, i.e., if vertices of degree smaller than 3 are present, then we choose v_0 such that it maximizes the distance to the boundary.

It seems at first glance as if Tutte's method would work much better for spatial embeddings, where there is no such thing as an outer face. Yet, if Equations 3.1 or 3.2 are applied to all vertices, then the embedded graph will eventually collapse to one point. A balancing force is necessary to prevent this collapse. One way to do this is to keep the vertices on the surface of a fixed sphere. But strictly spherical embeddings are too restricted for our purpose. An alternative is the classical approach of [**6**] of putting a repulsive force between each pair of vertices, as if they were equally charged particles. Here the problem is that we would need to calculate the Euclidean distances between each pair of vertices, resulting in quadratic time complexity for each iteration step. Although Fruchterman and Reingold present a potential remedy, we take a different approach here. To keep the running time linear in the number of edges (i.e., also in the number of vertices for, e.g., planar or 3-regular graphs), we use in Phase 1 a repulsive force between each vertex and the center of gravity, i.e. we set

$$r(v) = r_{f,G}(v) := f(v) - \frac{1}{n} \sum_{v \in V} f(v),$$

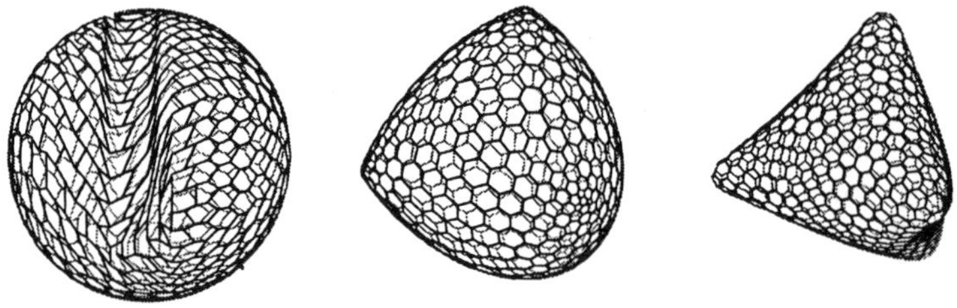

Initial Phase 1 Phase 2

FIGURE 5. Results of different embedding phases for the graph C_{916}.

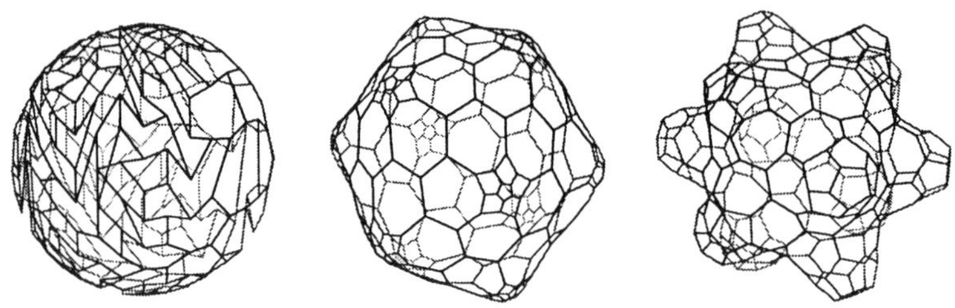

Initial Phase 1 Phase 2

FIGURE 6. Results of different embedding phases for a pentagon-heptagon fulleroid with icosahedral symmetry (c.f. [**4**]).

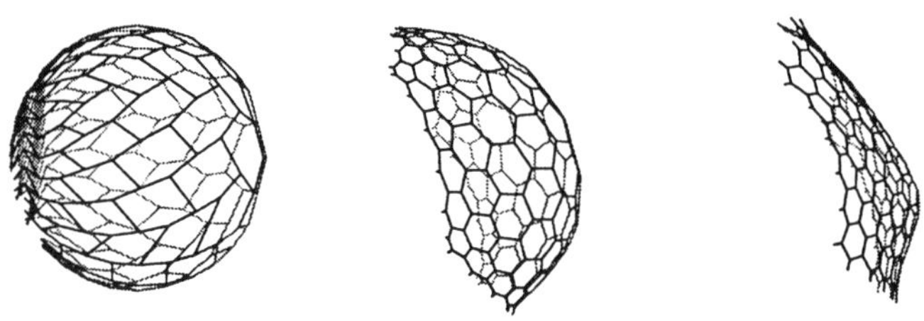

Initial Phase 1 Phase 2

FIGURE 7. Results of different embedding phases for a pericondensed polycyclic $C_{200}H_{32}$.

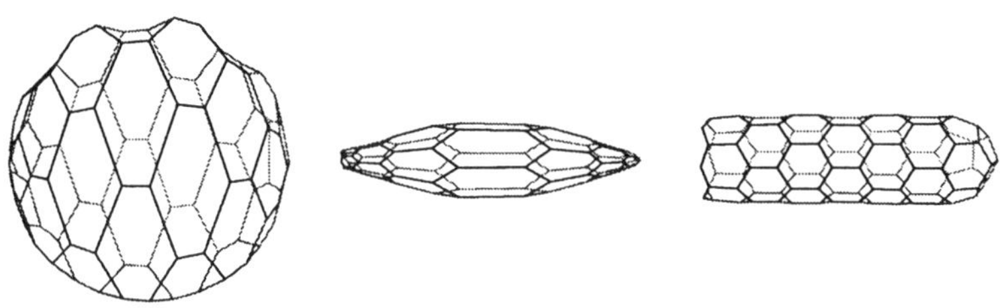

Initial Phase 1 Phase 2

FIGURE 8. Results of different embedding phases for a fullerene halftube.

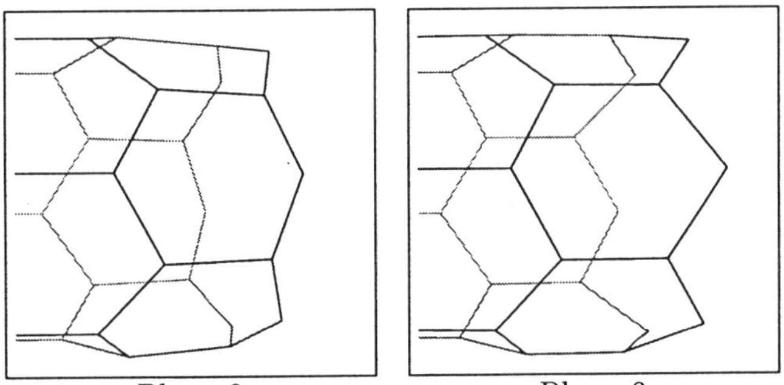

Phase 2 Phase 3

FIGURE 9. Effect of a third iteration phase on a fullerene halftube.

and

$$\delta_1(f,v) := \frac{\sqrt{n}}{2} \cdot \frac{r(v)}{||r(v)||^2} + \sum_{w \in \mathrm{adj}(v)} ||f(w) - f(v)|| \cdot (f(w) - f(v)),$$

where $n := \#V$ is the number of vertices in G.

In this phase, only a small number of iteration steps is performed for each graph, starting from the spherical initial configuration described above. In fact, the limit shape will be a planar sheet for the graph shown in Figure 7 and a straight line segment for the graph shown in Figure 8. In both cases, the best starting configuration for the following phase will be in between the spherical and the limit shape.

In Phase 2, we replace the repulsion from the center of gravity by repulsive forces between pairs of next-nearest neighbors. More precisely, we set

$$\mathrm{adj}^2(v) := \{ w \in V \mid v \neq w \text{ and } \mathrm{adj}(v) \cap \mathrm{adj}(w) \neq \emptyset \},$$

i.e., $\mathrm{adj}^2(v)$ is the set of next-nearest neighbors of a vertex $v \in V$, and

$$\delta_2(f,v) := \sum_{w \in \mathrm{adj}^2(v)} \frac{f(v) - f(w)}{||f(v) - f(w)||^2} + \sum_{w \in \mathrm{adj}(v)} ||f(w) - f(v)|| \cdot (f(w) - f(v)).$$

The effect is most obvious from Figure 6. Long edges are shrunk while short edges are stretched. The repulsion between next-nearest neighbors is still sufficient to keep the structure from collapsing, but allows more local variation in shape. On the other hand, the central repulsion of Phase 1 is more efficient in compensating for artifacts produced by the initial embedding.

The balancing between attractive and repulsive forces is adjusted as to keep embedded graphs slightly growing throughout Phase 1 and slightly shrinking throughout Phase 2. In other words, in Phase 1, the repulsive force is dominant in forming the shape, while in Phase 2, the attractive forces are dominant.

In our programs, we use an optional third phase with forces modelled after Hooke's law on both edge lengths and angles. The main objective here is to reduce the variation in edge lengths and angles while retaining the overall structure. Whether this is desirable depends on the application and sometimes on the given graphs. As an example, certain boundary effects occur for graphs which contain vertices of degree smaller than 3, as can be seen in the left part of Figure 9. The right part of the figure shows the result of applying Phase 3.

TABLE 1. Quality statistics for Schlegel diagrams

		bad (%)	s_e
FULL-100-IPR:	Tutte orig.	4.44	1.461
($n = 450$)	Tutte mod.	3.56	1.196
	CaGe	3.56	0.960
FULL-60:	Tutte orig.	1.77	1.154
($n = 1812$)	Tutte mod.	1.66	0.987
	CaGe	1.49	0.827
PLAN-28:	Tutte orig.	4.72	0.895
($n = 1802$)	Tutte mod.	0.83	0.817
	CaGe	0.17	0.698

5. Experimental Results

We found that the algorithms presented here work very well for fullerenes and still fairly well for general planar graphs.

Tables 1 and 2 show results for some typical sets of graphs as produced by CaGe. Here, *FULL-60* is the complete set of 60-vertex fullerenes, i.e. 3-regular 3-connected planar graphs with face sizes 5 and 6 only. Similarly, *FULL-100-IPR* is the set of all 100-vertex fullerenes which fullfill the so-called *IPR* or *isolated pentagon rule*, i.e. which do not contain any two pentagons sharing an edge. The set *PLAN-28* is a random sample from the set of 3-connected 3-regular planar graphs with 28 vertices (there are exactly 17 490 241 such graphs, see [**2**]).

For the Schlegel diagram algorithm, it is necessary to check whether a given vertex placement is an embedding at all or leads to edge crossings. A little more generally, we classify an alleged Schlegel diagram as bad if either two edges cross or some vertex gets very close to a non-incident edge. A further important criterion for readability is then the distribution of edge lengths.

Each embedding was scaled by the inverse of its average edge length. For each sample and each embedding method, we list in Table 1 the percentage of bad vertex placements produced and the average of the standard deviation s_e of the scaled edge lengths. For each graph and each algorithm, we performed 100 iteration steps. For graphs with more than 200 vertices, a larger number of steps will be necessary.

We see that the distribution of edge lengths is worst for the (simulation of the) original Tutte method and best for the CaGe method.

For spatial embeddings, the least that should be expected is that the connectivity of the original graph can be reconstructed from the embedding. More precisely, the closest neighbors to each vertex should be those with which it shares an edge. Moreover, distances between adjacent vertices should be in a relatively small range, whereas the shortest distance between nonadjacent vertices should be significantly larger.

Again, we scaled each embedding in order to obtain an average edge length of 1.0 and then computed for each embedded graph its minimal edge length $\min_e$, its maximal edge length $\max_e$ and the smallest distance between non-adjacent vertices $\min_s$. The averages of these quantities are displayed in Table 2. A graph is regarded as "bad", if $\min_s \leq \max_e$ or $\min_e < 0.8 \max_e$, as "excellent", if $\min_e \geq 0.95$, $\max_e \leq 1.05$ and $\min_s \geq 1.20$ and as "good" in all other cases. For each sample, we give the percentage of good, bad and excellent embeddings. Results for the first two and for all three phases of the algorithm are listed separately. In the three

TABLE 2. Quality statistics for spatial embeddings

		bad/good/exc. (%)			min_e	max_e	min_s
FULL-100-IPR:	2 phases	--	67.33	32.67	0.948	1.046	1.569
($n = 450$)	3 phases	--	--	100.00	0.995	1.008	1.541
FULL-60:	2 phases	--	83.83	16.17	0.947	1.053	1.561
($n = 1812$)	3 phases	--	--	100.00	0.992	1.008	1.545
PLAN-28:	2 phases	24.58	75.42	--	0.907	1.112	1.353
($n = 1802$)	3 phases	0.17	0.17	99.67	0.982	1.019	1.340

phases of the algorithm, $n/4$, 200 and 20 iteration steps were used, respectively. Again, more steps will be needed for larger graphs.

The table shows very similar behaviour for the two fullerene samples, while for the third sample, results are apparently much worse. We also see that Phase 3, as expected, significantly narrows the range of edge lengths. It should be interesting to perform more detailed experiments on the behaviour of the spatial embedder for different types of graphs.

Here are running times of our algorithms on a 233 MHz Pentium processor:

vertices	Schlegel	spatial
60	0.04 sec	0.15 sec
260	0.19 sec	1.21 sec
916	2.11 sec	4.50 sec

6. Conclusion

We have described fast embedding methods especially designed for 3-connected 3-regular planar graphs. Experiments show that even though the overall performance of these algorithms is very good, there is still room for improvement.

For the spatial embedder, we use a multi-phase iterative approach based on force models. The algorithm adjusts the global shape first and after that the local configuration. To this end, an assumption is made about the global shape of the best embedding. In this case, it is assumed to be vaguely spherical.

Future plans include application of a similar strategy to torus-graphs, where again we will assume that the facial rings of an embedding into the torus are given. We are also investigating embedding algorithms for periodic tilings in two and three dimensions.

References

[1] G. Brinkmann, O. Delgado Friedrichs, A. Dress, and Th. Harmuth, CaGe — a Virtual Environment for Studying some Special Classes of Large Molecules, *Match* **36** (1997) 233–237.
[2] G. Brinkmann, Problems and Scope of Spiral Algorithms and Spiral Codes for Polyhedral Cages, *Chemical Physics Letters* **272** (1997) 193–198.
[3] O. Delgado Friedrichs, and M. Deza, More Icosahedral Fulleroids, submitted to: *Proceedings of the Dimacs Workshop on Discrete Mathematical Chemistry*, 1998.
[4] A.W.M. Dress, and G. Brinkmann, Phantasmagorical Fulleroids, *Match* **33** (1996) 87–100.
[5] P. Eades, A Heuristic for Graph Drawing, *Congressus Numerantium* **42** (1984), 149–160.
[6] T. Fruchterman, and E. Reingold, Graph Drawing by Force-Directed Placement, *Software—Practice and Experience* **21**(11) (1991) 1129–1164.
[7] W.T. Tutte, How to Draw a Graph, *Proc. London Math. Soc.* **13** (1963) 743–767.

OLAF DELGADO FRIEDRICHS, FAKULTÄT FÜR MATHEMATIK, UNIVERSITÄT BIELEFELD, 33501 BIELEFELD, GERMANY.
E-mail address: delgado@Mathematik.Uni-Bielefeld.DE

DIMACS Series in Discrete Mathematics
and Theoretical Computer Science
Volume 51, 2000

More Icosahedral Fulleroids

Olaf Delgado Friedrichs and Michel Deza

ABSTRACT. Generalized fullerenes — or fulleroids — containing faces of degree larger than 6, are beginning to gain some interest in theoretical chemistry. Using the theory of *Delaney symbols*, we construct some relatively small fulleroids with icosahedral symmetries and without hexagons, which are unique with respect to their p-vectors and symmetries. We also obtain several infinite series of icosahedral fulleroids.

1. Introduction

Dress and Brinkmann [**3**] introduce *Fowler's Phantasmagorical Fulleroids* as the two unique planar polyhedral maps (or tilings of the sphere) with p-vector $(p_5, p_7) = (72, 60)$ and icosahedral symmetry. Their proof that no other such structures exist exemplifies the power and beauty of the method of Delaney symbols. Although for larger structures, the necessary case analysis becomes rather tedious and requires the help of a computer (cf. [**12**]), a number of similar classification results come up even more easily by the same general scheme.

The structures found by Dress and Brinkmann as well as the ones we will describe here can be constructed from certain well-known icosahedral fullerenes using simple subdivision operations. By applying such operations to infinite series of fullerenes, one obtains corresponding infinite series of fulleroids.

We will use the terminology of [**3**] plus the following definitions: a *fulleroid* is a tiling of the sphere such that all of its vertices have degree 3 while all of its faces have degree 5 or larger. A Γ-*fulleroid* is a fulleroid on which the group Γ acts as a group of symmetries. In particular, an I-fulleroid is one which has the same proper (i.e., orientation preserving) symmetries as the icosahedron. The *p-vector* (c.f. [**11**]) of a tiling is a vector $(p_i)_{i \in \mathbf{N}}$ such that p_i denotes the number of faces of degree i in that tiling.

Throughout this paper, we will assume that all faces are cells and that the intersection of any pair of faces is either empty, a single vertex or a single edge. This implies that the graph consisting of the vertices and edges of the given tiling is 3-connected, and therefore by Steinitz' famous theorem [**16**] (see also [**11**]), is the graph of a convex polytope. By a theorem of Mani [**14**], every 3-connected spherical map can be realized by a polytope with its full combinatorial symmetry. This allows us in the following to restrict our attention to combinatorial issues.

1991 *Mathematics Subject Classification.* Primary: 51M20. Secondary: 52B10, 52B15.

We will say that a given I-fulleroid F is of *type* (a, b) or an *$I(a, b)$-fulleroid*, if $p_i(F)$ is nonzero if and only if $i \in \{a, b\}$. See [**8**] for a more general study of such bifaced polyhedra. Let $a = 5$. The case $b < 5$ is not possible. For $a = b = 5$, the only possible I-fulleroid is the dodecahedron. The case $(a, b) = (5, 6)$ is the classical fullerene case. There is a well-known classification result for $I(5, 6)$-fulleroids, i.e. fullerenes of icosahedral symmetry ([**10**] and [**1**]). The smallest possible $I(5, 7)$-fulleroids are described in [**3**]. Here, we will look at the smallest $I(5, n)$-fulleroids for some numbers $n > 7$.

We will proceed as follows: in Section 2, we will describe and tabulate the new structures. All I-fulleroids known so far can be conveniently described as the result of regularly "decorating" certain well-known fullerenes. In Section 3, we will describe these decorations and introduce some infinite series of fulleroids which are obtained this way. Most remarkably, we will show that there are infinitely many $I(5, 7)$- and $I(5, 8)$-fulleroids. Then, in Section 4, we will derive a necessary condition on the p-vectors which implies that five of the new I-fulleroids are minimal for their respective values of n. Finally, in Section 5, we will apply the method of Delaney symbols due to A. Dress to establish the uniqueness of five of the new fulleroids and the minimality of the one not handled by Section 4.

2. Results

For later reference, we will start by describing the four smallest icosahedral fullerenes apart from the dodecahedron. The first one is $C_{60}(I_h)$, the bucky-ball or buckminsterfullerene. The second one is $C_{80}(I_h)$, the 'chamfered' dodecahedron, obtained from the dodecahedron by capping each face with a pentagonal prism, then removing all original edges. The other two are $C_{140}(I)$ and, finally, $C_{180}(I_h)$, the 'leap-frog' of the bucky-ball, obtained from the dual of $C_{60}(I_h)$ by truncating all the vertices. For reference, we show the first three of these well-known fullerenes in Figure 1. We use a simplified version of the conventional naming for fullerenes here, where $C_n(\Gamma)$ stands for a 'carbon molecule' with n atoms (i.e. a three-regular graph with n vertices) and full combinatorial symmetry group Γ. Although this notation is not generally unique, it will suffice for our purpose.

The four polyhedra mentioned above have parameters $(1, 1)$, $(2, 0)$, $(2, 1)$ and $(3, 0)$, respectively, as according to Goldberg [**10**], while the dodecahedron has parameters $(1, 0)$. These parameters refer to the distance vectors w.r.t. the hexagonal lattice A_2 between pairs of closest pentagons. We will sometimes use the notation $G_{p,q}$ for an icosahedral fullerene with Goldberg parameters (p, q).

Dress and Brinkmann show that the two $I(5, 7)$-fulleroids they describe are the only such structures with the (smallest possible) p-vector $(p_5, p_7) = (72, 60)$. We will generalize their result to $I(5, n)$-fulleroids, where $n = 8, 9, 10, 12, 14, 15$.

Note that because of Equation 4.1 below, the numbers p_a and p_b in an $I(a, b)$-fulleroid are linear functions of each other. This further implies that v, the number of vertices, is a linear function of any one of them. Therefore, we can use any one of these three quantities to measure the size of an $I(a, b)$-fulleroid. However, for $I(5, n)$-fulleroids, it will turn out to be most convenient to use p_n.

Table 1 shows the smallest possible p-vectors for $7 \leq n \leq 20$ according to Lemma 4.3 below. The first four columns of the table show the quantities n, p_5, p_n and v as defined above. Note that since all graphs described here are 3-regular (each vertex has valency three), the number of edges is simply $e = 3v/2$. The invariants

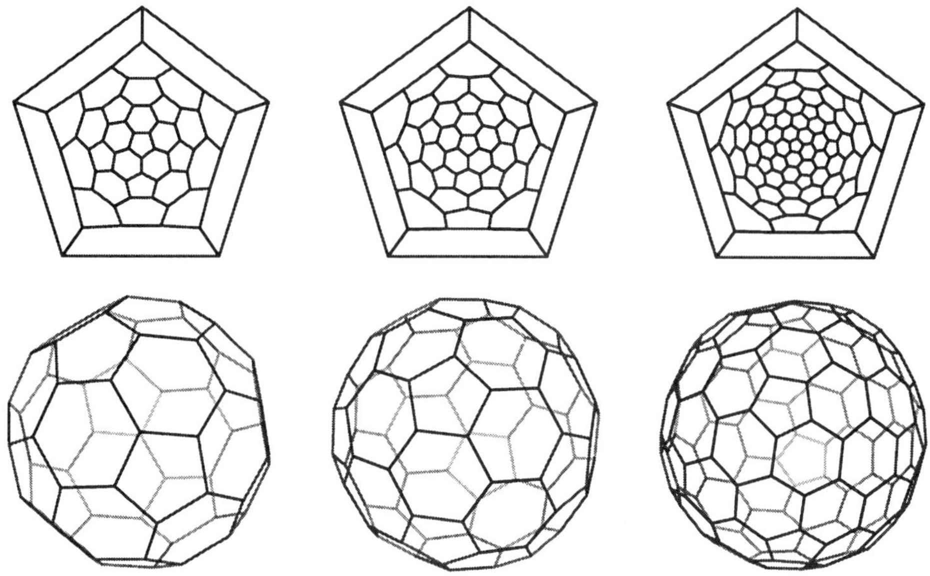

(a) $(1,1)$: $C_{60}(I_h)$ (b) $(2,0)$: $C_{80}(I_h)$ (c) $(2,1)$: $C_{140}(I)$

FIGURE 1. The icosahedral fullerenes with parameters $(1,1)$, $(2,0)$ and $(2,1)$

m_5, m_n, k_2, k_3 and k_5 are described in Section 4. The last three columns, if applicable, give the number (#) of structures with the given p-vector, the numbers of the figures in which they appear and the names by which we will refer to them. These names are of the form $F_{a,b}(\Gamma)$, where (a,b) is the type and Γ is the full combinatorial symmetry group of the fulleroid. Note that for $n = 12$, the smallest p-vector which fullfills the condition of the lemma is not realizable. The smallest $I(5,12)$-fulleroid has as its p-vector the second to smallest one predicted by that condition, which is also shown.

Figures 4 to 9 contain Schlegel diagrams and spatial embeddings with about constant edge lengths of the 6 new fulleroids. For reference, we also include pictures of the 2 structures from [**3**] as Figures 2 and 3. The coordinates for all of these figures where obtained by the algorithms described in [**7**].

3. The Decoration Operations

The six fulleroids shown in the previous section as well as the two described by Brinkmann and Dress have been found by a systematic investigation of *all* possible ways to assemble pentagons and, say, octagons into a structure with the desired properties. This will be explained in more detail in Section 5. However, it turns out that all eight structures can be conveniently described as the results of decorating small icosahedral fullerenes with symmetric patches from the dodecahedron. By applying these decorations to other icosahedral fullerenes, one obtains several infinite series of I-fulleroids. The two most interesting of these are an infinite series of $I(5,7)$-fulleroids and a similar series of $I(5,8)$-fulleroids.

We need the following four decoration operations producing pentagons:

TABLE 1. Potential p-vectors and invariants for certain $I(5,n)$-fulleroids.

n	p_5	p_n	v	m_5	m_n	k_2	k_3	k_5	#	Fig.	Name(s)
7	**72**	**60**	**260**	**1**	**1**	**3**	**2**	**1**	**2**	(2)	$\mathbf{F_{5,7}(I)a}$
										(3)	$\mathbf{F_{5,7}(I)b}$
8	**72**	**30**	**200**	**1**	**1**	**4**	**2**	**1**	**1**	(5)	$\mathbf{F_{5,8}(I_h)}$
9	**72**	**20**	**180**	**1**	**0**	**3**	**3**	**1**	**1**	(4)	$\mathbf{F_{5,9}(I_h)}$
10	**60**	**12**	**140**	**1**	**0**	**3**	**2**	**2**	**1**	(6)	$\mathbf{F_{5,10}(I_h)}$
11	312	60	740	5	1	3	2	1	?		
12	132	20	300	2	0	3	4	1	–		
12	**192**	**30**	**440**	**3**	**0**	**6**	**3**	**1**	**1**	(8)	$\mathbf{F_{5,12}(I_h)}$
13	432	60	980	7	1	3	2	1	?		
14	**252**	**30**	**560**	**4**	**0**	**7**	**2**	**1**	**1**	(9)	$\mathbf{F_{5,14}(I_h)}$
15	**120**	**12**	**260**	**2**	**0**	**3**	**2**	**3**	**1**	(7)	$\mathbf{F_{5,15}(I_h)}$
16	312	30	680	5	0	8	2	1	?		
17	672	60	1460	11	1	3	2	1	?		
18	252	20	540	4	0	3	6	1	?		
19	792	60	1700	13	1	3	2	1	?		
20	180	12	380	3	0	3	2	4	?		

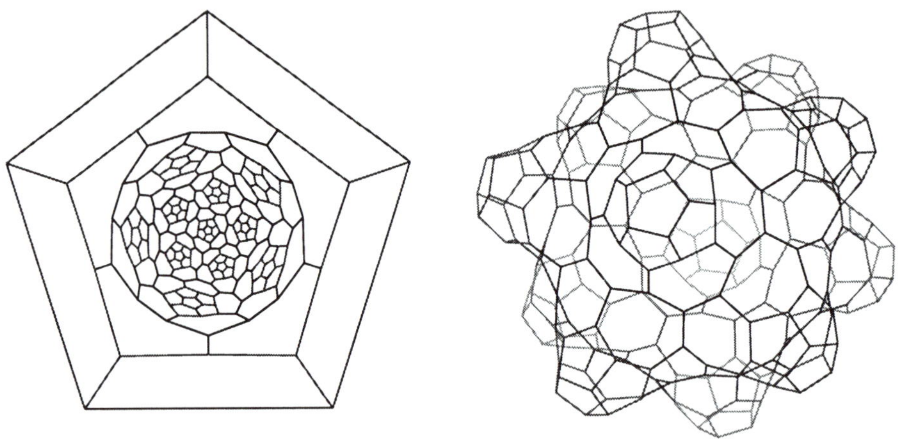

FIGURE 2. $F_{5,7}(I)a = P(C_{140}(I))$; $v = 260$

Triacon of first order: Split a hexagon into three pentagons by connecting midpoints of three pairwise nonadjacent sides to the face center.

Pentacon: Split a pentagon into five quadrangles by connecting midpoints of all sides to the face center, then truncate the central vertex to obtain six pentagons.

Triacon of second order: Split a hexagon into three quadrangles and a hexagon by three new edges parallel to pairwise nonadjacent sides, then apply the triacon of first order to the new hexagon, using midpoints of new edges, to obtain altogether six pentagons.

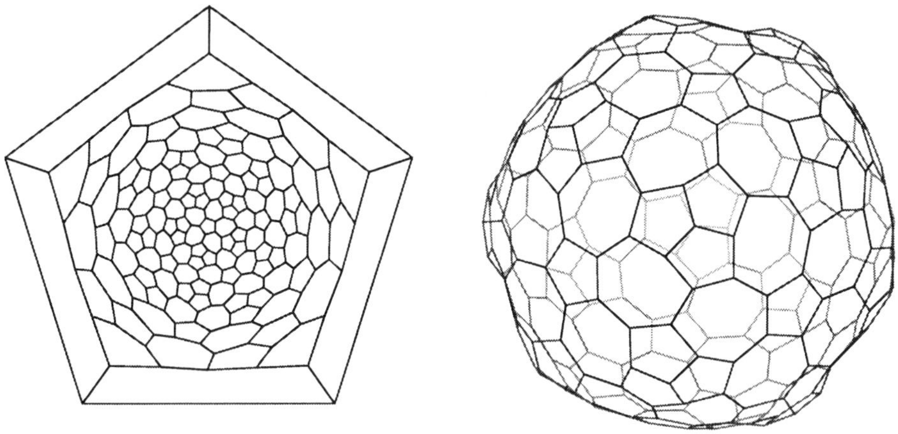

FIGURE 3. $F_{5,7}(I)b = T_1(C_{180}(I_h))$; $v = 260$

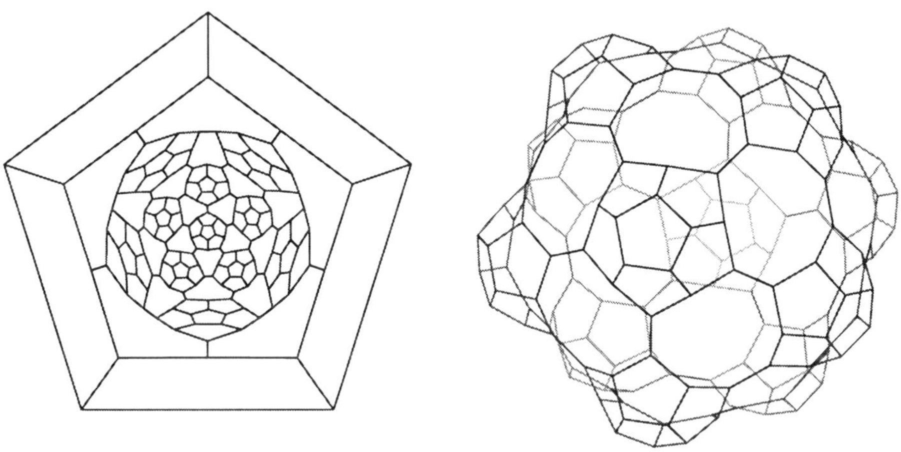

FIGURE 4. $F_{5,9}(I_h) = P(C_{60}(I_h))$; $v = 180$

Triacon of third order: Split a triangle into three new triangles and a hexagon by "cutting off" the vertices, then apply the triacon of second order to the hexagon to obtain altogether nine pentagons.

Note that these operations retain the full rotational symmetry of the face they are applied to, but that the first two triacon operations will destroy mirror symmetries in case a mirror plane intersects the decorated hexagon at two opposite vertices.

These and more elaborate decoration operations have long been known and used by specialists in the field. See for example [**15**].

In Table 2, we summarize the relations between small icosahedral fullerenes and fulleroids in terms of the four decoration operations. For that purpose, we denote the application of the pentacon operation to a complete set of 12 pentagons containing a 5-fold symmetry axis each by the letter P. We denote the application

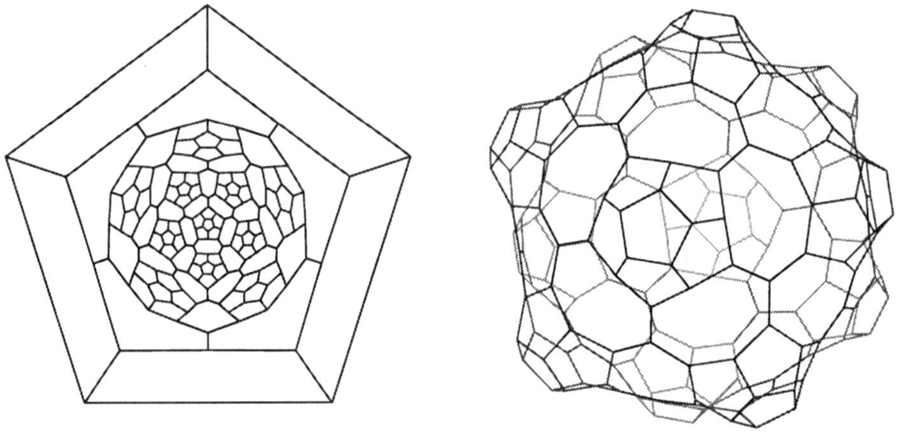

FIGURE 5. $F_{5,8}(I_h) = P(C_{80}(I_h))$; $v = 200$

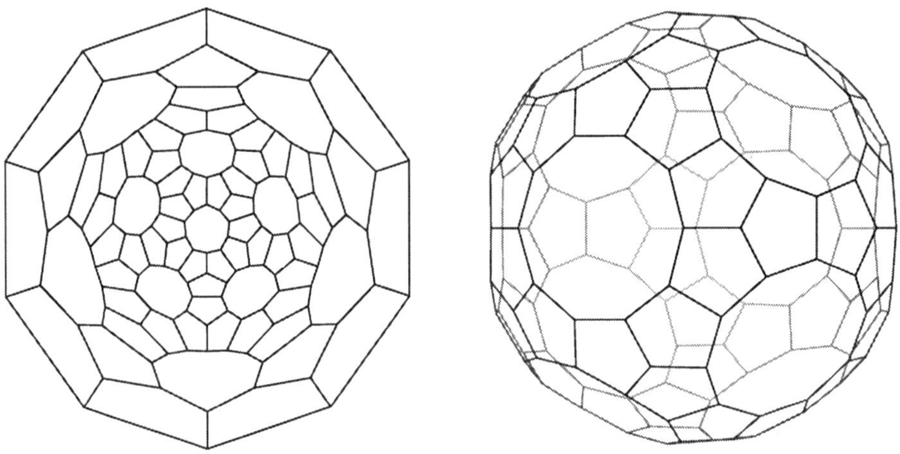

FIGURE 6. $F_{5,10}(I_h) = T_1(C_{60}(I_h))$; $v = 140$

TABLE 2. Description of fulleroids as decorated fullerenes

$$F_{5,7}(I)a = P(C_{140}(I))$$
$$F_{5,7}(I)b = T_1(C_{180}(I_h))$$
$$F_{5,8}(I_h) = P(C_{80}(I_h))$$
$$F_{5,9}(I_h) = P(C_{60}(I_h))$$
$$F_{5,10}(I_h) = T_1(C_{60}(I_h))$$
$$F_{5,12}(I_h) = T_3(C_{80}(I_h))$$
$$F_{5,14}(I_h) = P(F_{5,12}(I_h)) = T_3(F_{5,8}(I_h))$$
$$F_{5,15}(I_h) = T_2(C_{60}(I_h))$$

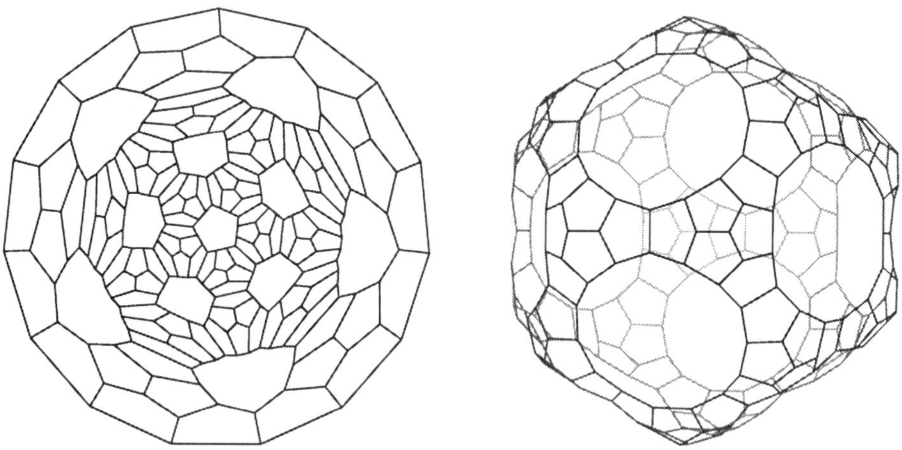

FIGURE 7. $F_{5,15}(I_h) = T_2(C_{60}(I_h))$; $v = 260$

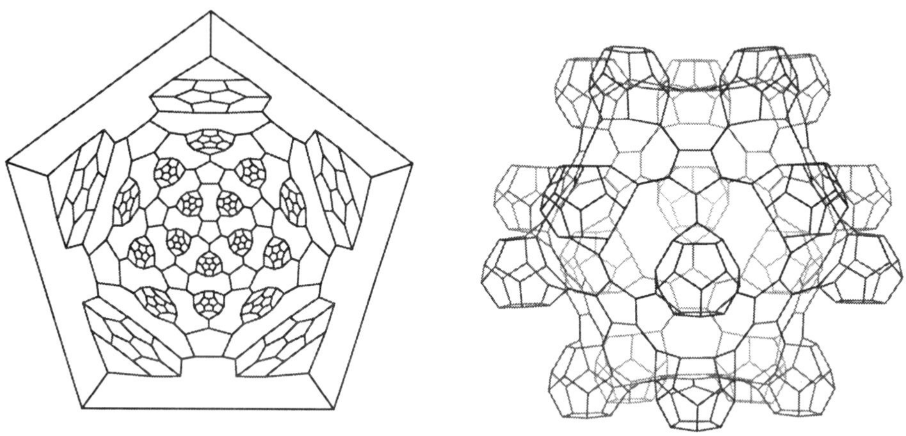

FIGURE 8. $F_{5,12}(I_h) = T_3(C_{80}(I_h))$; $v = 440$

of the first two triacon operations to sets of 20 faces containing a 3-fold rotational axis each by T_1 and T_2, respectively. If for some base structure the 3-fold axes meet vertices, then we denote the operation of truncating all 20 such vertices and applying the triacon of third order to the resulting triangles by T_3.

Note that the triacon of $C_{180}(I_h)$ has only I-symmetry due to the fact that the triacon operations do not always retain mirror symmetries. Also note that $F_{5,14}(I_h)$ is obtained from $C_{80}(I_h)$ by applying both P and T_3, in any order.

Infinitely many I-fulleroids can be obtained by applying the pentacon and triacon operations to the icosahedral fullerenes. In particular, the P operation can be applied to any icosahedral fullerene except $G_{1,0}$ (the dodecahedron). The resulting fulleroid will be of type $(5, 6, 7)$ except for $G_{1,1} = C_{60}(I_h)$, $G_{2,0} = C_{80}(I_h)$ and $G_{2,1} = C_{140}(I)$, namely the three cases appearing in Table 2. In a similar way, either one of T_1 and T_2 or T_3 can be applied to any icosahedral fullerene,

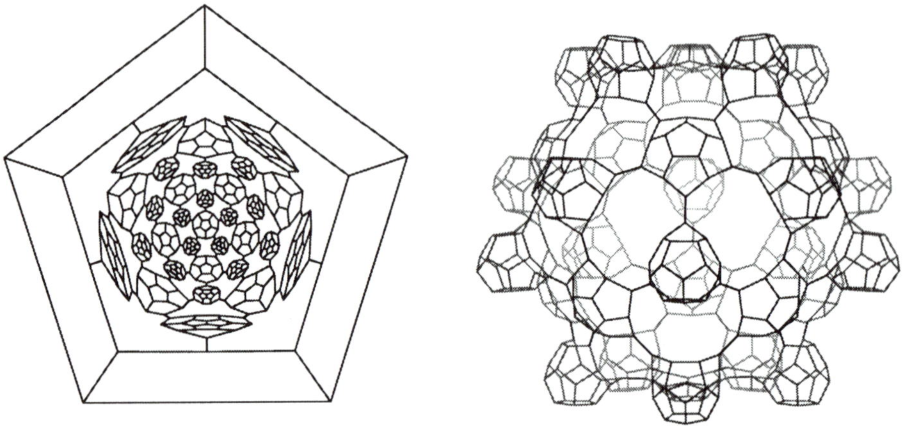

FIGURE 9. $F_{5,14}(I_h) = P(F_{5,12}(I_h))$; $v = 560$

depending on whether the 3-fold rotational axes meet (hexagonal) faces or vertices, respectively. This gives rise to three additional infinite families of fulleroids, but again all except finitely many of these fulleroids will be of type $(5, 6, n)$ for some $n > 6$.

On the other hand, an infinite series of $I(5,7)$-fulleroids $(H_n)_{n>0}$ is obtained from the series $(G_{2n+1,0})_{n>0}$ by applying T_1 in a regular pattern to one fourth of all hexagons (instead of just to those hexagons which contain a 3-fold rotational axis). The pattern is shown in Figure 10 together with fundamental domains for the first four of these $I(5,7)$-fulleroids. The fundamental domains are kite-shaped pieces of different sizes from the modified hexagonal grid. The axes of the 5- and the 3-fold rotations will go through the leftmost and rightmost vertex, respectively, of the kite. Note that the first $I(5,7)$-fulleroid of this series, H_1, is just $F_{5,7}(I_h)$.

A similar series $(O_n)_{n>0}$ of $I(5,8)$-fulleroids is obtained from $(G_{2n+1,0})_{n>0}$ by applying T_2 instead of T_1 in the same pattern as shown in Figure 10.

4. Restrictions on the p-Vectors

Here we will derive a simple necessary condition on the p-vector of an $I(a,b)$-fulleroid. We only use Euler's formula and some basic facts from group theory. Most of this material can be found in standard text books, see for example [**11**, **17**, **2**].

Let us first recall the well-known fact that, while the number of hexagons in a tiling with constant vertex degree 3 is arbitrary, the number of pentagons can be determined from the numbers of faces of all other degrees. Indeed, the number of edges is $e = \frac{1}{2}\sum i \cdot p_i$, the number of vertices is $v = \frac{1}{3}\sum i \cdot p_i$, thus by Euler's formula $f - e + v = 2$, we obtain

$$2 = \sum(1 - \frac{i}{2} + \frac{i}{3})p_i = \sum(1 - \frac{i}{6})p_i$$

or

$$0 = 12 + \sum(i-6)p_i. \tag{4.1}$$

For tilings with specified symmetries, much stronger conditions on the potential p-vectors are obtained by applying some elementary group theory. Let T be some

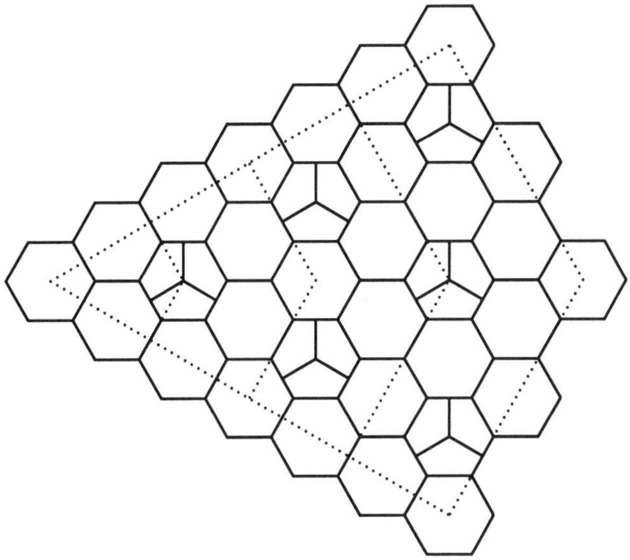

FIGURE 10. Fundamental domains for the $I(5,7)$-fulleroids H_1 to H_4 (dotted lines).

tiling of the sphere and let G be a subgroup of its full symmetry group. First, note that the image of an arbitrary face of T by an arbitrary element of G is again a face of T. Therefore, G can be interpreted as acting by permutations on the set of faces. By the same argument, G also acts on the set of edges and on the set of vertices. Recall that the *stabilizer* of some element x (i.e. some face, edge or vertex) is the set of operations in G which map it onto itself. The stabilizer of x is always a subgroup of G and is denoted by G_x. The orbit $G \cdot x$ of an element x is the set of images of x under all the operations in G. Obviously, the set of orbits forms a partition of the set of elements, thus, for example, the set of all face orbits under G forms a partition of the set of faces, and so on. The following elementary statement from group theory tells us how many elements these orbits can have.

LEMMA 4.1. *If G is a finite group acting on a finite set S and s is an arbitrary element of S, then $\#(G \cdot x) = \#G/\#G_x$.*

Now, consider a tiling T with symmetry group I, the group of orientation preserving symmetries of the icosahedron. All the elements of I are rotations. If some rotation maps a face f onto itself, then all the powers of this rotation do so, too. Moreover, no two rotations of different orders can fix the same face. Thus, the possible face stabilizers are exactly the groups generated by the rotations of orders 2, 3 and 5.

Another elementary fact is that the stabilizers of two elements in the same orbit are conjugate to each other, and that, moreover, if some subgroup H of G is conjugate to a stabilizer G_x, then H is the stabilizer of some element in the same orbit as x. Now, since in I, each pair of rotational subgroups of the same order are conjugate, the following statement holds.

THEOREM 4.2. *In an I-fulleroid, there is at most one orbit of faces with rotational symmetries of orders 2, 3 and 5 respectively. These orbits, if existent, contain*

TABLE 3. The possible face orbits of an I-fulleroid.

orbit size	60	30	20	12
number of orbits	any	≤ 1	≤ 1	1
face degree	any	2t	3t	5t

exactly 30, 20 *and* 12 *faces, respectively. All the other orbits have trivial stabilizers and contain exactly* 60 *faces, each.*

Table 3 gives an overview of the possible combinations of face orbits.

Now, a rotation of order 2 can never fix a face or vertex of odd degree, So, in this case, it has to fix either an edge or a face of even degree. Similarly, a rotation of order 3 must fix either a vertex or a face of some degree divisible by 3. On the other hand, a rotation of order 5 must fix a face, since there is no other element it could possibly fix. We can distinguish two broad classes of $I(5,n)$-fulleroids depending on whether or not the 5-fold rotations fix pentagons or larger faces. In the first case, the p-vector will be of the form

$$\mathbf{A}_{n,k}\colon (p_5, p_n) = \left(12 + 60k, \frac{60k}{n-6}\right),$$

where $k \geq 1$, $n \geq 6$ and $p_n \in \mathbf{N}$. In the second case, it will be of the form

$$\mathbf{B}_{n,k}\colon (p_5, p_n) = \left(60k, 12\frac{5k-1}{n-6}\right),$$

where $k \geq 1$, $n \geq 10$, $5|n$ and $p_n \in \mathbf{N}$. In the following, we will give a finer parametrization of potential p-vectors.

For a given degree i, let m_i denote the number of orbits of i-faces (i.e. faces of degree i) that are not mapped onto themselves by any symmetry but the identity. The total number of such faces is then $60m_i$ by the above. For $j = 2,3,5$, set $k_j := k$, if a $k \cdot j$-face exists which is fixed by a rotation of order j. If none such face exists for any k, then we set $k_j := 6/j$.

LEMMA 4.3. *For each I-fulleroid, the equality*

$$\sum_i m_i(i-6) + \sum_{j=2,3,5} k_j = 6$$

holds.

PROOF. Each face orbit has either a trivial stabilizer or is fixed by a rotation of order 2, 3 or 5. If some face f of degree m has a stabilizer of order $j = 2,3,5$, then we have $m = k_j \cdot j$, and the orbit of f contains exactly $60/j$ faces. Together with Equation 4.1, this implies

$$0 = \sum_i 60m_i(i-6) + \sum_{j=2,3,5} \frac{60}{j}(k_j \cdot j - 6) + 12.$$

Note that this equation remains true if for some j, there is no face fixed by a rotation of order j, because in that case we have $k_j = 6/j$, thus $k_j \cdot j - 6 = 0$.

We divide both sides by 60 and simplify the second sum, to obtain

$$\begin{aligned} 0 &= \sum_i m_i(i-6) + \sum_{j=2,3,5} k_j - \sum_{j=2,3,5} \frac{6}{j} + \frac{1}{5} \\ &= \sum_i m_i(i-6) + \sum_{j=2,3,5} k_j - \frac{6}{2} - \frac{6}{3} - \frac{6}{5} + \frac{1}{5} \\ &= \sum_i m_i(i-6) + \sum_{j=2,3,5} k_j - 6. \end{aligned}$$

□

We finish this section by giving formulae to calculate the p-vector and the number of vertices v from the invariants m_i and k_j. Since the invariants are insensitive to the number of hexagons, the following equations hold if and only if the fulleroid in question does not contain hexagons. For the p-vector, we have

$$p_i = 60 \cdot \left(m_i + \sum_{j \cdot k_j = i} \frac{1}{j} \right). \tag{4.2}$$

The number of vertices is

$$\begin{aligned} v = \frac{1}{3} \sum_i i \cdot p_i &= 20 \cdot \sum_i \left(i \cdot m_i + \sum_{j \cdot k_j = i} \frac{i}{j} \right) \\ &= 20 \cdot \sum_i \left(i \cdot m_i + \sum_{j \cdot k_j = i} k_j \right), \end{aligned}$$

thus

$$v = 20 \cdot \left(\sum_i i \cdot m_i + \sum_{j \cdot k_j \neq 6} k_j \right). \tag{4.3}$$

5. From the p-Vectors to the Structures

In the following, we will present a general method for the classification of I-fulleroids with given parameters m_i and k_j. This method easily generalizes to other, similar problems. We will present proofs for the individual classification results at different levels of abstraction, starting with an elementary proof of the uniqueness of $F_{5,10}(I_h)$ and gradually transforming our arguments into purely combinatorial ones expressed in the language of Delaney symbols. Our intention is to demonstrate thereby how Delaney symbols capture the essential structural information in a very natural and convenient way.

PROPOSITION 5.1. *The fulleroid $F_{5,10}(I_h)$ is the only I-fulleroid with p-vector $(p_5, p_{10}) = (60, 12)$ and the smallest $I(5,10)$-fulleroid.*

PROOF. First note that 12 is the smallest possible non-zero value of p_{10}, corresponding to the parameter values $m_{10} = 0$ and $k_5 = 2$. This means that there is exactly one orbit of decagons, each of which is fixed by a rotation of order 5, and one orbit of pentagons, which have trivial stabilizers. This gives us the additional information that all the rotation axes of order 2 meet edges and that all the axes

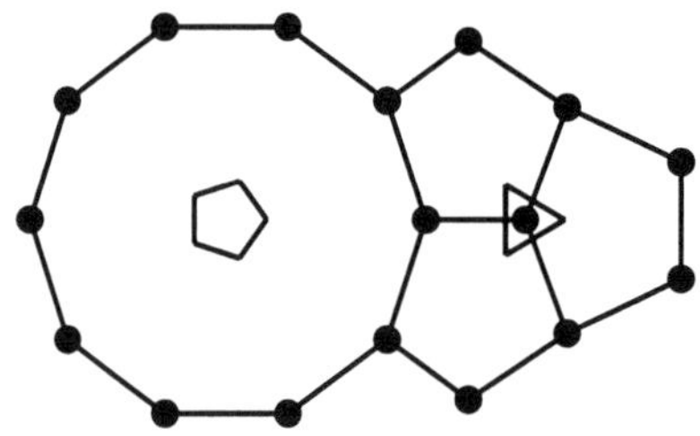

FIGURE 11. A forced subgraph for an I-fulleroid with p-vector $(p_5, p_{10}) = (60, 12)$.

of order 3 meet vertices. Let v_0 be one such vertex and let f_0 be one of the faces adjacent to it.

Assume f_0 is a decagon. Because of the 5-fold rotation mapping f_0 onto itself, every second vertex of f_0 must meet a 3-fold axis. The three faces containing such a vertex must then all be decagons, because they are rotated onto each other. This means that f_0 is completely surrounded by decagons, each of them in the same orbit as f_0. By induction, all the faces in the connected component which contains f_0 are decagons. There is only one connected component, so there are only decagons, a contradiction.

We conclude that v_0 must be adjacent to a pentagon. Again, this implies that all three faces adjacent to v_0 are pentagons. Let v_1 be a vertex adjacent to any two of these faces. The third face adjacent to v_1 cannot be a pentagon, since this pentagon would be in the same orbit as f_0 and thus contain a symmetric image v_2 of v_0, i.e., a vertex meeting a 3-fold rotation axis. There are three essentially different possible positions for v_2. In each case, by applying all symmetries one finds that the resulting structure must be the dodecahedron.

Consequently, the third face meeting v_2 is a decagon, thus the fulleroid in question must contain the substructure depicted in Figure 11, where the 3- and the 5-fold rotation centers are indicated by a small triangle and pentagon, respectively. By applying these rotations in a systematic way, one obtains the structure shown in Figure 6. □

Obviously, the argumentation in the above proof is straightforward but—with all the details worked out—rather tedious. This is partly due to the fact that instead of individual faces, edges and vertices, we have to deal with whole orbits, taking all symmetries known so far into account. It is much more convenient to work in the *orbit space* of the group I. The orbit space of some group Γ is defined as the image of a continuous function which maps two points p and q of the sphere onto the same image point if and only if there is some element of Γ which maps p to q. The orbit space of I is, topologically, just a sphere with three special points, namely the images of the three types of rotational centers. A simple way to obtain the orbit space is to take a fundamental domain and glue together boundary points which are mapped onto each other by some group element.

The degree of the image of a face in the orbit space depends on the stabilizer of that face. In general, an $(i \cdot j)$-face with a rotation center of order j is mapped to an i-face. Likewise, vertices with non-trivial stabilizers are mapped to vertices

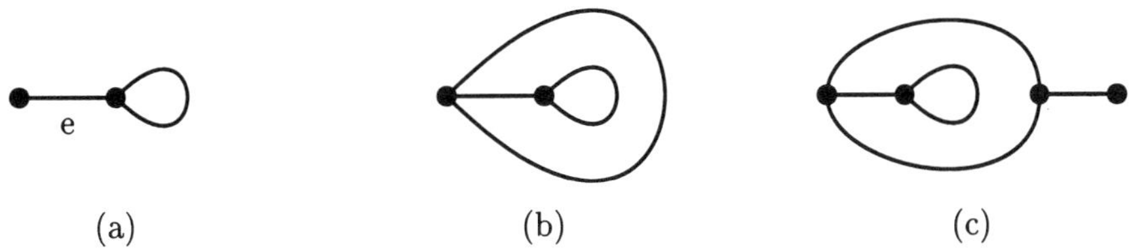

FIGURE 12. Forced subgraphs for the smallest $I(5,8)$-fulleroid.

of accordingly smaller degree. Therefore, in the orbit space, we have to consider special features like 1-gons and vertices of degree 1. Still more strangely, the image of an edge with 2-fold stabilizer is a "half-edge" with only one endpoint. A loop edge, one which has two identical endpoints, must be count twice when determining the vertex degree. Likewise, an edge adjacent to the same face on both sides has to be count twice when determining the face degree. In both cases, however, a half-edge is only count once.

Despite these difficulties, we can simplify our arguments considerably by working in the orbit space. We will call the image of a tiling in the orbit space its *orbit tiling.*

PROPOSITION 5.2. *The fulleroid $F_{5,8}(I_h)$ is the only I-fulleroid with p-vector $(p_5, p_8) = (72, 30)$ and the smallest $I(5,8)$-fulleroid.*

PROOF. Again, $p_8 = 30$ is the smallest possible value of p_8, since 2 is the largest rotational order occurring in I which divides 8. The p-vector $(p_5, p_8) = (72, 30)$ corresponds to the parameter values

$$(m_5, m_8, k_2, k_3, k_5) = (1, 0, 4, 2, 1).$$

The orbit tiling consists of a 1-gon (the image of a 5-fold symmetric pentagon), a 5-gon (the image of an asymmetric pentagon) and a 4-gon (the image of a 2-fold symmetric octagon). Exactly one degree 1 vertex and no half-edge occurs, because there are 2-fold symmetric, but no 3-fold symmetric faces.

The 1-gon gives rise to the subgraph depicted in Figure 12(a), because of the free valency at its vertex. The edge e is adjacent to the same face at both sides, thus if that face is a 4-gon, then the configuration of Figure 12(b) occurs. Now, both vertices have degree 3, so there is no possible continuation. We conclude that the 1-gon must be adjacent to the pentagon, leading to the configuration of Figure 12(c).

Now, the outer face is a 4-gon and there is exactly one vertex of degree 1, as required. The configuration is unique, as shown, and corresponds to the I-fulleroid $F_{5,8}(I_h)$. □

PROPOSITION 5.3. *The fulleroid $F_{5,9}(I_h)$ is the only I-fulleroid with p-vector $(p_5, p_9) = (72, 20)$ and the smallest $I(5,9)$-fulleroid.*

PROOF. As above, $p_9 = 20$ is smallest possible and the parameter values follow uniquely from the p-vector. The orbit tiling consists of a 1-gon and a pentagon as above and a triangle corresponding to an orbit of 3-fold symmetric 9-gons. We have no vertex of degree 1, but a half-edge. As above, the 1-gon must be adjacent to the pentagon, which in turn must be adjacent to the triangle. The only possible

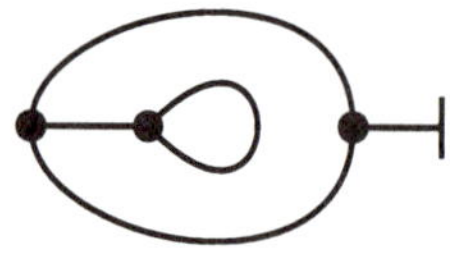

FIGURE 13. The orbit space of $F_{5,9}(I_h)$

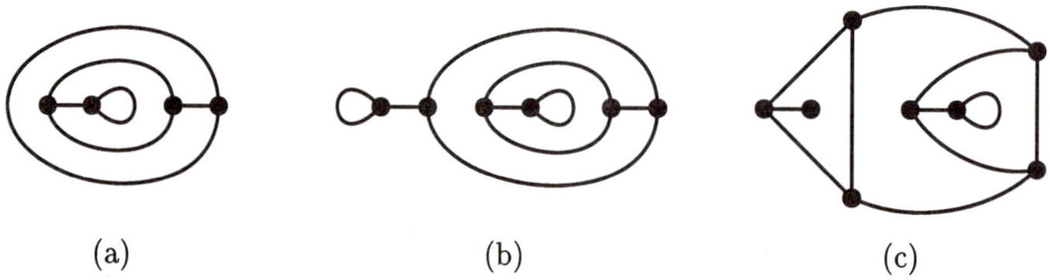

FIGURE 14. Forced subgraphs referred to in Proposition 5.4.

configuration is shown in Figure 13, where the half-edge is shown as a "T-shape" (note that a half-edge counts only once, so the outer face is indeed a triangle). □

PROPOSITION 5.4. *The fulleroid $F_{5,12}(I_h)$ is the only I-fulleroid with p-vector $(p_5, p_{12}) = (192, 30)$ and the smallest $I(5,12)$-fulleroid.*

PROOF. First, we must show that no I-fulleroid with the smaller p-vector $(p_5, p_{12}) = (132, 20)$ exists. For such a structure, the orbit tiling would contain a 1-gon, a quadrangle and two pentagon. As in the proof of Proposition 5.2, the 1-gon must be adjacent to a pentagon f_0. The other face adjacent to f_0 cannot be the quadrangle, since that would lead to the configuration of Figure 12(c). On the other hand, if f_0 is adjacent to the second pentagon, the configuration of Figure 14(a) results, which cannot be continued by a quadrangle. Thus, we have no I-fulleroid with face-vector $(p_5, p_{12}) = (132, 20)$ and come to check the next possible p-vector $(p_5, p_{12}) = (192, 30)$.

Here, in the orbit space, we have a 1-gon, a hexagon (the image of the 2-fold symmetric 12-gon) and three pentagons. We have no half-edge, but a degree 1 vertex. We already know that if the 1-gon is adjacent to a pentagon f_0, then f_0 can not be adjacent to a second pentagon. Let us assume that f_0 is adjacent to the hexagon, which we will denote by f_1. Now, f_1 must be adjacent to a second pentagon, but has only one vertex with a free valency on its boundary. This enforces a configuration as in Figure 14(b), in contradiction to the fact that we must have a third pentagon.

We conclude that the 1-gon must be adjacent to the hexagon. Again, there is only one possible continuation, leading to the configuration in Figure 14(c), which is legal. □

The reader is invited to proof the last 3 propositions in the straightforward manner of Proposition 5.1. Obviously, working in the orbit space simplifies things a lot. On the other hand, strange features like half-edges, loops and dead-end edges make the necessary case-analysis unintuitive and error-prone. Moreover, for

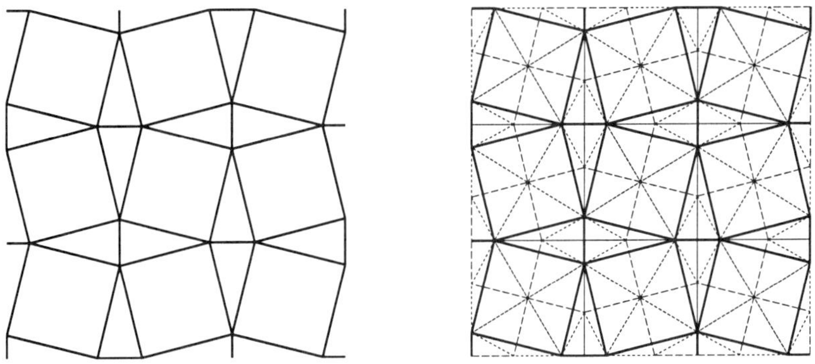

FIGURE 15. A tiling of the plane and its barycentric subdivision.

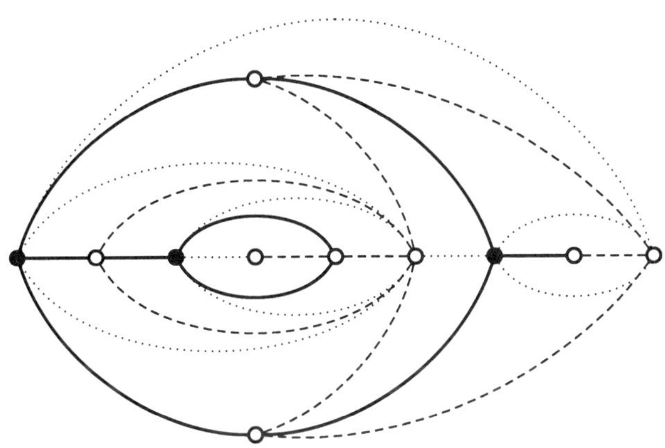

FIGURE 16. The barycentric triangulation of Figure 13.

more complicated tasks, the help of a computer will be indispensable. Therefore, a convenient combinatorial encoding which naturally includes all the singularities mentioned above should be most desirable. Such an encoding is provided by the Delaney symbol, as introduced by A. Dress [**9**] and inspired by M. Delaney [**6**].

Delaney symbols have been treated extensively in several publications (see for example [**5**, **4**], and, of course, [**3**]). We will only give an informal introduction here. For a given tiling, connect the center of each face to each of its vertices by a dotted line and to the midpoint of each of its edges by a dashed line as in Figure 15.

The result is a triangulation called the *barycentric subdivision*. Each of its triangles has exactly one dotted, one dashed and one solid edge. The subdivision can be performed in such a way that it is invariant under the symmetry group of the tiling. Then, obviously, the stabilizer of each triangle is trivial. The image of the barycentric subdivision in the orbit space is a subdivision of the orbit tiling into triangles such that again each triangle has exactly one dotted, one dashed and one solid edge. For example, the configuration of Figure 13 corresponds to the triangulation of the orbit space shown in Figure 16.

Admittedly, this looks much more complicated than the original figure. It has, on the other hand, the advantage that the whole structure can be conveniently

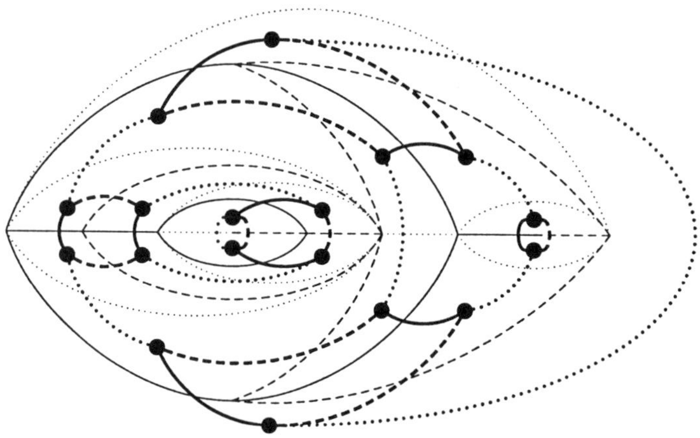

FIGURE 17. The triangulated orbit space (thin lines) and the Delaney graph of $F_{5,9}(I_h)$.

encoded by listing adjacent pairs of triangles sorted by the respective type (i.e. dotted, dashed or solid) of edge they are sharing. Since pairs of triangles can share more than one edge, a particular pair of triangles may occur more than once. A simple way to visualize the adjacency relations is to construct the dual graph of the orbit triangulation as in Figure 17. Here, to each triangle of Figure 16 (shown in gray), we assign a node, and we connect two nodes by a dotted, dashed or solid line, respectively, if the corresponding triangles share an edge of the same type. The resulting regularly edge-labelled graph is called the *Delaney graph* of the original tiling. The complete *Delaney symbol* is obtained by supplementing this graph with information on the images of the rotational axes. For details, see e.g. [**5**].

All we need to know in this context is how certain subgraphs of the Delaney graph correspond to the vertices, faces and edges of the orbit tiling. Note that if we remove all edges of a certain type, the remaining subgraph is a collection of cycles (this is only true if the symmetry group of the original tiling does not contain reflections). Each of these cycles corresponds to a constituent of the orbit tiling. In particular, dotted-dashed cycles correspond to faces, dotted-solid cycles to vertices and dashed-solid cycles to edges. The length of a dotted-dashed cycle is twice the degree of the corresponding face. The length of a dotted-solid cycle is twice the degree of the corresponding vertex. The length of a dashed-solid cycle is four if that cycle corresponds to an ordinary edge and two if it corresponds to a half-edge. An additional fact that helps to simplify the necessary case analysis is that the Delaney graph of a tiling with orientation preserving symmetry group is always *bipartite*, i.e. its nodes can be labelled by '+' and '-' in such a way that no edge connects two nodes with the same label.

PROPOSITION 5.5. *The fulleroid* $F_{5,15}(I_h)$ *is the only* I*-fulleroid with* p*-vector* $(p_5, p_{15}) = (120, 12)$ *and the smallest* $I(5,15)$*-fulleroid.*

PROOF. Obviously, the p-vector is smallest possible and corresponds to the parameter values given in Table 1. Thus, the orbit tiling consists of one triangle corresponding to an orbit of 3-fold symmetric 15-gons, and two pentagons. In effect, in the Delaney graph, we have one dotted dashed cycle of length 6 and two dotted-dashed cycles of length 10. Moreover, all dotted-solid cycles will have length 6 except for one of length 2 (corresponding to an orbit of 3-fold symmetric vertices). All dashed-solid cycles will have length 4 except for one of length 2 (corresponding to an orbit of 2-fold symmetric edges). Both nodes belonging to the small dotted-solid cycle $\mathcal{V}$ are contained in the same dotted-dashed cycle $\mathcal{F}$. There are two cases depending on the length of $\mathcal{F}$.

Let us first assume that $\mathcal{F}$ has length 6 (see Figure 18). Denote by a the solid edge in $\mathcal{V}$. The dashed-solid component to which a belongs consists of four nodes already, so it must be completed to a cycle by a solid edge b. Now, to make the Delaney graph connected, the cycle $\mathcal{F}$ must be connected to one of the dotted-dashed cycles of length 10 by a solid edge c. Again, to complete a dashed-solid cycle, a solid edge d is forced. But now, b, c and d belong to a common dotted-solid component with already 6 nodes, which has to be completed by a dotted edge e. This will produce a dotted-dashed cycle of length 2, which can not occur. The situation is depicted in Figure 18. We conclude that $\mathcal{F}$ can not have length 6.

Now let us assume that $\mathcal{F}$ has length 10. Again, denote by a the solid edge in $\mathcal{V}$. Again, a solid edge b is forced to complete a dashed-solid cycle. Let r be the dotted neighbour of one of the nodes incident to b. By considering legal lengths of dotted-solid and dashed-solid cycles again, it can be seen easily that r can not be connected by a solid line to any other node in $\mathcal{F}$. Therefore, r must be connected by an edge c to a different dotted-dashed cycle $\mathcal{G}$. We know that the length of $\mathcal{G}$ is at least 6, so we draw $\mathcal{G}$ as an incomplete cycle of length 6 in Figure 19. The solid edges d, e and f in this picture are forced by completion of a dashed-solid cycle, a dotted-solid cycle and a dashed-solid cycle, respectively.

The two remaining vertices of $\mathcal{F}$ can not be connected to each other by a solid edge, because this would lead to a dotted-solid cycle of length at least 8. Thus, there are edges g and h as in the picture connecting $\mathcal{F}$ to a dotted-dashed cycle $\mathcal{H}$, of which we don't know yet whether it is the same one as $\mathcal{G}$ or not. We only draw a portion of $\mathcal{H}$ containing 4 nodes. The solid edges i and j are enforced by the completion of dotted-solid cycles. Now, if $\mathcal{G}$ has only length 6, i.e. if the dashed edge k is present, then the dashed edge l is forced, thus $\mathcal{H}$ would have length 4, which is too short. If, on the other hand, $\mathcal{H}$ is equal to $\mathcal{G}$, then no connection to the third dotted-dashed cycle is possible. We conclude that $\mathcal{H}$ is different from $\mathcal{G}$ and has length 6, while $\mathcal{G}$ has length 10. This enforces edges k, l and m of Figure 20, which complete the Delaney graph. Obviously, this graph is legal (it contains, for example, exactly one dashed-solid cycle of length 2, as required), and our discussion above shows that it is unique. □

For $F_{5,14}(I_h)$, the uniqueness proof is rather lengthy, so we will not present it here, but it is still manageable without computer support.

6. Problems

We mention a few interesting problems regarding $I(5,n)$-fulleroids.

- Is there at least one $I(5,n)$-fulleroid for each $n > 6$?
- Is there at least one $I(5,n)$-fulleroid for infinitely many $n > 6$?
- Is there an infinite series of $I(5,n)$-fulleroids for each / infinitely many $n > 6$?
- For which n is there at least one $I(5,n)$-fulleroid realizing the smallest possible p-vector as according to Lemma 4.3?
- For which n is this smallest $I(5,n)$-fulleroid unique?

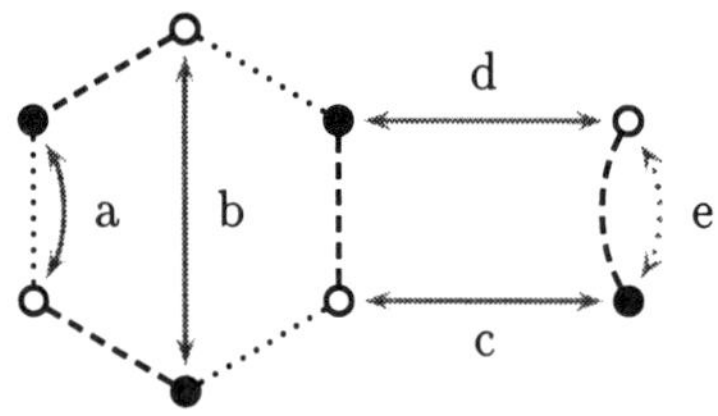

FIGURE 18. A forced partial Delaney graph.

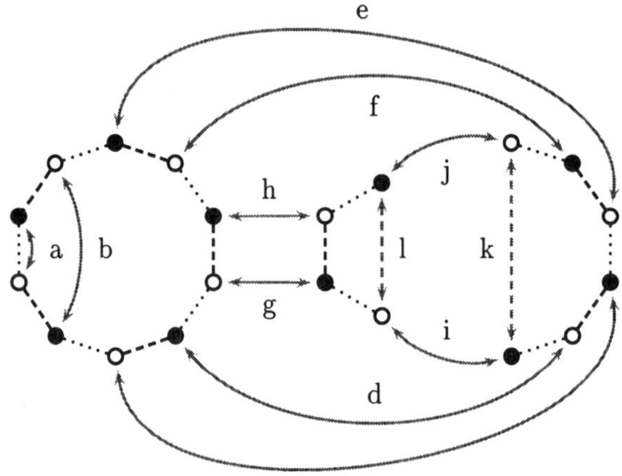

FIGURE 19. Another forced partial Delaney graph.

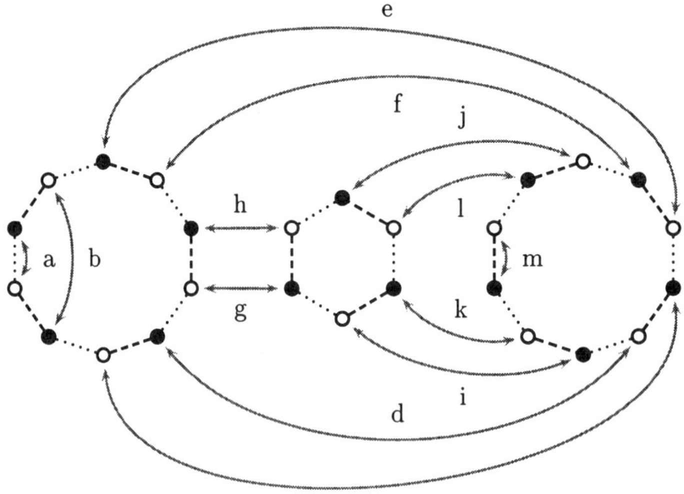

FIGURE 20. The Delaney graph of $F_{5,15}(I_h)$.

We think that the methods of [**15**] can be adapted to produce infinitely many $I(5,n)$-fulleroids for infinitely many n. Much work has been done on the classification of polytopes with a limited number of face or edge types (see also [**13**]), so we would not be surprised if even more of our problems could be answered easily by the experts.

References

[1] H.M.S. Coxeter, *Virus Macromolecules and Geodesic Domes*, A Selection of Mathematics (J.C. Butcher, ed.), Auckland University Press, 1971.

[2] H.M.S. Coxeter, *Regular Polytopes*, Dover, New York, 1973.

[3] A.W.M. Dress, and G. Brinkmann, Phantasmagorical Fulleroids, *Match* **33** (1996) 87–100.

[4] O. Delgado Friedrichs, A.W.M. Dress, and D.H. Huson, *Tilings and Symbols — A Report on the Uses of Symbolic Calculation in Tiling Theory*, Computer algebra in science and engineering (London) (J. Fleischer, J. Grabmeier, F.W. Hehl, and W. Küchlin, eds.), World Scientific, London, 1995, pp. 273–286.

[5] O. Delgado Friedrichs, A.W.M. Dress, A. Müller, and M.T. Pope, Polyoxometalates: A Class of Compounds with Remarkable Topology, *Molecular Engineering* **3** (1993) 9–28.

[6] M.S. Delaney, Quasisymmetries of Space Group Orbits, *Match* **9** (1980) 73–80.

[7] O. Delgado Friedrichs, Fast Embeddings for Planar Molecular Graphs, submitted to: *Proceedings of the Dimacs Workshop on Discrete Mathematical Chemistry*, 1998.

[8] M. Deza, and V.P. Grishukhin, *Face-regular bifaced polyhedra*, Preprint LIENS 97-10, École Normale Supérieure, Paris, 1997.

[9] A.W.M. Dress, *Regular Polytopes and Equivariant Tessellations from a Combinatorial Point of View*, Algebraic Topology (Göttingen), SLN 1172, Göttingen, 1984, pp. 56–72.

[10] M. Goldberg, A Class of Multisymmetric Polyhedra, *Tohoku Mathematical Journal* **43** (1937), 104–108.

[11] B. Grünbaum, *Convex polytopes*, John Wiley & Sons, Ltd., London, New York, Sydney, 1967.

[12] D.H. Huson, The Generation and Classification of Tile-k-Transitive Tilings of the Euclidean Plane, the Sphere and the Hyperbolic Plane, *Geometriae Dedicata* **47** (1993) 269–296.

[13] S. Jendrol, Convex 3-Polytopes with Exactly Two Types of Edges, *Discrete Mathematics* **84** (1990) 143–160.

[14] P. Mani, Automorphismen von Polyedrischen Graphen, *Mathematische Annalen* **192** (1971) 279–203.

[15] P.J. Owens, Simple 3-Polytopal Graphs with Edges of Only Two Types and Shortness Conditions, *Discrete Mathematics* **59** (1986) 107–114.

[16] E. Steinitz, Polyeder und Raumeinteilungen, *Enzykl. math. Wiss.* **3** (1922) 1–139.

[17] H. Zassenhaus, *The theory of groups*, second ed., Vandenhoeck & Ruprecht, Göttingen, 1956.

Olaf Delgado Friedrichs, Fakultät für Mathematik, Universität Bielefeld, 33501 Bielefeld, Germany.

E-mail address: `delgado@Mathematik.Uni-Bielefeld.DE`

URL: `http://www.mathematik.uni-bielefeld.de/~delgado`

Michel Deza, DMI-LIENS, Ecole Normale Superieure, 45 rue d'Ulm, 75005, Paris, France.

E-mail address: `deza@dmi.ens.fr`

DIMACS Series in Discrete Mathematics
and Theoretical Computer Science
Volume 51, 2000

Representation of Stereochemistry using Combinatorial Maps

A. Dietz, C. Fiorio, M. Habib, and C. Laurenço

ABSTRACT. N-dimensional maps are combinatorial models defined for representing the topology of subdivisions of n-dimensional geometric spaces, i.e., the topology of partitions of the spaces into i-dimensional cells (vertices, edges, faces, volumes, etc.; $0 \leq i \leq n$) [**1**]. They provide the basis for a simple, uniform, and general computer representation of the relative spatial arrangement of atoms in stereogenic entities. This article illustrates the use of 2-dimensional maps as a non-geometric computer representation of convex polyhedral and polygonal stereogenic entities, thus covering the vast majority of stereochemical atomic configurations, bond conformations, etc.

1. Introduction

The computer processing of chemical structural information requires a computer representation of stereochemistry which is adequate for a given application. A *geometric* model would seem to be the natural basis for the representation of stereochemistry. However, the manipulation of geometric chemical structures is computationally expensive [**2**]. Moreover, the chemist's reasoning is mostly based on structural formulae rather than on 3-dimensional molecular models. Therefore, it seems recommendable to use a *non-geometric* computer representation of stereochemistry for those applications which do not require exact atomic coordinates. Examples include systems for chemical information storage and retrieval or for computer-assisted organic synthesis design.[1] This non-geometric representation should be as expressive as the structural formulae which specify the *relative spatial arrangement* of the atoms in a molecule rather than *exact* atomic coordinates [**3**].

In general, a stereogenic entity is formed by atoms occupying the vertices of a convex polyhedron or polygon. Consequently, most of the non-geometric representations of molecular stereochemistry which have been proposed for the purpose of systematic nomenclature or the computer processing of chemical structures (e.g., [**4, 5, 6, 7, 8, 9**]) use convex polyhedral and polygonal templates in order to derive

1991 *Mathematics Subject Classification.* Primary 92E10; Secondary 52B10, 57N05.
The first author was supported by grant #96/10443 of the French State Department.

[1]However, some chemical information storage and retrieval systems (e.g., crystallographic databases) and some tasks of computer-assisted synthesis design (e.g., calculation of steric hindrance for reaction prediction) do require a geometric representation of stereochemistry.

stereodescriptors for stereogenic entities. For example, the configuration[2] of an asymmetric carbon atom is represented by placing its four neighbor atoms onto the vertices of a tetrahedral template and deriving a configurational descriptor. Analogously, templates are used to represent the conformation of a bond or to represent larger, non-local stereogenic entities which may encompass a whole molecule, e.g., the molecule in Figure 1.

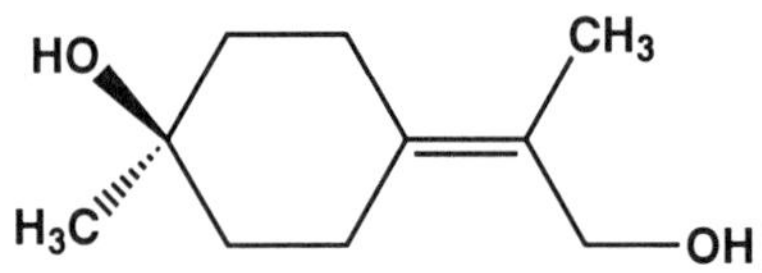

FIGURE 1. A non-local stereogenic entity

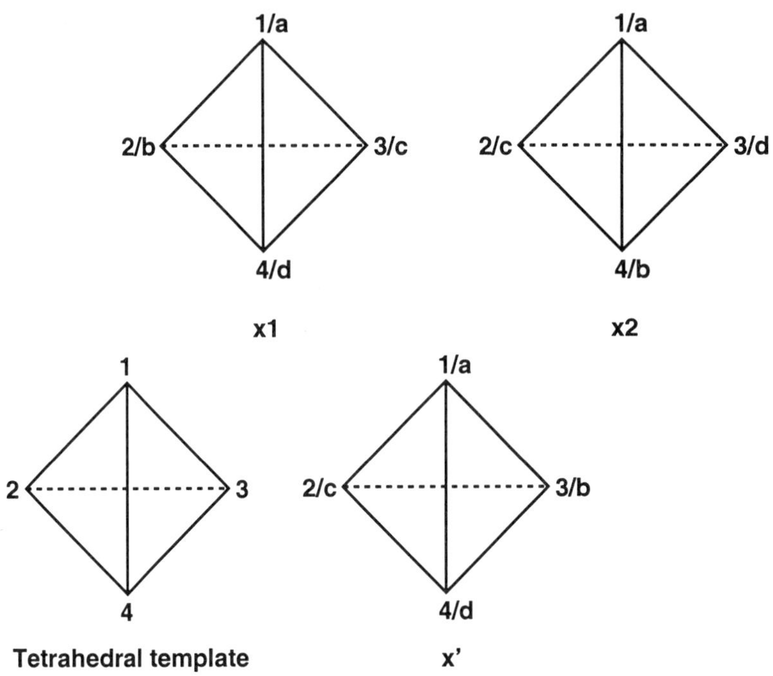

FIGURE 2. Template-based representation of asymmetric carbon atom configurations (the notation vx/at indicates that atom at occupies vertex vx of the template)

Let us consider the tetrahedral template in Figure 2 which serves to represent the configurations of asymmetric carbon atoms (we assume that the atoms a, b, c, and d are non-equivalent). Clearly, in order to determine stereodescriptors one must take into account the fact that there exist permutations of the atoms

[2] Informally, we define an "atomic configuration" as the relative spatial arrangement of an atom and its (chemically bonded) neighbor atoms, and the "conformation of a bond" between two atoms as the relative spatial arrangement of these atoms and their neighbor atoms.

which redistribute them on the vertices of a template without affecting their relative spatial arrangement. In our example, the application of the permutation $\{(a,a),(b,c),(c,d),(d,b)\}$[3] to *x1* results in *x2*. It can be interpreted as a rotation which leaves the relative spatial arrangement of the atoms a, b, c, and d unchanged. Consequently, *x1* and *x2* must give rise to the same stereodescriptor. However, *x'* represents the enantiomorphic configuration, thus resulting in a different stereodescriptor. Of course, the permutations of atoms are template-dependent and will not be the same, e.g., in the case of a tetrahedral template and in the case of a quadratic template.

In order to determine unique stereochemical descriptors, the approach of Ugi [**7, 8**] makes explicit use of groups of atom permutations which do not affect the relative spatial arrangement of these atoms.[4] The other methods [**4, 5, 6, 9**] use nomenclature rules which are, of course, also based on considerations concerning the invariance of the relative spatial arrangement of atoms under permutation.

Most of these methods have been devised for organic chemistry and their extensions for inorganic and organometallic chemistry cover only the most common templates, e.g., the octahedron. Hence, there is a need for a general and uniform approach to represent stereogenic entities.

An inconvenience of these approaches is that a different permutation group [**7, 8**] or set of nomenclature rules [**4, 5, 6, 9**] must be *manually* determined for and associated with each template, because so far no algorithm has been described for this purpose.

Moreover, these approaches have been developed to process full structures and cannot be used for the treatment of substructures, because the determination of stereodescriptors requires a priority value to be assigned to every atom that is involved in a stereogenic entity.

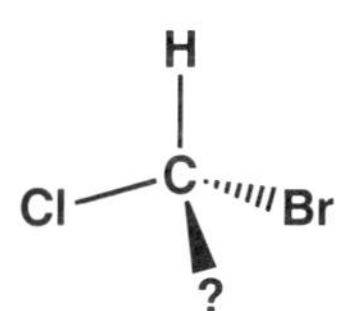

?: unknown substituent

FIGURE 3. The stereochemical aspect of this substructure cannot be represented by a stereodescriptor

For example, Figure 3 shows a substructure for which it is impossible to assign a stereodescriptor (e.g., R or S according to the CIP nomenclature [**4**]).

Of the approaches mentioned above, solely the group theoretical approach [**7, 8**] can be adapted to cope with the stereochemistry of substructures.

We have already pointed out [**3**] how to avoid the above problems by explicitly representing the relative spatial arrangement of the atoms involved in a stereogenic entity (or that of the vertices of the corresponding template). Moreover, we have

[3]We note a mapping μ from a set X to a set Y as a subset of the cartesian product $X \times Y = \{(x,y) | x \in X, y \in Y\}$.

[4]Ugi's original work [**7, 8**] deals with complete molecules instead of substructures. Here we have adapted the underlying ideas and the terminology which is somewhat different in [**7, 8**] ("ligands" are placed on "skeletal sites").

proposed a general solution for the representation of atomic configurations and bond conformations [**3**]. This representation is based on a formal definition of the notion of "relative spatial arrangement of atoms" (a procedure to translate a geometric arrangement of atoms into a non-geometric stereochemical representation). It allows to include the stereochemical aspect into full structure as well as substructure searches. It does not take into account, however, non-local stereogenic entities.

Here, we propose a simpler and more general representation of stereogenic entities. It is based on the mathematically well-formalized model of combinatorial maps [**1, 10, 11**]. The combinatorial map model is borrowed from geometric modeling: in computer aided design (CAD), simple *cells* (vertices, edges, faces, and volumes) are assembled to form complex geometric objects, e.g., a car body. In our stereochemical representation, combinatorial maps represent the boundary of convex polyhedral and polygonal stereogenic entities (an assembly of vertices, edges, and faces), but this topology-based approach is able to represent any relative spatial arrangement of atoms (stereogenic entities, molecular shapes, etc.). It has potential applications in every domain that relies on the computer processing of chemical structural information.

2. Using Combinatorial Maps for the Representation of Stereochemistry

We do not intend to give a mathematically rigorous presentation of combinatorial maps [**1, 10, 11**]. Instead, we informally describe the use of *2-dimensional maps* (2-maps) for the representation of the relative spatial arrangement of atoms forming a stereogenic entity.

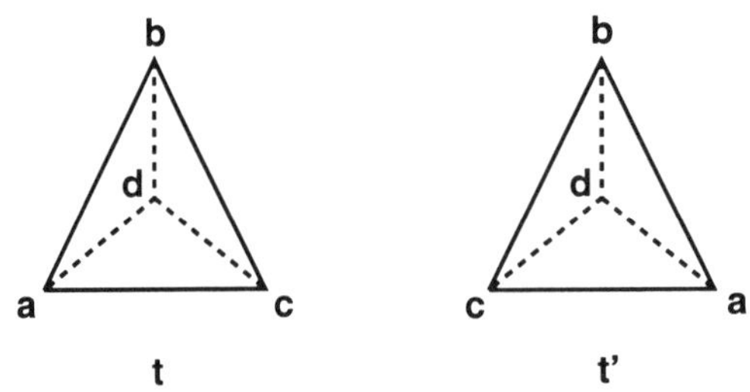

FIGURE 4. Enantiomorphic tetrahedra

Tetrahedral configurations are the most common ones encountered in organic molecules: for example, a carbon atom lies at the center of a tetrahedron defined by its four neighbor atoms; in a phosphane the relative positions of the phosphorus atom and its three neighbor atoms also define a tetrahedron, as do those of the sulfur atom and its three neighbor atoms in a sulfoxide.

We therefore illustrate the representation of atomic configurations via 2-maps with the tetrahedra t and t' (Figure 4) which are enantiomorphic if the atoms a, b, c, and d are non-equivalent. We shall see how the corresponding 2-maps allow both to distinguish t from t' and to determine their enantiomorphism relationship.

Our computer representation of stereochemistry sets out from the 3-dimensional coordinates of the atoms involved in a stereogenic entity (or the coordinates of the vertices of the corresponding template). These coordinates can be more or less precise. For example, they might have been determined experimentally or they might have been computed from a structural formula using standard bond lengths and bond angles. The first step is to generate the corresponding convex polyhedron or polygon from these coordinates. The second step is to generate the corresponding 2-map.

The convex polyhedron or polygon results from the computation of the convex hull [**12**] of the set of points specified by the 3-dimensional coordinates. In the following, we focus on the generation of the corresponding 2-map representing the relative spatial arrangement of the vertices of the convex polyhedron or polygon. Examples illustrate the comparison of stereogenic entities which are represented via 2-maps.

2.1. How to Obtain a 2-Map from a Polyhedron (and vice versa).

DEFINITION 1. 2-Dimensional map (2-map). *A 2-map is a quintuple* $M = (D, \beta_1, \beta_2, L, \lambda)$ *such that* D *is a finite nonempty set of* darts, β_1 *a permutation of* D, β_2 *an involution*[5] *of* D, L *a set of dart labels, and* λ *a surjection from* D *onto* L.

Let us consider the tetrahedron t in Figure 4. In order to represent the relative spatial arrangement of the vertices of a polyhedron, it is necessary to impose an orientation on its faces (symbolized by the arrows in Figure 5). We adopt the convention to view each polyhedron face along an axis which is perpendicular to that face and which is directed from the exterior of the polyhedron to its interior. Then, for each face the orientation is given through the circular sequence of vertices as they are encountered when traversing the boundary of the face in clockwise direction ("clockwise convention"). For example, in tetrahedron t (Figure 5) the orientation of the face containing the vertices a, b, and c is defined through the circular sequence (a, b, c).

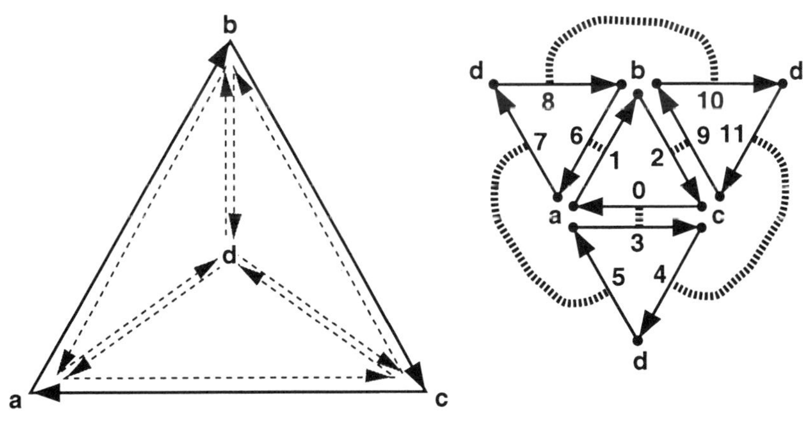

t m

FIGURE 5. Generating the 2-map m representing the relative spatial arrangement of the vertices of the tetrahedron t

[5]An involution is a permutation that is identical with its inverse.

We then cut the polyhedron along its edges, thus obtaining the set of its oriented faces (Figure 5). Each edge is thereby split up into two *darts*. A dart can be thought of as a "directed half-edge" whose starting point (symbolized by a disc) and end point (symbolized by an arrowhead) coincide with polyhedron vertices. A dart is identified by a natural number and labeled with the polyhedron vertex at its starting point. For example, vertex c is the label of darts 0, 4, and 9 in m.

The oriented polyhedron faces are represented by a permutation β_1 on the set of darts. In the graphical representation of the 2-map m in Figure 5 this permutation is symbolized by the circular dart sequences, where the end point of a dart x touches the starting point of a dart y if and only if β_1 maps x onto y. For example, in m we have $\beta_1 = \{(0,1),(1,2),(2,0),(3,4),(4,5),(5,3),(6,7),(7,8),(8,6),(9,10),(10,11),(11,9)\}$. This permutation induces the four circular dart sequences $(0,1,2)$, $(3,4,5)$, $(6,7,8)$, and $(9,10,11)$ as well as the corresponding circular label sequences (c,a,b), (a,c,d), (b,a,d), and (c,b,d), thus representing the oriented faces of the tetrahedron t.

An involution β_2 on the dart set represents how the oriented faces are assembled. In the graphical representation of the 2-map m in Figure 5 this involution is symbolized by thick dashed lines. For example, in m we have $\beta_2 = \{(0,3),(1,6),(2,9),(3,0),(4,11),(5,7),(6,1),(7,5),(8,10),(9,2),(10,8),(11,4)\}$. Two darts that are connected via β_2 correspond to a polyhedron edge. For example, in m the involution β_2 maps dart 0 onto dart 3 and vice versa, i.e., the dart set $\{0,3\}$ corresponds to the polyhedron edge ac.[6]

Note that a 2-map by itself represents only the topology of the boundary of a polyhedron, and that it does not contain any metric information (embedding of the boundary into 3-dimensional Euclidean space R^3). For a 2-map to represent the relative spatial arrangement of the vertices of a polyhedron, an embedding model into R^3 is required. The formerly mentioned "clockwise convention" and the fact that we consider only *convex* polyhedra define such an embedding model which enables us to (re)construct a polyhedron from a 2-map.

Informally, we can describe the process of (re)constructing a polyhedron from a 2-map as follows. The involution β_2 (thick dashed lines) might be viewed as a set of flexible connections between the oriented faces (circular dart sequences) of the 2-map. The oriented faces are now embedded into R^3 by shrinking these connections such that the starting point of each dart coincides with the end point of its image under β_2 and that we obtain a convex polyhedron in which the orientations of the faces satisfy the "clockwise convention". This embedding model does not specify *precise* 3-dimensional coordinates for the polyhedron vertices. However, it guarantees that the *relative* spatial arrangement of the polyhedron vertices is the same as in the polyhedron from which the 2-map has originally been generated.

Let us consider, for example, the 2-map m. If we embed the oriented face $(0,1,2)$ in the paper plane as drawn in Figure 5, then we must fold the remaining faces into the half space *below* the paper plane in order to obtain a tetrahedron having the same relative spatial arrangement of vertices a, b, c, and d as the original tetrahedron t.

[6]The correspondence between dart sets and polyhedron vertices is somewhat more complicated. For each dart d, the *orbit* of d related to the group generated by the permutation $\beta_2\beta_1^{-1}$ (composition of the inverse of β_1, followed by β_2) corresponds to a polyhedron vertex [**1**]. For example, in the 2-map m (Figure 5) the dart set $\{0,4,9\}$ represents the polyhedron vertex c.

2.2. Comparing Polyhedra via their Corresponding 2-Maps. Let V_P and V_Q be the vertex sets of two polyhedra P and Q and let $R_V \subseteq (V_P \cup V_Q) \times (V_P \cup V_Q)$ be an equivalence relation. *A priori*, we consider the vertices of a polyhedron as distinguishable from each other, that is, the default equivalence relation is the identity relation $R_V = \{(v,v) | v \in V_P \cup V_Q\}$. However, this equivalence relation may be redefined if necessary. For example, one might define two vertices as equivalent if the corresponding atoms belong to the same chemical element.

Let the 2-maps M_P and M_Q represent P and Q, respectively. Because each dart of the corresponding dart sets D_P and D_Q is labeled with a polyhedron vertex from V_P and V_Q, respectively, R_V is the natural equivalence relation on the union of the sets of dart labels.

We consider P and Q as equal if they are isomorphic, that is, if there exists a bijection $\sigma : V_P \to V_Q$ such that σ preserves the relative spatial arrangement of vertices and $\forall v \in V_P : (v, \sigma(v)) \in R_V$. Consequently, because a 2-map represents the relative spatial arrangement of the vertices of a polyhedron, P and Q are considered as equal if M_P and M_Q are isomorphic, i.e., if there exists a bijection from D_P to D_Q which respects the adjacency of darts as defined by the permutations β_1 and β_2 and which maps each dart from D_P onto an equivalently labeled dart of D_Q (according to the natural dart label equivalence relation R_V).

DEFINITION 2. Isomorphism of 2-maps. *Let* $M = (D, \beta_1, \beta_2, L, \lambda)$ *and* $M' = (D', \beta_1', \beta_2', L', \lambda')$ *be 2-maps and let* $R \subseteq (L \cup L') \times (L \cup L')$ *be an equivalence relation. A bijection* $\sigma : D \to D'$ *is an isomorphism which maps* M *onto* M' *if and only if* $\forall d \in D : \sigma(\beta_1(d)) = \beta_1'(\sigma(d)) \wedge \sigma(\beta_2(d)) = \beta_2'(\sigma(d)) \wedge (\lambda(d), \lambda'(\sigma(d))) \in R$.

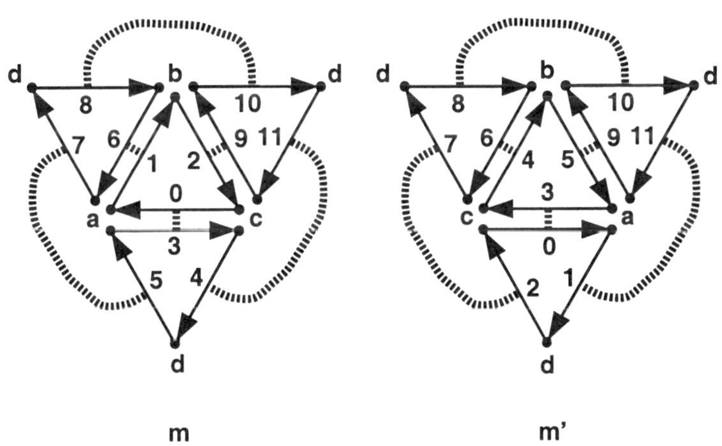

FIGURE 6. 2-maps representing the tetrahedra in Figure 4

For example, let us compare the tetrahedra t and t' (Figure 4) which are represented by the 2-maps m and m' (Figure 6). Accidentally, both polyhedra have the same vertex set $V = \{a, b, c, d\}$, and both 2-maps have the same dart set $D = \{0, 1, 2, 3, 4, 5, 6, 7, 8, 9, 10, 11\}$. Let the natural dart label equivalence relation be the identity relation $R_V = \{(v,v) | v \in V\}$. Under these conditions, there exists no isomorphism which maps m onto m', i.e., the corresponding tetrahedra t and t' are not equal (we shall see below how to determine that they are enantiomorphic).

However, if the natural dart label equivalence relation was $R_V = \{(v,v)|v \in V\} \cup \{(a,c),(c,a)\}$, the mapping $\{(0,3),(1,4),(2,5),(3,0),(4,1),(5,2),(6,6),(7,7),(8,8),(9,9),(10,10),(11,11)\}$ from the dart set D onto itself would be an isomorphism that maps m to m', i.e., the corresponding tetrahedra t and t' would be equal under these conditions.

2.3. Mirror Image of a Polyhedron. The clockwise convention relates the orientation of the faces of a polyhedron to the embedding of the polyhedron boundary in R^3 and thus determines the interior and exterior of the polyhedron. Therefore, inverting the orientation of the faces of a polyhedron amounts to turning the polyhedron boundary inside out and therefore leads to its mirror image.

DEFINITION 3. Inverse of a 2-map. *Let* $M = (D, \beta_1, \beta_2, L, \lambda)$ *be a 2-map. The inverse of* M *is* $M^{-1} = (D, \beta_1^{-1}, \beta_2, L, \lambda^{-1})$ *with* $\lambda^{-1} : D \to L, \lambda^{-1}(d) = \lambda(\beta_1(d))$.[7]

From this definition and the procedure for the (re)construction of polyhedra from 2-maps, it should be clear that a 2-map M^{-1} represents the mirror image of the polyhedron represented by the 2-map M. For example, *im* (Figure 7) is the inverse of m (Figure 6) and therefore represents the mirror image t' of t (Figure 4).

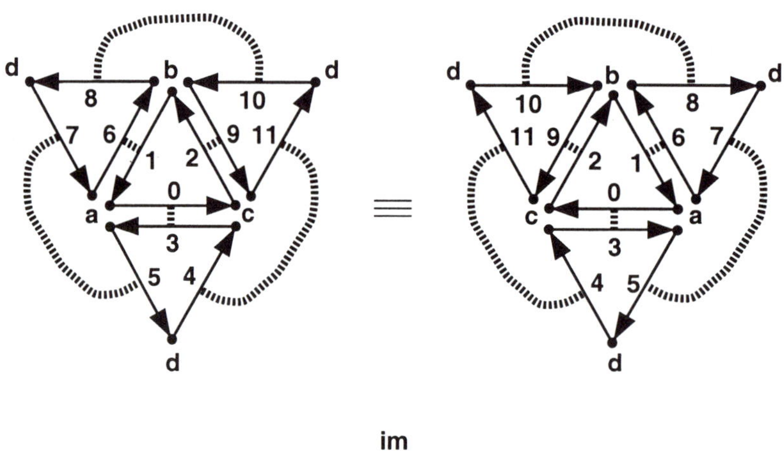

FIGURE 7. *im* (the inverse of m) represents t' (the mirror image of t)

2.4. Enantiomorphism of Polyhedra. Two polyhedra P and Q are enantiomorphic if P and its mirror image are not equal (that is, P is chiral) and if Q and the mirror image of P are equal.

In terms of 2-maps the above reads as follows. Let the 2-maps M_P and M_Q represent the polyhedra P and Q, respectively. P and Q are enantiomorphic if and only if there does not exist an isomorphism which maps M_P^{-1} onto M_P (i.e., P is chiral) and if there exists an isomorphism which maps M_P^{-1} onto M_Q (i.e., the mirror image of P and Q are equal).

Let us consider, e.g., the tetrahedra t and t' (Figure 4) and the corresponding 2-maps m and m' (Figure 6). The inverse of m is *im* (Figure 7). Accidentally, both polyhedra have the same vertex set $V = \{a,b,c,d\}$, and the three 2-maps have the

[7]It is easy to prove that M^{-1} is a 2 map, and that $(M^{-1})^{-1} = M$.

same dart set $D = \{0, 1, 2, 3, 4, 5, 6, 7, 8, 9, 10, 11\}$. Let the vertex equivalence relation and the natural dart label equivalence relation be the identity relation $R_V = \{(v, v)|v \in V\}$. Under these conditions, there exists no isomorphism mapping m onto im, i.e., t is chiral. However, the bijection $\{(0,3), (1,5), (2,4), (3,0), (4,2), (5,1), (6,9), (7,11), (8,10), (9,6), (10,8), (11,7)\}$ from D onto itself is an isomorphism which maps im onto m', i.e., the mirror image of t (represented by im) and t' are equal. Therefore, t and t' are enantiomorphic.

2.5. Extending the Scope of 2-Maps. The 2-maps as they have been described so far do not completely specify the relative spatial arrangement of the vertices of a polyhedron or a polygon. They only take into account the adjacency of the vertices as well as the orientation of the polyhedron faces.

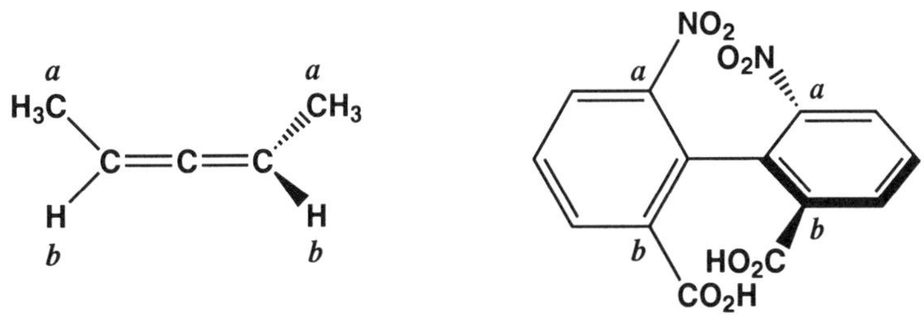

FIGURE 8. Irregular tetrahedral stereogenic entities

However, this is not sufficient to adequately represent, for example, the stereogenic entities encountered in allenes or biphenyls (Figure 8), where we consider as (constitutionally) equivalent the two atoms labeled a and those labeled b, respectively. These four atoms do form a stereogenic entity, because the tetrahedron whose vertices they occupy is an *irregular* one.

We therefore must be able to distinguish regular polyhedra (or polygons) from irregular ones. Consequently, it is necessary for the 2-maps to also take into account the *relative length* of the edges and the *relative size* of the angles of a polyhedron.

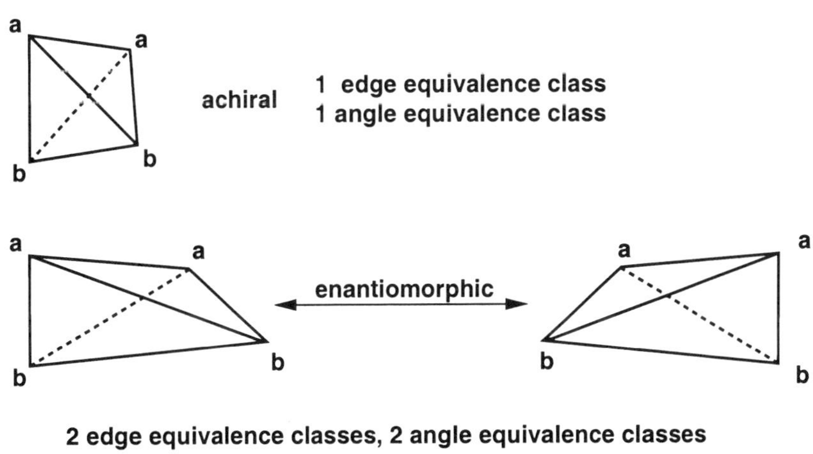

FIGURE 9. Tetrahedra which should be distinguishable

For example, we must be able to distinguish the regular tetrahedron in Figure 9 from the irregular ones, and to determine that the regular tetrahedron is achiral whereas the irregular ones are enantiomorphic.[8]

In the regular tetrahedron, all edges have the same relative length and all angles the same relative size. So, we have only one edge equivalence class and one angle equivalence class. In contrast to this, in the case of the irregular tetrahedra, we have two edge equivalence classes: one containing four long edges and another one containing two short edges. There are also two angle equivalence classes: one containing eight large angles and another one containing four small angles.

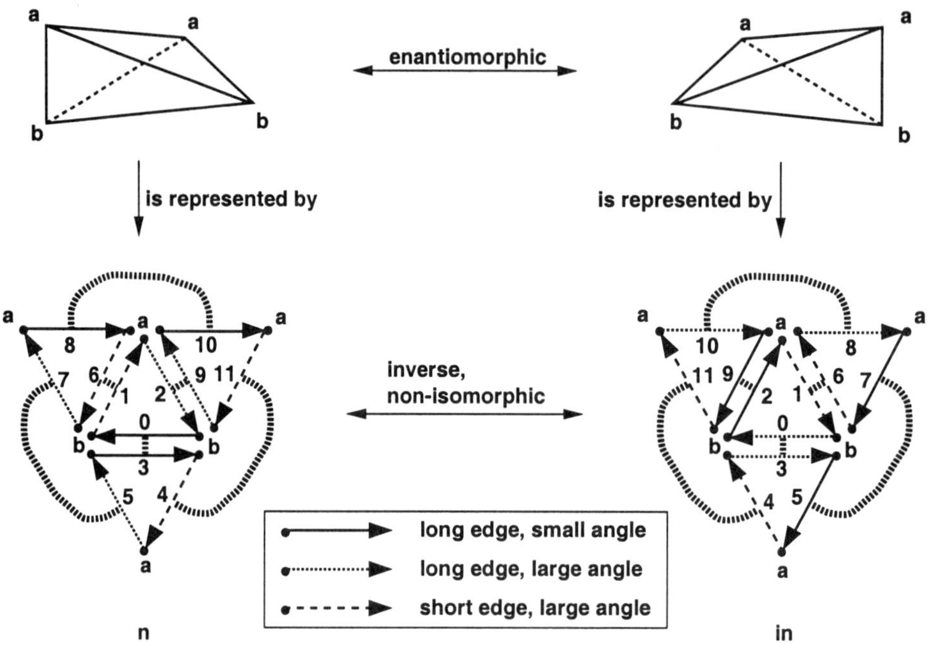

FIGURE 10. Dart labels including information about edge and angle equivalence classes allow to distinguish irregular tetrahedra from regular ones as well as to determine the enantiomorphism relationship between two irregular tetrahedra

This additional information is included into the dart labels of our 2-maps.[9] In Figure 10, this is visualized by using different line styles for the darts. For example, dart 0 in map n is labeled with the vertex equivalence class b, and with the *long edge* and *small angle*[10] equivalence classes (as indicated by the *solid* line).

The 2-map corresponding to the regular tetrahedron is not shown in Figure 10. It would have only a single dart line style, thus being clearly distinguishable from the 2-maps n and in where the different line styles reflect the irregularity of the tetrahedra on top of Figure 10.

[8]Note that, in order to simplify the presentation, the lowercase letters do not identify the polyhedron *vertices* but rather the *equivalence classes* to which these belong.

[9]Without this information, a 2-map would always represent the convex polyhedron of the highest possible symmetry.

[10]The label of a dart d contains the angle equivalence class of the interior angle φ of the polygonal face, where φ is formed by darts d and $\beta_1^{-1}(d)$.

Moreover, the extended dart labels allow to represent the chirality of these tetrahedra. The 2-map *in* is the inverse of the 2-map *n*. Thus, these 2-maps represent tetrahedra which are mirror images of each other. In *n*, the "solid dart" 0 starts from a vertex belonging to equivalence class *b*, and β_1 maps dart 0 onto dart 1 which is "widely dashed" and which also starts from a vertex belonging to equivalence class *b*. In *in*, there are two darts (2 and 5) which are compatible with dart 0 of *n*, that is, which are "solid darts" starting from a vertex belonging to equivalence class *b*. However, β_1 maps both dart 2 and dart 5 of *in* onto a dart (1 and 4, respectively) which is incompatible with dart 1 of *n* because it starts from a vertex belonging to equivalence class *a*. Clearly, these inverse maps are not isomorphic, thus representing enantiomorphic tetrahedra.

3. Conclusions

The 2-dimensional maps are a simple combinatorial model providing a uniform topology-based representation for convex polyhedral and polygonal stereogenic entities (a polygon can be considered as a degenerate polyhedron which has two faces but no interior). A 2-map describes *explicitly* the relative spatial arrangement of the atoms of a stereogenic entity, and there is no need to manually determine template-specific permutations in order to account for the symmetry of a given template. The test for map isomorphism allows to compare stereogenic entities, whether they are included in full structures or in substructures ("wildcard atoms" may be used to indicate the "free sites" of a stereogenic entity, if necessary). Moreover, the 2-map representations allow to construct the mirror image of a polyhedron, and to determine stereochemical properties and relationships such as chirality and enantiomorphism.

The scope of our representation is easily extended. The relative spatial arrangement of any set of points in R^3 (stereogenic entities, molecular shapes, etc.) can be represented by a set of 2-maps corresponding to a set of nested convex polyhedra. Non-orientable molecular surfaces (e.g., a molecular Möbius strip [**13**]) can be represented using generalized n-dimensional maps [**1, 10**].

References

[1] P. Lienhardt, N-Dimensional Generalized Combinatorial Maps and Cellular Quasi-Manifolds, Tech. report, Centre de Recherche en Informatique Université Louis Pasteur, Strasbourg, April 1993.

[2] P. Willett, Algorithms for the Calculation of Similarity in Chemical Structure Databases, in M.A. Johnson and G.M. Maggiora, Eds., *Concepts and Applications of Molecular Similarity*, Wiley, New York, 1990, ch. 3, pp. 43–63.

[3] A. Dietz, Yet Another Representation of Molecular Structure, *J. Chem. Inf. Comput. Sci.* **35** (1995), 787–802.

[4] R.S. Cahn, C.K. Ingold, and V. Prelog, Specification of Molecular Chirality, *Angew. Chem. Int. Ed. Engl.* **5** (1966), 385–415.

[5] A.E. Petrarca, M.F. Lynch, and J.E. Rush, A Method for Generating Unique Computer Structural Representations of Stereoisomers, *J. Chem. Doc.* **7** (1967), 154–165.

[6] A.E. Petrarca and J.E. Rush, Methods for Computer Generation of Unique Configurational Descriptors for Stereoisomeric Square-Planar and Octahedral Complexes, *J. Chem. Doc.* **9** (1969), 32–37.

[7] I. Ugi, D. Marquarding, H. Klusacek, G. Gokel, and P. Gillespie, Chemistry and Logical Structures, *Angew. Chem. Int. Ed. Engl.* **9** (1970), 703–730.

[8] I. Ugi, J. Dugundji, R. Kopp, and D. Marquarding, Perspectives in Theoretical Stereochemistry, Lecture Notes in Chemistry, no. 36, Springer-Verlag, Berlin, 1984.

[9] M.F. Brown, B.R. Cook, and T.E. Sloan, Stereochemical Notation in Coordination Chemistry. Mononuclear Complexes, *Inorg. Chem.* **14** (1975), 1273–1278.
[10] P. Lienhardt, Topological Models for Boundary Representation: A Survey, *Comput. Aided Design* **23** (1991), 59–81.
[11] R. Cori, Un code pour les graphes planaires et ses applications, *Astérisque* (1975), no. 27.
[12] F.P. Preparata and M.I. Shamos, Computational Geometry: An Introduction, Springer-Verlag, New York, 1985.
[13] D. M. Walba, Topological Stereochemistry, *Tetrahedron* **41** (1985), 3161–3212.

GDR 1093 CNRS, UMR 5506 CNRS/Université Montpellier II — LIRMM, 161 rue Ada, 34092 Montpellier Cedex 5, France
E-mail address: dietz@lirmm.fr

UMR 5506 CNRS/Université Montpellier II — LIRMM, 161 rue Ada, 34092 Montpellier Cedex 5, France
E-mail address: fiorio@lirmm.fr

GDR 1093 CNRS, UMR 5506 CNRS/Université Montpellier II — LIRMM, 161 rue Ada, 34092 Montpellier Cedex 5, France
E-mail address: habib@lirmm.fr

GDR 1093 CNRS, UPR 9023 CNRS — CCIPE, 141 rue de la Cardonille, 34094 Montpellier Cedex 5, France
E-mail address: cl@lirmm.fr

DIMACS Series in Discrete Mathematics
and Theoretical Computer Science
Volume 51, 2000

Geometric Symmetry and Chemical Equivalence

B.T. Fan, A. Panaye, J.H. Yao, S.G. Yuan, and J.P. Doucet

ABSTRACT. Symmetry is an essential concept in chemical graphs. But, generally, symmetry alone does not define either physical or chemical properties. We need often a vagueness and fuzziness of the broader interpretation of the symmetry. The absolute measure of molecular environment implicit in ^{13}C-NMR spectroscopic data is used in this paper to establish the magnitudes of topological, topographical and chemical equivalencies.

1. Introduction

Point symmetry and site equivalence are basic concepts in structural chemistry [**1, 2**]. Thus, the point symmetry of a molecular structure can be used to simplify many aspects of chemical information technology [**3, 4**], including especially molecular graph handling, analysis of physical properties, spectral simulation and structural elucidation, as well as management of chemical databases and computer-aided organic synthesis design.

Conventionally, molecular symmetry is distinguished, for a particular structure, by the assigning of the appropriate point groups, comprised of particular symmetry elements. Recently, Avnir *et al.* [**5, 6**] have suggested that descriptions of structure, based on the concept of continuous symmetry, may be particularly useful, in applications involving different states of a molecule, for instance, transition states, dynamic properties or conformational changes in large systems.

Mezey *et al.* have proposed, based on the observations of chemical and physical phenomena, a continuous extension of the discrete point symmetry concept to describe the approximate and quasi-symmetric systems, by using fuzzy-set theory [**7, 8**]. The "syntopy" fuzzy subsets generated from a defined membership function can be used to help to rationalize the statics or the dynamics of perturbed or intermediate quasi-symmetric systems.

Various approaches to the identification of chirality as a structural property have been suggested [**9**]–[**12**] based on the detection of chiral center and qualitative determinations of atom chirality. There is too, the need to distinguish the difference in behavior of sites, which look very similar but which exhibit non-equivalent environments (Figure 1). Thus, for example, such non-equivalence determines the

1991 *Mathematics Subject Classification.* 54-06.

This work was supported by Ministry of Foreign Affairs of France (PRA M95-01).

number of peaks observed in the ^{13}C NMR spectrum and the associated fine structure splitting.

Our purpose herein, is to examine the extent to which 2D-chemical graph theory can be used to distinguish such facets of chirality and equivalence. In particular, we are interested to evaluate whether 2D-theory alone, is sufficient to account for non-equivalence in three-dimensional structures.

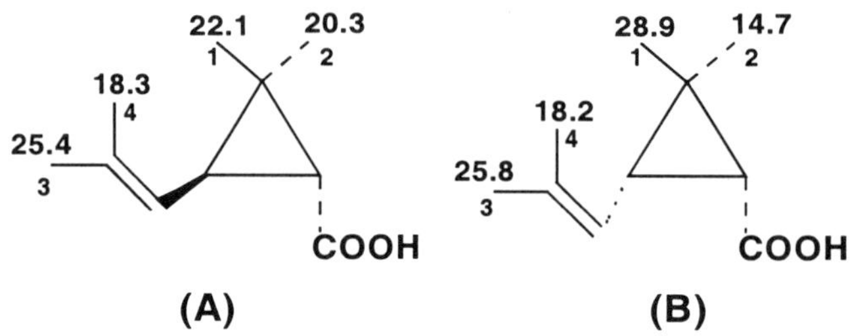

FIGURE 1. Influence of molecular configuration on NMR-13C chemical shift.

Numerous models are to be found in the literature concerning the evaluation of ^{13}C-NMR chemical shifts as a 2D problem [**13**]–[**16**] and there have been some attempts to develop 3D models [**17**]–[**19**] but, these latter are considerably more complex.

2. Site Equivalence in 2D and 3D

While the detection of equivalence, even as a 2D-problem can be difficult, generally it is straightforward if the analysis is based on automorphism, although, for many structures, this procedure is very time-consuming. As a development of other work [**20**]–[**22**], our MOST/ MINST (Maximum Overlapping Spanning Tree/ Minimum Non Overlapping Spanning Tree) procedure [**23**] has proved to be a very efficient procedure for the detection and classification of constitutionally equivalent sites from a topological connection table. The program successfully passed all the test molecules previously examined. The one failure reported to date by Faulon [**24**] involves a high symmetrical graph unlikely to be relevant in chemical application. However, for such highly symmetrical graphs where all the nodes have the same nature and connectivity but belong to cycles of different size, the problem can be easily solved by adding a simple test which can distinguish the sites linked to rings of different size [**25**]. It may be also possible to introduce a preliminary step of cycle detection using, for example, the algorithm, we proposed previously, for the determination of the Smallest Set of Smallest Rings (SSSR) [**26**].

Clearly, two carbon sites deemed non-equivalent in 2D, coincidentally, could exhibit distinct ^{13}C resonance. But, non-equivalence, alone, in a 3D analysis may not be a sufficient criterion to ensure that there are distinct peaks in the spectrum. Besides structural non- equivalence, the roles of conformational inter-conversions and their relative rates may need to be considered[1].

---

[1]Note that for ^1H-NMR spectra, the influence of coupling constants, e.g. in AA'BB' systems, is likely to lead to the need for a more complex analysis

For the cases considered below, our analyses have been carried out for one choice of conformation of each molecule. The choices are not based critically on complex energetic considerations, and so particular geometries considered may not be the most stable. However, some elementary rules can be constructed to cover common features e.g. local neighborhood of stereocenters: local asymmetry, occurrence of double bond, etc.

2.1. Detection of 3D Non Equivalence: Consider two sites deemed to be 2D equivalent and proceed as follows. First, determine their distance to the barycenter of the atomic positions of the structure. For faster processing, hydrogen atom positions are ignored in this calculation. If the distances to the barycenter are different, then the two sites are non equivalent in this structure. If the sites are equi-distant, or nearly equi-distant to the barycenter, if necessary refinements are possible by using dihedral angle calculations involving the positions of neighboring atoms. For geminal sites, a cutoff, also, can be introduced considering only atoms within a sphere of a given radius (usually 5-6 Å to incorporate proximate neighbors or chain folding).

For example, consider the structures A and B in Figure 1 and the distances, relative to the methyl groups, given as below (Schema 1):

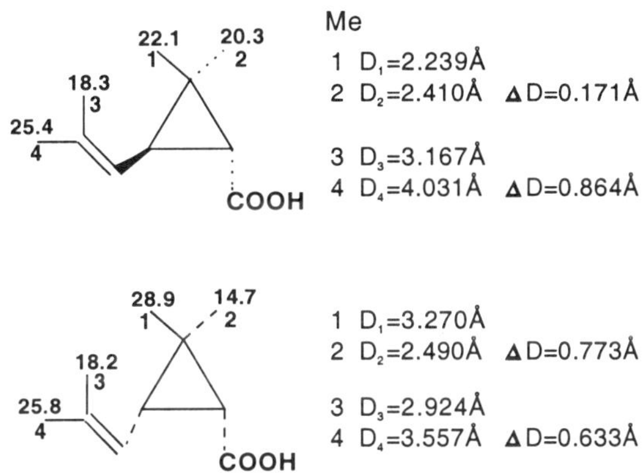

Schema 1.

It is attractive to associate differences in distance to differences in behavior. But the problem is much more complex. In fact, the dissimilarity in behavior is monitored by the dissimilarity of ***local*** environment. Consider, therefore, the main factors upon which the empirical rules can be established.

The properties of *gem*-dimethyl groups appear quite sensitive to structural details and, in particular, to their local environments. Elementary valence theory would suggest the *gem* methyls are practically at the same distances from the atoms at one or two bonds apart (A and B positions in the Schema 2) and so to be in similar local environments. While dissimilarity in their environment is to be expected for atoms at three or four bonds apart (C and D positions in the Schema 2).

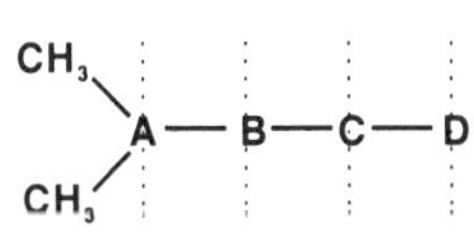

Schema 2.

The easiest rules to define arise for rigid moieties e.g. involving a double bond, or a ring structure such as is found in cyclopropyl systems. Based on an examination of literature results and subject to the caveat that caution may be required, we suggest the following rules.

Since we cannot consider every element independently, all hetero-atoms are gathered in a unique class (Z). So, we distinguish only 3 categories of substituents (H, C, Z). This is satisfactory for common functional groups (involving C, N, and O). But discrepancies may appear for very large atoms (Br, I) or groups where magnetic anisotropy is important (C≡C; C≡N). We distinguish methyl groups for sp^2 or sp^3 cases.

3. Empirical Rules for *Gem* Methyl Groups Adjacent to an sp^2 Carbon Site

The most common case corresponds to methyl groups adjacent to a carbon-carbon or carbon-nitrogen double bond. Such structural features are rigid and planar. A similar situation is found for methyls on a nitrogen alpha to a double bond carbon-oxygen or carbon-sulfur, since nitrogen adopts a sp^2 hybridization, due to conjugation.

3.1. sp^2 Carbon. For the typical structure in the diagram, the chemical shifts observed depend significantly on the natures of X and Y atoms on the neighboring site. We distinguish

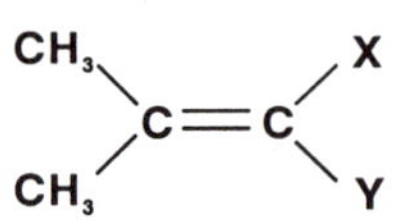

Schema 3.

Hydrogen H, Carbon C and hetero-atom Z.

The different couples X, Y which are observed lead to the following values for the difference in the ^{13}C chemical shifts of the two methyl groups.

	H, C	H, Z	Z, C	C, C
$\Delta\delta$	8 ppm	4.5 ppm	0.5 ppm	1 ppm

Indicated values correspond to estimated mean values on about twenty samples of double bonds extracted from ref. [**27**]. While we have not found an example of the Z, Z case, a shift magnitude of about 1 ppm seems appropriate. For the C, C case, the exact value seems to depend on the similarity of the two groups. Similarly, for the carbon- nitrogen double bond,

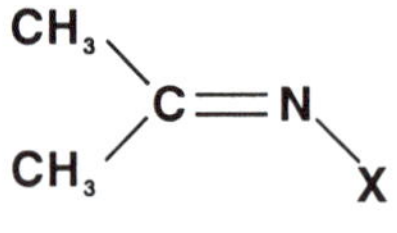

Schema 4.

If the nitrogen adjacent atom X is carbon, we estimate a $\Delta\delta$ value of 11 ppm. But $\Delta\delta$ is only about 7 ppm for $X = O$ or N.

3.2. sp2 Nitrogen. For the structure fragment in the following diagram, and a nitrogen atom alpha to a carbon-oxygen or carbon-sulfur double bond, local sp^2 hybridization is appropriate because of conjugation. So, while a less rigid structure is to be expected the lowest energy conformation should be planar.

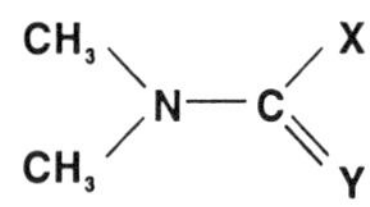

Schema 5.

Obviously, the difference in chemical shift $\Delta\delta$ depends on the nature of X and Y, and typical results are

X	$\Delta\delta_{(Y=O)}$	$\Delta\delta_{(Y=S)}$
H	5 ppm	8 ppm
C	3 ppm	2 ppm
O	0.5 ppm	

A strong correlation with the natures of adjacent atoms is evident. Atoms on more distant sites seem to be of little significance.

4. Empirical Rules for *Gem* Methyl Groups Borne by a sp^3 Atom

For local sp^3 hybridized environments the equilibrium conformation determines local geometrical structure. There are three cases, locally eclipsed, staggered or a mixture of these two extremes.

4.1. Eclipsed Bonds. Primary examples involve cyclopropyl and cyclobutyl derivatives. A three-membered ring can be only planar. Thus, each of two gem methyls always sees the same face of the triangle plane. So the main factors to be considered are the number and the location of other substituents (Figure 2).

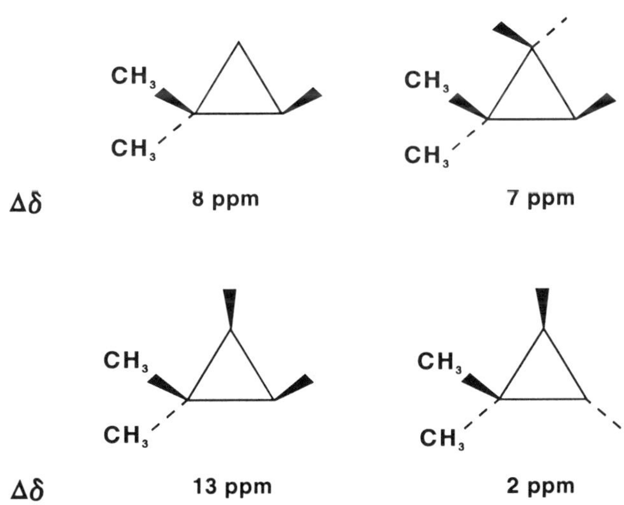

FIGURE 2. Gem methyl groups in three membered ring systems.

For one-or three-substituted rings (not counting the two methyls), and so the local environments differing only in one substituent similar chemical shifts are to be expected.

For two substituents in *cis*-geometry, chemical shifts are quite different (12 - 15 ppm), whereas for *trans*-substitution, (values are similar, since each *trans*-methyl is adjacent to one substituent, while the chemical shift difference observed (about 2 ppm) can be attributed to the dissimilarity of the substituents.

A four-membered ring is planar slightly buckled in shape, but there is a high inter-conversion rate (unless, the ring is embedded in a bridged structure). So, again, it is reasonable to consider only the case of planar geometry. However, very few data are available in the literature. The few examples found suggest differences of about 5 ppm to be typical.

Within a skeleton composed of single bonds, the dissimilarity is less dependent of the nature of atoms than in the preceding case of sp^2 atoms. On another hand, D atoms can modulate the effects.

4.2. Staggered Bonds in Acyclic Structures. One of the characteristics of such chains, is that they are very flexible and the rate of occupancy of the various spatial location depends on the atom connectivity. So, the results of topological analyses of the environments are valid. For example, isopropyl groups are found commonly in the structures of alkanes, alcohols and carboxylic acids bearing a chiral carbon, for which the two methyl chemical shifts differ by 0.2 to 7.2 ppm. For 21 such structures examined [**27**]–[**29**] a correlation between the connectivity of atoms located at two and three bonds from the methyls, i.e. C and D ranks of atoms (see Figure 3) can be established.

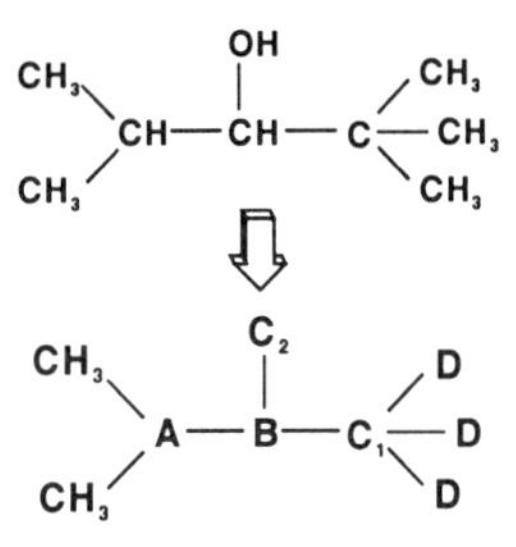

FIGURE 3. Staggered bonds in acyclic structures.

Because our survey did not include molecules wherein all the C sites are substituted, our rules are confined to structures for connectivity of sites B, C_1, C_2. Using a back-propagation neural network with a 3/2/1 architecture (3 input units, the connectivity of B, C_1, C_2; 2 units in the hidden layer and one output unit, the shift difference), cross-validation with a '*leave-one-out*' treatment leads to a mean error of 0.8 ppm on a total range of 7 ppm. This treatment accounts reasonably for the range of the shift difference found, and may be possibly introduced into ^{13}C spectral simulation system for improved accuracy.

4.3. Staggered Bonds in Cyclic Structures. Data for cyclohexyl based structures dominate for these cases, which are important since chemical shift differences of about 15 ppm are not uncommon. This large range for δ chemical shift differences originates in the diversity of the possible conformations depending on the structure of the systems: chair, boat, half-chair and inter-conversion. The most common cases correspond either to an isolated cycle or a polycyclic system (ring junction or bridged structure).

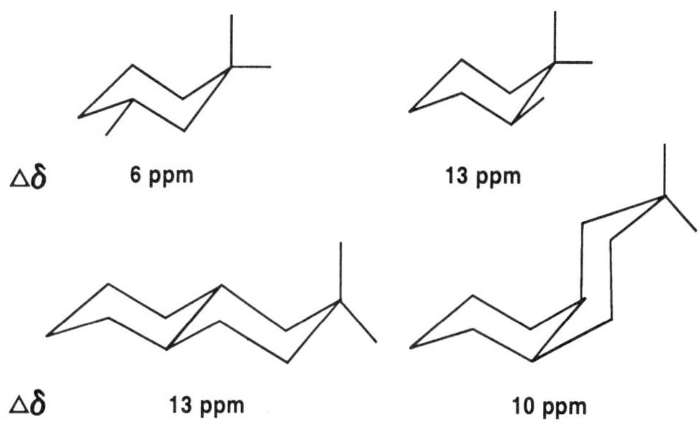

FIGURE 4. Staggered bonds in cyclic systems.

Based on the structures in Figure 4 appropriate rules seem to be for an isolated ring:

- if the cycle is substituted in alpha position, the shift difference is about 13 ppm;
- if the cycle is substituted in beta position, the difference decrease to 6 ppm.

for a polycyclic system :

- if the cycle junction is *trans*, the difference is often close to 13 ppm, (this value is nearly independent of the location of the gem di-methyl groups by respect to this junction).
- if the cycle junction is *cis*, where flexibility is greater, the difference decreases to near 10 ppm,

However, if the cycle geometry is distorted because of the presence of double bonds, either internal or external and so is nearly planar, the difference is smaller; about 4–5 ppm, although in case of rigid molecules or very disymmetrical faces, values up to 8–9 ppm have been observed (see Figure 5).

A seven-membered or larger rings is more flexible than a six-membered one. For less rigid structures, the chemical shift difference is smaller. What few examples were found suggest that the difference range is about one half of that for cyclohexyl analogs.

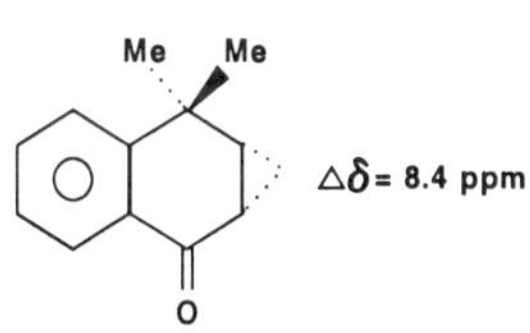

FIGURE 5. Distorted cyclic system by the presence of double bonds.

Structures involving the C and D neighbor types give rise to the maximum dissimilarity of behavior. Since conformational exchanges play a so important role, this maximum value must be weighted around the preceding observed values, which are related to the conformational flexibility.

4.4. Mixed Situations. For structures involving cyclopentyl components, many possibilities can be found with, on the one hand, largely flexible systems and, on the other, frozen rigid structures, involving different bonding modes. In the foregoing, the locations only of carbons and hetero-atoms was found to be sufficient to determine the positions of hydrogen atom and so their individual roles were not separated. But for five-membered rings, it would seem that the roles of all component atoms have to be included.

5. Conclusion

It has been shown that symmetry in chemistry is not always an absolute concept, using the example of chemical shift differences as a measure of local environmental effects. Thus, while 2D site equivalence can reflect symmetrical characteristics for many molecular structures, the consideration of only 2D symmetry leads often to an incorrect interpretation of chemical and physical properties. The concept of relative symmetry, which is geometrically interpreted, has been proposed. This geometrical symmetry can be correlated, in certain cases, with chemical equivalence in 2D. But it is often distinct from 2D equivalence in most cases. The influence on ^{13}C-NMR chemical shift is not only dependent upon the molecular geometrical information, and also upon the nature of substituents situated differently in spatial positions. Thus, attempts to establish a relationship between the difference in chemical properties, for instance the ^{13}C-NMR chemical shift, and the geometrical measurements (e.g. the spatial distance) require a more detailed interpretation. In this paper, we have examined the different cases, and propose a series of empirical rules for methyl-substituted compounds. Such empirical rules can be easily handled in ^{13}C-NMR spectral simulation systems. The combination of 2D equivalence with geometrical symmetry and the nature of atoms involved in local geometrical dissimilarity provides a more general perspective of chemical equivalence, and this can be applied to evaluate the properties of molecules such as ^{13}C-NMR chemical shifts.

References

[1] *Symmetries in Science. III*; Gruber B., Tachello I. Eds., Plenum Press, New York, 1988 and earlier volumes in that series.

[2] P.W. Atkins, *Physical Chemistry*, 3rd ed.;. Oxford University Press: Oxford, 1986, p406.

[3] P.G. Mezey, A Global Approach to Molecular Symmetry: Theorems on Symmetry Relations Between Ground- and Excited-State Configurations. *J. Am. Chem. Soc.*, **112** (1990) 3791–3802.

[4] P.G. Mezey, On the Allowed Symmetries of all Distorted Forms of Conformers, Molecules, and Transition Structures. *Can. J. Chem.*, **70** (1992) 343–347.

[5] H. Zabrodsky, S. Peleg, and D. Avnir, Continuous Symmetry Measures. *J. Am. Chem. Soc.*, **114** (1992) 7843–7851.

[6] 6. H. Zabrodsky, S. Peleg, D. Avnir, Continuous Symmetry Measures. 2. Symmetry Groups and the Tetrahedron. *J. Am. Chem. Soc.*, **115** (1993) 8278–8289.

[7] P.G. Mezey, and J. Maruani, The Concept of 'Syntopy', a Continuous Extension of the Symmetry Concept for Quasi-Symmetric Structures Using Fuzzy-Set Theory. *Mol. Phys.*, **69**(1) (1990) 97–113.

[8] J. Maruani, and P.G. Mezey, From Symmetry to Syntopy: A Fuzzy-Set Approach to Quasi-Symmetric Systems. *J. Chim. Phys.*, **87** (1990) 1025–1047.

[9] V.E. Kuz'min, I.B. Srel'makh, I.V. Yudanova, D.V. Pozigun, and M.B. Bekker, Quantitative Aspects of Chirality. II. Analysis of Dissymmetry Function Behaviour with Different Changes in the Structure of the Model Systems. *J. Phys. Org. Chem.*, **5** (1992) 299–307.

[10] H. Zabrodsky, S. Peleg, and D. Avnir, Continuous Symmetry Measures. 4. Chirality. *J. Am. Chem. Soc.*, **117** (1995) 462–473.

[11] J. Xu, GMA: A Generic Match Algorithm for Structural Homomorphism, Isomorphism, and Maximal Common Substructure Match and its Applications. *J. Chem. Inf. Comput. Sci.*, **36** (1996) 25–34.

[12] G. Moreau, Atomic Chirality, a Quantitative Measure of the Chirality of the Enviroment of an Atom. *J.Chem. Inf. Comput. Sci.*, **37** (1997) 929–938.

[13] J-E. Dubois, and M. Carabedian, Modeling of Alkyl Environment Effects on the ^{13}C Chemical Shift. *Org. Magn. Reson.*, **14** (1980) 264–271.

[14] A. Furst, and E. Pretsch, A Computer Program for the Prediction of ^{13}C-NMR Chemical Shifts of Organic Compounds. *Anal. Chim. Acta*, **229** (1990) 17–25.

[15] H.N. Cheng, and S.J. Ellingsen, Carbon-13 Nuclear Magnetic Resonance Spectral Interpretation by a Computerized Substituent Chemical Shift Method. *J. Chem. Inf. Comput. Sci.*, **23** (1983) 197–203.

[16] A. Panaye, J-P. Doucet, and B.T. Fan, Topological Approach of ^{13}C NMR Spectral Simulation : Application to Fuzzy Substructures. *J. Chem. Inf. Comput. Sci.*, **33** (1993) 258–265.

[17] H. Beierbeck, and J.K. Saunders, Conformational and Conformational Analysis of Hydrocarbons Chains Based on Time-Averaged Carbon-13 Chemical Shifts. *Can. J.. Chem.*, **55** (1977) 771–781.

[18] H. Beierbeck, and J.K. Saunders, Analysis of ^{13}C Nuclear Magnetic Resonance Chemical Shifts of Acyclic Hydrocarbons. *Can. J.. Chem.*, **58** (1980) 1258–1265.

[19] J.E. Dubois, J.P. Doucet, and A. Panaye, Similarités de comportement et formes locales. La topologie reflet de la topographie. *Bull. Soc. Chim. Belg.*, **98** (1989) 31–43.

[20] a — K. Varmuza, and H. Scsibrany, "Constitutionally Equivalent Atoms and Bonds in a Molecule: A Complete but Simple and Fast Solution". First European Conference on Computational Chemistry. May 23-27, 1994, Nancy, France.
b — Varmuza, K., and Scsibrany, H., "Determination of the Automorphism Group of Graphs by a Substructure Search Approach". ICCCRE, Jully 17-21, 1995, Paris, France.

[21] C. Jochum, and J. Gasteiger, Canonical Numbering and Constitutional Symmetry. *J. Chem. Inf. Comput. Sci.*, **17** (1977) 113–117.

[22] G. Rucker, and C. Rucker, On Using the Adjacency Matrix Power Method for Perception of Symmetry and for Isomorphism Testing of Highly Intricate Graphs. *J. Chem. Inf. Comput. Sci.*, **31** (1991) 123–126.

[23] B.T. Fan, A. Barbu, A. Panaye, and J.P. Doucet, Detection of Constitutionally Equivalent Sites from a Connection Table. *J. Chem. Inf. Comput. Sci.*, **36** (1996) 654–659.

[24] J.L. Faulon, Isomorphism, Automorphism Partitioning, and Canonical Labeling can be Solved in Polynomial-Time for Molecular Graphs. *J. Chem. Inf. Comput. Sci.*, **38** (1998) 432–444.

[25] B.T. Fan, A. Panaye, and J.P. Doucet, Comment on Isomorphism, Automorphism Partitioning and Canonical Labeling can be Solved in Polynomial Time for Molecular Graphs. *J. Chem. Inf. Comput. Sci.*, in press.

[26] B.T. Fan, A. Panaye, J.P. Doucet, and A. Barbu, Ring Perception: A New Algorithm for Finding Directly the SSSR from Connection Table. *J. Chem. Inf. Comput. Sci.*, **33** (1993) 657–662.

[27] H.O. Kalinowski, S. Berger, and S. Braun, *Carbon-13 NMR Spectroscopy.* Jon Wiley Ed., New York, 1988.

[28] L.P. Lindeman, and J.Q. Adams, Carbon 13 Nuclear Magnetic Resonance Spectrometry. *Anal. Chem.*, **43** (1971) 1245–1252.

[29] D.Z. Jiang, Effets de groupes alkyles en RMN ^{13}C d'acides alkylcarboxyliques encombrés. PhD thesis, Université Paris 7, Paris, 1982.

Université Paris 7, ITODYS-CNRS UPRES-A 7086, 1 rue Guy de la Brosse, 75005 Paris, France
E-mail address: fan@paris7.jussieu.fr

Université Paris 7, ITODYS-CNRS UPRES-A 7086, 1 rue Guy de la Brosse, 75005 Paris, France
E-mail address: panaye@paris7.jussieu.fr

Université Paris 7, ITODYS-CNRS UPRES-A 7086, 1 rue Guy de la Brosse, 75005 Paris, France, and Institute of Organic Chemistry of Shanghai, 354 Fenglin road, Shanghai, China
E-mail address: yao@paris7.jussieu.fr

Institute of Organic Chemistry of Shanghai, 354 Fenglin road, Shanghai, China
E-mail address: yuansg@pub.sioc.ac.cn

Université Paris 7, ITODYS-CNRS UPRES-A 7086, 1 rue Guy de la Brosse, 75005 Paris, France
E-mail address: doucet@paris7.jussieu.fr

DIMACS Series in Discrete Mathematics
and Theoretical Computer Science
Volume 51, 2000

(3,6)-Cages, Hexagonal Toroidal Cages, and their Spectra

Patrick W. Fowler, Peter E. John, and Horst Sachs

In memoriam François Jaeger

ABSTRACT. A (p,q)-*cage* is the graph of a simple polyhedron that has only p-gons and q-gons as its faces. A *hexagonal toroidal cage* (briefly, *ht-cage*) is the graph of a trivalent hexagonal tessellation of the torus.

The graphs of the molecules of simple carbon cages (fullerenes) and of (the hypothetical) toroidal graphitoids are $(5,6)$-cages and ht-cages, respectively. There are relations between $(5,6)$-cages and $(i,6)$-cages, $i \neq 5$.

In this paper, interrelations between various types of (generalized) $(3,6)$-cages, between these graphs and ht-cages, and between their spectra are discussed. In particular, a conjecture of the first author concerning "leapfrog" transforms of $(3,6)$-cages and their spectra is proved.

1. Preliminaries

Notation.

B, C, G, H:	graphs
V, E, L, ...:	elements of a graph (as a vertex, an edge, a loop,...)
$i, j, k, \ldots; 1, 2, 3, \ldots;$	
$x_1, x_2, x_3, \ldots$:	vertices of a (labelled) graph
$v(\mathrm{G})$, $h(\mathrm{G})$, $f_i(\mathrm{G})$:	number of vertices, of half edges, and of i-gons of graph G
$\mathbb{V}$, $\mathbb{B}$, $\mathbb{W}$:	sets of vertices
$\mathfrak{C}$, $\mathfrak{H}$:	sets of graphs
$\mathbf{A} = (a_{ij})$:	adjacency matrix of a graph G
$\mathbf{u}$, $\mathbf{v}$:	eigenvectors of $\mathbf{A}$
$f_{\mathrm{G}}(x) = \lvert x\mathbf{I} - \mathbf{A}\rvert$:	characteristic polynomial of graph G
specG:	spectrum of a graph G (family of zeros of $f_{\mathrm{G}}(x)$)
D, L:	graph operators

*

1991 *Mathematics Subject Classification.* Primary 05C10, 05C50, 52A25; Secondary 05C75, 81G55.

This work was supported by DAAD/ARC 313-ARC-XII-98/42.

A *cage* is defined as (the system of vertices, edges and faces of) a simple polyhedron; in order to enable certain degenerate cases to be included, we shall use the word *pseudocage* (briefly, *p*cage) for any finite connected cubic (i.e., trivalent) plane graph; in particular, a *p*cage may have digons, monogons and "zerogons" (see below) among its faces (a face is considered an open region). A $(q_1, q_2, \dots, q_k)$-cage is a *p*cage each face of which is a q_1-gon, q_2-gon, ..., or q_k-gon.

A *loop* is an edge whose end vertices coincide. We distinguish between *heavy loops* of weight 1 and *light loops* of weight $\frac{1}{2}$. Let L be a loop attached to vertex i of *p*cage C. Then the contribution of L to the valency of i and, simultaneously, to the entry a_{ii} in the adjacency matrix $\mathbf{A} = (a_{ij})$ of C is 1 if L is a light loop and 2 if L is a heavy loop. Every light loop is assumed to form the full boundary of some face (namely, a zerogon: see below).

Drawing figures it is convenient to suppress the interior Z of a light loop L and identify the pair (L, Z) with a *half edge* H: see Figure 1.1 where L_h is a heavy loop, L_l is a light loop, and the end point Π of H is a point of H but not a vertex.

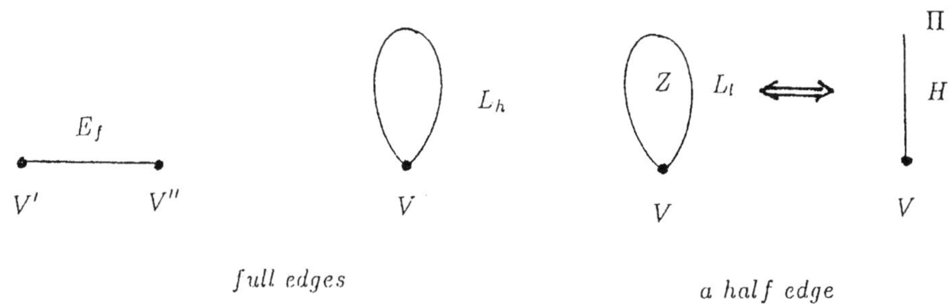

Figure 1.1

We may assume that all edges (full or half) are realized by smooth arcs. Consider Figure 1.2 where curves $c(E, F)$ (E standing for E_f, H, L_h) are also smooth arcs, close enough to E, such that their interiors lie entirely in the face F and that the interiors of $c(E', F)$ and $c(E'', F)$ intersect (in precisely two points if F is a digon and in precisely one point otherwise) if and only if E' and E'' are consecutive on the boundary of F. Then the union of E (without its end vertex or end vertices) and the interior of the region $R(E, F)$ is a realization of "the F-*bank* of E" if E separates two distinct faces or is a half edge, and of "one of the two F-banks of E" otherwise.

Each full edge has two distinct banks, a half edge (or light loop) has only one bank (Fig. 1.2). A face is a *q-gon* iff it contains the interiors of precisely q banks in its interior; it is a *monogon* iff its boundary is a heavy loop, and it is a *zerogon* iff its boundary is a light loop.

Banks B', B'' of a *p*cage C are considered *neighbouring with multiplicity* μ iff their intersection consists of precisely μ components; they are *neighbouring* iff $\mu > 0$, i.e.,

- if they are the two banks of a full edge; in this case, $\mu = 1$

or

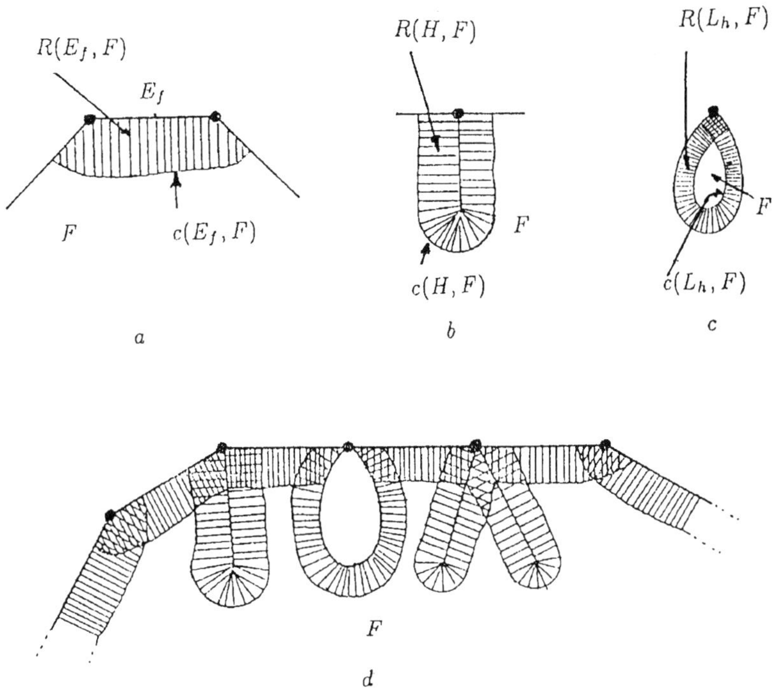

Figure 1.2

- if they belong to edges consecutive on the boundary of some face F and their interiors lie in F; in this case, $\mu = 2$ if F is a digon and $\mu = 1$ otherwise.

A bank neighbours itself if and only if its interior lies in a monogon; in this case, $\mu = 1$ (see Fig 1.2c).

Let $v = v(\mathrm{C})$, $f_i = f_i(\mathrm{C})$ and $h = h(\mathrm{C})$ denote the number of vertices, the number of i-gons and the number of half edges (light loops) of $p$cage C, respectively; note that $h = f_0$ and that $v+h$ is even. Counting the banks, as a simple consequence of Euler's formula we obtain

$$3f_0 + \sum_{i=1,2,\dots} (6-i)f_i = 12. \tag{1.1}$$

Now consider an $(i,6)$-cage C. From (1.1) we have

$$f_i(\mathrm{C}) = \begin{cases} \dfrac{12}{6-i} & \text{for} \quad i > 0, \\ 4 & \text{for} \quad i = 0 \end{cases} \tag{1.2}$$

implying that $(i,6)$-cages cannot exist for $i = 1$ and $i \geq 6$; for $i = 0,2,3,4,5$ they do exist. For the smallest $(0,6)$-cages see Figure 1.3.

Let g_i denote the number of i-gons in an $(i,6)$-cage. By (1.2),

$$g_0 = 4, g_2 = 3, g_3 = 4, g_4 = 6, g_5 = 12. \tag{1.3}$$

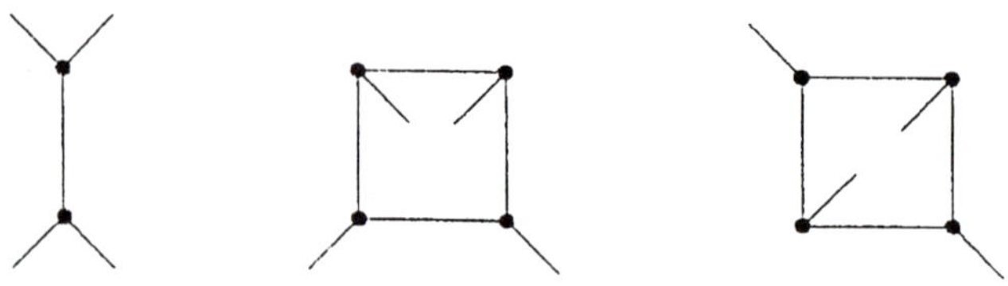

Figure 1.3

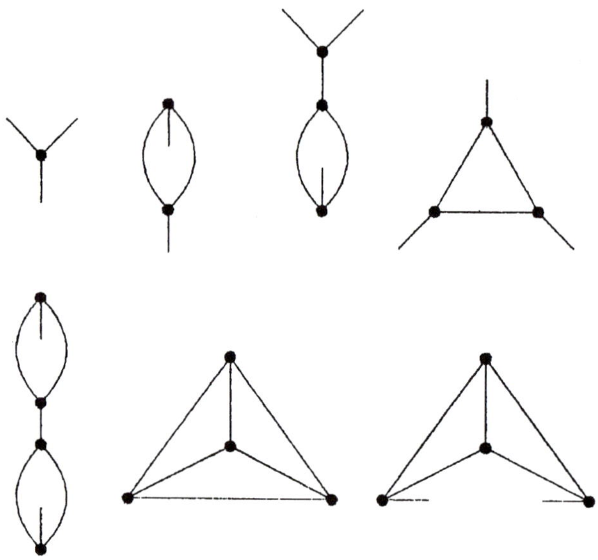

Figure 1.4

One of the main objects of our investigations will be $(0,3,6)$-cages; for these we obtain from (1.1) $3(f_0 + f_3) = 12$, thus

(1.4) $$f_0 + f_3 = 4 \quad \text{for all } (0,3,6)\text{-cages.}$$

For the smallest $(0,3,6)$-cages with no more than four vertices see Figures 1.3, 1.4.

REMARK 1.1. A $(0,3,6)$-cage with precisely one half edge does not exist. This follows immediately from Theorem 2(3) (Section 13.4, page 272) in Grünbaum's book on convex polytopes [**4**].

*

The diagrams of fullerene molecules being $(5,6)$-cages, the chemist is especially interested in the theory of $(5,6)$-cages. There are, however, three reasons that justify the study of $(i,6)$-cages for values of i other than 5 also from the chemist's point of view.

1. $(5,6)$-cages are known to be hard to deal with: therefore, it makes sense first to consider some classes of related objects with a less complex structure for gaining insight that may also be useful for the investigation of $(5,6)$-cages.

2. An isolated i-gon ($i = 0, 2, 3, 4$) can easily be replaced by a certain cluster of pentagons (Figure 1.5), thus $(i, 6)$-cages can be transformed into $(5, 6)$-cages; in a subsequent transformation (e.g. by "leapfrogging", see Section 3) the pentagons can be separated from one another.
3. There are close relations between $(0, 3, 6)$-cages (especially, $(3, 6)$-cages) and "toroidal 6-cages", i.e. hexagonal tessellations of the torus representing the (hypothetical) "torene" molecules (= toroidal graphitoids = toroidal fullerenes = toroidal polyhexes), and between the spectra of the corresponding graphs (see Sections 10 – 12).
Added in proof. There is evidence that toroidal fullerenes have been made, see [**9**] (1997).

*

In some cases it would require an excessive amount of space to spell out the many (but simple) details of the proofs; we hope that the figures will make the situation clear.

2. Spectrum and Divisors of a Graph

Let G be a graph with adjacency matrix $\mathbf{A} = (a_{ik})$. The eigenvalues and eigenvectors of G are defined to be the eigenvalues and eigenvectors of $\mathbf{A}$. The spectrum and the characteristic polynomial of G (i.e., of $\mathbf{A}$) are denoted by specG and $f_{\mathrm{G}}(x)$, respectively.

Let G have vertex set $\mathbb{V} = \{1, 2, \dots, m\}$. Assume that H is a graph on n vertices that can be assigned labels $1, 2, \dots, m$ such that, for all $i, k \in \mathbb{V}$, every vertex of H labelled i has among its neighbours precisely a_{ik} vertices (counting multiplicities if multiple edges or heavy loops are admitted) labelled k. This means that the labelling of H defines a locally homeomorphic homomorphism from H onto G, in other words, H is a locally homeomorphic cover of G; graph G is also called a *divisor* of H, in symbols:

$$\mathrm{G} \mid \mathrm{H}$$

(see [**1**]). Note that the divisor relation is transitive.

Now let $\mathbf{u} = (u_1, u_2, \dots, u_m)$ be an eigenvector of G belonging to the eigenvalue λ. For $i \in \mathbb{V}$, assign u_i, as a component of a vector $\mathbf{v}$ on n components to be formed, to all vertices of H labelled i. Then, clearly, $\mathbf{v}$ is an eigenvector of H belonging to λ. As in this process linear independence is preserved, we conclude that every eigenvalue λ of G with multiplicity μ is also an eigenvalue of H with multiplicity at least μ. Thus we have proved

THEOREM 2.1 (see [**1**]). G | H *implies* specG $\subseteq$ specH; *equivalently,*

$$\mathrm{G} \mid \mathrm{H} \textit{ implies } f_{\mathrm{G}}(x) \mid f_{\mathrm{H}}(x).$$

Set $\mathbb{V} = \mathbb{B}$ and introduce an additional vertex set $\mathbb{W} = \{1', 2', \dots, m'\}$. Replace every full edge with end vertices i, j of G by a pair of distinct edges with end vertices i, j' and j, i', and replace every half edge of G attached to vertex k by a full edge with end vertices k, k'. The resulting graph on $n = 2m$ vertices is called the *duplex* of G and denoted by DG (see [**10**]). Clearly, DG is bipartite with bipartition $\mathbb{B}, \mathbb{W}$; if G is connected then DG is connected if and only if G is non-bipartite, otherwise it decomposes into two disjoint parts both isomorphic to G.

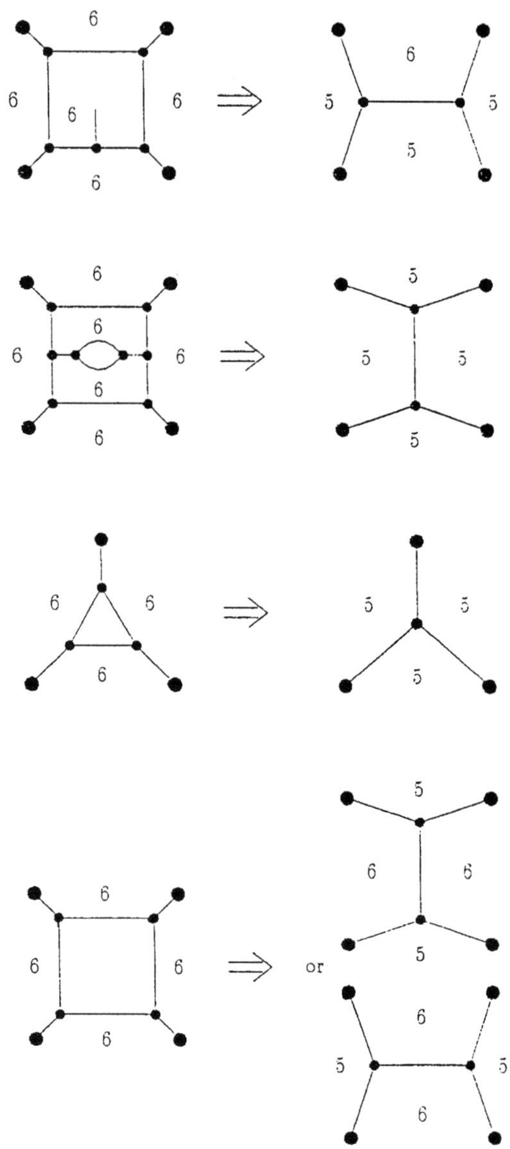

Figure 1.5

As an immediate consequence of the construction,

$$\text{G} \mid \text{DG}.$$

Modifying the above assignment procedure, assign $v_i^* = u_i$ to vertex i and $v_{i'}^* = -u_i$ to vertex i' of DG to obtain a new vector $\mathbf{v}^*$ on n components: clearly, $\mathbf{v}^*$ is an eigenvector of DG belonging to the eigenvalue $-\lambda$.

Define $\{\lambda_1, \lambda_2, \ldots, \lambda_m\}^- = \{-\lambda_1, -\lambda_2, \ldots, -\lambda_m\}$. We conclude that

$$\text{spec}\,\text{DG} = \text{spec}\,\text{G} \cup (\text{spec}\,\text{G})^- ;$$

equivalently,

$$f_{\mathsf{DG}}(x) = (-1)^m f_{\mathsf{G}}(x) \cdot f_{\mathsf{G}}(-x). \tag{2.1}$$

3. The Leapfrog Transformation

The leapfrog transform of a polyhedral graph P, denoted by LP, is defined as the truncation of the dual of P, see [**2**],[**7**] (see also [**4**] p. 266, Fig. 13.3.21 which shows the leapfrog transform of the pentagonal prism). This definition can be extended to the class of all connected plane graphs where digons, monogons and half edges are allowed. However, we prefer to give a more direct definition based on the simple observation that LP is precisely the same as the plane graph defined by the banks of P and their neighbourhood relation: this concept extends immediately to the general case.

We restrict our considerations to the class of *p*cages.

Construction of the leapfrog transform LC *of a* p*cage.*

Into the interior of each bank B of C, insert a new vertex $V(B)$. For any pair of banks B' and B'' neighbouring with multiplicity μ ($\mu > 0$), connect $V(B')$ and $V(B'')$ by μ edges. In addition, for any half edge H of C add a half edge H^* attached to $V(B(H))$ and touching H (Fig. 3.1).

Note that each edge may be realized by a smooth arc lying in the interior of the union of the corresponding (one or two) banks.

The resulting graph is LC.

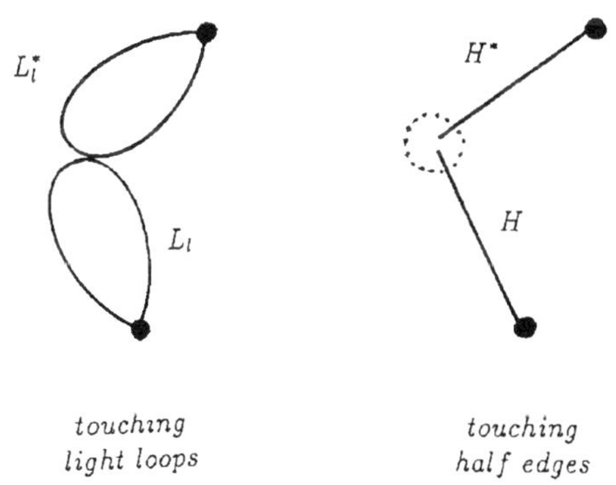

Figure 3.1

LC is again a *p*cage,

$$v(\mathsf{LC}) = 3v(\mathrm{C}) \tag{3.1}$$

and, for all $i \neq 6$, there is a $(1,1)$ - correspondence between the i-gons of C and those of LC.

For an example, see Figure 3.2: for each i-gon F of C, $i > 0$, its image LF - which is again an i-gon - is contained in F; the image LH of a half-edge H is again a half edge touching H. The image LV of a vertex V is a hexagon containing V (and no other vertex of C) in its interior.

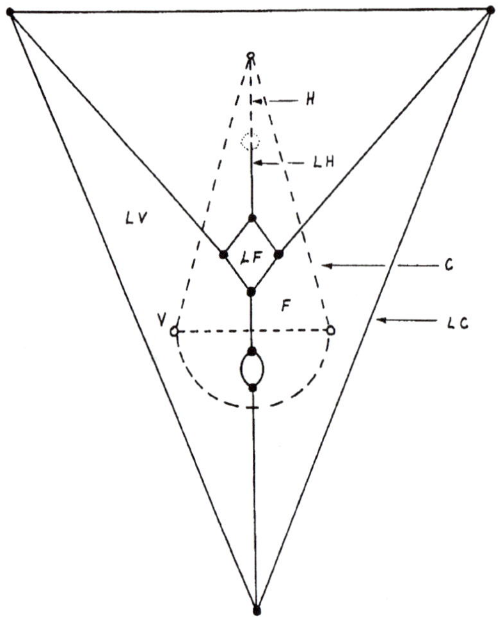

Figure 3.2

The leapfrog transform of an $(i, 6)$-cage, or of a $(0, i, 6)$-cage, is again an $(i, 6)$-cage, or a $(0, i, 6)$-cage, respectively.

4. The Structure of (3,6)-Cages

The basic facts concerning the structure of a $(3, 6)$-cage C are well known [**3**],[**4**],[**5**]. Select in C an arbitrary triangle T and an arbitrary boundary edge E of T. Starting from an interior point of T, draw a transversal τ intersecting E and leaving each hexagon it enters through the edge opposite to the edge of its entrance (as in Fig. 4.1). By means of Euler's formula it can easily be shown that τ can neither intersect itself nor return to T: thus it must be a simple arc terminating after the intersection of a certain number S of "steps" in one of the three triangles different from T (Figure 4.1).

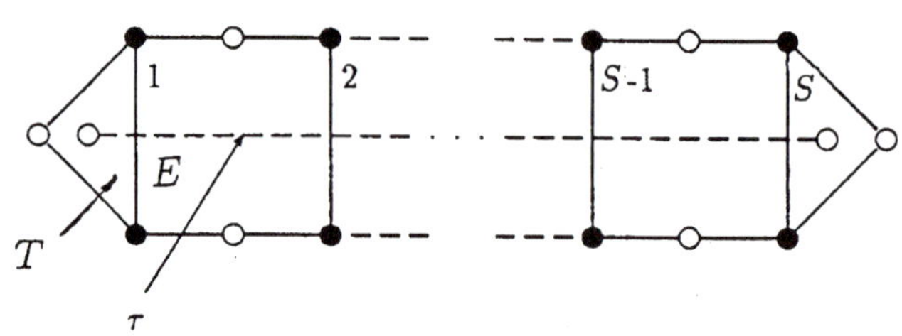

Figure 4.1

This observation implies that C has a "polar" representation of the form depicted in Figure 4.2. C is uniquely determined by only 3 parameters, namely,

r:	the **r**adius (number of **r**ings),
$s = 2S$:	the **s**ize (number of **s**pokes = twice the number of **s**teps),
t:	the **t**wist (**t**orsion)

where, if t is taken to be non-negative and minimum, $0 \leq t < s$ and, necessarily, $t \equiv r$, mod 2 (see Figs. 4.2, 4.3).

Conversely, any triple of integers r, s, t satisfying

$$r \geq 1;\ s \equiv 0, \text{mod } 2;\ s \geq 2;\ 0 \leq t < s;\ t \equiv r, \text{mod } 2$$

is the parameter triple of some $(3,6)$-cage C. We shall use the notation

$$\mathrm{C} = (r, s, t).$$

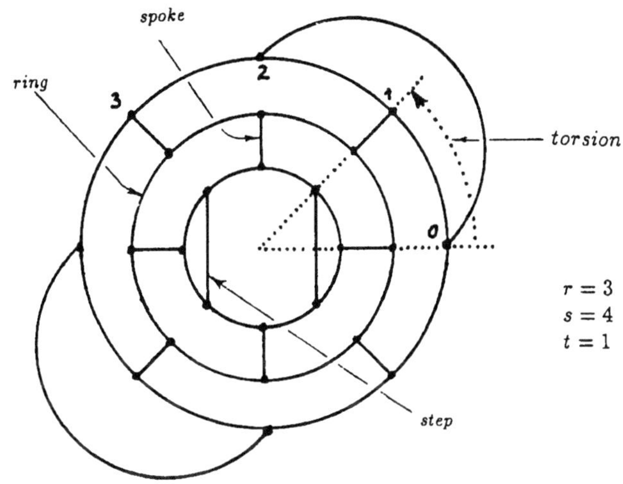

Figure 4.2

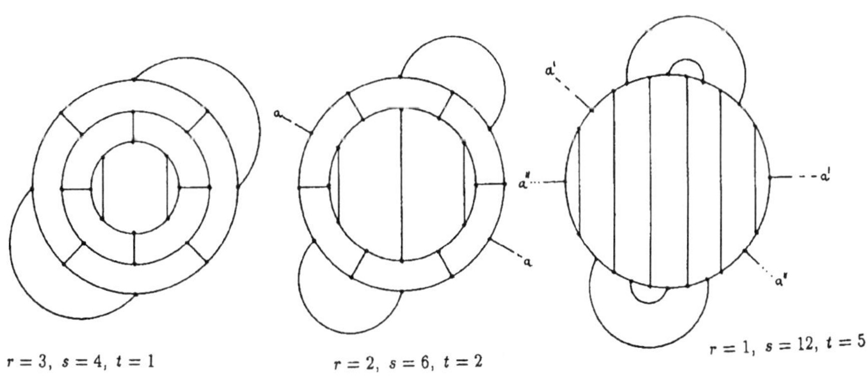

Figure 4.3

However, the parameter triple is not unique: it depends on the choice of the initial edge E (Fig. 4.1). Thus, for any $(3,6)$-cage C we find three parameter triples which, in general (though not necessarily), are distinct: for an example, see Figures 4.3 which represent the same cage.

Whenever convenient, we shall allow t to take any value of its residue class mod s.

Note that $2rs$ is the number of vertices of C.

The formulæ that describe the transformations of one parameter triple into the two others (see [**6**]) will be discussed elsewhere.

The representation of C given in Figure 4.2 can easily be transformed into a 3-dimensional representation revealing the symmetries of C: it consists of a cylindrical "hollow column" - the *body* - composed of $r-1$ layers of "bricks", with s bricks in each layer, and two "lids", one at the top and one at the bottom of the column, as depicted in Figure 4.4.

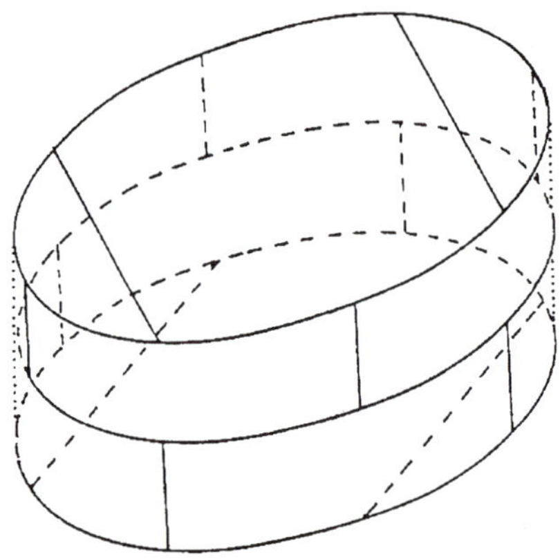

Figure 4.4

Note that the definition of the body makes sense also for odd values of s.

However, for plane drawings it is more convenient to use one of the representations given in Figure 4.5 where the labels attached to some of the vertices and edges indicate identification.

5. A Special Class of (0,3,6)-Cages

Next we define a class $\mathfrak{C}$ of $(0,3,6)$-cages comprising the class of $(3,6)$-cages.

fig5.1

Figure 5.1 shows five types of lids "fitting" on some body (see Fig. 5.2). Class $\mathfrak{C}$ consists of all $(0,3,6)$-cages that can be represented by a body together with a pair of lids each of one of the types α, β, β', γ, γ'.

There are $(0,3,6)$-cages that are not in $\mathfrak{C}$: for examples see Figures 1.3, 1.4 and 5.3 (However, these may be considered degenerate members of $\mathfrak{C}$ obtained by "squeezing" the (only) ring of a lid: see the next two paragraphs).

Note that a member of $\mathfrak{C}$ may have two body-lid representations with different pairs of lid types (Fig. 5.4). Let $\mathfrak{R}$ be the set of all body-lid representations of the members of $\mathfrak{C}$.

For each lid, its circumference (the length of its outer circuit) is $2s$ where s is even for lids of types α, β, γ and odd for lids of types β', γ'. Therefore, the possible combinations of the types of lids that can occur in a member of $\mathfrak{R}$ are $\alpha\alpha$, $\alpha\beta$, $\alpha\gamma$, $\beta\beta$, $\beta\gamma$, $\gamma\gamma$; $\beta'\beta'$, $\beta'\gamma'$, $\gamma'\gamma'$. According to these combinations, $\mathfrak{R}$

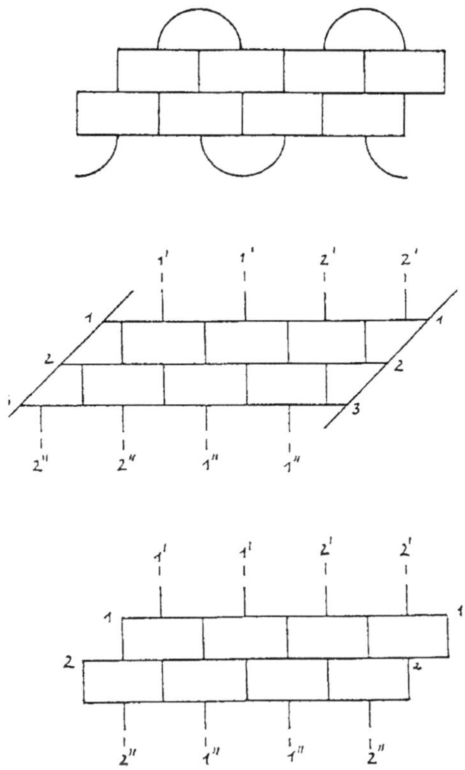

Figure 4.5

decomposes into 9 subclasses $\mathfrak{R}_{\alpha\alpha}, \dots, \mathfrak{R}_{\gamma'\gamma'}$. Note that $\mathfrak{R}_{\alpha\alpha}$ is the class of all body-lid representations of $(3,6)$-cages.

We shall also say that a $(0,3,6)$-cage represented in $\mathfrak{R}_{\alpha\alpha}$ is of *type* $\alpha\alpha$, etc.

*

We need the concept of a "squeezed circuit".

Let Q be a circuit of length $2q$ $(q \geq 1)$ with a pair of opposite edges E, E^* distinguished (Fig. 5.5). Replace E and E^* by half edges H and H^* and identify pairwise the remaining vertices and edges as indicated in Figure 5.5. The resulting *squeezed circuit* S has q vertices and $2q$ banks, its outer face is a $2q$-gon. As its total edge weight is q, we shall say that S has length q.

*

If in a lid of type γ or γ' the vertices on its outer circuit are deleted then what remains is a squeezed circuit of length s. Thus a lid of type γ or γ' at the top (bottom) of a body on ρ rings can be produced by squeezing the top (bottom) ring of a body on $\rho + 1$ rings. Therefore, it is reasonable to define the parameter r for any member of $\mathfrak{R}$ as $r = \rho + \frac{\lambda}{2}$ where ρ is the number of rings of the body and λ is the number of lids of type γ or γ'.

Note that, for any $C \in \mathfrak{C}$, the integer $2rs$ (which may be odd) is independent of the representation of C in $\mathfrak{R}$ since it equals the number of vertices of C.

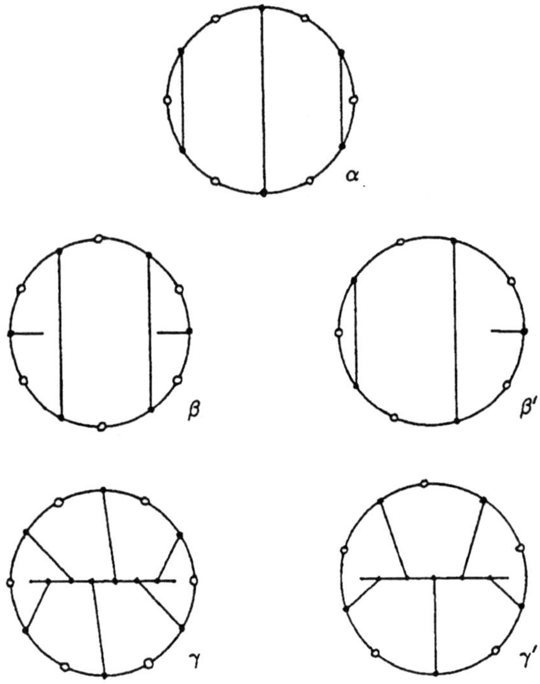

s is even in cases α, β, γ and odd in cases β', γ'

Figure 5.1

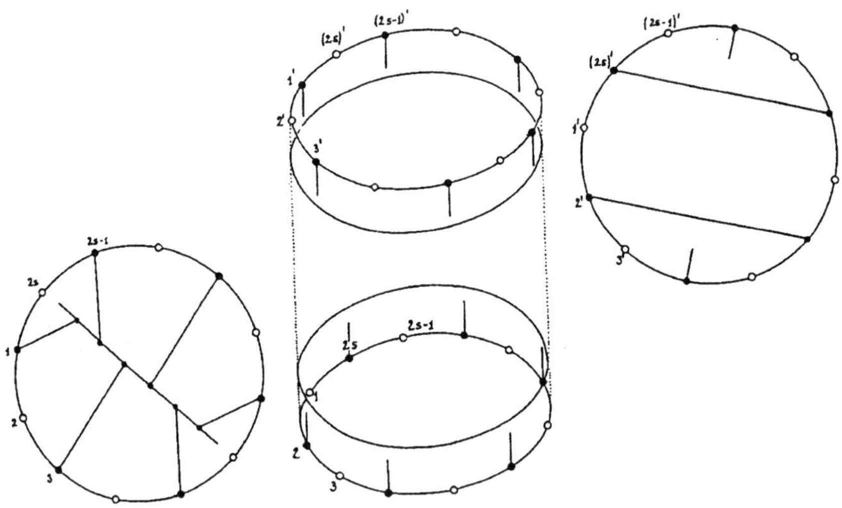

Figure 5.2

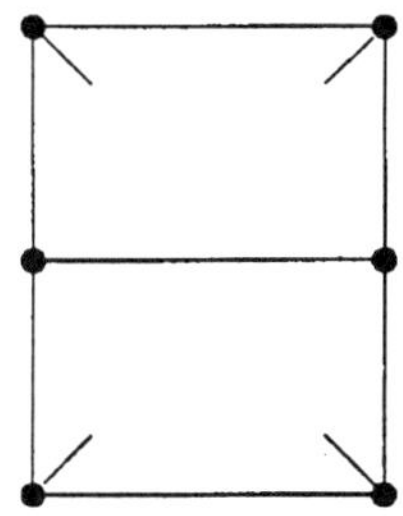

Figure 5.3

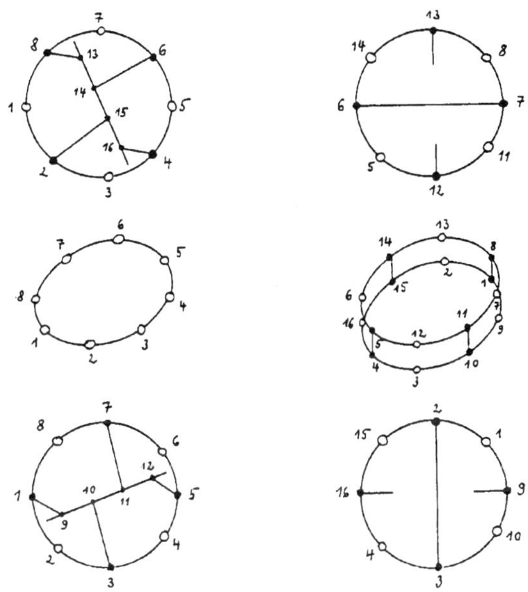

Different representations of the same (0, 6)-cage

Figure 5.4

6. Divisor Relations for (0,3,6)-Cages

Next we shall show that the $(0,3,6)$-cages of class $\mathfrak{C}$ are closely interrelated by their divisibility properties.

For each type of p-cage to be treated we shall give an example big enough to represent the general case. As we have defined a measure for the torsion only for cages of type $\alpha\alpha$ (and not for p-cages in $\mathfrak{C}$ of other types - which could easily be done, if desired), we shall not discuss the "torsion" here (except for cages of type $\alpha\alpha$).

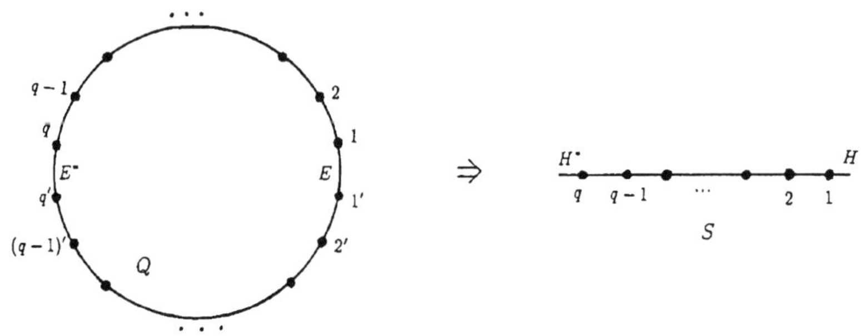

Figure 5.5

6.1 Multiples of cages in $\mathfrak{C}$

Let $\mathrm{C} \in \mathfrak{C}$ have radius r and size s. For any positive integers ρ, σ we shall construct $(0,3,6)$-cages C^*, $\mathrm{C}^{**} \in \mathfrak{C}$ with parameters $r^* = \rho r$, $s^* = s$ and $r^{**} = r$, $s^{**} = \sigma s$, respectively, such that $\mathrm{C} \mid \mathrm{C}^*$ and $\mathrm{C} \mid \mathrm{C}^{**}$.

Multiplication of the radius

Let C have a bottom lid of type ξ and a top lid of type η, where $\xi, \eta \in \{\alpha, \beta, \gamma; \beta', \gamma'\}$, and assume that its vertices are labelled 1 through $v(\mathrm{C}) = 2rs$ (which may be odd). Let $\mathrm{C}', \mathrm{C}'', \dots, \mathrm{C}^{(\rho)}$ be copies of C, labelled the same way as C; as labels for $\mathrm{C}^{(i)}$ we shall use $1^{(i)}, 2^{(i)}, \dots, (2rs)^{(i)}$ $(i = 1, 2, \dots, \rho)$. "Gluing" these copies together we shall construct $\mathrm{C}^* \in \mathfrak{C}$ with $r^* = \rho r$, $s^* = s$ and of type $\xi\xi$ if ρ is even and $\xi\eta$ if ρ is odd.

Let us describe in detail the first step. We "place C'' upside down upon C' and glue the two copies together": this instruction is made precise by Figures 6.1 which show the top lid of C and the two or three rings of C^* that correspond to the top lids of C' and C''.

If $\rho = 2$ then we are done: the resulting p-cage is of type $\xi\xi$. If $\rho > 2$ then we continue the gluing process with the top lid of the part already constructed and the bottom lid of C''' (both of type ξ), etc.. After $\rho - 1$ steps we have created a p-cage C^* of type $\xi\xi$ if ρ is even and of type $\xi\eta$ if ρ is odd. Dropping the superscripts of the labels we obtain a labelling of C^* which shows that $\mathrm{C} \mid \mathrm{C}^*$. Further, the "torsion" (if correctly defined and taken mod s for types α, β, γ and mod $2s$ for types β', γ') of C^* is ρ times the torsion of C.

Reduction of the radius

The above results can be used for searching for divisors of a given member of $\mathfrak{C}$. In particular, if $\xi = \eta = \alpha$ then both C and C^* are $(3,6)$-cages. Let $\widehat{\mathrm{C}} = (\hat{r}, \hat{s}, \hat{t})$ be any $(3,6)$-cage: then we can always find in $\mathfrak{C}$ a divisor C of $\widehat{\mathrm{C}}$ such that $r = \dfrac{\hat{r}}{2}$, $s = \hat{s}$ and $\widehat{\mathrm{C}}$ coincides with C^*.

To this end, consider the body-lid representation of $\widehat{\mathrm{C}}$. Select a pair of triangles, T^0 and T_0 say, T^0 at the top and T_0 on the bottom. Denote the "angle difference" between T^0 and T_0 by d, as depicted in Figure 6.2; note that $d = \hat{t}$ or $d = \hat{t} + \hat{s}$ (we need only d mod $2\hat{s}$).

Case 1: $\hat{r}$ is even. Label the vertices 1 through $\hat{r}\hat{s}$ as indicated in Figures 6.3.1a. This labelling defines a divisor C of $\widehat{\mathrm{C}}$ with parameters $r = \dfrac{\hat{r}}{2}$ and $s = \hat{s}$ of

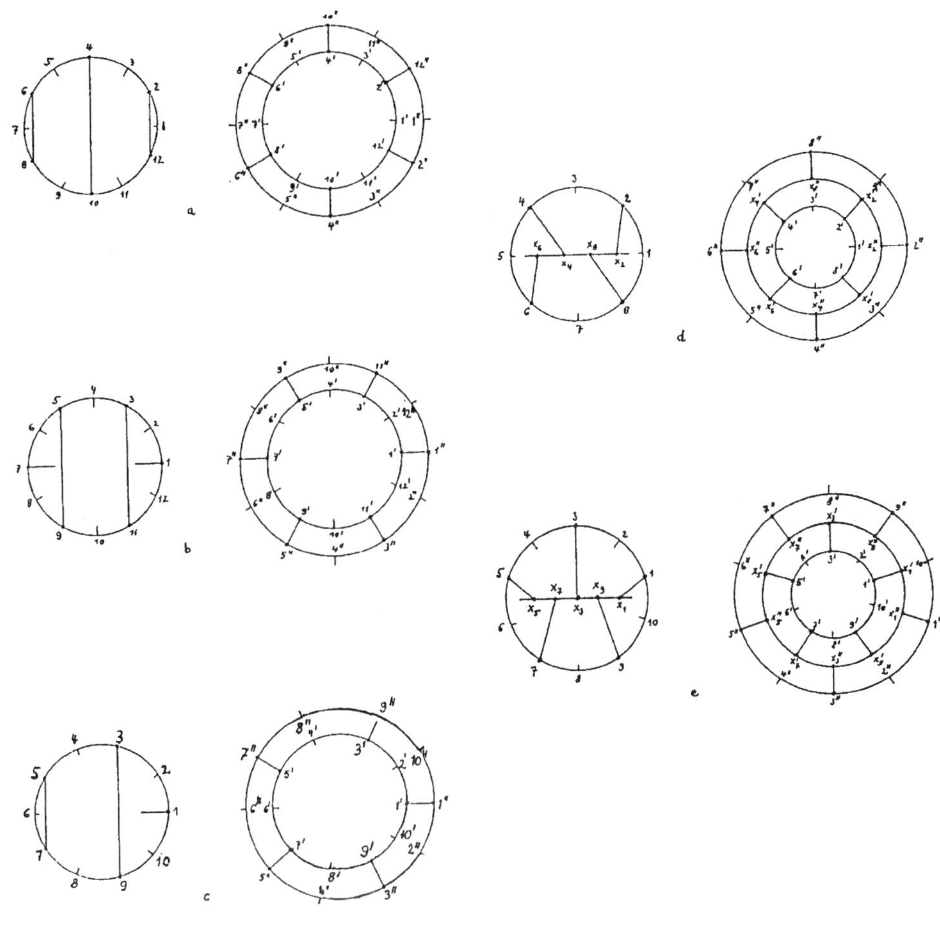

Figure 6.1

type $\alpha\alpha$ if $\hat{r}+d \equiv 0$, mod 4 and of type $\alpha\beta$ if $\hat{r}+d \equiv 2$, mod 4: see Figures 6.3.1b. A label attached to a "dangling" edge E indicates the vertex to which E goes.

Case 2: $\hat{r}$ is odd. Label the vertices 1 through $\hat{r}\hat{s}$ as indicated in Figure 6.3.2a. This labelling defines a divisor C of $\widehat{\mathrm{C}}$ with parameters $r = \dfrac{\hat{r}}{2}$ and $s = \hat{s}$ of type $\alpha\gamma$: see Figure 6.3.2b.

Multiplication of the size

As above assume that the vertices of C are labelled 1 through $2rs$. We enlarge the size of the body B of C from s to $s^{**} = \sigma s$ and label the vertices of the new body B^{**} - which is to become the body of C^{**} - such that the labelling of B is locally preserved but repeated σ times in a circular manner; in order to be able to distinguish between vertices having the same label, we shall use - as above - superscripts $i = 1, 2, \ldots, \sigma$. Figures 6.4.1,2 present for each type of lid L of C a new (labelled) lid L^{**} such that B^{**} and the corresponding top and bottom lids $\mathrm{L}^{**}_{\text{top}}$, $\mathrm{L}^{**}_{\text{bottom}}$ form a new p-cage $\mathrm{C}^{**} \in \mathfrak{C}$ whose labelling (after omission of the

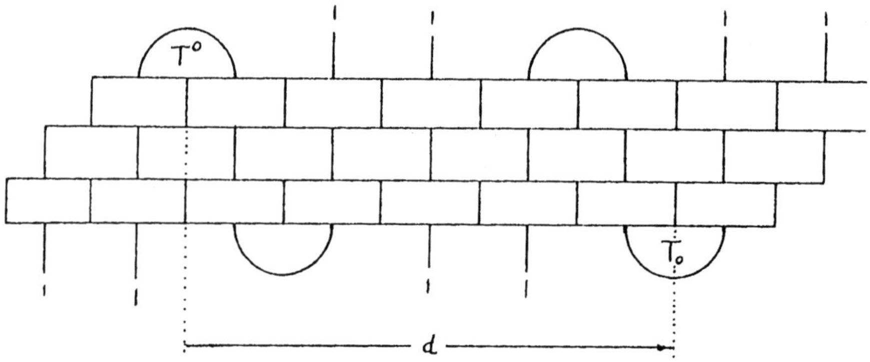

$r = 4,\ s = 8,\ t = 2,\ d = 10$

Figure 6.2

superscripts) shows immediately that C | C**. In Figures 6.4.1,2 the circumferences (outer circuits) of the lids are represented by straight line segments with their end vertices identified.

Note that the lid L** is not unique: there may be several possibilities of connecting the vertices, and any shift by a multiple of s in cases α, β, γ and of $2s$ in cases β', γ' is allowed. In any case, the number of different lids we obtain is equal to σ.

Reduction of the size

Let $\widehat{\mathrm{C}} = (\hat{r}, \hat{s}, \hat{t})$ be any $(3,6)$-cage and consider its polar representation (Figs. 4.2, 6.5.1a,2a). Label the vertices 1 through $\hat{r}\hat{s}$ as indicated in Figures 6.5.1a,2a. This labelling defines a divisor C of $\widehat{\mathrm{C}}$ with parameters $r = \hat{r}$, $s = \frac{\hat{s}}{2}$. If $\hat{s} \equiv 0$, mod 4 then s is even and C is of type $\alpha\alpha$: see Figures 6.5.1; if $\hat{s} \equiv 2$, mod 4 then s is odd and C is of type $\beta'\beta'$: see Figures 6.5.2.

6.2 A divisibility theorem for $(3,6)$-cages

Applying the results about the multiplication of the radius and the size to $(3,6)$-cages we immediately obtain

THEOREM 6.1. *Let*

$$\mathrm{C} = (r, s, t), \quad \mathrm{C}' = (\rho r, s, t'), \quad \mathrm{C}'' = (r, \sigma s, t'')$$

where ρ, σ are positive integers.
If

(i) $t' \equiv \rho t$, *mod* s

then C | C′;

if

(ii) $t'' \equiv t$, *mod* s

then C | C″. ◻

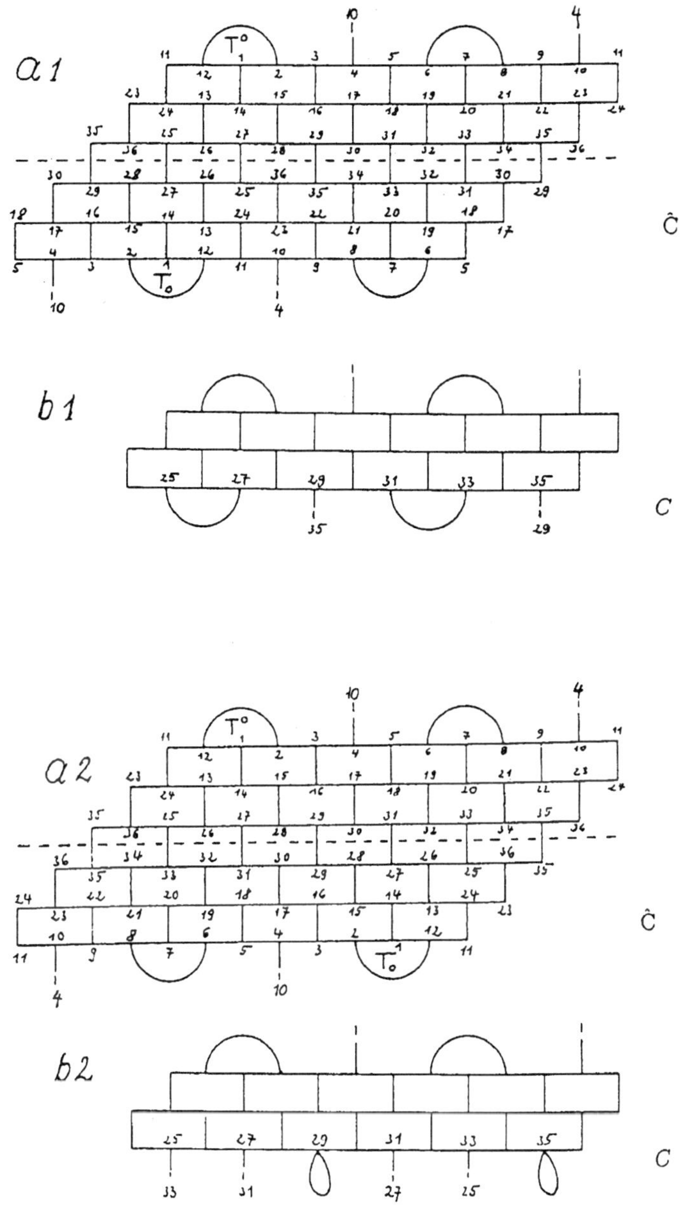

Figure 6.3.1

COROLLARY 6.2. *For any positive integer q and any $(3,6)$-cage (r,s,t),*

$$(r,s,t) \mid (qr,qs,qt).$$

□

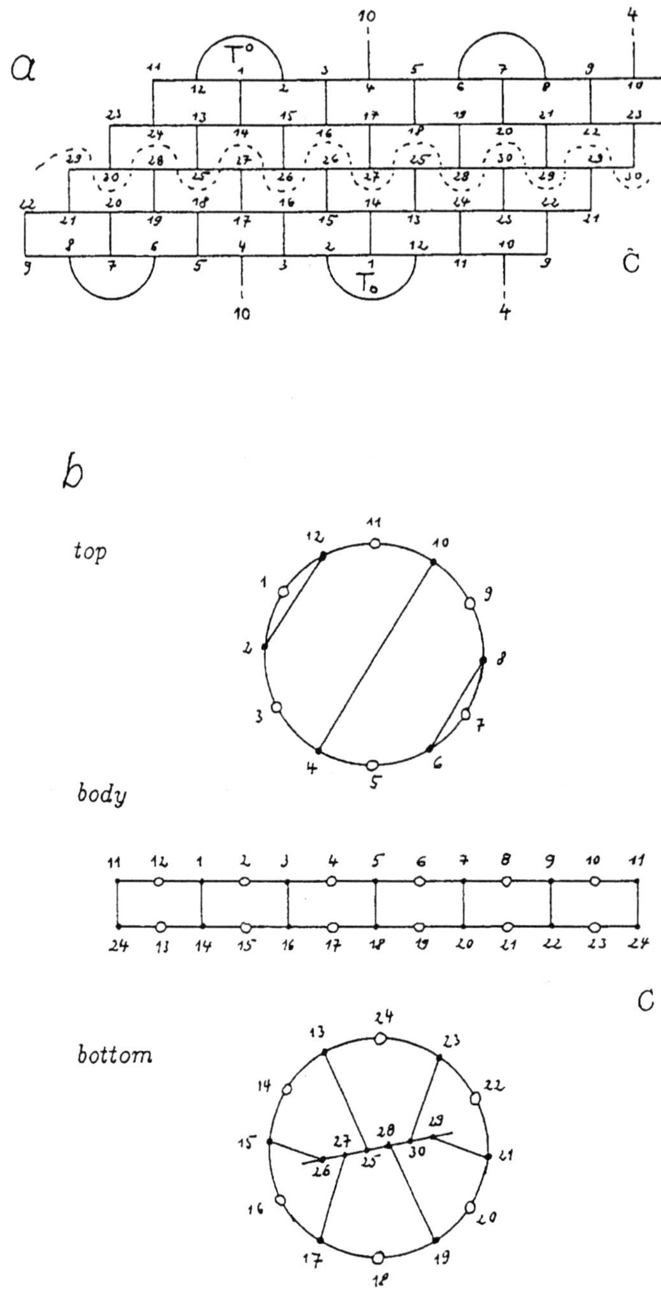

Figure 6.3.2

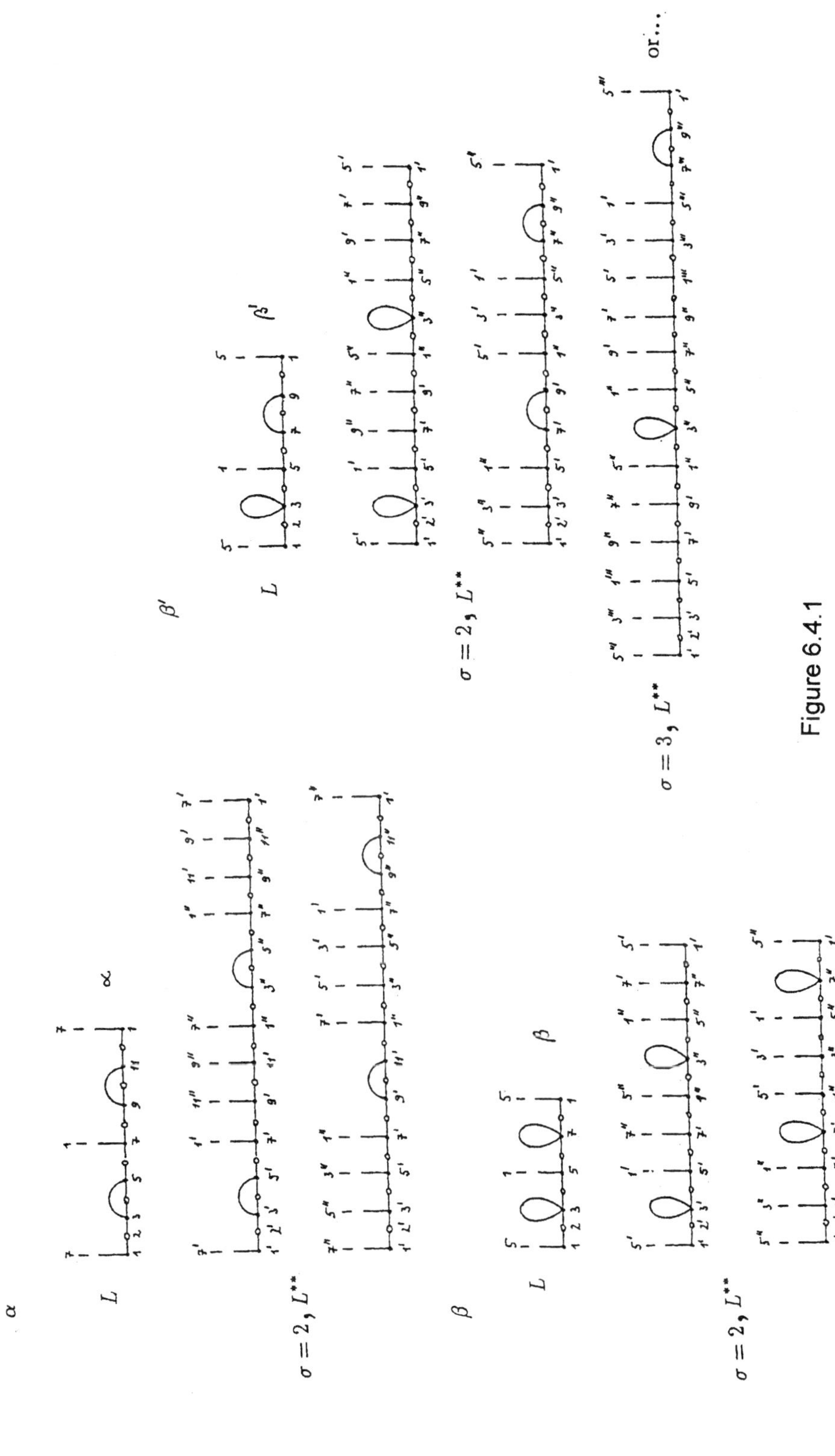

Figure 6.4.1

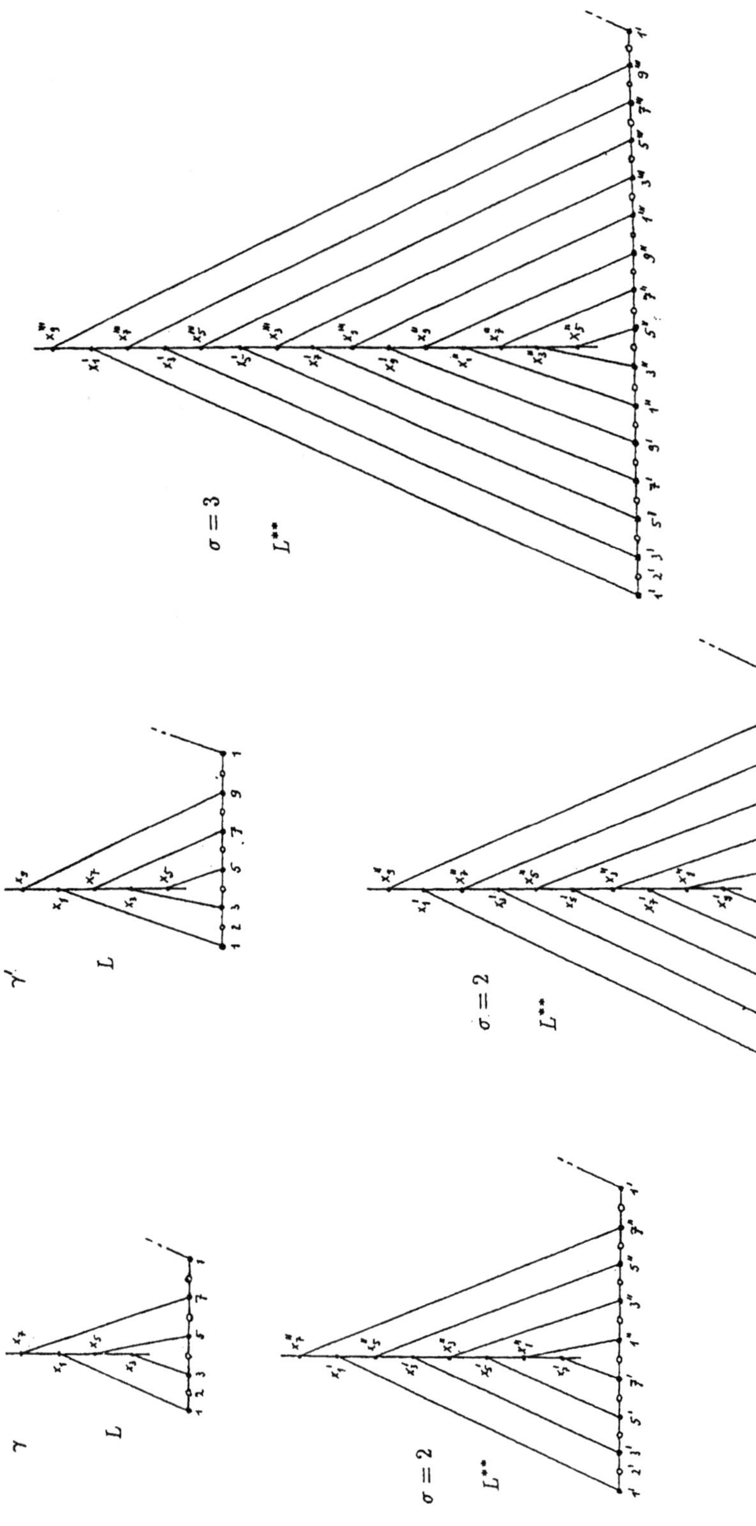

Figure 6.4.2

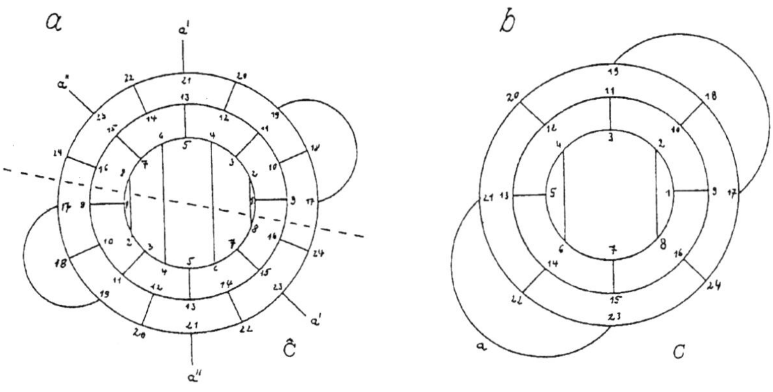

Figure 6.5.1

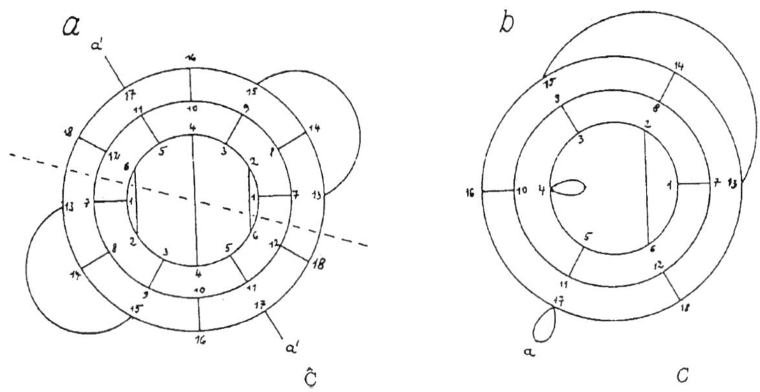

Figure 6.5.2

7. Three Conjectures

In 1995 the first author formulated three conjectures about the spectra of $(3,6)$-cages.

- The first conjecture.

  The spectrum of any $(3,6)$*-cage* C *has the following form:*

  $\mathrm{spec}C = \{3, -1, -1, -1; \lambda_1, \lambda_2, \dots, \lambda_{2(\nu-1)}; -\lambda_1, -\lambda_2, \dots, -\lambda_{2(\nu-1)}\}$

  where $\nu = \frac{1}{4}v(\mathrm{C})$.

  Equivalently, $f_{\mathrm{C}}(x) = (x-3)(x+1)^3 \cdot g(x^2)$ *where* g *is a polynomial of degree* $\frac{1}{2}\,(v(\mathrm{C}) - 4)$.

Note that the spectrum of the graph T of the tetrahedron is

$$\mathrm{spec}T = \{3, -1, -1, -1\}.$$

It can easily be proved that T | C, therefore, by Theorem 2.1, $\mathrm{spec}T \subseteq \mathrm{spec}C$. Thus what remains is the conjecture that the "reduced spectrum" $\mathrm{spec}C - \mathrm{spec}T$ is symmetric with respect to the zero point of the real axis.

We have been able to prove this conjecture for all $(3,6)$-cages on $4q \cdot 2^k$ vertices where k is any non-negative integer and q is an odd integer between 0 and 100. We shall not pursue this problem here; let us just mention that we are inclined to extend this conjecture to the class of all $(0,3,6)$-cages as follows.

CONJECTURE 7.1. *Let* C *be any* $(0,3,6)$*-cage with precisely* h *half edges. Then*

$$f_{\mathrm{C}}(x) = (x-3)(x+1)^{\alpha}(x-1)^{\beta} \cdot g(x^2)$$

where $0 \le \alpha \le 3$, $0 \le \beta \le 1$, $\alpha \cdot \beta = 0$, $3 - \alpha + \beta = h$ *and* g *is a polynomial of degree* $\frac{1}{2}(v(\mathrm{C}) - \alpha - \beta - 1)$.

The motivation for α and β is the following: the sum of the eigenvalues must equal the trace of the adjacency matrix, i.e., the number of half edges.

- The second conjecture.
  For any $(3,6)$*-cage* C,

$$\mathrm{spec}\mathrm{C} \subset \mathrm{spec}\mathsf{L}^2\mathrm{C}$$

(recall that L denotes the leapfrog transformation, see Section 3).

We shall prove this conjecture in Section 8 . Again we are inclined to extend this conjecture to the class of all $(0,3,6)$-cages.

- The third conjecture.
  For certain classes of $(3,6)$-cages with a higher degree of symmetry (for details see Section 9), not only $\mathrm{spec}\mathrm{C} \subset \mathrm{spec}\mathsf{L}^2\mathrm{C}$ but also

$$\mathrm{spec}\mathrm{C} \subset \mathrm{spec}\mathsf{L}\mathrm{C} \subset \mathrm{spec}\mathsf{L}^2\mathrm{C}.$$

We shall prove this conjecture in Section 9.

8. Proof of the Second Conjecture

Let L denote the leapfrog transformation (Section 3) and let C be any $(3,6)$-cage.

THEOREM 8.1.

$$\mathrm{C} \mid \mathsf{L}^2\mathrm{C}.$$

PROOF. Let $\mathrm{C} = (r,s,t)$. Figures 8.1 show the effect of the transformation L^2 if applied to a hexagon H and to a triangle T of C, respectively. From these figures we infer that L^2 triples the values of all three parameters r, s, t, i.e.,

$$\mathsf{L}^2\mathrm{C} = (3r, 3s, 3t).$$

By Corollary 6.2, $\mathrm{C} \mid \mathsf{L}^2\mathrm{C}$. □

Theorems 8.1 and 2.1 imply the validity of the second conjecture (Section 7):

$$f_{\mathrm{C}}(x) \mid f_{\mathsf{L}^2\mathrm{C}}(x)$$

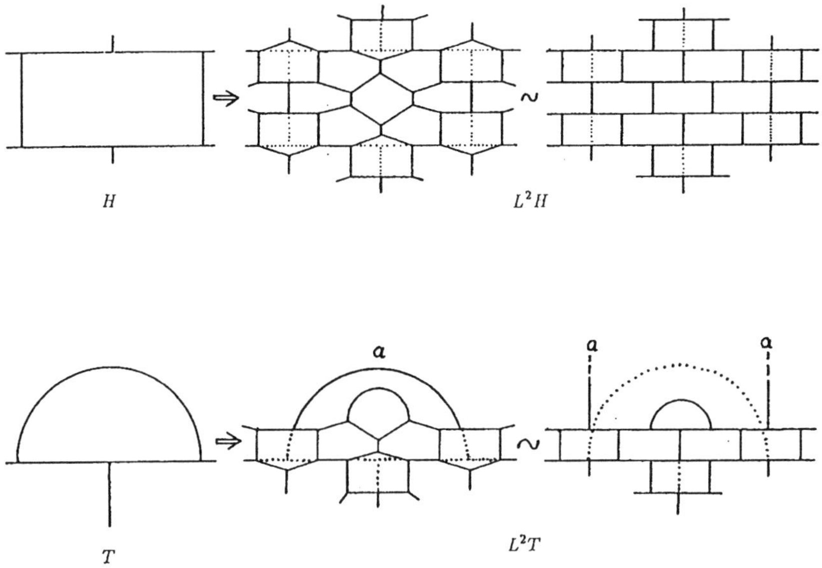

Figure 8.1

9. Proof of the Third Conjecture

Let C be a $(3,6)$-cage and let B be one of its body-lid representations with parameters r, s, t. Having in view the pairs of triangles at the top and on the bottom of B, we shall call B a *parallel representation* if $t = 0$ and an *orthogonal representation* if $t = S = \frac{s}{2}$ (see Figures 9.1).

The third conjecture says that if C has a parallel or orthogonal representation and its parameters satisfy a certain additional condition then

$$\mathrm{spec}C \subset \mathrm{spec}LC \subset \mathrm{spec}L^2C \subset \mathrm{spec}L^3C \subset \dots .$$

We shall prove the following theorem which implies the truth of this conjecture.

THEOREM 9.1. *Let C^* be one of the following $(3,6)$-cages:*

$$C_1^{=}(q) = (2q, 2q, 0), \quad C_2^{=}(q) = (2q, 6q, 0)$$

$$C_1^{\perp}(q) = (q, 2q, q), \quad C_2^{\perp}(q) = (q, 6q, 3q)$$

where q is a positive integer. Then

$$C^* \mid LC^* \mid L^2C^* \mid L^3C^* \mid \dots .$$

PROOF. Let $C = (r, s, t)$ be any $(3,6)$-cage that has a parallel (case **A**) or orthogonal (case **B**) representation. Recall that $r \equiv t$, mod 2.

In case A, $t = 0$; set $r = 2q$, so

in **case A:** $\quad C = (2q, s, 0);$

in case B, $s = 2t$, so

in **case B:** $\quad C = (r, 2t, t).$

It is not difficult to show (we omit the details) that

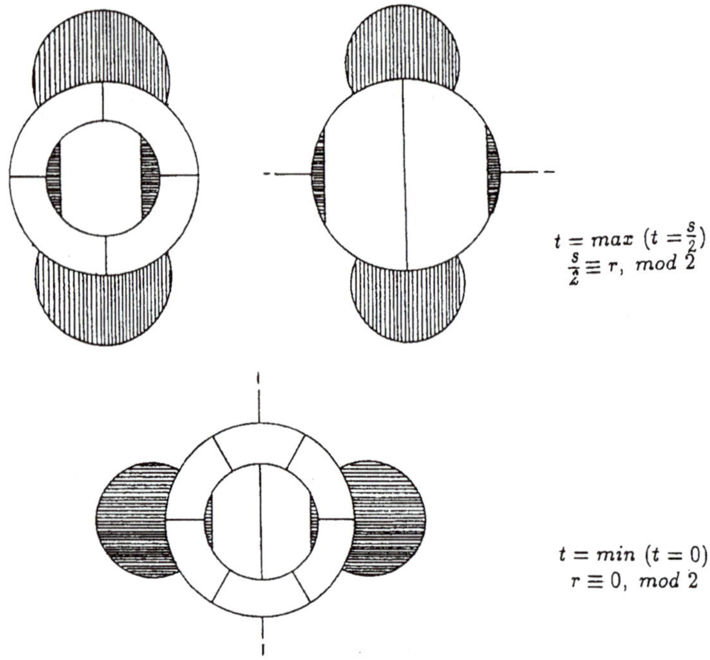

Figure 9.1

in case A: $\mathsf{L}\mathrm{C} = (s, 6q, 0)$,

in case B: $\mathsf{L}\mathrm{C} = (t, 6r, 3r)$.

Case A. If $s = 2q$ then $\mathrm{C} = \mathrm{C}_{\bar{1}}^{=}(q)$, $\mathsf{L}\mathrm{C} = \mathrm{C}_{\bar{2}}^{=}(q)$, and C, LC satisfy condition (ii) of Theorem 6.1 ; if $s = 6q$ then $\mathrm{C} = \mathrm{C}_{\bar{2}}^{=}(q)$, $\mathsf{L}\mathrm{C} = \mathrm{C}_{\bar{1}}^{=}(3q)$, and C, LC satisfy condition (i) of Theorem 6.1.

We conclude that

$$\begin{aligned} & \mathrm{C}_{\bar{1}}^{=}(q) \mid \mathsf{L}\mathrm{C}_{\bar{1}}^{=}(q) = \mathrm{C}_{\bar{2}}^{=}(q) \mid \mathsf{L}\mathrm{C}_{\bar{2}}^{=}(q) = \mathsf{L}^2\mathrm{C}_{\bar{1}}^{=}(q) \\ = \; & \mathrm{C}_{\bar{1}}^{=}(3q) \mid \mathsf{L}\mathrm{C}_{\bar{1}}^{=}(3q) = \mathsf{L}^3\mathrm{C}_{\bar{1}}^{=}(q) = \mathsf{L}^2\mathrm{C}_{\bar{2}}^{=}(q) \quad \text{etc.} \end{aligned}$$

Case B. If $r = t = q$ then $\mathrm{C} = \mathrm{C}_{\bar{1}}^{\perp}(q)$, $\mathsf{L}\mathrm{C} = \mathrm{C}_{\bar{2}}^{\perp}(q)$ and C, LC satisfy condition (ii) of Theorem 6.1; if $r = q$, $t = 3q$ then $\mathrm{C} = \mathrm{C}_{\bar{2}}^{\perp}(q)$, $\mathsf{L}\mathrm{C} = \mathrm{C}_{\bar{1}}^{\perp}(3q)$, and C, LC satisfy condition (i) of Theorem 6.1.

We conclude that

$$\begin{aligned} & \mathrm{C}_{\bar{1}}^{\perp}(q) \mid \mathsf{L}\mathrm{C}_{\bar{1}}^{\perp}(q) = \mathrm{C}_{\bar{2}}^{\perp}(q) \mid \mathsf{L}\mathrm{C}_{\bar{2}}^{\perp}(q) = \mathsf{L}^2\mathrm{C}_{\bar{1}}^{\perp}(q) \\ = \; & \mathrm{C}_{\bar{1}}^{\perp}(3q) \mid \mathsf{L}\mathrm{C}_{\bar{1}}^{\perp}(3q) = \mathsf{L}^3\mathrm{C}_{\bar{1}}^{\perp}(q) = \mathsf{L}^2\mathrm{C}_{\bar{2}}^{\perp}(q) \quad \text{etc.} \end{aligned}$$

□

10. (0,3,6)-Cages and ht-Cages

A *hexagonal toroidal cage* (briefly, *ht-cage*) is the graph of a trivalent hexagonal tessellation of the torus. Let $\mathfrak{H}$ denote the set of all ht-cages. $\mathrm{H} \in \mathfrak{H}$ can be represented in the plane by a diagram Δ as depicted in Figure 10.1 with the usual boundary identification: elements at the boundary of the diagram with the same label are to be identified. Graph H is bipartite, with bipartition $\mathbb{B}$, $\mathbb{W}$, say, where $|\mathbb{B}| = |\mathbb{W}|$, thus $v(\mathrm{H})$ is even.

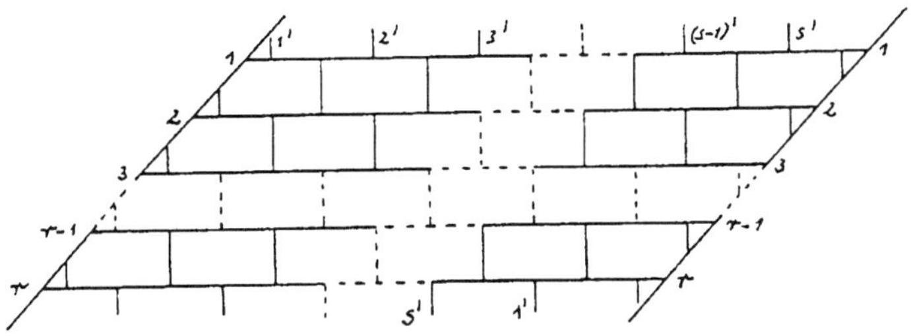

or

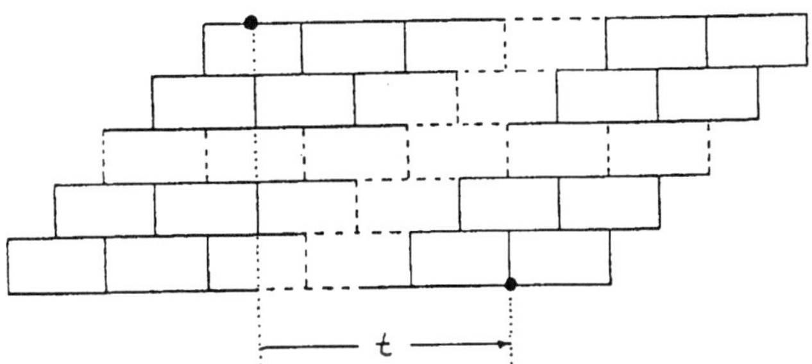

Δ

Figure 10.1

The representation Δ of H depends on three parameters:

r: the number of **r**ings represented in Δ by horizontal straight line segments,

s: the **s**ize, i.e. the number of hexagons in each horizontal row of Δ,

t: the **t**wist (**t**orsion), in Δ characterized by the shift of the dotted vertex from its position on the upper boundary to its position on the lower boundary of Δ.

We shall briefly write

$$\mathrm{H} = \langle r, s, t \rangle.$$

Note that, in general, the parameter triple is not unique: as from a hexagon there are three directions to proceed, we obtain for H also three (not necessarily distinct) parameter triples.

*

First we will show that the duplex of any $(0,3,6)$-cage can be realized as an ht-cage.

Let $\mathbb{P}$ denote the plane, homeomorphic to the sphere, and let C be any $(0,3,6)$-cage embedded in $\mathbb{P}$. Consider the pair $\mathbf{C} = (\mathrm{C}, \mathbb{P})$ and two identical copies $\mathbf{C}' = (\mathrm{C}', \mathbb{P}')$, $\mathbf{C}'' = (\mathrm{C}'', \mathbb{P}'')$ of $\mathbf{C}$. In the interior of each triangle and each zerogon of C distinguish a point so that we obtain a quadruple of points, Q_1, Q_2, Q_3, Q_4, say. Connect points Q_1, Q_2 and points Q_3, Q_4 by smooth arcs α_{12} and α_{34}, respectively, that do not intersect, do not meet a vertex and do not touch an edge (w.l.o.g., α_{12} and α_{34} may be taken to be straight line segments); let α'_{pq} and α''_{pq} be the arcs corresponding to α_{pq} in $\mathbf{C}'$ and $\mathbf{C}''$, respectively. Cut planes $\mathbb{P}'$ and $\mathbb{P}''$ along α'_{12}, α'_{34} and α''_{12}, α''_{34} and, following the cuts, glue $\mathbb{P}'$ and $\mathbb{P}''$ crosswise together (in the same way as in the elements of the theory of elliptic functions): the result is a Riemann surface of genus 1 (i.e., homeomorphic to the torus) with a cover of C, say $\widehat{\mathrm{C}}$, embedded in it. Clearly, every face of $\widehat{\mathrm{C}}$ is a hexagon (see Figures 10.2a,b; the zerogons of C have no image in $\widehat{\mathrm{C}}$), thus $\widehat{\mathrm{C}} \in \mathfrak{H}$. Recall that the members of $\mathfrak{H}$ are bipartite; let $\mathbb{V}(\widehat{\mathrm{C}}) = \mathbb{V}(\mathrm{C}') \cup \mathbb{V}(\mathrm{C}'')$ have bipartition $\mathbb{B}$, $\mathbb{W}$. Let $i_0 \in \mathbb{V}(\mathrm{C})$ be a vertex that lies on the boundary of a triangle or zerogon of C: then one of i'_0, i''_0 belongs to $\mathbb{B}$ and the other one belongs to $\mathbb{W}$. This implies that the analogue is true for any vertex $i \in \mathbb{V}(\mathrm{C})$.

Set

$$i' = \begin{cases} i^* & \text{if} \quad i' \in \mathbb{B} \\ i^{**} & \text{if} \quad i' \in \mathbb{W} \end{cases}, \quad i'' = \begin{cases} i^* & \text{if} \quad i'' \in \mathbb{B} \\ i^{**} & \text{if} \quad i'' \in \mathbb{W}. \end{cases}$$

Then to an edge (i,j) $(i \neq j)$ (to a light loop (half edge) attached to vertex k) in C there corresponds in $\widehat{\mathrm{C}}$ the pair of edges (i^*, j^{**}) and (i^{**}, j^*) (the edge (k^*, k^{**})). This means that $\widehat{\mathrm{C}}$ is the duplex of C (see Section 2).

*

Next we show that every graph $\mathrm{H} \in \mathfrak{H}$ is the duplex of some $(0,3,6)$-cage.

Given an ht-cage $\mathrm{H} = \langle r, s, t \rangle$, we shall construct (one or more) $(0,3,6)$-cages on $\frac{1}{2}v(\mathrm{H})$ vertices that are divisors of H.

Consider Figures 10.3 (compare also Figures 10.4): we shall use one of the patterns **AA**, **BB** if r is even and pattern **AB** if r is odd. First we label the vertices of part P, 1 through rs, from bottom to top and from left to right in each row, as indicated in Figures 10.3; then we select in part P′ an arbitrary vertex $V \in \{V_1, V_2, \ldots, V_s\}$; eventually we label the rs vertices in part P′, 1 through rs, such that

(i) V is labelled 1,

(ii) in P′, the labelling proceeds from top to bottom and in each row from right to left,

(iii) the adjacency relation for the labelled vertices in P′ is the same as in P.

In each case, this labelling defines a divisor C of H on $rs = \frac{1}{2}v(\mathrm{H})$ vertices which is a member of $\mathfrak{C}$ that has a representation in $\mathfrak{R}$ with radius $\dfrac{r}{2}$ and size s.

The types of the resulting divisors are displayed in Table 1.

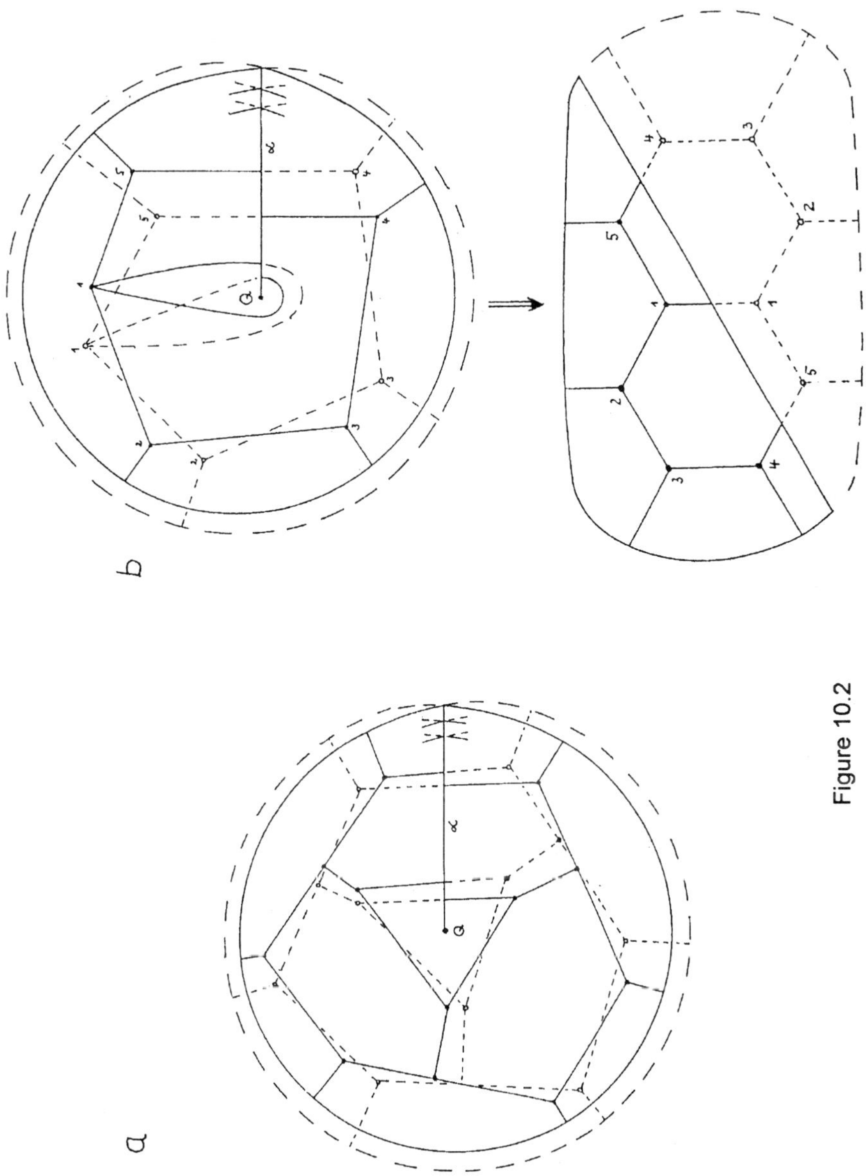

Figure 10.2

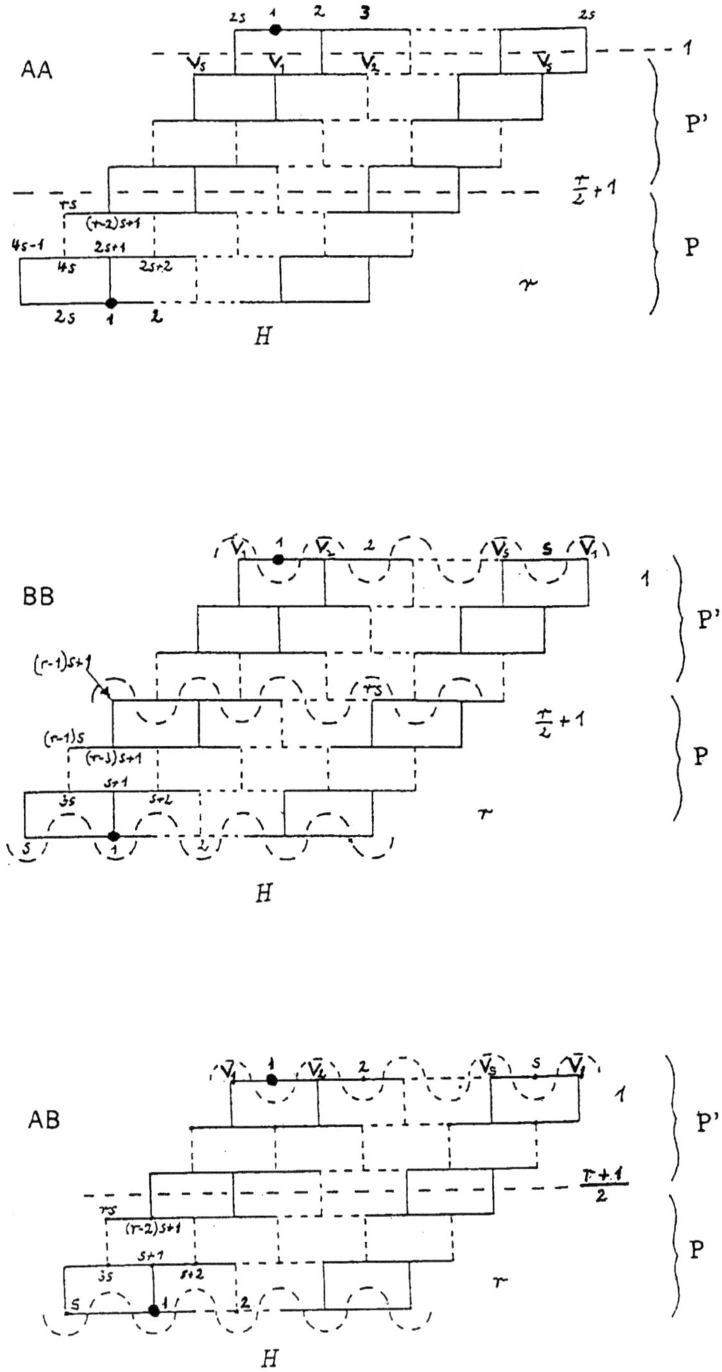

Figure 10.3

TABLE 1

r	even				odd	
Pattern	AA		BB		AB	
s	even	odd	even	odd	even	odd
type of divisor	$\alpha\alpha$ $\alpha\beta$ $\beta\beta$	$\beta'\beta'$	$\gamma\gamma$	$\gamma'\gamma'$	$\alpha\gamma$ $\beta\gamma$	$\beta'\gamma'$
Figures 10.4	1; 2, 3	4	5	6	7, 8	9

Let $\mathrm{H} \in \mathfrak{H}$ with bipartition $\mathbb{B}$, $\mathbb{W}$ and let C be one of its divisors constructed according to the above instructions. Then, for $i = 1, 2, \ldots, rs$, one of the two vertices labelled i is in $\mathbb{B}$ and the other one is in $\mathbb{W}$. Mark every vertex in $\mathbb{W}$ by a dash. Then to an edge (i, j) $(i \neq j)$ (to a half edge attached to vertex k) in C there corresponds in H the pair of edges (i, j') and (i', j) (the edge (k, k')). This means that H is the duplex of C.

In particular, by formula (2.1),

$$f_{\mathrm{H}}(x) = (-1)^{rs} f_{\mathrm{C}}(x) f_{\mathrm{C}}(-x). \tag{10.1}$$

*

Summarizing the results of this Section, we have proved

THEOREM 10.1. (I) *The duplex of any* $(0, 3, 6)$*-cage has a realization as an ht-cage.*
(II) *Every ht-cage is a realization of the duplex of some* $(0, 3, 6)$*- cage.*

□

REMARK 10.2. Let H denote an ht-cage. Set $v(\mathrm{H}) = 4k + 2(\gamma + 1)$ where $\gamma \in \{0, 1\}$.

If Conjecture 7.1 is true then Theorem 10.1(II) in connection with formula (2.1) implies the truth of the following proposition.

The characteristic polynomial of H *has the form*

$$f_{\mathrm{H}}(x) = (x^2 - 9)(x^2 - 1)^{\gamma} h^2(x^2) \tag{10.2}$$

where h is a polynomial of degree k.

This proposition is indeed true: this was proved in 1993 by Kirby, Mallion and Pollak [**8**].

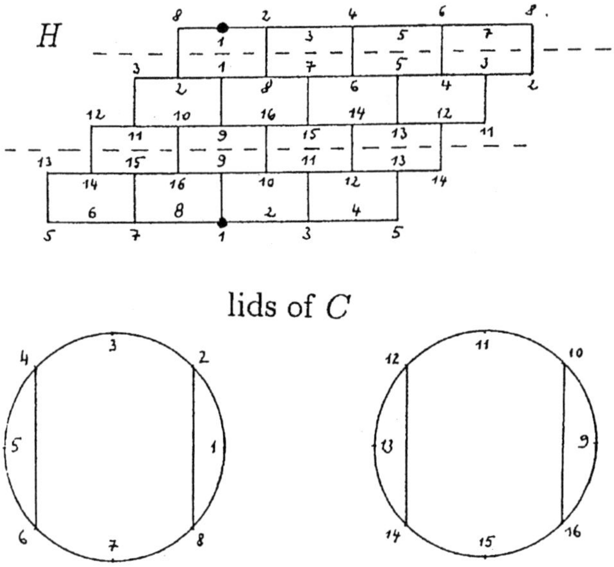

Figure 10.4.1

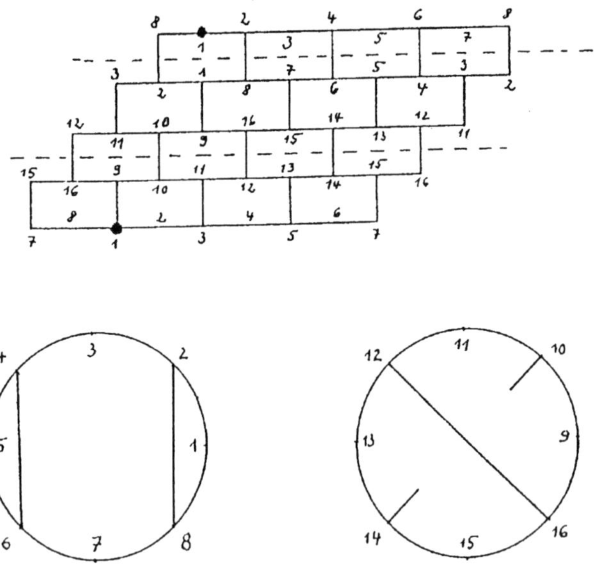

Figure 10.4.2

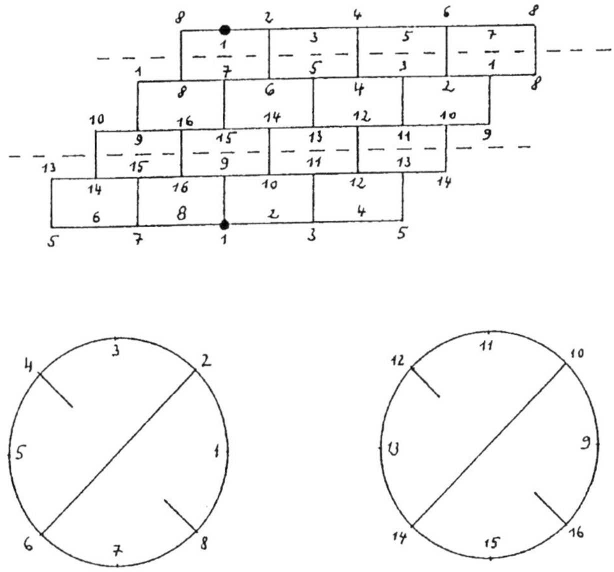

Figure 10.4.3

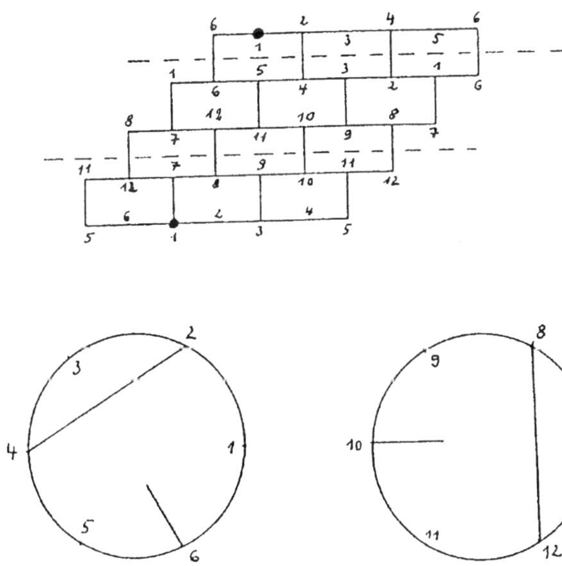

Figure 10.4.4

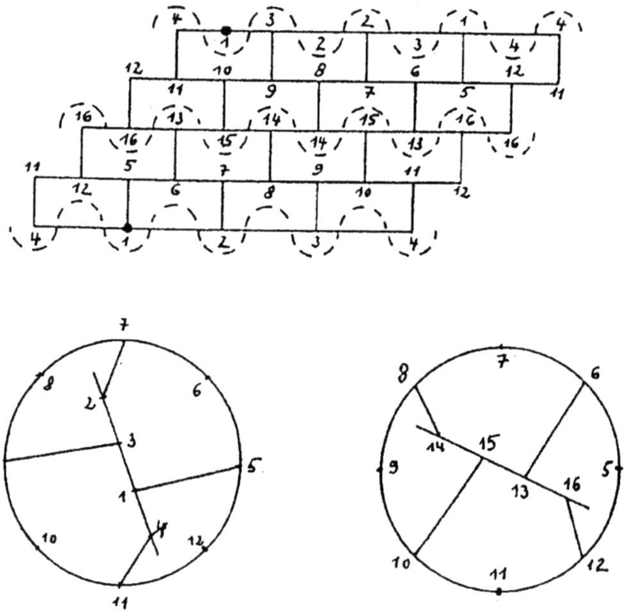

Figure 10.4.5

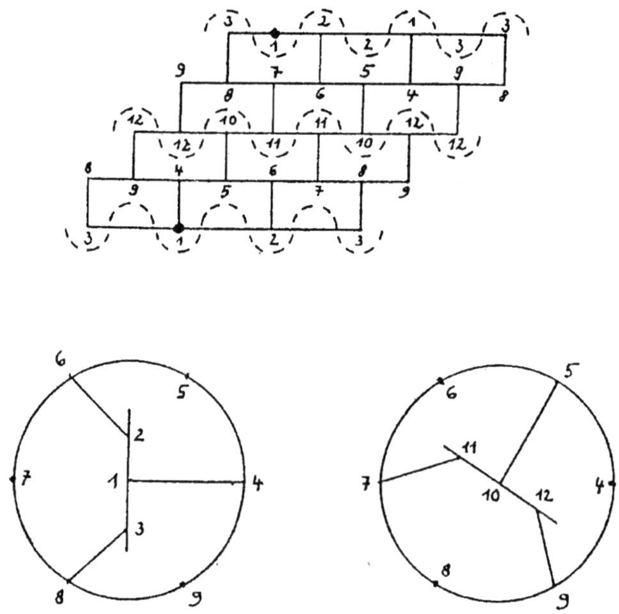

Figure 10.4.6

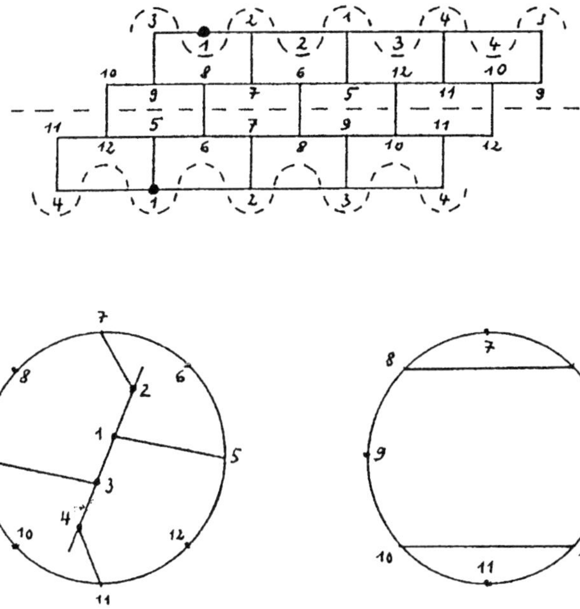

Figure 10.4.7

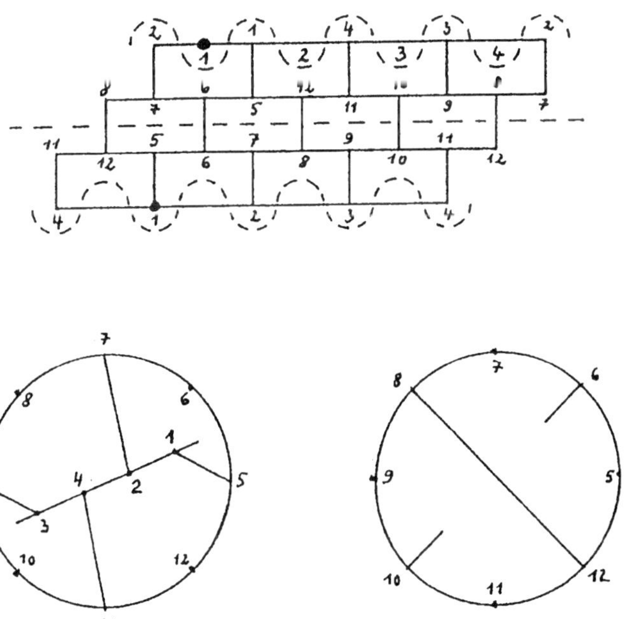

Figure 10.4.8

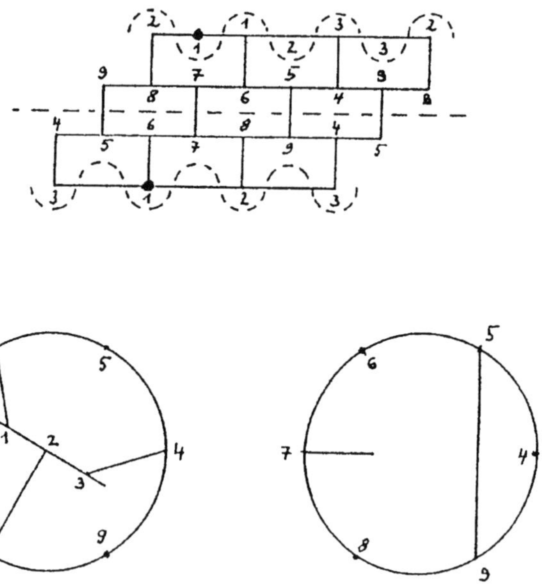

Figure 10.4.9

11. Divisor Relations for ht-Cages

THEOREM 11.1. *Let* $\mathrm{H} = \langle r,s,t\rangle$ *and* $\mathrm{H}' = \langle r',s',t'\rangle$ *be ht-cages and assume that one of the following cases applies.*

Case A (A1) $r \mid r',\ s' = s,$
(A2) $rt' \equiv r't, \mathrm{mod}\ v(\mathrm{H});$

Case B (B1) $r' = r,\ s \mid s',$
(B2) $t' \equiv t, \mathrm{mod}\ 2s.$

Then

$$\mathrm{H} \mid \mathrm{H}' .$$

COROLLARY 11.2. *For any ht-cage* $\langle r,s,t\rangle$ *and any positive integer* q,

$$\langle r,s,t\rangle \mid \langle r,qs,t\rangle \mid \langle qr,qs,qt\rangle.$$

PROOF of Theorem 11.1.

Case A. Set $\rho = \dfrac{r'}{r}$. Note that $v(\mathrm{H}) = 2rs$, thus (A2) is equivalent to

$$\text{(A2}')\quad t' \equiv \rho t, \quad \mathrm{mod}\ 2s.$$

Assign labels $1,2,\dots,v(\mathrm{H}) = 2rs$ to the vertices of H, as shown in Figure 11.1.

Let $\mathrm{H}_1 = \mathrm{H}, \mathrm{H}_2, \dots, \mathrm{H}_\rho$ be copies of H, labelled the same way as H and represented in such a way that the vertex labelled 1 in the bottom line of the representation of H_i coincides with the vertex labelled 1 in the top line of the representation of H_{i+1} ($i = 1,2,\dots,\rho-1$), as depicted in Figure 11.1. The resulting diagram, with t' as in Figure 11.1, is a (labelled) representation of H′. The labelling shows that H | H′.

Case B. Set $\sigma = \dfrac{s'}{s}$. The proof should now be clear from Figure 11.2. □

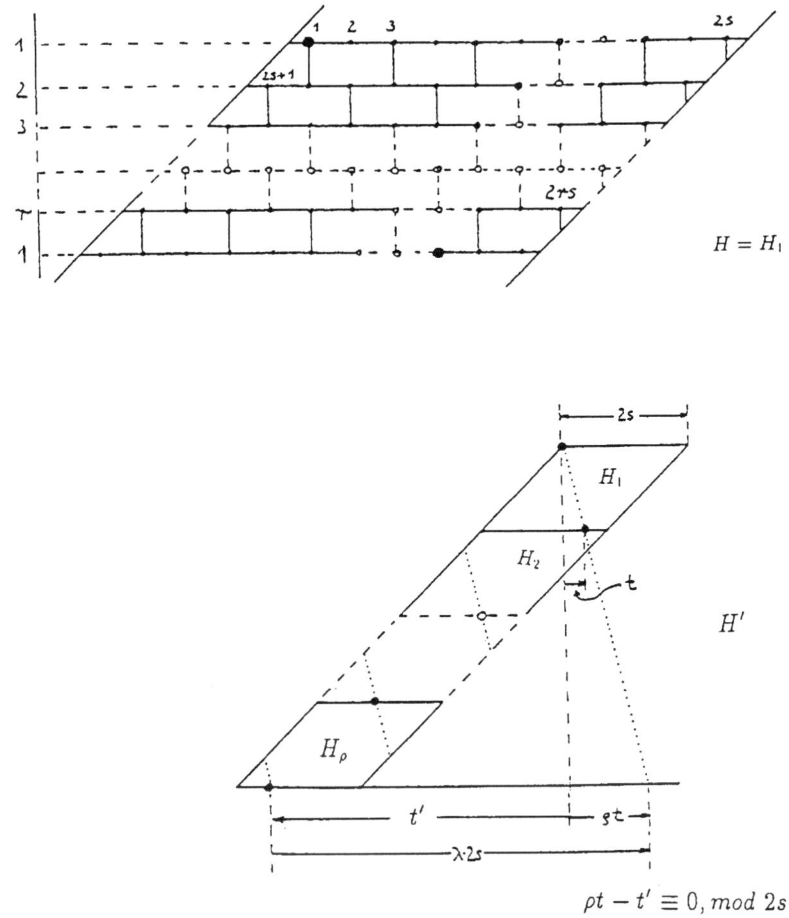

Figure 11.1

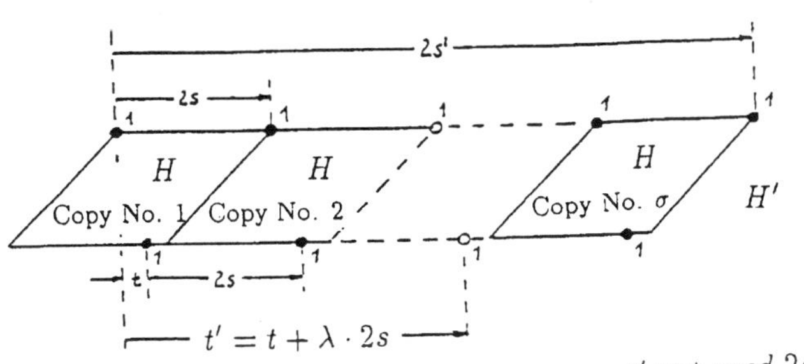

Figure 11.2

12. The Second Conjecture Extended to ht-Cages

The leapfrog transformation L can in a natural and direct way be extended to connected graphs embedded in any orientable closed surface, in particular, to ht-cages [**11**]. Clearly, if H is an ht-cage then so is LH.

For the second conjecture see Sections 7 and 8; its analogue for ht-cages is:

$$\mathrm{spec}\mathrm{H} \subset \mathrm{spec}\mathsf{L}^2\mathrm{H} \quad \textit{for any ht-cage } \mathrm{H}.$$

The validity of this statement follows immediately from

THEOREM 12.1. *Every ht-cage* H *satisfies*

$$\mathrm{H} \mid \mathsf{L}^2\mathrm{H}.$$

The proof which runs along the same lines as the proof of Theorem 8.1 is omitted. □

*

The authors thank the referees for their valuable advices.

References

[1] D. M. Cvetković, M. Doob and H. Sachs, *Spectra of Graphs*, Deutscher Verlag der Wissenschaften, Berlin 1980/81; Academic Press, New York 1980; J.A. Barth Verlag, Heidelberg-Leipzig 1995. Chapter 4, pp.116–133.

[2] P. W. Fowler and D.E. Manolopoulos, *An Atlas of Fullerenes*, Clarendon Press, Oxford 1995. Section 3.4, pp. 51–59.

[3] P. R. Goodey, *A class of Hamiltonian polytopes*, J. Graph Theory **1** (1977), 181–185.

[4] B. Grünbaum, *Convex Polytopes*, Interscience Publishers (J. Wiley & Sons), London-New York-Sydney 1967. Section 13.4, Exercise No. 2, p. 281.

[5] B. Grünbaum and T. S. Motzkin, *The number of hexagons and the simplicity of geodesics on certain polyhedra*, Can. J. Math. **15** (1963), 744–751.

[6] P. E. John, *Canonical representation of (3,6)-cages*, Poster, "Clusters, Cages and Giant Molecules", Bielefeld, Dec. 10-14, 1997.

[7] R. L. Johnston, J. Chem. Soc. Faraday Trans. **87** (1991), 3353.

[8] E. C. Kirby, R. B. Mallion and P. Pollak, *Toroidal Polyhexes*, J. Chem. Soc. Faraday Trans. **89** (12) (1993), 1945–1953.

[9] J. Liu, H. Dai, J. H. Hafner, D. T. Colbert, R. E. Smalley, S. J. Tans and C. Dekker, *Fullerene 'crop circles'*, Nature **385** (1997), 780–781.

[10] R. E. Merrifield and H. E. Simmons, *Topological Methods in Chemistry*, J. Wiley & Sons, New York 1989. Chapter 11, pp. 167–177.

[11] M. Yoshida, M. Fujita, P. W. Fowler and E. C. Kirby, *Non-bonding orbitals in graphite, carbon tubules, toroids and fullerenes*, J. Chem. Soc. Faraday Trans. **93** (1997), 1037–1043.

UNIVERSITY OF EXETER, DEPARTMENT OF CHEMISTRY, EXETER, ENGLAND, U.K.
E-mail address: `P.W.Fowler@exeter.ac.uk`

TECHNICAL UNIVERSITY OF ILMENAU, INSTITUTE OF MATHEMATICS, ILMENAU, GERMANY
E-mail address: `Peter.John@Mathematik.TU-Ilmenau.DE`

TECHNICAL UNIVERSITY OF ILMENAU, INSTITUTE OF MATHEMATICS, ILMENAU, GERMANY
E-mail address: `Horst.Sachs@Mathematik.TU-Ilmenau.DE`

DIMACS Series in Discrete Mathematics
and Theoretical Computer Science
Volume 51, 2000

A Generalized Ring Spiral Algorithm for Coding Fullerenes and other Cubic Polyhedra

Patrick W. Fowler, Tomaž Pisanski, Ante Graovac, and Janez Žerovnik

ABSTRACT. The so-called *ring spiral* algorithm is a convenient means for generating and representing certain fullerenes and some other cubic polyhedra. In 1993 Manolopoulos and Fowler presented a fullerene on 380 vertices without a spiral. No smaller unspiralable fullerene is known. In the spring of 1997, using computer, Gunnar Brinkmann found the smallest cubic polyhedron without a spiral. It has only 18 vertices. Here we generalize the ring spiral approach in order to obtain a canonical representation for arbitrary planar cubic polyhedra. Some other questions are addressed: for instance possible generalization of this method to polyhedra with vertices of arbitrary valency.

1. Introduction

Quite apart from the many mathematical reasons for interest in the cubic polyhedra, there are topical chemical reasons for a renewed investigation of their properties. Polyhedral carbon cages with pentagonal and hexagonal faces (*'fullerenes'*) have been discovered **[12]** and have a rich and growing chemistry. Problems of enumeration, construction, interconversion and relative energetics are still subjects of intensive research but it appears that much of the general behaviour of fullerenes can be understood in terms of simple graph-theory-based treatments of π-electronic structure coupled to curvature-based discussion of steric strain in the σ-electronic framework **[9]**. Much early systematic work concentrated on the construction and enumeration of fullerene isomers, and made heavy use of the ring spiral algorithm **[14]**. The present paper returns to this algorithm, its successes, failures and generalisations.

The *ring spiral algorithm* for construction of a polyhedron of f faces starts with an oriented disk with a polygon of size r_1 and then attaches polygons of size $r_2, r_3, \ldots, r_k, \ldots r_f$ to the disk in a uniquely determined spiral in such a way that the

1991 *Mathematics Subject Classification.* 05 C 12, 92 E 10.

Part of this paper was delivered as a talk by T. P. and J. Ž. at a conference in Kansas City in 1993. The second author was supported in part by "Ministrstvo za znanost in tehnologijo Slovenije", proj. no. J2-6193-0101-97 and J1-6161-0101-97. The fourth author was supported in part by "Ministrstvo za znanost in tehnologijo Slovenije", proj.no. J1-7035-0101-97.
The authors would like to thank Gunnar Brinkmann for interesting discussion and for providing some vital data for the paper, and Bojan Mohar for pointing out that by considering triangulations rather than fullerenes, one can simplify the proof of Proposition 3.2.

graphs $P_1, P_2, \ldots, P_k, \ldots, P_f$, that are produced in this process, have all maximum valency ≤ 3. At each step of the algorithm, we can view P_k as a map (polyhedron) on the sphere with a disk removed. If flattened, P_k has a shape of a disk (with no holes). All vertices in the interior are trivalent whereas the vertices on the boundary are either 3- or 2-valent. If the boundary ∂P_k is traced according to the orientation of the disk, we can record the valencies of the s_k boundary vertices. In this way we can produce a *word* $w_k = x_{k,1}x_{k,2}\ldots x_{k,s_k}$ over the symbols $\{0,1\}$, where 0 represents valency 3 and 1 represents valency 2. The word w_k can be cyclically permuted. Its actual form depends on the starting vertex on the boundary. Such a word is sometimes called the *boundary code*. For a discussion on the history, variations and improvements on the boundary code for polyhexes, see [**4**].

Clearly, $w_1 = 11\ldots1 - (r_1 - \text{times})$. If the graphs $G_{f-1} = G_f$ are trivalent then the disk closes up to the sphere and $w_{f-1} = 00\ldots0 - (r_f - \text{times}), w_f = \emptyset$. In order to avoid ambiguity we will choose at each step of the spiral algorithm a particular cyclic permutation of the word w_k according to the following rule:

For a given ring size r one can choose any pair (z, o) as indices of entries **z**eroes and **o**nes, such that $r = z + o + 2$. The general rule in an obvious exponent notation reads as follows:

$$R(z,o) = 10^z1 \to 01^o0$$

This means that the word of the form

$$w_k = 10^z1v$$

is replaced by

$$u_k = v01^o0$$

If the resulting word consists of zeros only, the procedure stops and $w_{k+1} = u_k$. Otherwise, u_k is cyclically shifted to the right until the rightmost 1 arrives on the first position (leftmost position) in order to obtain w_{k+1}.

It may happen that a step is not possible. In such a case the sequence $s = (r_1, r_2, \ldots, r_f)$ is not a legitimate ring-spiral sequence. If however, the algorithm succeeds then the sequence $s = (r_1, r_2, \ldots, r_f)$ is called a *ring spiral* of the trivalent polyhedron. If we choose from all possible ring spirals the one that is lexicographically first, we get the *lexicographic generalized spiral code* of the polyhedron. If there is no danger for confusion we will simplify the code to the form $s = (r_1, r_2, \ldots, r_f) = r_1r_2\ldots r_f$. We also use the notation ∂s_k to denote the boundary code w_k of a partial polyhedron P_k defined by the first k steps of the spiral code. Note that lexicographic generalized spiral code falls into the category of *canonical codes*. Roughly speaking a code is canonical if it is the same for all isomorphic polyhedra and different for non-isomorphic ones.

EXAMPLE 1.1. Let us show how one can produce the trigonal prism given its spiral code

$$s = 34443.$$

The corresponding sequence of words is as follows:

$$s_1 = 3, \partial s_1 = w_1 = 111$$

$$s_2 = 34, \partial s_2 = w_2 = 10101$$

$$s_3 = 344, \partial s_3 = w_3 = 10010$$

$$s_4 = 3444, \partial s_4 = w_4 = 000$$

$$s = s_5 = 34443, \partial s_5 = \emptyset$$

When we attach a triangle we apply one of the following rules:

$$R(0,1) = 11 \to 010$$
$$R(1,0) = 101 \to 00$$

For a quadrilateral the rules would be:

$$R(0,2) = 11 \to 0110$$
$$R(1,1) = 101 \to 010$$
$$R(2,0) = 1001 \to 00$$

The whole process can be thus described more concisely:

$$(11)1 \overset{R(0,2)}{\to} (0110)1 = (101)01 \overset{R(1,1)}{\to} (010)01 = (1001)0 \overset{R(2,0)}{\to} 000.$$

We should mention that the code

$$s = 34443$$

contains superfluous information. The very last symbol 3 represents the length of the all-zero string at the end. The reason that we distinguish between the code

$$s = 3444$$

and

$$s = 34443$$

is topological. While

$$s = 34443$$

represents the polyhedron,

$$s = 34443$$

represents the same polyhedron without the last triangular face. For the sake of completness we should add an extra rule

$$R(3,-) = 000 \to \emptyset$$

which is applicable only if $s_k = 000$ for some k.

In the exponent notation r^0 represents the empty string.

EXAMPLE 1.2. Hence we can abbreviate

$$55555655655555 = 5^5 65^2 65^5$$

and

$$55555656665655555 = 5^5 656^3 565^5$$

If $r_k \in \{5,6\}$, for each k, that is, if the polygons are pentagons and hexagons, we get in such a way a *fullerene*.

To generate fullerenes, we start with one of the two initial words:

$$S(5) = 11111 = 1^5$$
$$S(6) = 111111 = 1^6$$

and we apply a sequence of the following rules in any order

Pentagon:

$$R(0,3) = 11 \to 01110$$
$$R(1,2) = 101 \to 0110$$
$$R(2,1) = 1001 \to 010$$
$$R(3,0) = 10001 \to 00$$

Hexagon:

$$R(0,4) = 11 \to 011110$$
$$R(1,3) = 101 \to 01110$$
$$R(2,2) = 1001 \to 0110$$
$$R(3,1) = 10001 \to 010$$
$$R(4,0) = 100001 \to 00$$

We obtain a fullerene if the rules may be applied and we finish the process in one of the following two terminal words:

$$T(5) = 00000 = 0^5$$
$$T(6) = 000000 = 0^6$$

The spiral code has several advantages.

1. It is easy both for a human and a computer program to generate (or reconstruct) a polyhedron from its spiral code.
2. It is a concise description of a cubic polyhedron.
3. From the code one can immediately recognise the sizes of faces of the polyhedron.
4. Initial sub-sequences of the code define an increasing sequence of partial polyhedra, such that each polyhedron is homeomorphic to a disk.

Unfortunately not all trivalent polyhedra (not even fullerenes!) admit ring spirals. This was shown by D.E Manolopoulos and P.W. Fowler in [**14**]. They produced a fullerene on 380 vertices that has no spiral. Over 400 larger fullerenes with ≤ 1000 vertices have since been found, but none smaller than the 380 vertex example [**3, 20, 21**].

Recently Gunnar Brinkmann [**2**] found the smallest trivalent planar polyhedron without a spiral (see Figure 1) while the problem of finding the smallest fullerene without a spiral remains open.

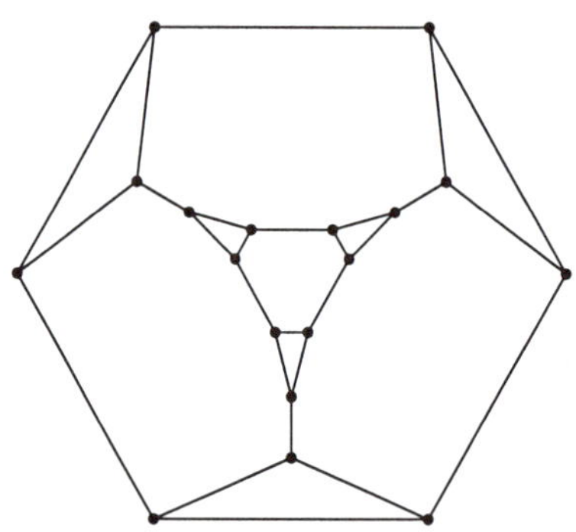

FIGURE 1. The smallest trivalent planar polyhedron without a ring spiral, determined by Gunnar Brinkmann, is a truncated trigonal prism.

It is natural to ask whether it is possible to generalize the spiral code in such a way as to encode all planar trivalent polyhedra and such that the conventional ring spiral - if it exists - would be lexicographically the best among all descriptions.

In [2] Brinkmann proposed an efficient generalization of such a code which shares almost all advantages of the original ring-spiral code.

The smallest trivalent planar polyhedron without a spiral turned out to be a truncated trigonal prism. It was shown by Brinkmann and Fowler [3] that this is not a coincidence.

THEOREM 1.3 (Brinkmann and Fowler, [3].). *The truncate of any trivalent planar polyhedron P does not have a spiral unless P is the tetrahedron.*

Obviously it would be interesting to explore spiral properties of truncations of four-valent polyhedra, in particular those that arise as truncations of medials of other polyhedra. For a definition of the *medial transformation*, see for instance [8].

2. Generalized Spiral Code

The generalization that we propose here is different from the *Brinkmann code*, first published in [2]. While both generalizations have comparable characteristics our approach has one minor advantage. Since our partial polyhedra during the reconstruction process are simply connected, the generation of the polyhedron from our code may be more suitable for hand calculations.

As in the conventional ring spiral algorithm, we choose positive orientation of the disk. We will assume that the ring spiral step attaches the polygon to the disk at *position* 1. If, instead, we slide the polygon along the perimeter of the disk to the next position, we call that *position* 2, etc. We are considering the gap between two consecutive 1's on the boundary; see Figure 2.

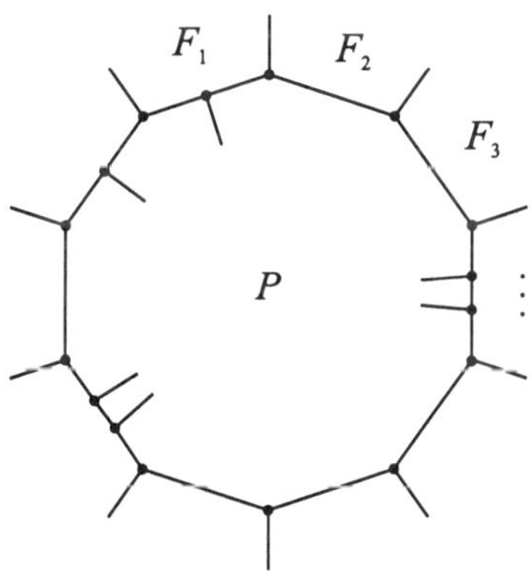

FIGURE 2. A partial polyhedron P and positions along the boundary ∂P where other faces $F_1, F_2, \ldots,$ can be attached. If $F_1 \neq F_i$, for $i \neq 1$, the spiral step is possible.

Formally, the *generalized ring spiral* is a sequence:

$$s_f = (p_1, p_2, \ldots, p_f; r_1, r_2, \ldots, r_f)$$

with the following interpretation: At step k the polygon r_k is attached to the partial polyhedron P_{k-1} at the position p_k. By convention $p_1 = p_2 = p_f = 1$.

We can simplify the notation if there is no danger of confusion. If a face of size r is attached at position p, we can write this as r_p. We will also omit subscript 1 in order to shorten the notation. Hence $r_1 = r$. If $p = 1$ we say that k is a *spiral step*.

TABLE 1. A rewriting of a part of Table 2 of [**10**], giving the canonical codes of all fullerenes on ≤ 30 vertices.

C_{20}	5^{12}
C_{24}	$5^5 6 5^2 6 5^2$
C_{26}	$5^5 (65)^3 5^4$
C_{28}	$5^3 (65)^3 5^6 6$
C_{30}	$5^6 6^4 5^6$
C_{30}	$5^5 6 5 6^3 5 6 5^5$
C_{30}	$5^4 (6^2 5)^2 5^6 6$

In order to compare the sequences, we consider the full form of s_f:

$$s_f = (p_1, p_2, \dots, p_f, r_1, r_2, \dots, r_f)$$

and the sequences are ordered lexicographically.

This means that if a polyhedron admits a spiral then its canonical form will reflect this fact.

However, not every sequence is realizable. For instance, the sequence $433_2 3_p, \dots$ is not realizable for any position p.

Some codes for small fullerenes are listed in Table 1, rewritten from [**10**] in our shorthand notation.

The C_{380} fullerene without a spiral can be described as follows:

$$C_{380} = 5^3 6^{133} (5^2 6^8)^2 5^2 6^7 5_2 6^6 5_3 6^6 5_3 6^{12}$$

The Tutte graph [**1**] is:

$$T = 5^2 4 5^3 10 9 10 4_2 5^3 5_2 (54)^2 5_2 5 4 5_3 5 10$$

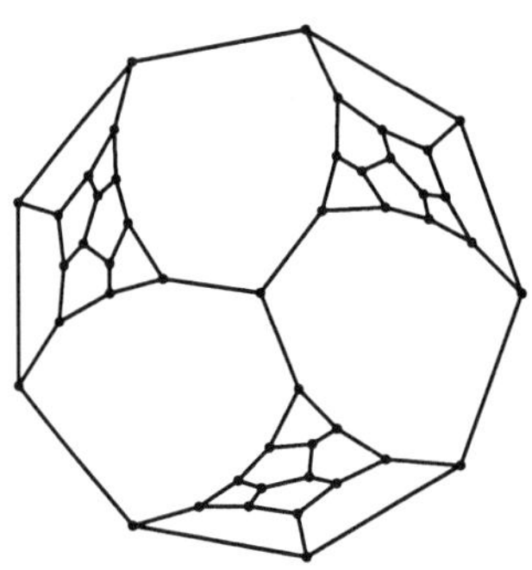

FIGURE 3. The Tutte graph

An undergraduate student, Darja Gartner, from Ljubljana wrote a computer program for generating a cubic polyhedron from its generalized spiral code. Her program has formed part of the Vega package since February 1994; see [**16**].

THEOREM 2.1. *Every convex cubic polyhedron can be described by a generalized ring spiral.*

PROOF. We give only a sketch. If false, there must be a smallest counterexample. Let us build a partial polyhedron P using the ordinary spiral. The inability to continue results from the fact that at some point adding a polygon F to the boundary ∂P would make the new partial polyhedron Q non-simply connected, see Figure 4. Let $(p_1 = 1, p_2, \ldots, p_k)$ be positions at which F meets P. There must be a gap between p_i and p_{i+1}. Take the boundary component of ∂Q that is the first after the edge where the face F was attached to P. This boundary component is composed of positions $p_1 + 1, p_1 + 2, \ldots, p_2 - 1$. Now proceed with the next face F' in a spiral-like step within this boundary component. This means F' touches P at step $p_1 + 1$. If the resulting partial polyhedron $P + F'$ is simply connected, then we succeeded to make another step in our generalized procedure, otherwise we repeat the argument but now the boundary component has the part belonging to ∂P shorter than before since P and F' may intersect only at some of the positions: $p_1 + 1, p_1 + 2, \ldots, p_2 - 1$. This follows from the renowned Jordan curve theorem from topology. Since we are working with integers, there are only a finite number of steps possible before we have to find a solution and make a step forward. After filling all but the last boundary component of Q by disks, we attach face F to the resulting partial polyhedron. □

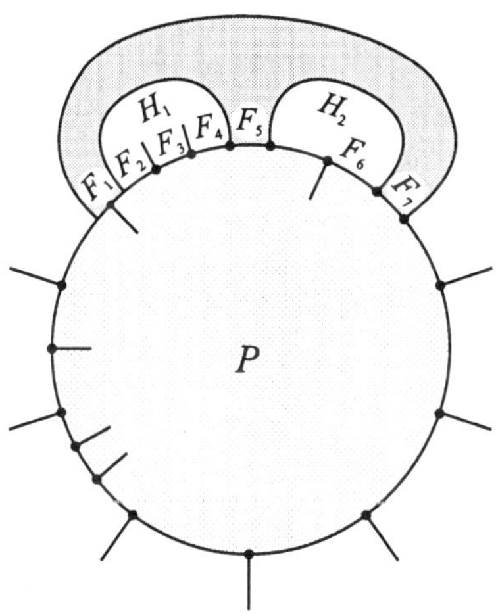

FIGURE 4. Attaching a face to a partial polyhedron may result in a multiply connected region.

This result can be used as the basis of a straight-forward polynomial-time algorithm of complexity $O(n^3)$ for computing a lexicographic generalized ring spiral code for a cubic polyhedron. Namely, at each step attach to the partial polyhedron the face at the first position that does not produce a multiply connected region. If the polyhedron admits a spiral, this algorithm (call it A) will find the lexicographically smallest. Algorithm A was implemented in June 1998 by Drago Bokal, another undergraduate from Ljubljana. A more sophisticated algorithm (call it P), presented in the proof of the following result attains the optimal time complexity. The canonical codes produced by the Algorithms A and P are not necessarily the same.

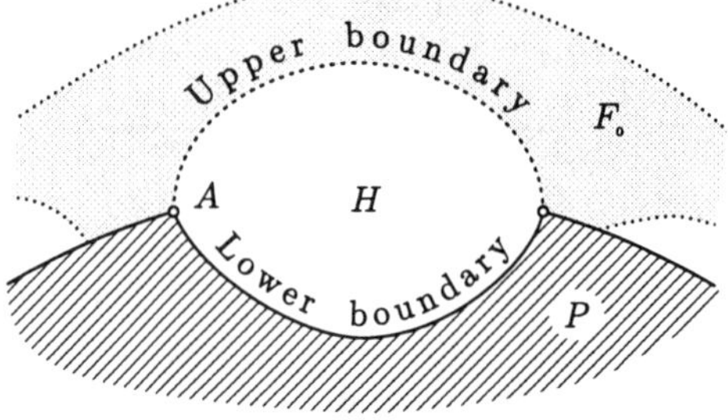

FIGURE 5. The hole H prevents the face F_0 from being attached to the partial polyhedron P.

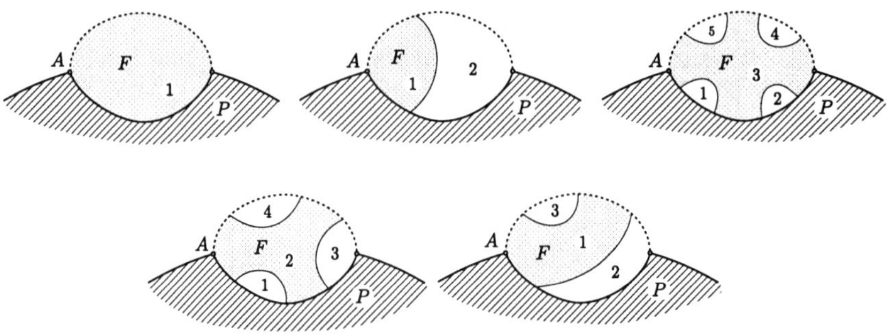

FIGURE 6. When trying out the face F at A, the remainder of the hole H breaks down into some lower and upper holes. An *upper hole* is a hole touching the upper boundary. A *lower hole* is a hole that is not an upper hole. The order of patches in typical cases is shown by the numbers inside the faces.

A canonical generalized ring spiral code for a convex cubic polyhedron can be computed in time $O(n^2)$. A formal proof of this will be presented in another paper. Here we only sketch the idea. Since there are $O(n)$ starting positions we have only to show that one code - not necessarily the canonical one - can be found in linear time.

The key idea lies in efficient data structures, for instance cross-linked doubly-linked lists connecting the edges. In addition we present here the Algorithm P, a recursive procedure for patching holes.

Algorithm P. [Patching a hole].

- Input: A hole H with lower and upper boundaries (+ global information about the unused faces)
- Recursive step: Add a face F at A.
    1. Try out the face F in order to determine new lower and upper holes.
    2. Patch each lower hole.
    3. Patch F [Print out the size of F and its relative position along the current boundary].
    4. Patch each upper hole.

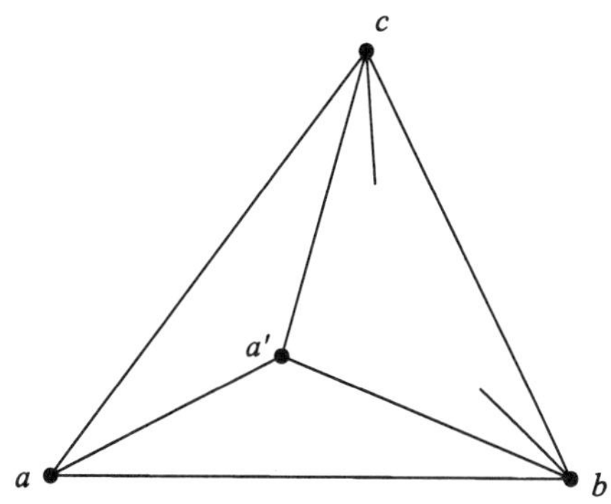

FIGURE 7. The cases 122 and 123 result in a smaller separating triangle $a'bc$.

The missing step of the proof, which we postpone for another paper, is the precise description of how the relative position of F is determined in step 3. This is very technical and not relevant for chemical applications. In August 1998 Drago Bokal programmed and extensively tested a variation of algorithm P. It turns out that the practical time complexity of Algorithm P is indeed linear.

From the proof of the previous theorem we can deduce both the correctness of the algorithm and the fact that there is no need for backtracking.

3. Proper Generalized Spirals

The generalized spiral algorithm generates all sequences that are of interest here, in the planar case.

A generalized spiral is called *proper* if the face r_{k+1} shares an edge with the face r_k. Note, that a ring spiral is always a generalized proper spiral.

There is an old result of Tutte that each four-connected planar graph is hamiltonian; see [**18**]. This means that a planar triangulation (deltahedron) without separating triangles is four-connected and hence hamiltonian as shown even earlier by Whitney; see [**19**]. Since the dual of a planar triangulation is a cubic map, a proper generalized spiral in a cubic map corresponds to a hamilton path in its dual. Hence, if a cubic map has the property that any three faces that are mutually adjacent are adjacent only along three edges with a common vertex, then such a map has a proper generalized spiral.

Actually we have:

PROPOSITION 3.1. *A cubic map has a proper generalized spiral if and only if its dual has a Hamilton path.*

PROPOSITION 3.2. *No dual of a fullerene has a separating triangle.*

This is mentioned in the fullerene literature[**9**] but we include a proof in order to convince the reader.

PROOF. Let F^* be a triangulation, dual to a fullerene. Then F^* has vertices of valency 5 and 6[**15**]. Let T, with vertices a, b, c be a separating triangle in F^*. At each corner of T we put a label, denoting the valency of that vertex that leads inside the triangle. Since T is separating, each label is at least 1. Since F^* is a

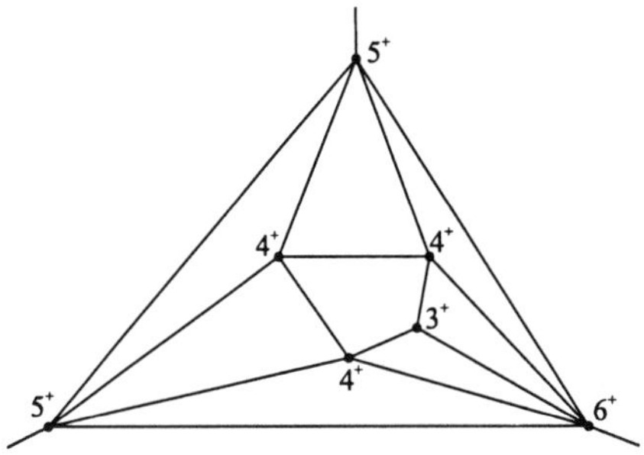

FIGURE 8. The case 223 would lead to a vertex of valency < 5 since the vertex 3^+ can be completed either to a 3-valent or 4-valent vertex.

fullerene, each label is at most 3. The following are all possible cases:

$$111, 112, 113, 122, 123, 133, 222, 223, 233, 333.$$

Note that the label 11x implies 111 and the interior of T is capped. Since the interior vertex is of valency 3, F^* would not be a fullerene. The following cases remain:

$$122, 123, 133, 222, 223, 233, 333.$$

A label 3 inside implies a label 1 outside. Reversing the roles of inside and outside we rule out all cases 33x. The following cases remain:

$$122, 123, 222, 223.$$

Without loss of generality we may assume that T does not have a separating triangle in its interior.

This rules out 222. Case 12x leads to a smaller separating triangle as shown in Figure 7. The only remaining case is 223 and is impossible since it would yield a vertex of valency < 5 inside T; compare Figure 8. □

COROLLARY 3.3. *The dual of a fullerene is 4-connected.*

This gives us the following result.

COROLLARY 3.4. *A fullerene has a proper generalized spiral.*

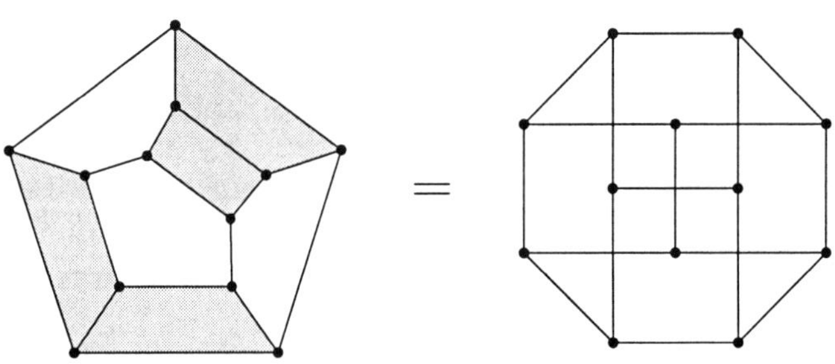

FIGURE 9. The smallest non-trivial omnispiral polyhedron has 12 vertices.

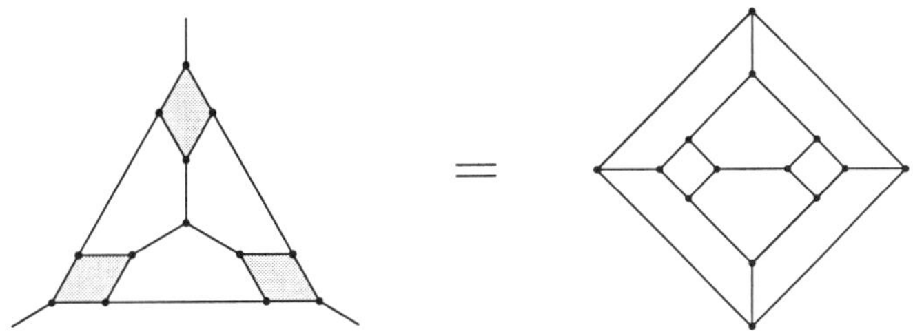

FIGURE 10. The next smallest non-trivial omnispiral polyhedron on 14 vertices has isolated quadrilaterals.

TABLE 2. The fraction of omnispiral cubic polyhedra on a given number of vertices seems to be small.

Number of vertices	4	6	8	10	12	14	16	18	20	22
Number of omnispirals	1	1	1	1	2	2	3	3	3	1
Number of nonomnispirals	0	0	1	4	12	48	230	1246	7592	49565

4. Omnispiral Polyhedra

One may think of polyhedra without a spiral as *the worst possible case* when each starting position leads to at least one non-spiral step. Here we take a brief look at the opposite extreme: the case where each starting position leads to a spiral. We call such a polyhedron an *omnispiral polyhedron.*

For instance, every n−sided prism Π_n is an omnispiral polyhedron. In general, there are three different starting positions, leading to the following codes:

- $n4^n n$
- $4^2 n4n4^{n-3}$
- $4n4n4^{n-2}$

For a trigonal prism the first code is canonical while for $n > 3$ the second code is canonical.

The smallest possible case is the tetrahedron T with the (unique) code 3^4. Figures 9 and 10 depict the two smallest non-trivial omnispiral polyhedra which were found by Gunnar Brinkmann. His preliminary computations suggest that omnispiral polyhedra are very scarce and that their number does not grow in a regular monotonic manner (see Table 2.)

5. Non-Cubic Polyhedra

The generalized ring spiral code can be used to describe any regular planar polyhedron, but for non-cubic polyhedra the interpretation of the code has to be changed. For instance, a four-valent polyhedron can be built by repeatedly attaching polygons to a partial polyhedron. In case of cubic polyhedra, we used only 0 and 1 on the boundary of the disk. If we keep the meaning of 0, ie. a vertex whose valencies have been used, and interpret 1 as a vertex with one free valency, then we may introduce 2 to denote a vertex with two free valencies. Figure 11 shows how the elongated octahedron, ie. the polyhedron obtained by capping two opposite sides of a cube, can be built by a spiral.

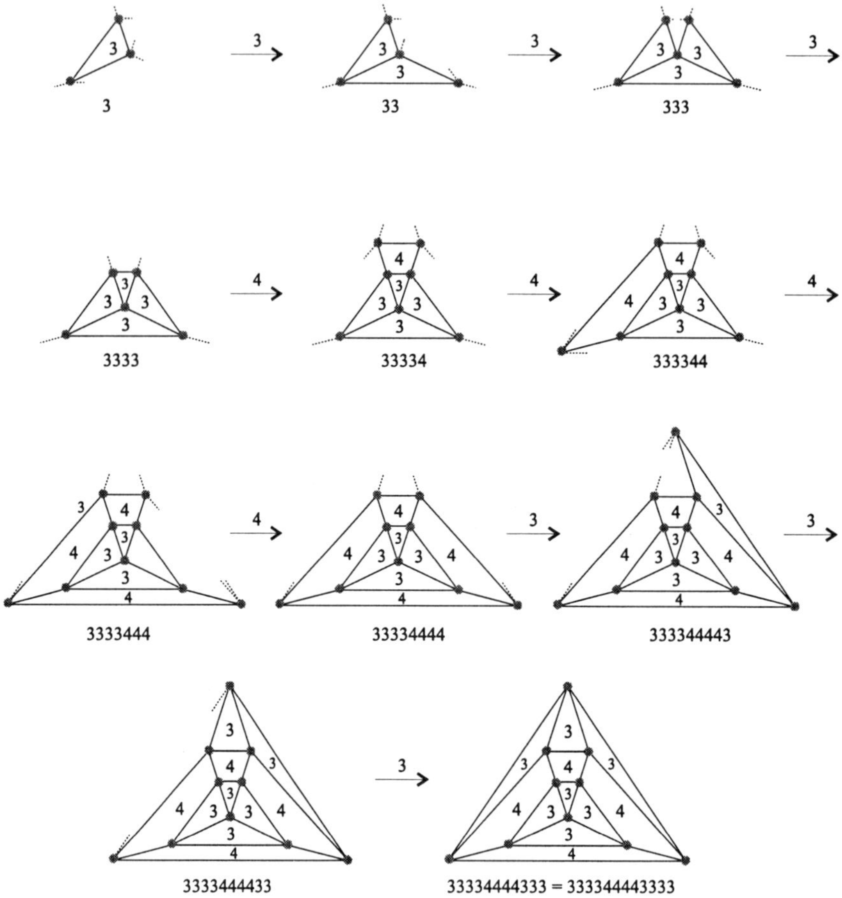

FIGURE 11. The elongated octahedron can be described by a generalization of the spiral code for 4-valent planar polyhedra. The sequence of face additions is shown. An obvious extension gives $3^4 4^{4m} 3^4$ for all members of the family of 4-valent planar polyhedra obtained by capping a tower of m cubes at both ends.

There are several ways in which one can extend this approach to non-planar polyhedra, which could be useful for describing, say toroidal cages. However, the extensions we have found so far are not very natural and the time needed to find the canonical code may grow. Probably the most reasonable extension would be to attach faces as in the planar case but cutting the polyhedron along all those edges that would result in a non-planar polyhedron. This means that at the end one would have a disk with all faces attached correctly but on the boundary all edges would have to be pairwise identified. In this respect our code would become similar to the code proposed by Gunnar Brinkmann [**2**].

In this paper we have presented an algorithm for computing the lexicographic generalized spiral code in time $O(n^3)$. We also presented an algorithm that can compute a differently canonicalised generalized spiral code faster (in practice the time complexity is $O(n^2)$.) It would be interesting to see if it is possible to find the lexicographic generalized spiral code in time $O(n^2)$.

References

[1] M. Behzad, G. Chartrand, and L. Lesniak-Foster, *Graphs & Digraphs*, Prindle, Weber & Schmidt International Series, 1979.
[2] G. Brinkmann, Problems and Scope of the Spiral Algorithm and Spiral Codes for Polyhedral Cages. *Chemical Physics Letters*, **272** (1997) 193–198.
[3] G. Brinkmann, and P.W. Fowler, Spiral coding of leapfrog polyhedra, *J. Chem. Inf. Comp. Sci.*, **38** 463–468.
[4] G. Caporossi, and P. Hansen, Enumeration of Polyhex Hydrocarbons to $h = 21$. *Journal of Chemical Information and Computer Sciences*, 38:610–619, 1998.
[5] H.S.M. Coxeter, *Regular Polytopes*, 3rd ed. Dover Publications Inc., New York 1973.
[6] H.M. Cundy, and A.R. Rollett, *Mathematical Models*, 3rd ed. Tarquin Publications, Norfolk, 1981.
[7] P.W. Fowler, and J.E. Cremona, Fullerenes Containing Fused Triples of Pentagonal Rings, *J. Chem. Soc. Faraday Trans.* **93** (1997) 2255–2262.
[8] P. Fowler, and T. Pisanski, Leapfrog Transformations and Polyhedra of Clar Type, *J. Chem. Soc. Faraday Trans.* **90**(19) (1994) 2865–2871.
[9] P.W. Fowler, and D.E. Manolopoulos, *An Atlas of Fullerenes*, Oxford Univ. Press 1995.
[10] P.W. Fowler, D.E. Manolopoulos, and R.P. Ryan, Isomerisation of the Fullerenes, *Carbon* **30** (1992) 1235–1250.
[11] J.L. Gross, and T.W. Tucker, *Topological Graph Theory*, Wiley Interscience, 1987.
[12] H.W. Kroto, J.R. Heath, S.C. O'Brien, R.F. Curl, and R.E. Smalley, C_{60} - Buckminsterfullerene, *Nature*, **318**(6042) (1985) 162–163.
[13] D.E. Manolopoulos, D.R. Woodall, and P.W. Fowler, Electronic Stability of Fullerenes: Eigenvalue Theorems for Leapfrog Carbon Clusters, *J. Chem. Soc. Faraday Trans.*, **88**(17) (1992) 2427–2435.
[14] D.E. Manolopoulos, and P.W. Fowler, A Fullerene without a Spiral, *Chem. Phys. Lett.* **204** (1993) 1–7.
[15] T. Pisanski, On Planar Graphs with 12 Vertices of Degree Five, *Glasnik Mat.* **12** (1977) 233–235.
[16] T. Pisanski (ed.), *Vega Version 0.2 Quick Reference Manual and Vega Graph Gallery*, Ljubljana, 1995, `http://vega.ijp.si/`
[17] T. Saaty, and P. Kainen, *The Four Color Problem; Assaults and Conquest*, McGraw-Hill 1977.
[18] W.T. Tutte, A Theorem on Planar Graphs, *Trans. Am. Math. Soc.* **82** (1956) 99–116.
[19] H. Whitney, A Theorem on Graphs, *Ann. Math.* **32** (1931) 378–390.
[20] M. Yoshida, and P.W. Fowler, Dihedral Fullerenes of Threefold Symmetry with and without Face Spirals, *J. Chem. Soc. Faraday*, **93** (1997) 3289–3294.
[21] M. Yoshida, and P.W. Fowler, Systematic Relationships between Fullerenes without Spirals, *Chem. Phys. Lett.* **278** (1997) 256–261.

DEPT OF CHEMISTRY, UNIVERSITY OF EXETER, STOCKER ROAD, EXETER, EX4 4QD, UK
E-mail address: `P.W.Fowler@exeter.ac.uk`

IMFM/TCS UNIVERSITY OF LJUBLJANA, JADRANSKA 19, SI-1111 LJUBLJANA, SLOVENIJA
E-mail address: `Tomaz.Pisanski@fmf.uni-lj.si`

RUGJER BOŠKOVIĆ INSTITUTE, HR-41001 ZAGREB, POB 1016, CROATIA
E-mail address: `Graovac@rudjer.irb.hr`

UNIVERSITY OF MARIBOR, TECHNICAL FACULTY, SMETANOVA 17, SI-2000 MARIBOR, SLOVENIJA
E-mail address: `Janez.Zerovnik@fmf.uni-lj.si`

DIMACS Series in Discrete Mathematics
and Theoretical Computer Science
Volume 51, 2000

An Efficient Algorithm for Determining Fixed Bonds and Normal Components in a Bipartite Graph

Xiaofeng Guo and Milan Randić

ABSTRACT. Determining fixed bonds and normal components of bipartite molecular graphs is closely related to the enumeration of Kekule structures, and calculations of LM-conjugated circuit polynomials, sextet polynomials, and generalized molecular bond orders. Some algorithms for determining fixed bonds and normal components of benzenoid hydrocarbons have been reported. In this paper, we will consider general bipartite graphs. The concept of maximal alternating growing subgraphs of a conjugated bipartite graph G in two directions starting from a root vertex is introduced. Some interesting properties of maximal alternating growing subgraphs and normal bipartite graphs are given. Based on the important properties, we establish an efficient algorithm for determining fixed bonds and normal components of a bipartite graph G. The complexity of the algorithm is $O(mn^{1/2})$ in general cases, where m and n are the numbers of edges and vertices of G respectively. Particularly, if G is a benzenoid hydrocarbon, the algorithm becomes linear.

1. Introduction

A graph G is said to be a conjugated bipartite graph if it is bipartite and has perfect matchings (or Kekule structures). A circuit C in a conjugated bipartite graph G is said to be conjugated if the graph obtained from G by deleting all the vertices on C has Kekule structures. An edge of G is said to be a fixed double (resp. single) bond if it is (resp. is not) contained in any Kekule structure of G. If G has no fixed bond, it is called a normal bipartite graph. If G has fixed bonds, a maximal connected subgraph of G with no fixed bond is called a normal component of G. The concept of fixed bonds has been shown to be very useful in the enumeration of Kekule structures [**1**]–[**4**], and in calculations of LM-conjugated circuits polynomials [**5, 6**], and sextet polynomials [**7**]–[**12**] of benzenoid hydrocarbons, etc. In fact, fixed bonds are not in any conjugated circuit, so if we delete all fixed single bonds and the end vertices of all fixed double bonds from a benzenoid hydrocarbon B, the resultant graph consisting of normal components of B has the same number

1991 *Mathematics Subject Classification.* 05, 92.

This work is supported by NSFC (National Natural Science Foundation of China). We also acknowledge the support of the CSC (China Scholarship Council) for making the visit of Xiaofeng Guo to Drake University possible. We also thank Dr. B.D. Gute (Duluth, MN) for suggesting a number of improvements in the presentations of the material.

of Kekule structures, and the same LM-conjugated circuit polynomial and sextet polynomial as B. Therefore, it can simplify the above mentioned enumeration and calculations to determine fixed bonds and normal components of conjugated bipartite molecular graphs. In recent years, many investigations on the properties and structures of benzenoid hydrocarbons with (or without) fixed bonds have been made [**13**]–[**19**]. Some $O(n^3)$ and $O(n^2)$ algorithms for determining fixed bonds and normal components of benzenoid hydrocarbons have been given [**20, 21**]. In this paper, we will consider general bipartite graphs, which include several classes of molecular graphs such as generalized polyhex graphs, helicenes, etc. The concept of maximal alternating growing subgraphs in two directions of a conjugated bipartite graph G starting from a root vertex is introduced. It is proved that if G is a normal bipartite graph then any vertex or any edge of G is contained in the intersection of maximal alternating growing subgraphs of G in two directions starting from a root vertex, and that any two edges or any two vertices of G are contained in a conjugated circuit of G. Based on the above important properties, an efficient algorithm for determining all fixed bonds and normal components of a conjugated bipartite graph G is established. The complexity of the algorithm depends on the complexity of another algorithm for finding a Kekule structure of G, which is $O(mn^{1/2})$ in general cases [**22, 23**], and is linear if G is a benzenoid hydrocarbon [**24, 25**], where m and n are the numbers of edges and vertices of G respectively. After a Kekule structure has been given, the complexity of the other parts of the algorithm is $O(m) = O(n^2)$, and if G is a planar bipartite graph or a non-planar bipartite graph with the maximum degree being not greater than a constant, $O(m) = O(n)$. Therefore the complexity of the algorithm is $O(mn^{1/2})$ in general cases, and, if G is a benzenoid hydrocarbon, the algorithm becomes linear, that is the complexity is $O(m) = O(n)$. We conjecture that if G is a planar bipartite graph or a non-planar bipartite molecular graph there is a linear algorithm for finding a Kekule structure of G and a linear algorithm for finding fixed bonds and normal components of G.

2. Properties of Conjugated Bipartite Graphs and Maximal Alternating Growing Subgraphs

Let G be a conjugated bipartite graph, and let K be a Kekule structure of G. A path (resp. circuit) in G is said to be a K-alternating path (resp. K-alternating circuit) if its edges alternately appear in K and $E(G) \setminus K$ (the difference of edge sets $E(G)$ and K). Since G is bipartite, we can color vertices of G black and white so that any two adjacent vertices have different colors.

Definition 2.1. Let G be a conjugated bipartite graph, and K a Kekule structure of G. A maximal K-alternating growing subgraph of G is obtained by the following growing procedure:

1. Take a vertex v_0 of G. Let $G_0 = v_0$, let G_1 be the subgraph of G induced by all the single bonds (resp. the double bond) incident with v_0, and let G_2 be the subgraph of G obtained from G_1 by adding all the double (resp. single) bonds, each of which is incident with a vertex in $V(G_1) \setminus V(G_0)$, together with their end vertices.
2. Suppose G_i has been given and all the edges of G_i not in G_{i-1} are single (or double) bonds. Then G_{i+1} is obtained from G_i by adding all the double (or

single) bonds, each of which is incident with a vertex in $V(G_i) \setminus V(G_{i-1})$, together with their end vertices.

3. If either $V(G_{i+1}) \setminus V(G_i) = \phi$ or there is no edge incident with a vertex in $V(G_{i+1}) \setminus V(G_i)$ then go to Step 4. Otherwise set $i := i+1$, and go to Step 2.
4. The so obtained subgraph G_{i+1} of G is a maximal K-alternating growing subgraph of G rooted at v_0, which is of type I (resp. of type II) and is denoted by $G_1(v_0)$ (resp. $G_2(v_0)$).

Figure 1 shows a conjugated bipartite graph G and maximal K-alternating growing subgraphs $G_1(v_0)$ and $G_2(v_0)$ of two types of G, where the arrows indicate the growing process of $G_i(v_0)$, $i = 1, 2$. Note that a maximal K-alternating growing subgraph of G can be easily obtained, because the rules of the growth are very simple.

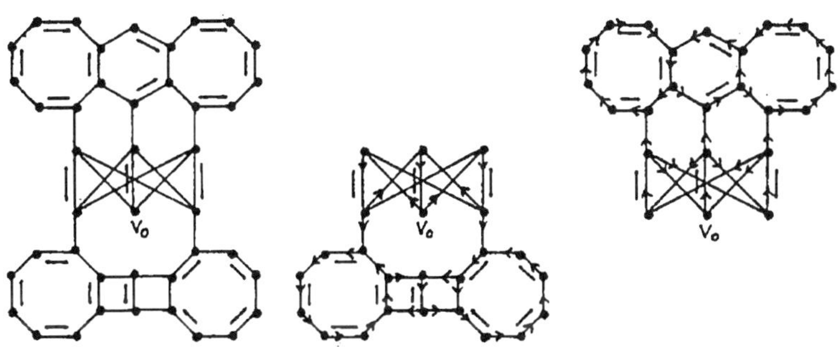

G $\quad$ $G_1(v_0)$ $\quad$ $G_2(v_0)$

FIGURE 1. A conjugated bipartite graph G and maximal K-alternating growing subgraphs of two types of G, where the arrows indicate the growing process.

By Definition 2.1, the following lemma obviously holds.

LEMMA 2.2. *Let G be a conjugated bipartite graph, K a Kekule structure of G, and $G_i(v_0)$ a maximal K-alternating growing subgraph of G with a root vertex v_0. Then, for any vertex v_j in $G_i(v_0)$, there is a K-alternating path from v_0 to v_j.*

LEMMA 2.3. *Let G be a conjugated bipartite graph, K a Kekule structure of G, and $G_1(v_0)$ (resp. $G_2(v_0)$) a maximal K-alternating growing proper subgraph of G with a root vertex v_0. Let E_1 (resp. E_2) be the set of all the edges in $E(G) \setminus E(G_1(v_0))$ (resp. $E(G) \setminus E(G_2(v_0))$ incident with a vertex in $G_1(v_0)$ (resp. $G_2(v_0)$), let E_1^* the set of all the edges in $E(G) \setminus E(G_1(v_0) - v_0)$ incident with a vertex in $G_1(v_0) - v_0$, and let e_0 be the double bond incident with v_0. (i) If e_0 does not belong to $E(G_1(v_0))$, then all edges in E_1^* are fixed single bonds of G; otherwise all edges in E_1 are fixed single bonds. (ii) All edges in E_2 are fixed single bonds.*

PROOF. We first consider the case that e_0 does not belong to $E(G_1(v_0))$.

For any edge e_k in E_1^* not incident with v_0, let v_k be the end vertex of e_k in $G_1(v_0) - v_0$. By Lemma 1, there is a K-alternating path starting from v_0 and a single bond to v_k in $G_1(v_0)$. In addition, any such K-alternating path from v_0 to

v_k in $G_1(v_0)$ must terminate at a K-single bond, otherwise e_k would be added into $G_1(v_0)$ by the growing procedure in Definition 2.1, a contradiction. By the same reasoning e_k is also a K-single bond. Then any K-alternating path starting from v_0 and a single bond to v_k in $G_1(v_0)$ is of odd length, and so v_0 and v_k must have different colors. Note that v_0 and any of its adjacent vertices also have different colors and all the edges in E_1^* incident with v_0 are K-single bonds. Hence all the end vertices of all edges in E_1^* belonging to $G_1(v_0) - v_0$ must have the same color and any edge in E_1^* is a K-single bond. Then, since $G_1(v_0) - v_0$ itself has a Kekule structure, all edges in E_1^* are fixed single bonds.

By a similar argument as above, all edges in E_2 (resp. E_1, if e_0 belongs to $E(G_1(v_0))$,) are also fixed single bonds. □

THEOREM 2.4. *Let G be a normal bipartite graph, and K a Kekule structure of G. Then, for any two vertices v_i, v_j of G, (i) v_j belongs to both K-alternating growing subgraphs $G_1(v_i)$ and $G_2(v_i)$; (ii) v_i and v_j are on a conjugated circuit of G.*

PROOF. (i) Assume that v_j does not belong to $G_1(v_i)$. Let E_i be the set of all the edges with only one end vertex in $G_1(v_i)$. Clearly, E_i is an edge cut of G. If there is an edge e_k in E_i not incident with v_i, then, by Lemma 2, e_k would be a fixed single bond of G, contradicting that G is normal. If all edges in E_i are incident with v_0, the double bond incident with v_0 would be a cut edge of G, and so were a fixed double bond of G, again a contradiction.

By a similar argument, v_j belongs to $G_2(v_i)$.

(ii) By (i), v_j belongs to both $G_1(v_i)$ and $G_2(v_i)$. Thus there are two K-alternating paths P_1 and P_2 from v_i to v_j respectively in $G_1(v_i)$ and $G_2(v_i)$. From the definition of $G_1(v_i)$ and $G_2(v_i)$, the initial edges of P_1 and P_2 are different. We can prove that the terminating edges of P_1 and P_2 are also different. Otherwise, $P_1 \cup P_2$ would be a trail of odd length, which contains a circuit of odd length, contradicting that G is bipartite.

Each component in $P_1 \cap P_2$ is a K-alternating path with the end edges being K-double bonds, and $P_1 \Delta P_2$ is a regular graph of degree 2 containing both v_i and v_j. We now prove that $P_1 \Delta P_2$ is a circuit. Otherwise, there would be a circuit C_1 in $P_1 \Delta P_2$ which contains v_i but not v_j. C_1 is divided by v_i and the end vertices of $P_1 \cap P_2$ on C_1 as $2t$ segments which alternately appear on P_1 and P_2. Each of the $2t-1$ end vertices of $P_1 \cap P_2$ on C_1 is incident with two single bonds on C_1, and any other vertex on C_1 is incident with a double bond and a single bond on C_1. This means that C would be an odd circuit, again a contradiction.

Since $P_1 \cup P_2 - P_1 \Delta P_2$ consists of paths of odd length, it had Kekule structures. $G - P_1 \cup P_2$ also has Kekule structures. Now it follows that $G - P_1 \Delta P_2$ has Kekule structures, that is, $P_1 \Delta P_2$ is a conjugated circuit containing v_i and v_j. □

THEOREM 2.5. *Let G be a normal bipartite graph, and K a Kekule structure of G. Then, (i) for any vertex v_i and any edge e_j of G, e_j belongs to both K-alternating growing subgraphs $G_1(v_i)$ and $G_2(v_i)$; (ii) any two edges e_i, e_j of G are on a conjugated circuit of G.*

PROOF. (i) By Theorem 2.4 (i), any end vertex of e_j belongs to both $G_1(v_i)$ and $G_2(v_i)$, that is, it can be reached by a K-alternating path starting from v_i in both $G_1(v_i)$ and $G_2(v_i)$. Without loss of generality, let v_i be colored black, and let u_b (resp. u_w) denote the end vertex of e_j black (resp. white). If e_j is a K-double bond, any K-alternating path from v_i to u_b in $G_1(v_i)$ must contain e_j. If e_j is a K-single bond, any K-alternating path from v_i to u_b in $G_1(v_i)$ will terminate at a K-double bond other than e_j, and the K-alternating path can further grow to pass through e_j.

By the same reasoning, e_j also belongs to $G_2(v_i)$.

(ii) Since e_i is not a fixed bond, there is a Kekule structure K of G such that e_i is a K-double bond. Let v_i be the black end vertex of e_i, and let u_b and u_w be the black end vertex and the white end vertex of e_j, respectively. By (i), there is a K-alternating path P_1 in $G_1(v_i)$ from v_i to u_b (or u_w), which contains e_j. By Theorem 2.4 (i), there is a K-alternating path P_2 in $G_2(v_i)$ from v_i to u_b (or u_w), which must not contain e_j. By a similar argument as in the proof of Theorem 2.4 (ii), we have that $P_1 \Delta P_2$ is a conjugated circuit containing e_i and e_j. □

The proof is thus completed.

THEOREM 2.6. *Let G be a conjugated bipartite graph, let K be a Kekule structure of G, and let G_i be a proper normal component of G. Then all the edges in $E(G) \setminus E(G_i)$ with an end vertex in G_i are fixed single bonds.*

PROOF. From the maximality of G_i, we assert that all the edges in $E(G)\setminus E(G_i)$ with an end vertex in G_i are fixed bonds. In addition, they must be fixed single bonds, otherwise an edge in G_i adjacent to them would be a fixed single bond, a contradiction. □

3. An Efficient Algorithm for Determining Fixed Bonds and Normal Components of a Conjugated Bipartite Graph

Based on the above Theorems, we can establish an efficient algorithm for determining fixed bonds and normal components of a conjugated bipartite graph.

THEOREM 3.1. *Let G be a conjugated bipartite graph, K a Kekule structure of G, and v_b and v_w the end vertices of a K-double bond e_i of G. Then $G_2(v_b) \cap G_2(v_w)$ is either a normal component of G or a fixed double bond (that is, e_i itself) of G.*

PROOF. Assume that $G_2(v_b) \cap G_2(v_w)$ is not a fixed double bond.

Let G_i be a normal component of G which contains e_i. By Theorems 2.4, 2.5, any vertex and any edge of G_i belong to $G1(v_b) \cap G_2(v_b)$. The edge e_i itself is also contained in $G_1(v_b)$. Then $G_2(v_w) = G_1(v_b)$, and so any vertex and any edge of G_i belong to $G_2(v_b) \cap G_2(v_w)$.

Conversely, by a similar argument as in the proof of Theorems 2.4 (ii) and 2.5 (ii), any vertex and any edge in $G_1(v_b) \cap G_2(v_b) = G_2(v_w) \cap G_2(v_b)$ belong to the same normal component of G with the K-double bond e_i. Therefore, we have $G_i = G_2(v_b) \cap G_2(v_w)$. □

ALGORITHM 1. Let G be a conjugated bipartite graph.

1. Find a Kekule structure K.
2. Set $i = 1$.
3. Take a K-double bond e_i with end vertices v_{ib} and v_{iw}.
4. Find $G_2(v_{ib})$ and $G_2(v_{iw})$ in $G - \bigcup_{j \le i-1} G_j$ by using the growing procedure of maximal K-alternating growing subgraphs in Definition 1, which can be accomplished by drawing arrows on the bipartite graph directly (note that, if

e_i has obtained an arrow in a previous step, then one of $G_2(v_{ib})$ and $G_2(v_{iw})$ has also been given and we only need to find the other).

5. Take $G_i = G_2(v_{ib}) \cap G_2(v_{iw})$ (that is, the subgraph of G induced by all the edges with two arrows in two directions).
6. If every K-double bond in $G - \bigcup_{j\le i} G_j$, has two arrows in two directions, go to Step 9.
7. Set $i = i+1$. If there is an edge with an end vertex of degree one in $G - \bigcup_{j\le i-1} G_j$, take it as G_i, and draw the second arrow in a different direction at the edge (note that any edge in $G - \bigcup_{j\le i-1} G_j$ with an end vertex adjacent to a vertex in $\bigcup_{j\le i-1} G_j$ has obtained an arrow in previous steps). Go to Step 7.
8. Take a K-double bond $e_i = v_{ib}v_{iw}$ in $G - \bigcup_{j\le i-1} G_j$ with an end vertex adjacent to a vertex in $\bigcup_{j\le i-1} G_j$. Go to Step 4.
9. Take all the graphs in $\bigcup_{i\ge 1} G_i$ having only one edge as fixed double bonds of G, and take all the edges with only one arrow as fixed single bonds of G. The other connected components in $\bigcup_{i\ge 1} G_i$ are normal components of G. Stop.

The algorithm is based on the above Theorems 2.6, 3.1 and the fact that any K-double bond of G is either a fixed double bond or contained in a normal component of G, and so all the fixed double bonds and normal components found by the algorithm will cover all vertices of G, and the other edges must be fixed single bonds of G. In Steps 7, 8, the choice of e_i may ensure that a fixed single bond can only obtain an arrow.

THEOREM 3.2. *Let G be a conjugated bipartite graph with n vertices and m edges. Then the complexity of Algorithm 1 is $O(mn^{1/2})$. Particularly, if G is a benzenoid hydrocarbon, the complexity of the algorithm becomes linear, that is $O(n)$.*

PROOF. The complexity of Step 1 of the algorithm is $O(mn^{1/2})$ in general cases [**22, 23**]. If G is a benzenoid hydrocarbon, the complexity is $O(n)$ [**20**]. Note that in the process in finding $G_1 = G_2(v_{1b}) \cap G_2(v_{1w})$, each edge in G_2 will also obtain an arrow, that is, one of $G_2(v_{2b})$ and $G_2(v_{2w})$ is a subgraph of one of $G_2(v_{1b})$ and $G_2(v_{1w})$ and has been found in a previous step. So we need only to find the other one. In general cases, one of $G_2(v_{ib})$ and $G_2(v_{iw})$ is a subgraph of one among $\{G_2(v_{1b}), G_2(v_{1w}), G_2(v_{2b}), G_2(v_{2w}), \dots, G_2(v_{(i-1)b}), G_2(v_{(i-1)w})\}$. To finish Steps 4, 5, and 7, we need to draw at most $2m = O(m)$ arrows on edges of G, at most two arrows on every edge. The number of edges of G is at most $\frac{n^2}{4}$, that is, $m \le \frac{n^2}{4} = O(n^2)$. If G is a benzenoid hydrocarbon, then $O(m) = O(n)$. Now the conclusion of the Theorem holds. □

For a planar bipartite graph G or a bipartite (molecular) graph G with the maximum degree not greater than some constant, we have that $O(m) = O(n)$. If there is an algorithm for finding a Kekule structure of G which is linear, then the complexity of Algorithm 1 will also be linear. Hence we have the following.

CONJECTURE 3.3. *Let G be a planar bipartite graph or a non-planar bipartite molecular graph with Kekule structures. Then there is a linear algorithm for finding a Kekule structure of G, and so the complexity of Algorithm 1 is $O(n)$.*

In the following we give two examples to demonstrate the process of finding all fixed bonds and all normal components of conjugated bipartite graphs by Algorithm 1.

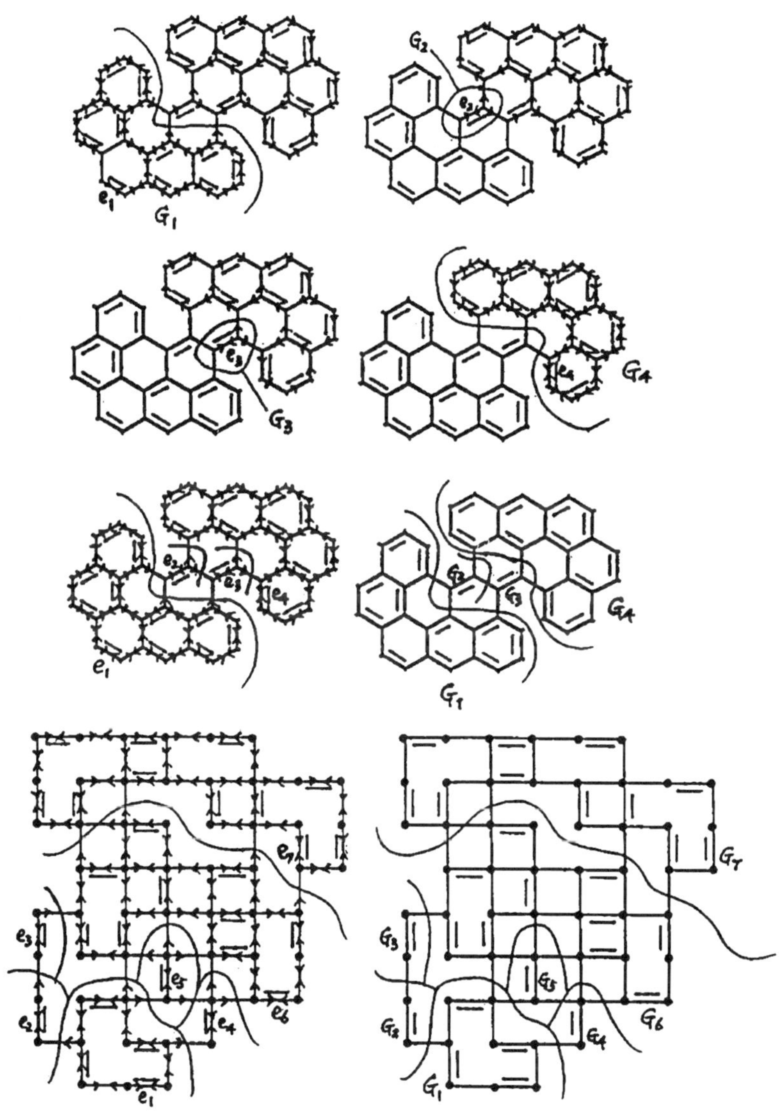

FIGURE 2. Two examples for demonstrating Algorithm 1.

Appendix

For the reader's convenience we have included definitions of LM-Conjugated circuits polynomial, sextet polynomial, and generalized molecular bond orders.

DEFINITION A.1. A set S of conjugated circuits of a Kekule structure K_i of a benzenoid hydrocarbon B is said to be linearly independent and minimal if S consists of a maximum number of linearly independent circuits of B and has the minimum total circuit length. We denote a circuit of size $4n+2$ in S by R_n, and the summation expression of circuits in S with respect to K_i by

$$R(K_i) = \sum_{R_i \in S} R_i = \sum_{n=1,2,\dots} r_n(K_i) R_n,$$

where $r_n(K_i)$ is the number of circuits of size $4n+2$ in S. The summation expression of all sets of LM-conjugated circuits of all Kekule structures of B is called the LM-conjugated circuit polynomial of B and is denoted by

$$R(B) = R = \sum_{K_i} R(K_i) = \sum_{n=1,2,\dots} r_n R_n,$$

where $r_n = \sum_{K_i} r_n(K_i)$.

DEFINITION A.2. Let G be a graph representing a polycyclic aromatic hydrocarbon. The sextet polynomial $B_G(x)$ is defined as follows:

$$B_G(x) = \sum_{i=0}^{m} r(G,i) x^i,$$

where $r(G,i)$ is the number of ways in which i mutually resonant sextets are chosen from G, and m is the maximum number of i for which $r(G,i)$ is non-zero.

We assign weights $R(K_i)$ to the individual Kekule structure K_i of a benzenoid hydrocarbon B. Let

$$R_e(B) = R_e = \sum_{K_i \ni e} R(K_i)$$

denote the weight sum of all the Kekule structures of B that contain an edge e of B. A generalized VB bond order (GBO) is defined as follows:

DEFINITION A.3. For an edge e of a benzenoid hydrocarbon B, the generalized bond order $GBO(e)$ of e is equal to the ratio of the weight sum of all the Kekule valence structures that contain e to the weight of all Kekule structures of B, that is: $GBO(e) = Re(B)/R(B)$.

References

[1] J. Brunvoll, S.J. Cyvin, B.N. Cyvin, and I. Gutman, Essentially Disconnected Benzenoids—Enumeration and Classification of Benzenoid Hydrocarbons—IX, *Match*, **23** (1988) 209–227.

[2] J. Brunvoll, S.J. Cyvin, B.N. Cyvin, and I. Gutman, Essentially Disconnected Benzenoids—Distribution of K, the Number of Kekule Structures, in Benzenoid Hydrocarbons—VII, *Match*, **23** (1988) 239–243.

[3] J. Brunvoll, S.J. Cyvin, B.N. Cyvin, and I. Gutman, Essentially Disconnected Benzenoids—Distribution of K, the Number of Kekule Structures, in Benzenoid Hydrocarbons—VIII, *Match*, **24** (1989) 51–57.

[4] S.J. Cyvin, and I. Gutman, Kekule Structures in Benzenoid Hydrocarbons, Lecture Notes in Chemistry 46 (Springer, Berlin, 1988).

[5] X. Guo, and M. Randić, Recursive Method for Enumeration of Linearly Independent and Minimal Conjugated Circuits of Benzenoid Hydrocarbons, *J. Chem. Inf. Comput. Sci.*, **34** (1994) 339–348.
[6] X. Guo, M. Randić, and D.J. Klein, Analytical Expressions for the Count of LM-Conjugated Circuits of Benzenoid Hydrocarbons, *International Journal of Quantum Chemistry*, **60** (1996) 943–958.
[7] X. Guo, and F. Zhang, Mathematical Properties and Structures of Sets of Sextet Patterns of Generalized Polyhexes, *Journal of Mathematical Chemistry*, **9** (1992) 279–290.
[8] F. Zhang, and X. Guo, Directed Tree Structure of the Set of Kekule Patterns of Generalized Polyhex Graphs, *Discrete Applied Mathematics*, **32** (1991) 295.
[9] He Wenjie and He Wenchen, *Theor. Chim. Acta*, **70** (1986) 43.
[10] N. Ohkami, *Journal of Mathematical Chemistry*, **5** (1990) 23.
[11] H. Hosoya, and T. Yamaguchi, *Tetrahedron Lett.*, (1975) 4669.
[12] N. Ohkami, A. Motoyama, T. Yamaguchi, H. Hosoya, and I. Gutman, *Tetrahedron*, **37** (1981) 1113.
[13] R. Chen, S.J. Cyvin, and B.N. Cyvin, Recognition of Essentially Disconnected Benzenoids, *Match*, **25** (1990) 71–81.
[14] I. Gutman, Covering a Hexagonal System with Hexagons, in: *Proceedings of the Fourth Yugoslavia Seminar on Graph Theory*, Novi Sad University (1983) 151–160.
[15] I. Gutman, and S.J. Cyvin, Topological Property of Benzenoid Systems XXXV—Number of Kekule Structures of Multiple Chain, *Aromatics Monatsh. Chem.*, **118** (1987) 541–548.
[16] F. Zhang, and R. Chen, The Construction Method of Kekulean Hexagonal Systems with Each Hexagon being Resonant, *Match*, **24** (1989) 323–332.
[17] F. Zhang, and R. Chen, When Each Hexagon of a Hexagonal System Covers It, *Discrete Applied Mathematics*, **30** (1991) 63–75.
[18] F. Zhang, X. Guo, and R. Chen, Z-Transformation Graphs of Perfect Matchings of Hexagonal Systems, *Discrete Mathematics*, **72** (1988) 405–415.
[19] F. Zhang, and X. Li, Hexagonal Systems without Fixed Double Bonds, *Match*, **25** (1990) 251–259.
[20] X. Li and F. Zhang, A Fast Algorithm to Determine Fixed Bonds in Hexagonal Systems, *Match* **25** (1990) 151–163.
[21] F. Zhang and X. Li, Hexagonal Systems with Fixed Bonds, *Discrete Mathematics* **47** (1993) 285–296.
[22] J.A. Bondy and U.S.R. Murty, Graph Theory with Applications, American Elsevier Publishing Co. Inc., 1976.
[23] L. Lovasz and M.D. Plummer, Matching Theory, North-Holland Mathematics 121, Annals of Discrete Mathematics 29, 1986.
[24] P. Hansen and M. Zheng, A Linear Algorithm for Perfect Matching in Hexagonal Systems, *Discrete Mathematics* **122** (1993) 179–196.
[25] P. Hansen, B. Jaumard, H. Sacks and M. Zheng, Finding a Kekule Structure in a Benzenoid System in Linear Time, *J. Chem. Inf. Comput. Sci.* **35** (1995) 561–567.

INSTITUTE OF MATHEMATICS AND PHYSICS, XINJIANG UNIVERSITY WULUMUQI XINJIANG 830046, P. R. CHINA AND DEPARTMENT OF MATHEMATICS AND COMPUTER SCIENCE, DRAKE UNIVERSITY, DES MOINES, IOWA 50311-4505

E-mail address: `xfguo@xju.edu.cn`

DEPARTMENT OF MATHEMATICS AND COMPUTER SCIENCE, DRAKE UNIVERSITY, DES MOINES, IOWA 50311-4505

E-mail address: `milan.randic@drake.edu`

DIMACS Series in Discrete Mathematics
and Theoretical Computer Science
Volume 51, 2000

Numerical Solutions of the Laplace Equation in Chemical Spaces

Ray Hefferlin

ABSTRACT. The triad principle states that numerical values of properties of atoms and molecules can be obtained by interpolation in their chemical spaces. This principle is associated with the numerical solution to the Laplace equation.

1. Chemical Spaces

Chemical spaces are invoked in the large-scale classification of molecules to forecast new numerous applications including molecular design. Numerous coordinate systems have been used to place and locate atoms and molecules in these spaces and to employ multiple-regression methods [**1**] and neural networks [**2, 3**] data.

2. Coordinate Systems, the Triad Principle, and the Laplace Equation

2.1. Addressing all Possible Molecules with Fixed Numbers of Electrons. The prototype coordinate system has atomic-number axes $(Z_I, Z_2, \ldots)$, one for each atom in the molecules; along each axis are enumerated H, He, Li, Be, ... The common use of this system is to have a lattice point for each of all possible molecules with a given number of atoms.

From this system can be derived [**4, 5**] another system that has one axis for the period number (R_i) and one for the group number (C_i) of each of the atoms in the molecules with a given number of atoms. The periodic system of the elements, with axes (R, C), is the simplest and the most well known example.

The triad principle has been associated with the entire history of the periodic system; it was used by Mendeleev to predict properties of three elements [**6**]. The triad principle is also a simple numerical solution to Laplace's equation [**7**]. Figure 1 shows its use to obtain approximate values for ionization potentials [**8**] for elements on the interior of a portion of the Periodic Table. Other properties can be fitted as well.

Periodic charts of diatomic molecules exist in a space (R_1, C_1, R_2, C_2), with one axis for the period number and one for the group number of each of the two atoms in the molecules [**9, 10**]. Figure 2 shows that the Laplace equation can yield approximate values of dissociation potentials [**11**] in this space.

1991 *Mathematics Subject Classification.* Primary 92E10; Secondary 65N50.

Li **5.39**	**Be** **9.32**	**B** **8.3**	**C** **11.3**	**N** **14.5**	**O** **13.6**	**F** **17.4**	**Ne** **21.6**
Na **5.14**	**Mg** 7.43 *7.246*	**Al** 5.98 *8.329*	**Si** 8.15 *10.17*	**P** 10.5 *12.02*	**S** 10.4 *12.99*	**Cl** 13 *14.76*	**Ar** **15.8**
K **4.34**	**Ca** 6.11 *6.2*	**Ga** 6 *7.608*	**Ge** 7.88 *9.058*	**As** 9.81 *10.4*	**Se** 9.75 *11.56*	**Br** 11.8 *12.86*	**Kr** **14**
Rb **4.18**	**Sr** 5.69 *5.608*	**In** 5.79 *6.847*	**Sn** 7.34 *8.065*	**Sb** 8.64 *8.941*	**Te** 9.01 *10.02*	**I** 10.5 *11.13*	**Xe** **12.1**
Cs **3.89**	**Ba** **5.21**	**Tl** **6.11**	**Pb** **7.42**	**Bi** **7.29**	**Po** **8.43**	**At** **9.5**	**Rn** **10.7**

FIGURE 1. Numerical solution of the Laplace equation for atomic radii (Angstrom units) in the interior of the rectangular, main-group, rows 2 to 6, portion of the periodic chart of the elements. Tabulated values (boldface) are entered on the boundary of the rectangle. In each interior cell is the average of the data above, below, and to the left and right, of the cell (italics). Tabulated data for the interior region are shown for comparison. The necessary iterations were performed by Excel^{TM}.

						0	Be
					7.81	**0**	B
				11.09	5.67 *5.71*	**0**	C
			9.76	6.5 *7.44*	3.5 *3.92*	**0**	N
		11.09	6.5 *7.44*	5.12 *5.00*	2.23 *2.55*	**0**	O
	7.81	5.67 *5.71*	3.5 *3.92*	2.23 *2.55*	1.6 *1.27*	**0**	F
0	**0**	**0**	**0**	**0**	**0**	**0**	Ne
Be	B	C	N	O	F	Ne	

FIGURE 2. Numerical solution for dissociation potentials (in eV) for the triangular portion of the space of row-2 diatomic molecules possessing ten or more valence electrons. Tabulated dissociation potentials for the ten-electron species are shown along the diagonal in boldface. These data are maximum values compared to those for adjacent molecules in the direction transverse to the ten-electronic sequence. Tabulated data for the inert-gas species along the two other boundaries, also in boldface, are all zero within the accuracy of the method. In each interior cell are the interpolated value, in italics, and the tabulated value for comparison. The agreement is far better than that to be expected from least-squares and neural-network methods, as the terrain in this part of the space is very rugged. The method is limited for diatomics by the fact that known data for all ten-electron and inert gas molecules do not exist for all properties nor for analogous portions of the space with heavier (isovalent) molecules. Nevertheless, there should be portions of some chemical spaces where it has been or can be used effectively.

2.2. Addressing Special Collections of Molecules with Various Numbers of Electrons. Completely different kinds of coordinate systems are used for studies of all members (regardless of the numbers of atoms) of special classes of molecules. Benzenoid polycyclic aromatic hydrocarbons [**12**] and indacene-related polycyclic hydrocarbons [**13**], cases in point, require only two axes: one to enumerate the carbon atoms and one to enumerate the hydrogen atoms. The triad principle applies quite faithfully to the space with these axes [**14**] and it follows that data in the interior spaces of the periodic systems of these species can be obtained from numerical solutions of Laplace's equation.

References

[1] C Carlson, J. Gilkeson, K. Linderman, S. LeBlanc, R. Hefferlin, and B. Davis, Estimation of Properties of Triatomic Molecules from Tabulated Data Using Least-Squares Fitting. *Croatica Chem. Acta*, 70:479–508, 1997.

[2] T.R. Cundari, and E.W. Moody, A Comparison of Neural Networks versus Quantum Mechanics for Inorganic Systems. *J. Chem. Inf. Comp. Sci.*, 37:871–875, 1997.

[3] J. Wohlers, B. Laing, and R. Hefferlin, Least-Squares and Neural-Network Forecasting from Critical Data: Diatomic Molecular r_e and Triatomic ΔH_a and IP. Chapter for proceedings volume of the 7th *International Conference on Mathematical Chemistry*, JAI Press, Greenwich, CN. In press.

[4] R.A. Hefferlin, R. Campbell, D. Gimbel, H. Kuhlman, and T. Cayton, The Periodic Table of Diatomic Molecules — I. An Algorithm for the Retrieval and Prediction of Spectrophysical Properties. *J. Quant. Spectrosc. Radiat. Transfer*, 21:315–336, 1979.

[5] R. Hefferlin, *Periodic Systems and their Relation to the Systematic Analysis of Molecular Data.* The Edwin Mellen Press. Lewiston, New York, pages 256–293, 1989.

[6] J.W. Van Spronsen, *The Periodic System of Chemical Elements: A History of the First Hundred Years.* Elsevier, New York, pages 63–96, 1969. Horizontal and vertical interpolations are described separately.

[7] H.D. Young, and R.A. Freedman, *University Physics.* Addison-Wesley Publishing Compmly, New York, page 757, 1996.

[8] R.E. Weast, editor. *Handbook of Chemistry and Physics.* CRC Press, Boca Raton, FL, page E-67, 1973.

[9] R.A. Hefferlin, G.V. Zhuvikin, K.E. Caviness, and P.J. Duerksen, *J. Quant. Spectrosc. Radiat. Transfer*, 32:257–268, 1984.

[10] R. Hefferlin, Matrix-Product Periodic Systems of Molecules. *J. Chem. Inf. Comp. Sci.*, 34:314–317, 1994.

[11] K.P. Huber, and G. Herzberg, *Molecular Spectra and Molecular Structure. IV Constants of Diatomic Molecules.* Van Nostrand Reinhold Company, New York, 1979.

[12] J.R. Dias, A Periodic Table for Polycyclic Aromatic Hydrocarbons. *Acc. Chem. Res.*, 18:241–248, 1985.

[13] J.R. Dias, Deciphering the Information Content of Chemical Formulas: Chemical and Structural Characteristics and Enumeration of Isomers. *J. Chem. Inf. Comp. Sci.*, 32:203–209, 1992.

[14] J.R. Dias, Constant-Isomer Benzenoid Series and their Topological Characteristics. *Theor. Chim. Acta*, 77:143–162, 1990.

PHYSICS DEPARTMENT, SOUTHERN ADVENTIST UNIVERSITY, COLLEGEDALE, TN 37315
E-mail address: hefferln@southern.edu

DIMACS Series in Discrete Mathematics
and Theoretical Computer Science
Volume 51, 2000

Chemical Frameworks and Hyperbolic Tilings

S. T. Hyde and S. Ramsden

ABSTRACT. Many complex crystalline 3-dimensional graphs are of interest to solid state chemists, in order to relate atomic and molecular crystal structures. Some of those graphs can be generated from tessellations of the 2*d* hyperbolic plane, and then mapped onto triply periodic hyperbolic surfaces. The properties of such graphs, including their densities, can be analysed in terms of topological and non-euclidean geometric concepts. Examples include a variety of zeolite frameworks and novel graphitic carbon structures.

1. Introduction

Three-dimensional (3*d*) euclidean networks (or just "nets"), such as ball and stick models of bonded chemical frameworks, are deceptively difficult objects to construct and describe *ab initio*. The nets for ideal atomic crystals are infinite graphs, although many of their properties can be determined from the associated finite graphs, formed by dividing the infinite graph by its translations. Studies of finite graphs, characteristic of molecules, spawned many of the earliest developments in graph theory. Sylvester, Cayley and other founders of the field considered (finite) molecular nets in some detail [**2**]. Graph theory and chemistry have diverged since then, and many of the simpler properties of infinite nets remain unexplored by graph theorists.

Solid state chemists, who deal with structures of infinite crystals, have generated a corpus of results and conjectures that deserve attention. The work of Wells [**53**], Smith [**50**] and later O'Keeffe *et al.* [**40, 41, 43, 44**] are rich sources for the curious graph theorist. Some of these data are used below for our own purposes, namely the reduction of euclidean 3-d (E^3) nets to non-euclidean 2*d* nets, using results from topology and non-euclidean (particularly hyperbolic) geometry. The underlying 2*d* surface geometries used to support many of the infinite nets are themselves of relevance to packing problems of interest to chemists, including sphere[**9, 10, 44**] and rod [**42, 37, 44**] packings, and some of those connections will also be mentioned.

1991 *Mathematics Subject Classification.* Primary 57M50; Secondary 57N25,57M15.

STH thanks the organisers of the DIMACS meeting for arranging a stimulating conjunction of mathematicians, chemists and physicists, and the editors for their extreme patience.

2. Regular Nets and Periodic Hyperbolic Surfaces

The simplest nets known to crystal chemists are "regular". All vertices and edges of regular z-connected nets are symmetrically identical and at least z vertex angles, of the $\frac{(z)(z-1)}{2}$ angles, are equal [**44**]. (Note that this sense of "regular" is distinct from that of graph theory.) Regular nets offer favorable geometries for homo-atomic crystals, given a preferred bonding geometry, specified usually in terms of bond lengths and bond angle. While this paper is partly concerned with geometric embedding of nets (resulting in specific symmetries), the net topology, characterised by its rings and connectivity, is our first concern.

Clearly the simplest non-trivial regular 3-d nets are three-connected. The only regular three-connected net that can be realised in E^3 is the chiral Y^* graph, also called the Laves' graph, or Wells' $(10,3)-a$ [**53**] (Figure 1) (Enantiomers are known as the $+Y^*$ and $-Y^*$ nets). In contrast to the better-known $2d$ regular nets, this is a surprisingly complex net to the uninitiated eye. Its nodes occupy invariant sites of the space group $I4_132$ [**18**]; they are connected via equivalent edges to form a network of 10-rings. [1] The chiral net contains 4_1 and 3_2 helices, shown in Figure 1. These helices are arranged in regular arrays: they wrap along the rods of two cubic cylinder (or "rod") packings: the $\beta - Mn$ (4_1) and $SrSi_2$ (3_2) rod packings [**10, 37, 42, 44**]. The net is known to chemists: a slight distortion from

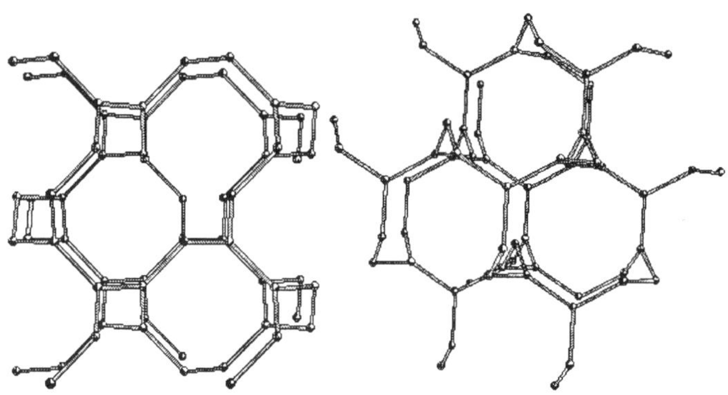

FIGURE 1. Views of a finite portion of the $+Y^*$ net (left) along the four-fold [100] axis (note 4_1 helices along these directions) and (right) almost along a three-fold [111] axis (note 3_2 helices)

regularity (which does not alter the net topology) results in the reduced symmetry embedding ($P4_332$). These nodes describe the array of silicon atoms in the solid $SrSi_2$ crystal [2][**42**].

The Y^* net is of low vertex density, in the sense that the shortest distances between unlinked vertices are far larger than those between linked vertices. Indeed, there is so much free space that a pair of enantiomeric $+Y^*$ and $-Y^*$ nets can be intergrown to give the Y^{**} lattice complex, shown in Figure 2, whose symmetry in E^3 is described by the space group ($Ia\underline{3}d$) [**18**]. Minor distortion of this array (to

[1] "Rings" in a net are defined for each vertex angle (defined by a pair of edges sharing a vertex). They are the shortest cycles free of short-cuts, *cf.* [**44**].

[2] Abstract nets will be *italicized*, so as not to confuse them with real chemical frameworks, which may be of lower symmetry.

$Ia\bar{3}$) gives the crystal structure of $\gamma - Si$ [**44**]. The Y^* and $-Y^*$ nets are dual, called "maximally interpenetrating", in the sense that all rings of one net enclose an edge of the other [**48, 53**]. Pairs of maximally interpenetrating nets can be derived from the following construction. Imagine net edges of one labyrinth to be a network of inflatable tubes, of vanishing tube radius. Now increase the pressure within the tubes, so that they swell. Eventually, they define the complementary network, closing down about its edges. Interpenetrating nets define the channels of

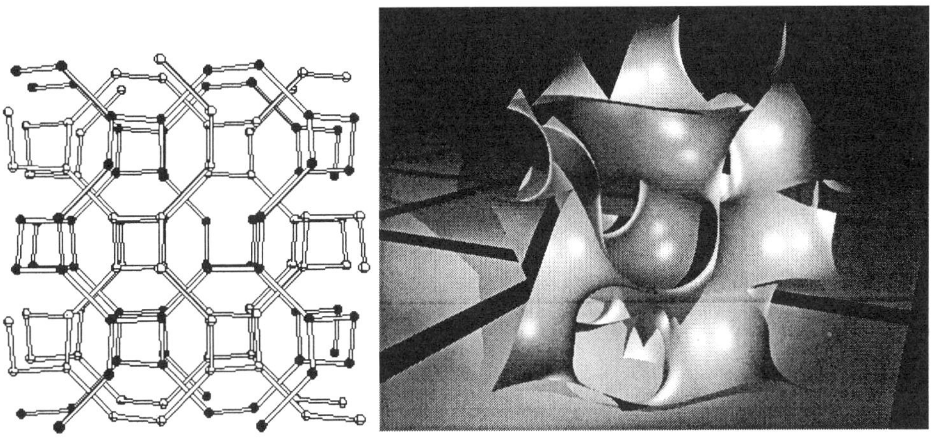

FIGURE 2. (Left) View of a Y^{**} net, consisting of intergrown $+Y^*$ and $-Y^*$ nets (light and dark nodes resp.). (Right) The gyroid, a cubic ($Ia\bar{3}d$) triply periodic minimal surface

one of the (topologically) simplest intersection-free triply periodic minimal surfaces (IPMS), the *gyroid*, discovered by Schoen [**48**]. This surface, shown in Figure 2, has zero mean curvature, and results if equal pressure is maintained in the interwoven inflated $+Y^*$ and $-Y^*$ labyrinth graphs. (That follows from the link between Laplace pressure and mean curvature).

Minimal surfaces are synclastic, or saddle-shaped; their zero mean curvature means that the surface concavity at any point is balanced exactly by its convexity. The simplest examples are embedded (intersection-free), in which case the surface carves space into two continuous sub-volumes, defined by the pair of maximally interpenetrating nets. More complex intersecting examples can carve space into more than two subvolumes, including combinations of closed, $1d$, $2d$ or $3d$ domains [**12, 11**].

The Gaussian curvature necessarily varies from point to point on minimal surfaces, remaining everywhere non-positive. The underlying $2d$ geometry of minimal surfaces is thus hyperbolic. All known minimal surfaces display some symmetries, although it is unclear whether that is a prerequisite or a product of the parametrisation procedures adopted.

Among all minimal surfaces known to date (in E^3), IPMS have the most homogeneous distribution of Gaussian curvature, with isolated flat points of vanishing Gaussian curvature (that can be considered as singularities of the embedding). These surfaces lie on an underlying lattice containing three independent translations. The topology of IPMS (and topologically equivalent, or homeomorphic triply periodic hyperbolic surfaces) is complex: all these surfaces are of unbounded genus. Nevertheless, they can be characterized by their genus per primitive unit cell of the oriented surface, the "lattice fundamental region" of the surface. That genus is formally defined in terms of the integral (Gaussian) curvature of the lattice fundamental region (lfr), using the Gauss-Bonnet theorem. For IPMS,

$$\chi_{lfr} = \frac{1}{2\pi} \iint_{lfr} K\, da \tag{2.1}$$

where χ_{lfr} is the Euler-Poincaré characteristic per lattice fundamental region. The surface genus g is given by:

$$\chi =: j\,(1 - g) \tag{2.2}$$

where $j = 1$ for non-orientable surfaces, and $j = 2$ for orientable surfaces. The genus is thus equal to that of a "compactified" lattice fundamental region of the triply periodic surface, whose boundary is glued according to the lattice translations. This genus must exceed two. Data so far suggest that the most homogeneous IPMS are cubic examples of genus three and cubic symmetry. Three such surfaces are known; the gyroid is one example.

Consider next regular four-connected nets in E^3. As for three-connected nets, only one example is known: the *diamond* net, which in its most symmetric E^3 embedding is cubic ($Fd\bar{3}m$), Figure 3. It too is an important structure type in the solid state: decoration of the vertices by four-valent sp^3 hybridised carbon atoms gives the structure of real diamond. Silicon, germanium and ordinary (cubic) ice (ice_{Ic}) also crystallize in this array. Coloring of alternative vertices by cations and anions gives the ZnS (sphalerite) structure-type.

Like Y^*, two *diamond* nets can be maximally interpenetrated to give a pair of (identical) complementary nets, describing the structures of *cuprite* and ice_{VII} [**1, 53**], Figure 3. This pair of four-connected regular nets are partitioned by another genus-three, cubic ($Pn\bar{3}m$) triply periodic minimal surface, the D (sometimes denoted F) surface [**49, 48**], shown in Figure 3. Like the *gyroid*, this surface is also "quasi-homogeneous", with a nearly uniform distribution of Gaussian curvature. Lower symmetry variants of the D surface,which are less homogeneous, also lead to chemically interesting nets. For example, tetragonal distortion gives the tD (including the T) surface ($I4_1/amd$, oriented symmetry $I4_1/acd$) [**29, 13**], which consists of a pair of maximally interpenetrating $ThSi_2$ nets (whose nodes describe the arrangement of Si atoms in $ThSi_2$) [**44**]. Regular five-connected nets cannot be realized in E^3.

Among the regular six-connected nets in E^3, the simplest is the PC net, that resembles a "jungle-gym", with vertices forming a primitive cubic lattice, (space group $Pm\bar{3}m$), shown in Figure 4. This describes, for example, the array of Na and Cl ions in common salt, $NaCl$. A maximally interpenetrating pair of these PC nets gives the body-centred cubic ($Im\bar{3}m$) arrangement (e.g. $CsCl$), Figure 4. This arrangement is partitioned by the third quasi-homogeneous triply periodic minimal surfaces of genus three and cubic symmetry, the P surface. In addition to these cubic minimal surfaces, lower symmetry variants exist for certain deformations, including tetragonal, rhombohedral, ... (tP, tD, tG, rPD, rG, $rPDG$, ... IPMS) [**30, 16**]. These surfaces display a less homogeneous distribution of Gaussian curvature than the parent cubic surfaces. A wealth of higher genus, and less homogeneous, IPMS are also known, and new examples are published regularly

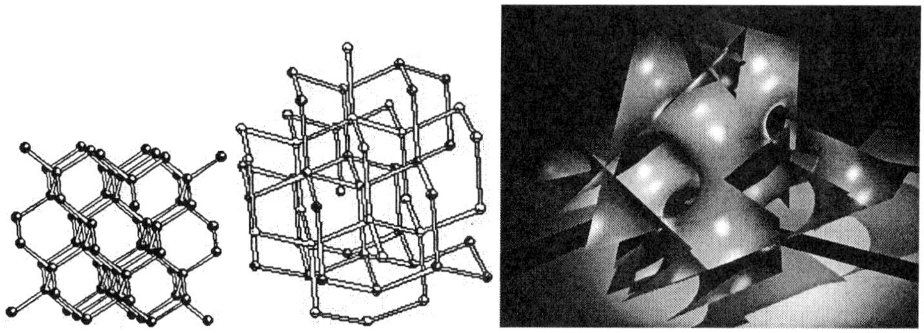

FIGURE 3. (Top left) View of a *diamond* net. (Top right) A pair of maximally interpenetrating *diamond* nets. (Bottom) The D surface, a cubic ($Pn\underline{3}m$) IPMS, whose labyrinth graphs are interpenetrating *diamond* nets

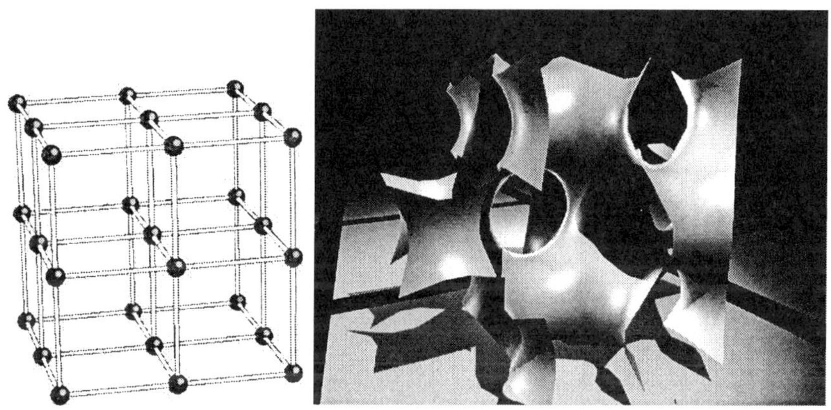

FIGURE 4. (Left) View of a PC net which can clearly be interpenetrated by an identical net, shifted to the centre of the cubes. (Right) The P surface, a cubic ($Pn\bar{3}m$) triply periodic minimal surface.

[**30, 25, 13, 14, 15, 11, 16**]. For example, the cubic ($Im\bar{3}m$) genus-four $I-WP$ triply periodic minimal surface separates distinct, but maximally interpenetrating, graphs [**31**]. One labyrinth is defined by the eight-connected I (body-centred cubic) graph, the other by the four-connected NbO graph [**44**]

3. Nets in E^3 and H^2

The simpler regular nets in E^3 thus form labyrinth graphs of the most homogeneous IPMS. That connection can be fruitfully exploited further to effectively map the problem of nets in E^3 onto that of tilings of H^2 ($2d$ hyperbolic space), the universal covering space of triply periodic hyperbolic surfaces. Any genus g surface, where g exceeds two, can be unfolded to form a $4g$-gonal hyperbolic polygon, which tiles H^2, with $4g$ polygons sharing each vertex [**47, 8**]. (Actually, some distortion of the metric of H^2 is required for the IPMS and related triply-periodic hyperbolic surfaces, due to their inhomogeneity. That distortion is unimportant for the topological analysis that follows here, but can be analysed in detail to provide a quantitative map from E^3 to H^2 (and *vice versa* for IPMS.) We label the family of homeomorphic surfaces sharing the same set of complementary labyrinth graphs as an IPMS (P, say) by a prime, (P').

The map from labyrinth graphs in E^3 onto H^2 can be effected by a translation in E^3 of the labyrinth net(s) relative to its surface, so that all nodes lie *in* the surface, and edges form geodesics between nodes. This embedding distorts the conformal structure of the E^3 nets, but results in topologically equivalent H^2 nets. All the rings (or cycles) of the labyrinth graphs in E^3 surround catenoidal tunnels of the hyperbolic surface, and are not contractible on the surface. (We call such rings - the shortest non-null-homotopic cycles - "collar rings".) Thus the labyrinths graphs are trees in H^2. Indeed, this map sometimes produces two disjoint trees in H^2. In other cases, the labyrinth graphs reticulate the surface with crossings, but they can be "disentangled", *cf.* Figure 5. Clearly, these labyrinth graphs are

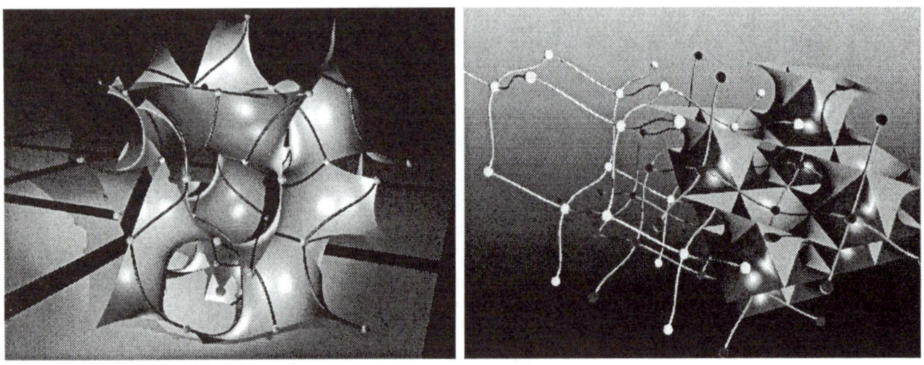

FIGURE 5. Embedding regular 3-, and 4-connected labyrinth graphs onto the *gyroid* and D IPMS. (Left) The *gyroid* cannot be reticulated by its racemic pair of labyrinth graphs, $+Y^*$ and $-Y^*$ without crossings of the graphs, but it can be reticulated by a pair of chiral $+Y^*$ (or $-Y^*$) graphs. (Right) The D surface can be reticulated by a pair (light and dark vertices) of disjoint *diamond* labyrinth graphs.

not planar in E^3. However, the representation within the hyperbolic plane is a planar one, and the usual tools of graph theory for planar graphs can be applied. Of course, this universal cover dramatically overcounts vertices. That is clear from analysis of the coordination sequence (CS) in H^2 and E^3. The CS is the number of vertices, $n(k)$, in the graph displaced from a central reference node by a shortest path of k edges [**3, 35, 51, 45, 44**]. The CS of the H^2 tree of connectivity z (a Bethe lattice) is exponential:

$$n(k) = z(z-1)^{k-1} \tag{3.1}$$

(For example, $z = 3$ gives the Bethe lattice, the H_2 map of a Y^* net.) The E^3 net, however, exhibits an asymptotically quadratic CS, since the k-th shell of the net is a surface in E^3. If the net nodes form a lattice, the CS adopts the form [**44**]:

$$n(k) = (z-2)k^2 + 2 \tag{3.2}$$

where, for example, $z = 6$ for the PC net.

Evidently, a number of nodes must be excised to recover the original E^3 net. The construction was introduced by Sadoc and Charvolin to generate the simplest IPMS [**47**]. Infinite slices of the hyperbolic plane are removed, and the resulting boundaries are glued pairwise, to yield the E^3 embedding. An example of the surgery for the P' surface, represented in the Poincaré disc model of H^2, is shown in Figure 6. Planar representations of the labyrinth graphs can be imposed on this

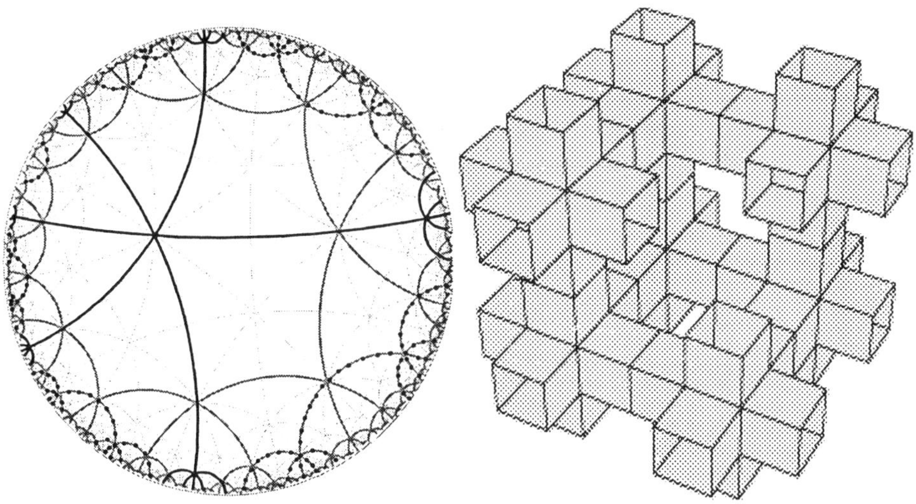

FIGURE 6. (Left) Mapping of the genus three P' surface into the hyperbolic plane, drawn in the Poincaré disc model. (The surface is recovered from the map by gluing edges, with sectors excised to give a single-sheeted cover, *cf.* [**47**]). The pair of labyrinth nets (PC graphs) of the P' surface are mapped onto H^2 to give an unbounded number of disjoint six-connected trees. One set of graphs that fold in E^3 to form a single labyrinth graph is marked by the dotted lines, a single example of the other labyrinth is marked by the full lines. (Right) The infinite polyhedron $\{4, 6|4\}$, a regular sponge.

map, e.g. two disjoint 6-connected trees from labyrinth graphs of the P', shown in Figure 6. The first few terms in the CS for the PC graph can be read off the Poincaré disc representation of the E^3 net, ignoring vertices lying in the excised sectors. (Alternatively, it could be determined algebraically by computing the number of distinct words of length k, within the subgroup of the $\{4, 6\}$ hyperbolic reflection group induced by the P surface topology.) The density of nodes is clearly far lower than that of the nodes of the hyperbolic $\{4, 6\}$ tiling (underlying the genus-three P surface). The combined map of both labyrinth graphs is, however, just sufficient to define a tiling of the hyperbolic surface separating those graphs. Euler's Theorem (for a regular $\{n_2, z\}$ tiling of n_2-rings, z around each of the N vertices),

$$\frac{\chi}{N} = 1 - \frac{z}{2} + \frac{z}{n_2} \tag{3.3}$$

can be adopted in modified form, where $n_2 = \infty$. χ refers to the integral curvature of those portions of the periodic minimal surface that is associated with a single labyrinth. The value of χ is thus a half of the true Euler-Poincaré characteristic (per crystallographic unit cell,χ_{uc}), so that:

$$\frac{\chi_{uc}}{N_i} = 2 - z_i \tag{3.4}$$

for labyrinth $i(=1,2)$, containing N_i nodes within the unit cell, and (average) net connectivity z_i. In many cases, the two labyrinth graphs are distinct, and $z_1 \neq z_2$. For example, the genus-four I-WP surface, has ($2d$) labyrinth graph topologies:

$$\chi_I = -6N_I = -12$$

(since the 8-connected I net has two nodes per conventional cell) and the four-connected NbO labyrinth has topology:

$$\chi_{NbO} = -2N_{NbO} = -12$$

(six nodes) as expected. Further, the CS of both interpenetrating graphs exhibit nearly the same functional form, differing only by the ratio of connectivities,

$$\frac{n(k)_I}{n(k)_{NbO}} \approx \frac{z_I}{z_{NbO}} = 2$$

(with minor deviations). Whether this is a general feature of maximally interpenetrating nets (labyrinth graphs of a single hyperbolic surface) remains unknown, and deserves further investigation.

König's Theorem guarantees that we can embed any 3d euclidean net in a surface [**32**], and that surface may be hyperbolic if the net is crystalline and infinite. The H^2 embedding of an E^3 net is far from unique. For example, the mapping of a (labyrinth) graph onto the surface that is formed by inflating the graph's edges results in a tree - with unbounded hyperbolic ring-size (n_2) - we call this embedding of the graph into H^2 "maximal". Topologically leaner embeddings can be found. Among these embeddings, there is one "minimal embedding", namely the densest $2d$ (hyperbolic) packing of nodes, giving the smallest value ofn_2 [**8**]. [3]

This distinction between embeddings is particularly simple for the primitive cubic PC net in E^3. Since this net is a labyrinth net of the P surface, the P surface is a maximal embedding, with a single node of the PC net per lattice fundamental region, of the genus-three surface ($\chi = -4$, *cf.* Figure 3.4). This open reticulation can be replaced by a far denser one, again on the P surface, which forms the network of edges in the infinite polyhedron (also called a "skew polyhedron" or "regular sponge") constructed exclusively of regular 4-gons, with 6 sharing each vertex. We call this regular tessellation of H^2, $\{4, 6\}$. Its folding on the P surface gives extra 4-gonal collar rings, and we denote this E^3 net $\{4, 6 \mid 4\}$ [**5**]. (Extending somewhat the notation of Coxeter [**5**], we denote a generic embedding $\{n_2, z \mid t\}$, with t-gonal collar rings.) In that minimal embedding, the hyperbolic ring-size, n_2, is equal to 4, and it follows from Euler's Theorem (3.3) that the Euler characteristic for the minimal embedding (per conventional cubic unit cell of the PC net, containing $N = 1$ node) is:

$$\chi_{min} = \frac{-1}{2}$$

[3] A "minimal" embedding is a topological notion, unrelated to the geometric sense of "minimal" surfaces.

compared with

$$\chi_{max} = -4$$

for the maximal embedding (see Figure 6).

These examples indicate the wealth of embeddings of 3-d euclidean nets into $2d$ hyperbolic analogues. The construction can be reversed, and reticulations of H^2 employed to generate nets in E^3. The range of E^3 embeddings of a hyperbolic (n_2, z) net should scale according to the ratio:

$$\frac{\chi_{min}}{\chi_{max}} = \frac{1}{2} + \frac{z}{n_2(2-z)} \tag{3.5}$$

(where the values are χ are normalized to include the same number of vertices in each case.)

Consider the dual net in H^2 of the $\{4,6\}$ net in H^2 (the minimal embedding of the PC net): the $\{6,4\}$ net. That hyperbolic net can be realised in a number of ways in E^3, $\{6,4 \mid t\}$. One of those is the *sodalite* net (Figure 7), which contains planar square and hexagonal faces [**36**]. Its edges describe the edges of a space-filling array of truncated octahedra, arranged in a body-centred cubic array ($Im\bar{3}m$ symmetry), sharing square faces. This E^3 net is named after the zeolite of the same name, whose tetrahedral T atoms (Si or Al) form a (somewhat distorted) *sodalite* net. If all hexagons of *sodalite* are covered by faces, and all squares empty, forming collar rings, the $\{6,4 \mid 4\}$ infinite polyhedron results, whose topology is that of the P' surface, Figure 7. There are 12 vertices per conventional (genus three) unit cell, or

$$\frac{\chi\{6,4\}_P}{N} = -\frac{1}{3}$$

(the same result follows from Equation 3.3, noting that $n_2 = 6$ and $z = 4$.) Since the P', D' and G' surfaces are homeomorphic, topologically identical $\{6,4 \mid 4\}$ nets can be realized on the D' and $G(yroid)'$ surfaces (as well as the other genus-three surfaces; the CLP' and H' surfaces). The particular E^3 embedding is dependent on the exact $3d$ net configuration, i.e. net edge lengths, vertex angles and edge torsion - all measured in E^3. The P' embedding preserves the exact geometry of the vertices in the *sodalite* net; other genus-three embeddings require some relative shifts of vertex positions.

Alternatively, a half of the hexagons can be punctured by labyrinth graph edges - to form a distinct infinite euclidean polyhedron of D' surface topology $\{(6.4.6.4), 4|6\}$, Figure 7 (where the first (bracketed) entries denote the cyclic arrangement of surface rings about each vertex). This (minimal) embedding of *sodalite* is denser in H^2 than the P' reticulation. It has ring-size:

$$n_2 = \frac{4}{\frac{2}{4} + \frac{2}{6}} (= 4.8\ldots)$$

which means that

$$\frac{\chi\{6,4\}_D}{N} = -\frac{1}{6}$$

The maximal embedding of *sodalite* has $2d$ topology:

$$\frac{\chi\{6,4\}_{max}}{N} = -2$$

The reticulation on the P surface lies - as expected - between minimal and maximal embeddings. A second embedding of the $2d$ $\{6,4\}$ net is possible in the D surface: the $\{(6,4) \mid 6\}$ NbO net (Figure 8) [**44**].

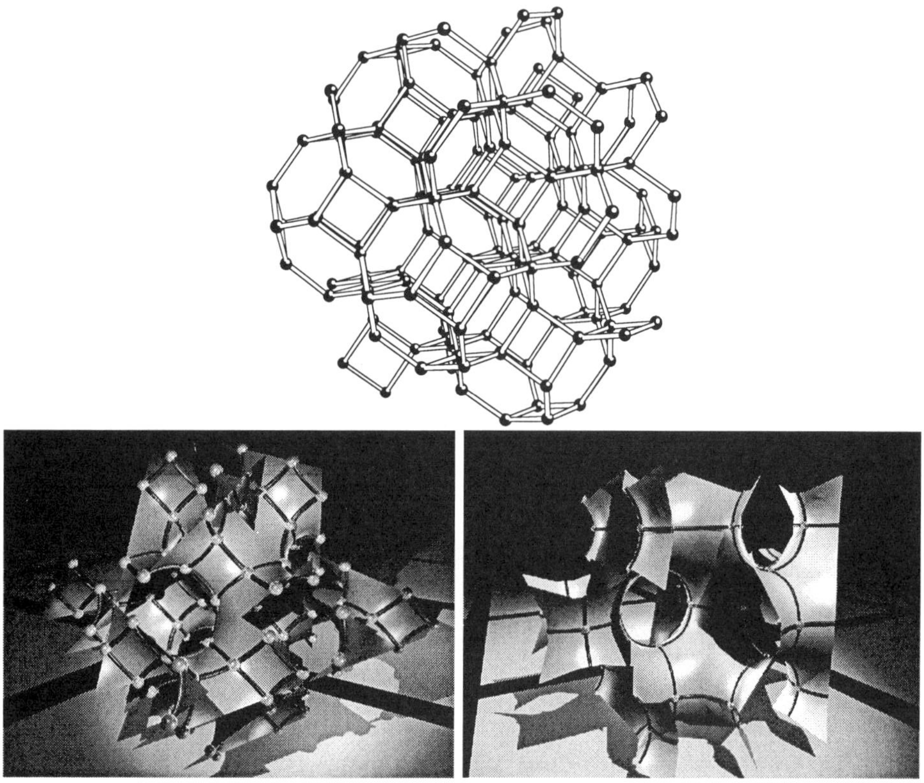

FIGURE 7. (Top) The *sodalite* graph in E^3, containing 4- and 6-rings. (Lower left) Reticulation of the P surface by a hyperbolic $\{6,4\}$ net; forming a *sodalite* graph, $\{(6.6.6.6),4 \mid 4\}$. (Lower right) Reticulation of the D surface by $\{(6.4.6.4),4\}$, forming *sodalite*, $\{(6.4.6.4),4 \mid 6\}$.

In addition to their topological character, these nets can be geometrised by demanding that all net edges are totally geodesic on the IPMS. In that case, the embedding geometry of NbO differs from all previous examples; the edges now lie along the least curved asymptotic directions of the hyperbolic surface, rather than the most concave and convex directions, the lines of curvature.

The curvilinear geometries of net edges follows from differential geometry. In the case of an asymptotic net, such as NbO, the underlying surface curvature is encoded in the net in a torsional sense, and the curvature of the edges vanishes. At the other extreme, where all net edges lie along the lines of curvature, the torsion of edges vanishes and they are necessarily curved. Intermediate cases are also possible, for which the Gaussian curvature of the underlying hyperbolic surface is manifested in both curvature and torsion of the net edges. The geometric density of nets embedded in hyperbolic surfaces is also sensitive to the net orientation relative

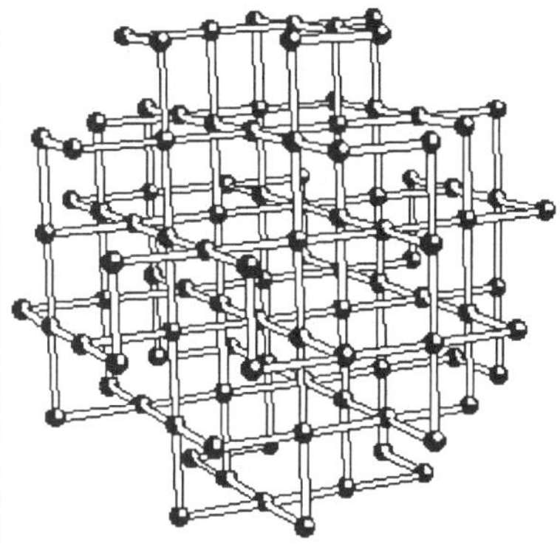

FIGURE 8. (Left) Reticulation of the D surface by the $\{6,4\}$ net. (Right) The NbO net.

to the principal directions, discussed in more detail in Section 6. The *sodalite* embedding on the D surface, a $\{(6.4.6.4),4\}$ net in H^2, can also reticulate the gyroid, forming a denser decoration of that surface than the Y^* nets introduced above. The resulting E^3 net, $\{(6.4.6.4),4 \mid 8\}$, is also known to zeolite chemists. It describes the framework of the natural zeolite, *analcime*, shown in Figure 9 [**1, 36**]. The embedding topology of the ANA framework is:

$$\frac{\chi\{(6.6.4.4),4\}_{gyr.}}{N} = \frac{-1}{6}$$

Crystallographers adopt a larger conventional cubic unit cell of the non-oriented gyroid ($Ia\bar{3}d$), of topology $\chi = -8$. It follows that each such cell contains 48 vertices. (The oriented cell, of genus three, has 24 vertices.)

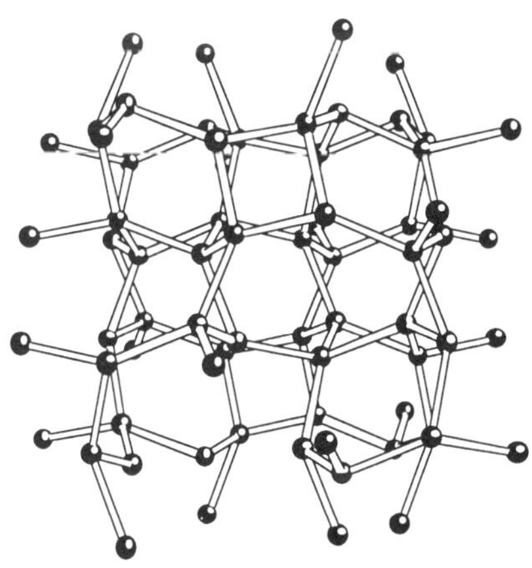

FIGURE 9. (Left) The H^2 $\{(6.4.6.4),4\}$ net reticulating the gyroid. (Right) Fragment of the resulting *analcime* E^3 net.

4. Branched Embeddings and Non-Planar Graphs

The choice of a minimal embedding for an arbitrary four-connected net is, at this stage, unknown for a number of chemically relevant nets. For example, while the maximal embedding of the *diamond* net is the 4-connected Bethe lattice on the D surface described above, the minimal embedding surface is unknown. In some cases, the minimal embedding surface is expected to contain first-order branch points, resulting in self intersections. These branches are often manifested geometrically by the presence of edges along axes of three-fold symmetry in the most symmetric embeddings of the nets in E^3.

That condition is met along some of the edges of the *melanophlogite* net (in its most nearly regular $Pm\bar{3}n$ embedding in E^3), Figure 10 [**36**]. This is a fascinating graph, containing five- and six-rings, that is known to chemists as a silicate framework ("clathrasil") and the *Type I clathrate* framework found in gas hydrates (as well as Si and Ge clathrates) [**44**]. In this case, no $2d$ embedding is known, but the graph can be simply derived a related "collapsed" version, that embeds a branched triply periodic minimal surface, the $WI-10$ surface, recently discovered by Koch and Fischer [**11**]. A single element of this surface is shown in Figures 11.

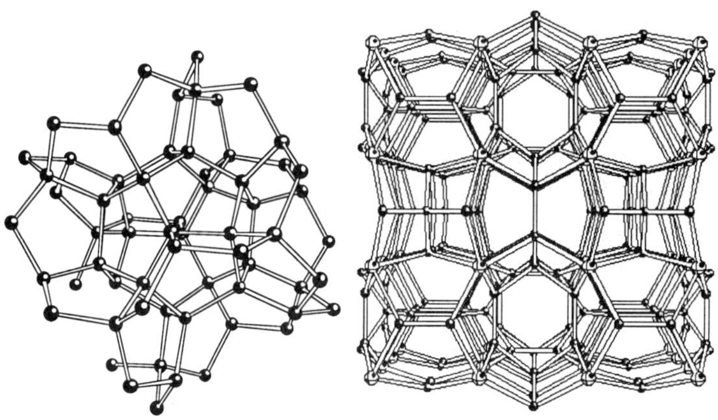

FIGURE 10. (Left) The E^3 *melanophlogite* net, containing 5- and 6-rings. (Right) A collapsed $(4,6)$-connected net, the *melanophlogite** net.

The embedding, like *diamond*, is more complex than those described earlier. The related network, *melanophlogite**, is generated from the *melanophlogite* net as follows. Those edges lying in three-fold axes of the net are shrunk on the WI-10 minimal surface, forming six-connected vertices on the first-order branch points of the surface, illustrated in Figure 10. The H^2 embedding of the $(4,6)$-connected *melanophlogite** net contains three types of vertices, resulting in an average $2d$ ring-size close to five. Note that this ring-size is close to that expected from the $3d$ ring-sizes (five- and six-rings). The *melanophlogite* net is regenerated by splitting the six-connected vertex into a pair of vertices, joined by an edge that runs through the branch point, *not* on the surface.

This operation – embedding four-connected nets via the related contracted net of higher connectivity, formed by contraction of edges lying in three-fold axes – is one way to study the embedding problem for such graphs. For example, the

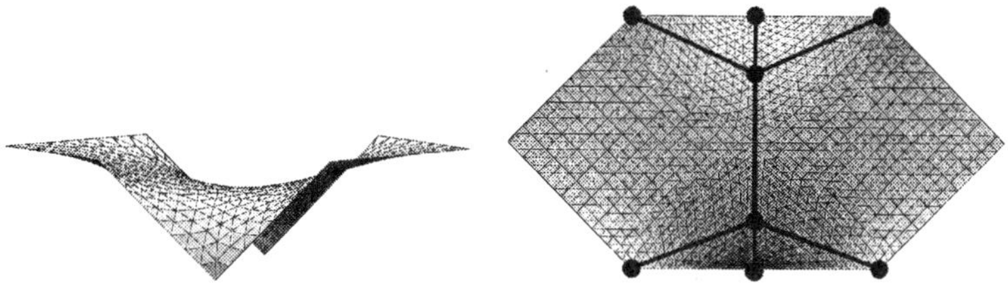

FIGURE 11. Minimal embedding of the *melanophlogite** net on the self-intersecting triply periodic minimal surface containing order-one branch points, the $WI - 10$ surface. (Left) Surface element, bounded by straight lines. (Right) Plan view of surface with *melanophlogite** edges in the surface. (The four outermost vertices are on branches of the surface, and form six-connected vertices.) Images are courtesy of R. Corkery.

minimal embedding of *diamond* is unknown. But we can form variants, *diamond** Since all edges in the *diamond* graph are along 3-fold axes, a number of distinct graphs can be formed, depending on the set of collapsed edges. One contraction gives the T lattice complex ($Fd\bar{3}m$), that is a minimal embedding of an (unknown) surface containing branch points distributed on the T vertices. A second pairing gives the PC graph, whose minimal embedding is in the P surface.

5. The Bonnet Isometry Links the P, D and G IPMS

The simplest triply periodic minimal surfaces - the cubic P, D and G surfaces - have common topology, and they can be readily transformed into each other by a simple surface distortion. For example, the P and D surfaces share a one-parameter family of rhombohedral rPD surfaces ($R\bar{3}m$); if the cubic surface are transformed to the rhombohedral unit cells, the D and P surfaces differ only in their axial $\frac{c}{a}$ ratios (the P is double that of the D) [**30, 13, 16**]. Further, a rhombohedrally distorted gyroid (a member of the rG family, space group $R\bar{3}c, \frac{c}{a}$ ratio $0.7975\ldots$) is also a member of the rPD family, so that all three surfaces can be linked via a continuous family of embedded triply periodic surfaces, all of zero mean curvature [**16**]. That transformation path between the P, D and G surfaces is continuous, and reticulations on the surface are transformed likewise, without any change in surface or collar rings.

A second, more restrictive, transformation path is possible, which preserves both mean curvature (so that all intermediates are minimal surfaces) *and* Gaussian curvature at corresponding points on the transformed surfaces. The transformation, known as the "Bonnet transformation" is thus isometric - and is achieved by pure bending of the surface without any stretching. Bonnet-related surfaces have identical hyperbolic geometry (they are intrinsically identical), only their embedding in E^3 changes. (In fact, the gyroid was discovered empirically by Schoen by subjecting the D and P surfaces to the Bonnet transformation [**48**]). There is an important distinction between the Bonnet isometry and the crystallographic transformation route described in the previous paragraph. Generic intermediate surfaces of the Bonnet transformation are not embedded, and some surface surgery is necessary to realize the transformation, involving cutting the surface and regluing, although the intrinsic topology of the surfaces remains fixed throughout. That means that reticulations of isometric surfaces, although identical in their $2d$ hyperbolic geometry, and thus containing an identical distribution of surface rings, can exhibit very different E^3 embeddings, with different collar rings.

Minimal embeddings of the *analcime* and *sodalite* nets, described above, are examples of locally identical reticulations of Bonnet-related surfaces. Both nets are $\{(6.6.4.4), 4\}$ on the G and D surfaces respectively, the former has eight-rings, the latter six-rings as collar-rings. Fragments of both reticulations on their underlying surfaces indicate the isometric correspondence between both graphs. This example illustrates the topological flexibility of realizing nets in E^3 from H^2: despite apparent differences in E^3, nets may be identical in a $2d$ sense.

6. Geometric Density of Nets

The H^2 description can be used to derive useful relations between the net's $2d$ topology and its density in E^3 [**20, 22**]. The Gauss-Bonnet theorem and Euler's relation allows a connection to be drawn between the H^2 net density, ρ_2, and the density of the net in E^3, ρ_3. Denote the area of the surface per unit cell (of topology χ and volume V) by A. Then

$$\rho_3 := \frac{A}{V}\rho_2 \tag{6.1}$$

Rewriting the surface area in terms of its topology and (surface-averaged) Gaussian curvature,

$$\rho_3 = \frac{2\pi\chi}{V\langle|K|\rangle}\rho_2 \tag{6.2}$$

It is convenient to introduce a global geometric characteristic of hyperbolic surfaces, called the homogeneity index, h:

$$h := \frac{A^{3/2}}{(2\pi|\chi|)^{1/2}V} \tag{6.3}$$

This number is exactly $\frac{3}{4}$ for idealized homogeneous minimal surfaces of constant negative Gaussian curvature, and close to that value for the nearly homogeneous triply periodic surfaces used here for net embeddings.

This allows the $2d$ and $3d$ densities to be simply related:

$$\rho_3 = h\rho_2^{3/2}(\frac{2\pi|\chi|}{N})^{1/2} \tag{6.4}$$

where the net contains N vertices per unit cell volume (and $\frac{\chi}{N}$ is related to the $\{n_2, z\}$ by Equation 3.3).

Estimates of the density, ρ_2, (the reciprocal area per node in the H^2 reticulation), can be made from purely geometric considerations. These estimates reveal slow variation of the area with increasing ring-size n_2, (or negative Gaussian curvature of the underlying hyperbolic surface), and as a first approximation, assume ρ_2 is constant (suitably scaled by the edge length), and choose the minimal embedding, thereby minimising the ring-size, n_2. That allows the E^3 net density to be expressed solely in terms of the global homogeneity index (h) of the underlying hyperbolic surface, and the ring-size of the minimally embedded hyperbolic net, n_2, via Equation 6.4. The homogeneity index can be estimated from the (average) mean curvature of the underlying hyperbolic surface.

Plots of net density versus hyperbolic ring-size for a range of low density nets including the zeolite nets described above, indeed support the validity of the constant $2d$ density approximation. Actual density data for a range of zeolites fits well Equation 6.4 above, suggesting that the hyperbolic area per T-atom (assumed here to be Si) remains constant, regardless of the underlying surface curvature over a range of nets. The average ring-size (n_2) is unknown for many other silicates. However, if these are estimated from the 3-d rings alone (a procedure that is known to be reasonable for the case of melanophlogite), Equation 6.4 leads to conserved surface density for an array of nets, from low density zeolites to higher density cases [**21**].

These results are of immediate relevance to the solid state. For example the *analcime* and *sodalite* frameworks are expected to be related by homothetic scaling factor of lattice parameters, a,

$$\frac{a_{ana}}{a_{sod}} \approx (\frac{N^3_{ana}C^2_D\chi_D}{N^3_{sod}C^2_G\chi_G})^{1/6} = (\frac{(48)^3\cdot(0.7498)^2 2}{(12)^3\cdot(0.7665)^2 8})^{1/2} = 1.58$$

in reasonable agreement with the measured lattice parameter ratio of silicon-rich examples of real-world analcime and sodalite, 1.54 (13.7Åand 8.9Åcells for analcime and sodalite respectively [**36**]).

This example is particularly interesting, given the isometry between the underlying D and G surfaces due to the Bonnet transformation. (The corresponding $\{(6.6.4.4), 4\}$ net on the P surface is that of another zeolite, Linde A [**1, 36**]. However, this embedding is not minimal, since the collar rings (4-gons are smaller than surface rings, $\{(6.4.6.4) \mid 4\}$,and so the net density is not expected to scale according to the Bonnet isometry.)

6.1. Wide-Pore Nets. Net densities are of particular relevance to zeolite catalysis, a process that is dependent in part on the pore size of the matrix zeolite material, in part on the incoming petroleum precursor. The density can be defined in both topological and geometric senses. The former is gauged by the coordination sequence, the latter by the curvilinear bonding geometry of the framework.

First, note that the density in E^3 scales with the $2d$ rings size. Somewhat paradoxically then, large-pore materials, which are necessarily of low density, require the presence of small ($2d$) rings in the H^2 net. A correlation between density and ring-size has been noticed by others within the setting of E^3, leading to the suggestion that wide-pore zeolites and related (four-connected) catalytic materials necessarily contain significant number of three-rings [**3, 4, 38, 51**].

As pores defined by channels of the embedding of the hyperbolic net grow, the curvature of the underlying IPMS decreases. Therefore, the bonding geometry must alter. The ring-size (n_2) is coupled to the bonding geometry within the framework, whose preferred dimensions are decided by chemists. (Those dimensions are almost invariably referred to the flat metric of E^3, with some exceptions [**1**].

Although no definitive bounds can be placed on accessible bond angles for *Si-O-Si* angles (although we can assume that these are more flexible than the relatively rigid tetrahedral *O-Si-O* bonds), they are observed to be constrained. We impose the constraint fulfilled by most framework silicates, that this angle be fixed at 145°. The effective T-T-T angle, θ, is then dependent on the disposition of intermediate -O- atoms with respect to the euclidean T-T edges (if the O atoms lie within a T-T ... ring, the T-T-T angle is about 150°, if they all lie outside the T-T ... rings, (θ)

falls to about 70°. The resulting curvature depends on the planarity of the ring, and the underlying surface curvature is itself dependent on the tangential direction of the edges with respect to principal directions of the surface. (For edges embedded in a minimal surface subtending an angle of ϕ with the principal directions, it follows from differential geometry that the edge curvature and torsion (κ and τ) are related to the Gaussian curvature of the underlying surface, K, by:

$$-K = \kappa^2 + \tau^2$$

$$\kappa = (\sqrt{-K})(cos^2(\phi) - sin^2(\phi))$$

$$\tau = (2\sqrt{-K})(cos(\phi)sin(\phi))$$

Assume, for simplicity, planar collar rings surrounding channels, which are necessarily along principal directions. (That assumption is convenient, but relevant, as it results in the least dense embedding.) Assuming the usual Si-Si distance (l) of 3.1Å,

$$\kappa\ell = 2cos(\frac{\theta}{2})$$

so that the largest open rings are *ca.* 6Åradius, and the resulting framework density (ρ_3) can be determined from this average radius of curvature:

$$\frac{2\pi||chi|^{1/2}}{N} = (|K|\Omega)^{1/2} = \frac{2\Omega^{1/2}}{\ell}cos(\frac{\theta}{2})$$

For a range of silicates, the nodal area on the surface, Ω is close to 12Å^2. That implies a lower bound of about 12.3 Si atoms per 1000Å^3 for the framework density. To date, the largest pores found in fully four-connected silicas are those of silica-rich faujasite, with a framework density of 12.7. In order to achieve still-

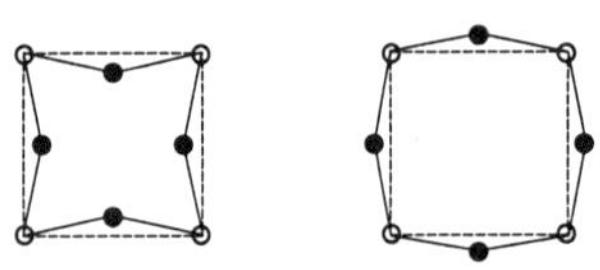

FIGURE 12. T-O-T ... ring in a four-connected net, such as SiO_2 (T atoms are light and two-connected O atoms dark). The T-T-T bond angles (dashed lines) depend on torsion of the T-O-T bonds (full edges) to the inside (left) or outside (right) of collar rings.

wider pores, any of the variables determining ρ_3 can be adjusted (ρ_2, θ, ℓ). One alternative is to vary the surface density, substituting larger ($2d$) atoms. That route goes towards explaining the number of wider-pore aluminum phosphate ($AlPO$) materials compared with their more silicon-rich analogues. (and can be enhanced by substitution of Si atoms by Ge or O by S). Alternatively, the effective net connectivity (z) can be lowered by disconnecting some edges from the net (*e.g.* formation of silanol groups in the silica). The formation of silica double layers on swollen hyperbolic surfaces is also feasible, since each layer then carries an effective connectivity of three (although the bridging -O- links connecting layers are themselves likely to be strained). A combination of both routes is responsible for the ultra-wide pore materials (typically 20 – 300Å) found in more recently discovered mesoporous silicates (and other inorganics) [**26**].

7. Three-Connected Nets, Fullerenes, Schwarzites and Sphere Packings

Three-connected crystalline nets are also of great interest to chemists. The best-known example[4] is surely the {5.6.6} network of sp^2 carbon-carbon bonds in buckminsterfullerene, C_{60}. Many other "elliptic" nets (reticulating surfaces of positive Gaussian curvature) have been described [**17**]. Hyperbolic analogues of the C_{60} net (as models for hypothetical "schwarzite" graphitic carbon structure) have been canvassed by others [**52, 28, 39, 34**], following the suggestion that hyperbolic carbon nets (such as {8.6.6}) are energetically stable with respect to fullerenes (and often far more stable) [**33**]. The hyperbolic analogue of C_{60}, {7.6.6} is readily described in terms of its H^2 tiling, and both P' and D' embeddings have been described. The {8.8.6} "polybenzene" tilings of the P and D surfaces are believed to be particularly favorable hypothetical structures for condensed graphite [**39**], Figures 13, 14. Duals of these three-connected nets, $\{n_2, 3\}$, whose $2d$ rings are exclusively triangular, $\{3, z\}$, are of likely relevance to crystalline materials. Closest (hexagonal) packing of equal discs in the (euclidean) plane is realised with touching discs forming a $\{3, 6\}$ net. If extra discs are inserted around each vertex,

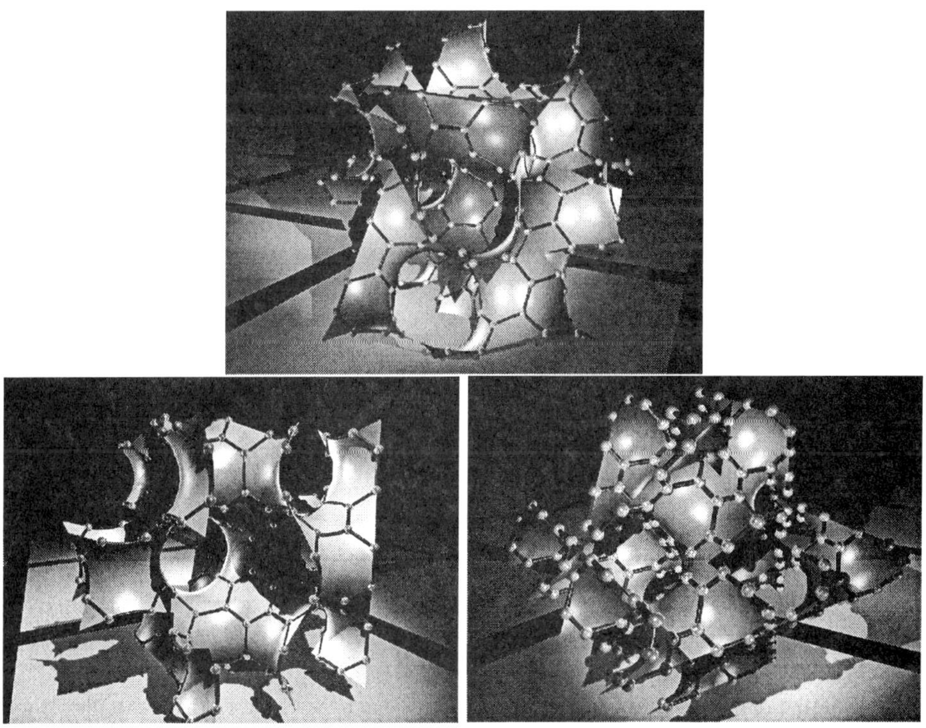

FIGURE 13. The three-connected "schwarzite" {8.8.6} net on the (top) *gyroid* surface (lower left) *P* and (lower right) *D* surfaces.

[4]The connectivity, z, is understood to be three from here on, and the net symbol is simplified accordingly.

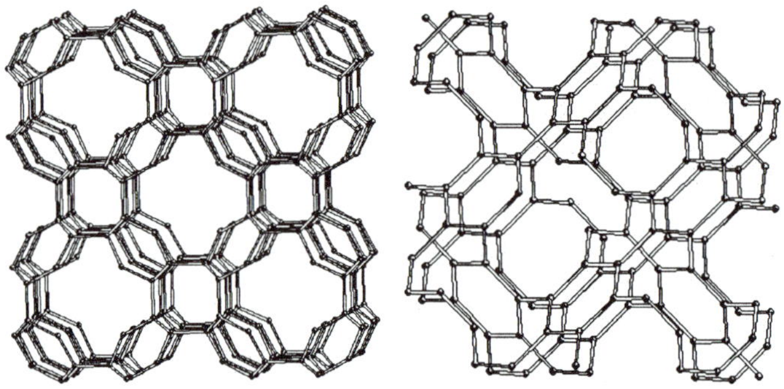

FIGURE 14. (Left) The three-connected "schwarzite" $\{8.8.6\}_P$ net in E^3 and (right) schwarzite $\{8.8.6\}_D$.

$\{3, z\}$ ($z > 6$) nets result. These tile H^2, and their embeddings in E^3 (via IPMS) correspond to a number of known sphere packings in E^3. The nets are triangulations of the IPMS, and offer the densest surface packings (ρ_2) of all reticulations of these surfaces. Since all faces of the tiling are triangular, polyhedral decompositions of these nets consist of the "triangular" polyhedra or deltahedra. (Those geodesic nets in H^2 are hyperbolic analogues of tetrahedra, octahedra, icosahedra and semi-regular polyhedra). Many of the simpler embeddings in E^3 are found in large metallic alloy arrays, (and, one could speculate, virus crystals)[5]. For example, $\{3,7\}$ on the D' gives a very low density regular sphere packing [**44**]. The $\{3,8\}$ net describes Zn sites in $NaZn_{13}$ on the P', *pyrochlore* on the D' and Th sites in Th_3P_4 [**44**] (the S lattice complex) on the gyroid. {3,9} describes cubic close packing on the D, body-centred cubic packing on the G and simple cubic packing on the P. The same $2d$ net defines Al sites in WAl_{12} [**44**] on the $I - WP'$ surface). Wells noted the possibility of semi-regular infinite polyhedra $\{3, 10\}$ (on the $I - WP'$) and $\{3, 12\}$ (on the $I - WP'$) [**53**].

The link between H^2 and E^3 descriptions of E^3 sphere packings can be developed in more generality. In particular, the packing efficiency (e), defined as the fraction of space covered by equivalent discs, of equal circular discs can be readily derived from non-euclidean ($2d$) geometry (as a function of sphere coordination number, z), that exceeds six for hyperbolic packings.

$$e = \frac{6(\frac{1}{2}cosec(\frac{\pi}{z} - 1)}{z - 6} \tag{7.1}$$

The packing efficiency increases from a minimum of $(2(1 - \frac{1}{\sqrt{3}}))$ (*ca.* 85%) for the elliptic $\{3,3\}$ (tetrahedral) sphere tiling of discs, to a theoretical maximum of $\frac{3}{\sqrt{\pi}}$ for the $\{3, \infty\}$ hyperbolic tree tiling (the Bethe lattice), *cf.* Figure 15. One way to achieve dense packings of ($2d$) nodes, o,r in chemical terms, increase the coordination of atoms, (increasing z), is to warp the plane hyperbolically (inserting interstitial "disclinations"). An apparently $3d$ structure can be comprehended then as a realization of that warping, produced in response to local $2d$ coordination requirements alone. The upper limit (for a $(3, \infty)$ tiling) is inaccessible in E^3 if

[5]Though many of the simpler alloy networks have not yet been identified as minimal embeddings of $2d$ structures, including the Frank-Kasper structures. The embedding for these structures is likely to be closer topologically to a closed cell foam.

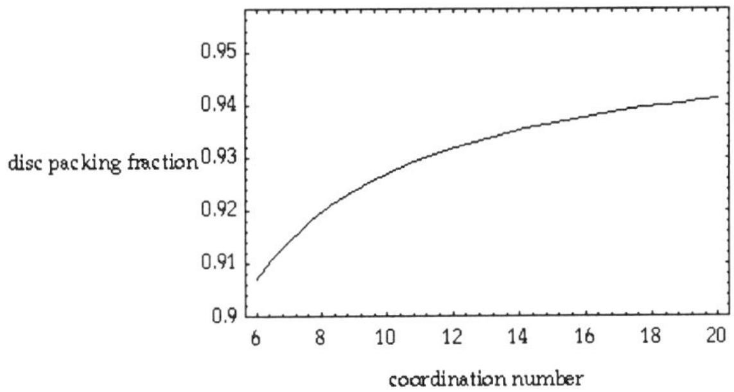

FIGURE 15. Disc packing efficiency on H^2 as a function of the sphere coordination.

we demand an intersection-free embedding (indeed, it is only realisable in H^2). The limit is dependent on the singular structure of the surface normals (the Gauss map) around a flat/umbilic point. Equilateral triangulations have been found for coordination numbers up to $z = 12$ [**53**] (unrealisable on the genus-three surfaces, but reticulating the genus-four $I - WP'$ surface).

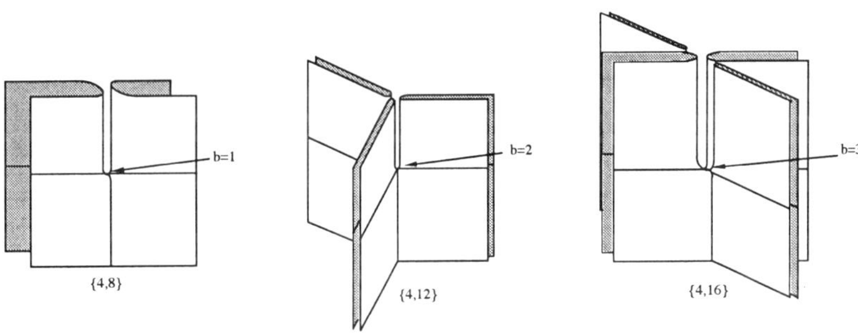

FIGURE 16. Maximally warped arrangements of $\{4, z\}$ nets, forming flat points of increasing order. The arrows point to domains on the surface whose Gauss maps contain branch points of order$b = 0, 1, 2$.

The branch point order of the Gauss map around a flat point sets the topology of the hyperbolic surface.

$$\chi = \frac{-1}{2} \sum_i b_i \tag{7.2}$$

For example, the (eight distinct) flat points on the genus-three IPMS are all of first order. In that case, an upper bound for the connectivity is:

$$z \leq \frac{2n_2(b+1)}{n_2 - 2} \tag{7.3}$$

so that $\{3, 12\}$ is the densest possible triangulation on intersection-free triply periodic surfaces of genus-three (per compactified unit cell).

Some limit cases for $\{4, z\}$ tilings are shown in Figure 16.

Once that upper bound is exceeded, self-intersecting surfaces result, which may also contain surface branch points. Though these too are hyperbolic structures, they can partition space into closed domains (as in convex polyhedral arrays), one-dimensional channels, etc. For example, the $WI-10$ surface described above partitions space into $1d$ channels (forming the $CrSi_3$ rod packing) and isolated polyhedra. Another example of a self-intersecting triply periodic minimal surface found by Koch and Fischer partitions space into eight distinct $3d$ labyrinths [**11**]. (Embeddings of many interwoven disjoint graphs, have recently been derived by decoration of these IPMS with close-packed trees [**24**].)

Finally, note that consideration of tilings of H^2 also allows exploration of still higher-dimensional euclidean spaces,E^n, by increasing the coordination number (z) beyond its upper limit for intersection-free surfaces (of fixed topology per unit cell), equation 7.3. For example, the $\{4, 2n\}$ tiling has nodes at lattice sites of the generalized simple cubic lattice in E^n, and $\{4, 4n\}$ faces partition E^n into closed cells, the "measure polytopes" (generalized cubes) (*cf.* Figure 17). For example, the $\{4, 12\}$ tiling partitions E^3 into closed cubes.) This example suggests the possibility of tiling E^3 with appropriate graphs, subdividing the space into closed polyhedra (and thereby generating the polyhedral packings).

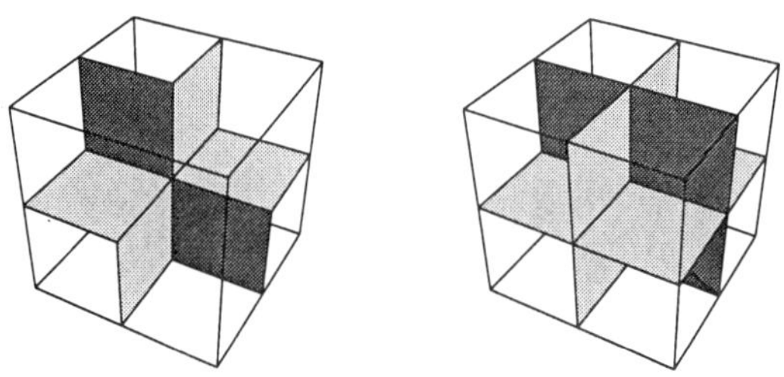

FIGURE 17. Patch of a $\{4,6\}$ tiling, a part of the P' surface. (Right) The $\{4,12\}$ tiling, showing the dissection of E^3 into cubes.

8. Future Developments

This paper outlines a possible approach to tilings of E^3 that are relevant to condensed atomic structures, and advocates study of the rich variety of H^2 tilings along the way. The examples derived here are the simplest possible ones, derived from reticulations of embedded triply periodic hyperbolic surfaces, S, whose compactified surfaces, S/T, where T are translational symmetry elements of the surface, has been confined to low genera (generally three or four).

Many open questions remain. It is natural to generalize the analysis to more complex triply periodic hyperbolic surfaces, of higher genus. Chemists will note that some of the more common graphs found in crystals are not canvassed here, though the variety of structures generated already by these examples is encouraging. Many of those cases require more complex embedding surfaces, including self-intersecting examples. Those examples span an array of morphologies, including the intersection-free "sponges" discussed here, "foams", "honeycombs" and combinations thereof. Generic features of intersecting triply periodic hyperbolic surfaces need to be better established before further progress can be claimed. Some progress has been made recently towards a topological description of novel "polycontinuous" triply periodic hyperbolic surfaces, that partition E^3 into more than two labyrinth graphs [**7, 24**]. These partitioning surfaces must contain line singularities, and lie between the simpler IPMS and the classical hexagonal honeycomb.

Further, the constraints imposed by the shape of E^3 remain unknown. For example, how constrained is the topology of a given $\{n_2, z|t\}$ embedding? Further, how is t dependent on the underlying hyperbolic net, $\{n_2, z\}$? (Examples are known of distinct E^3 embeddings of a hyperbolic net on the *same* surface (within a homeomorphism), so this question is a subtle one).

At a more philosophical level, this account calls into question the widely-held implicit assumption that chemical networks are $3d$ euclidean structures. [6]

References

[1] S. Andersson, S.T. Hyde, and H.-G. von Schnering, The Intrinsic Curvature of Solids, *Z. Kristallogr.*, **168** (1984) 1–17.
[2] N.L. Biggs, E.K. Lloyd, and R.J. Wilson, *Graph Theory*, Oxford University Press, 1976.
[3] G.O. Brunner, The Properties of Coordination Sequences and Conclusions Regarding the Lowest Possible Density of Zeolites, *J. Solid St. Chem.*, **29** (1979) 41–45.
[4] G.O. Brunner, and W.M. Meier, Framework Density Distribution of Zeolite-Type Tetrahedral Nets, *Nature*, **337** (1989) 146–147.
[5] H.S.M. Coxeter, *Regular Skew Polyhedra in Three and Four Dimensions and their Topological Analogues*, **Ser. 2, 43**, (1937) 33–62.
[6] H.S.M. Coxeter, On Laves' Graph of Girth Ten, *Can. J. Math.*, **7** (1955) 18–23.
[7] V. Elser, A Cubic Archimedean Screw, *Phil. Trans. Roy. Soc. Lond. A*, **354** (1996) 2071–2075.
[8] P.A. Firby, and C.F. Gardiner, *Surface Topology*, Ellis Horwood Ltd., Chichester, 1991.
[9] W. Fischer, Existenzbedingungen homogener Kugelpackungen zu kubischen Gittercomplexen mit weniger als drei Freiheitsgraden, *Z. Kristallogr.*, **138** (1973) 129.
[10] W. Fischer, Existenzbedingungen homogener Kugelpackungen zu kubischen Gittercomplexen mit drei Freiheitsgraden, *Z. Kristallogr.*, **140** (1974) 50–74.
[11] W. Fischer, and E. Koch, Spanning Minimal Surfaces, *Phil. Trans. Roy. Soc. Lond. A*, **354** (1996) 2105–2142.
[12] A. Fogden, Weierstrass Representation of Periodic Minimal Surfaces, *J. Physique (Colloque)*, **C7** (1990) C7–149.
[13] A. Fogden, Parametrisation of Triply Periodic Minimal Surfaces. 3. General Algorithm and Specific Representations for the Irregular Class, *Acta Cryst.*, **A49** (1992) 409
[14] A. Fogden, Parametrisation of Triply Periodic Minimal Surfaces. 3. General Algorithm and Specific Representations for the Irregular Class, *Acta Cryst.*, **A48** (1992) 442–451; 575–591.
[15] A. Fogden, Description of a 3-Periodic Minimal Surface Family with Trigonal Symmetry, *Z. Kristallogr.*, **209** (1994) 22–31
[16] A. Fogden, and S.T. Hyde, Continuous Transformations of Cubic Minimal Surfaces, *Eur. Phys. J. B*, **7** (1999) 91–104.
[17] P.W. Fowler, and D.E. Manolopoulos, *An Atlas of Fullerenes*, Oxford University Press, (1995).
[18] T. Hahn (ed.), *International Tables for Crystallography*, Kluwer Academic Publishers, 1992.
[19] H. Heesch, and F. Laves, Über dünne Kugelpackungen, *Z. Kristallogr.*, **85** (1933) 443–453.
[20] S.T. Hyde, B.W. Ninham, and Z. Blum, The Density Bof Silicon-Rich Zeolite Frameworks, *Acta. Cryst.*, **A49** (1993) 586–589.
[21] S.T. Hyde, *Crystalline Frameworks as Hyperbolic Films*, in *Defects and Processes in the Solid State. Some Examples in Earth Sciences*, J.D. FitzGerald and J.N. Boland (eds.), Elsevier, Amsterdam, (1993).
[22] S.T. Hyde, The Density of Three-Dimensional Nets, *Acta. Cryst.*, **A50** (1994) 753–759.

[6] Full-colour $3d$ models of many of these hyperbolic embeddings can be accessed from: http://rsphysse.anu.edu.au/ sth110/sth.html

[23] S.T. Hyde, and M. O'Keefe, Elastic Warping of Graphitic Carbon Sheets: Relative Energies of Some Fullerenes, Schwarzites and Buckytubes, *Phil. Trans. Roy. Soc. Lond. A*, **354** (1996) 1999–2008.
[24] S.T. Hyde, C. Oguey, and S. Ramsden, *Packings of Polycontinuous Graphs*, in press.
[25] H. Karcher, The Triply Periodic Minimal Surfaces of Alan Schoen and their Constant Mean Curvature Companions, *Manuscr. Math.*, **64** (1989) 291–337.
[26] G. Kresge, M.E. Leonowicz, W.J. Roth, J.C. Vartuli, and J.S. Beck, Ordered Mesoporous Molecular Sieves Synthesized by a Liquid Crystal Template Mechanism, *Nature*, **359** (1992) 710–712.
[27] F. Laves, Zur Klassifikation der Silikate, *Z. Kristallogr.*, **82** (1932) 1–14.
[28] T. Lenosky, X. Gonze, M. Teter, and V. Elser, Energetics of Negatively Curved Graphitic Carbon, *Nature*, **355** (1992) 333–335.
[29] S. Lidin, and S.T. Hyde, A Construction Algorithm for Minimal Surfaces, *J. Phys. France*, **48** (1987) 1585–1590.
[30] S. Lidin, Ring-Like Minimal Surfaces, *J. Phys. France*, *49* (1988) 421–427.
[31] S. Lidin, S.T. Hyde, and B.W. Ninham, Exact Construction of Periodic Minimal Surfaces: The I-WP Surface and its Isometries, *J. Phys. France*, *51* (1990) 801–813.
[32] J.H. Lindsay, An Elementary Treatment of the Imbedding of a Graph in a Surface, *Am. Math. Monthly*, (1959) 117–118.
[33] A.L. Mackay, and H. Terrones, Diamond from Graphite, *Nature*, **352** (1991) 762.
[34] A.L. Mackay, and H. Terrones, Hypothetical Graphite Structures with Negative Gaussian Curvature, *Phil. Trans. Roy. Soc. Lond.*, **A343** (1993) 113–127.
[35] W.M. Meier, *Pure Appl. Chem.*, **58** (1986) 1323–1328.
[36] W.M. Meier, and D.H. Olson, *Atlas of Zeolite Structure Types*, Butterworth-Heinemann, London, (1992), (also accessible on the WWW).
[37] M. O'Keeffe, and S. Andersson, *Acta Cryst.*, **A33** (1997) 914–923.
[38] M. O'Keeffe, Dense and Rare Four-Connected Nets, *Z. Kristallogr.*, **196** (1991) 21–37.
[39] M. O'Keeffe, G.B. Adams, and O.F. Sankey, Predicted New Low Energy Forms of Carbon, *Phys. Rev. Lett.*, **68** (1992) 2325–2328.
[40] M. O'Keeffe, and N.E. Breese, Uninodal 4-Connected 3d Nets. I. Nets Without 3- or 4-Rings, *Acta Cryst.*, **A48** (1992) 663–669.
[41] M. O'Keeffe, Uninodal 4-Connected 3d Nets. II. Nets With 3-Rings, *Acta Cryst.*, **A48** (1992) 670–673.
[42] M. O'Keeffe, Cubic Cylinder Packings, *Acta Cryst.*, **A48** (1992) 879–884.
[43] M. O'Keeffe, Uninodal 4-Connected 3d Nets. III. Nets With Three or Four 4-Rings at a Vertex, *Acta. Cryst.*, **A51** (1995) 916–920.
[44] M. O'Keeffe, and B.G. Hyde, *Crystal Structures. Patterns and Symmetry Theory*, Mineralogical Society of America, Washington, 1996.
[45] M. O'Keeffe, Coordination Sequences for Lattices, *Z. Kristallogr.*, **210** (1995) 905–908.
[46] M. O'Keeffe, and S. Andersson, *Acta Cryst.*, **A33** (1997) 914–923.
[47] J.-F. Sadoc, and J. Charvolin, Infinite Periodic Minimal Surfaces and their Crystallography in the Hyperbolic Plane, *Acta Cryst.*, **A45** (1989) 10–20.
[48] A.H. Schoen, *Infinite Periodic Minimal Surfaces without Self-Intersections*, N.A.S.A. Technical Note No. D5541, 1970.
[49] H.A. Schwarz, *Gesammelte Mathematische Abhandlungen*, Springer, Berlin, 1890.
[50] J.V. Smith, Topochemistry of Zeolites and Related Materials, *Chem. Rev.*, **88** (1988) 149–182.
[51] L. Stixrude, and M.S.T. Bukowinski, Rings, Topology and the Density of Tectosilicates, *Am. Mineral.*, **75** (1990) 1159–1169.
[52] D. Vanderbilt, and J. Tersoff, Negative-Curvature Fullerene Analog of C_{60}, *Phys. Rev. Lett.*, **68** (1991) 511–513.
[53] A.F. Wells, *Three-Dimensional Nets and Polyhedra*, John Wiley and Sons, 1977.

Department of Applied Mathematics, Research School of Physical Sciences, Australian National University, Canberra 0200 Australia.
E-mail address: `stephen.hyde@anu.edu.au`

Visualisation Laboratory,ANU Supercomputer Facility, Australian National University, Canberra 0200 Australia
E-mail address: `stuart.ramsden@anu.edu.au`

DIMACS Series in Discrete Mathematics
and Theoretical Computer Science
Volume **51**, 2000

Discrete Mathematics for Combinatorial Chemistry

Adalbert Kerber, Reinhard Laue, and Thomas Wieland

ABSTRACT. The aim is a description of discrete mathematics used in a project[1] devoted to the implementation of a software package for the simulation of combinatorial chemistry.

1. Combinatorial Chemistry

Rapid technical screening methods have been developed recently which allow quickly to test thousands of chemical substances if they are helpful or not. Thus the results of mass synthesis (in contrast to classical synthesis of a single substance) can rapidly be searched for an efficient pharmaceutical drug, say. Correspondingly, automatic devices for controlled mass synthesis are available now.

This is a challenge for mathematicians to provide methods that allow a simulation of such experiments in order to see *in advance* if a planned mass synthesis can deliver an interesting substance, at least in principle. The history of mass synthesis is described in the review article [**14**].

The aim of an experiment in combinatorial chemistry is the synthesis of a suitable *library* of molecules. There are single and multistep methods to do that, and they all use well defined *building blocks* and known *reactions* for the generation of such a library which is then *screened* in order to find synthesized molecules with the required medical or biological activity.

A mathematical model for a chemical reaction between two molecules (there are, of course, much more difficult cases, but for the purposes here, it suffices to consider two-component syntheses of the form $A + B \to C$) is a *reaction scheme*, say

$$((\eta_1, \beta_1),\ (\eta_2, \beta_2),\ \rho),$$

which is a pair of labeled molecular graphs, together with a mapping ρ that describes the reaction[1], say. For example, peptides are built by joining amino acids via condensation of the carboxyl group (COOH) and the amino group (NH_2). Consider, for example, the amino acids

1991 *Mathematics Subject Classification.* 05C90, 68R10, 90C35, 94C15.

[1]This is a simplification of the situation and is only used for a formalization of the construction problem discussed below. For more sophisticated purposes more comprehensive approaches like the algebra of *be*- & *r*-matrices of [**2**] are necessary.

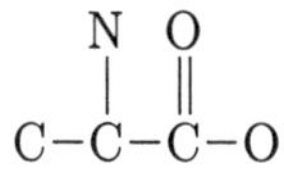

alanine

and

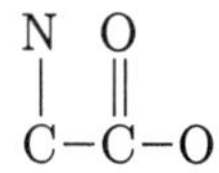

glycine

(For the sake of simplicity we took off the numbering of the atoms which is supposed to be — in the common substructure — the very same as indicated above.) Taking alanine first gives, via the well-known condensation,

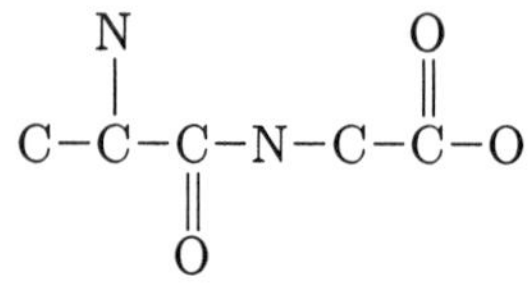

While glycine as η_1 yields

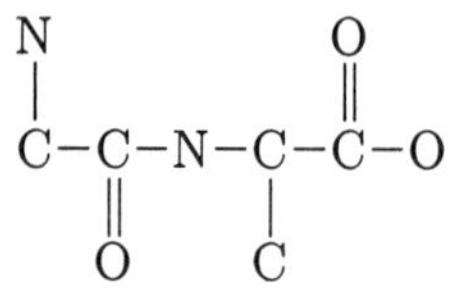

This was a case of so-called *single-attachment.* The case of *multiple attachment* is, of course, more complicated. Consider the following core with several reaction sites and a number of ligand compounds, assuming, for the sake of simplicity, that each of them contains exactly one substructure isomorphic to (η_2, χ_2) and that the sites are all equivalent. Here is a sketch of such a situation where there are exactly four equivalent sites:

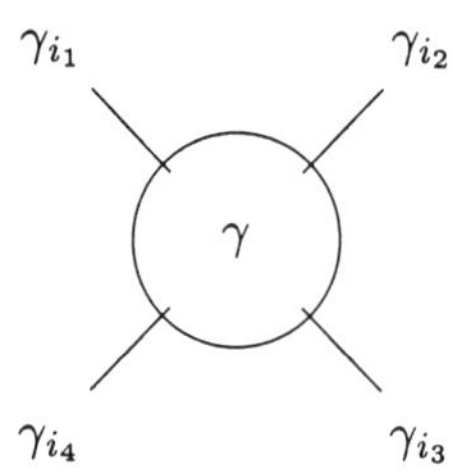

From the mathematical point of view this situation is the very same situation as in the case when we want to obtain *permutational isomers,* say the 22 isomers of dioxin which has the skeleton

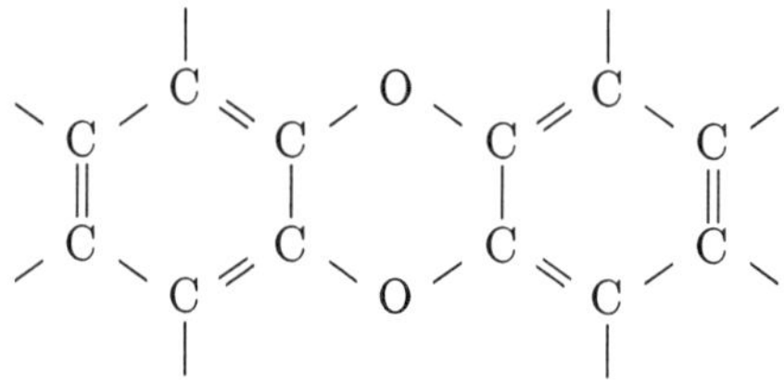

Ruch, Hässelbarth and Richter have shown [**10, 11**], how permutational isomers are bijectiveley related to double-cosets. The set of permutational isomers of the dioxin is bijectively to the set of double cosets

$$V_4\backslash S_8/S_4 \oplus S_4.$$

We shall explain the reason for that later. (The Kleinian four group V_4 comes from the fact that this is the symmetry group of the skeleton, while the subgroup $S_4 \oplus S_4$ is due to the fact that we have to distribute 4 chloride and 4 hydrogen atoms among the 8 free valences on which the symmetric group S_8 acts transitively.) This set of double cosets is of order 22, but even more: Using double cosets we can *construct* and not just count, a complete system of representatives of these orbits, i. e. we can construct the molecular graphs corresponding to these 22 permutational isomers of dioxin!

2. A Mathematical Model

Slightly more abstract than the consideration of permutational isomers is the approach used by Pólya [8] in order to consider connectivity–isomers and many other situations. He considered sets

$$Y^X := \{f : X \to Y\}$$

of mappings together with actions of a group G on X and the corresponding action of G on Y^X :

$$G \times Y^X \to Y^X : (g, f) \mapsto f \circ g^{-1}.$$

The set of *orbits*

$$G(f) := \{f \circ g^{-1} \mid g \in G\}$$

of G on Y^X will be denoted by

$$G\backslash\!\backslash Y^X.$$

This model covers the case of permutational isomers, since we need only identify the set X with the set of free valences, Y with the set of admissible ligands and G with the symmetry group of the skeleton. Then $f \in Y^X$ means a particular attachment of ligands, while the orbit of f consists of all the attachments that arise from f by applying a symmetry operation. Thus the essentially different attachments of ligands are in 1-1–correspondence with the orbits. Moreover, as Ruch, Hässelbarth and Richter have shown, these orbits are bijectively related to double cosets in the following way. The set of orbits

$$G\backslash\!\backslash_f Y^X$$

consisting of mappings with the same *weight* (or *content*, as Pólya called it), i.e. which take the values $y \in Y$ with the multiplicities $|f^{-1}(y)|$ (for example in the case of the dioxin $|f^{-1}(H)| = |f^{-1}(Cl)| = 4$), can be mapped bijectively onto a set of double cosets as follows:

$$G\backslash\!\backslash_f Y^X \longrightarrow G\backslash S_X/(S_X)_f : \; G(f \circ \pi) \longmapsto G\pi(S_X)_f,$$

where $(S_X)_f$ denotes the stabilizer of f in the symmetric group S_X. In fact, the stabilizer of f in the symmetric group S_X is the direct sum (or direct product, if you prefer) of the symmetric groups $S_{f^{-1}(y)}$ on the inverse images of the values which f takes:

$$(S_X)_f = \oplus_y S_{f^{-1}(y)}.$$

for each element π of the symmetric group S_X on X. In the dioxin case we obtain the bijection that was mentioned above already.

The crucial point is now that we can obtain a transversal of this set of orbits $G\backslash\!\backslash_f Y^X$ by constructing a transversal of this set of double cosets and then retranslating its elements, which are permutations, into mappings which are attachments of the ligands and hence a model for the desired permutational isomers which are molecules.

Let us apply this approach to some of the famous examples from combinatorial chemistry, taken from [**1**]. Here are the cores that are considered:

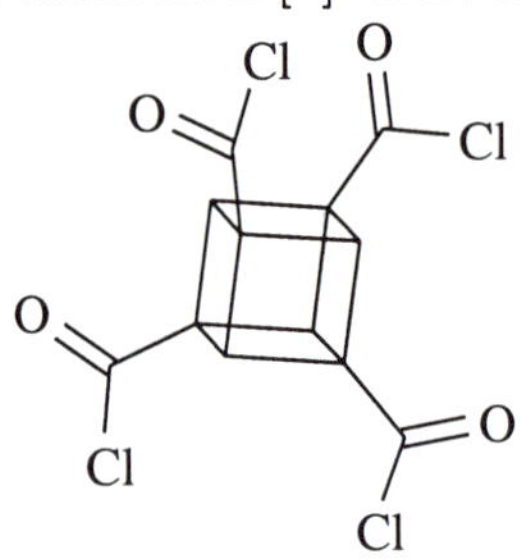

1

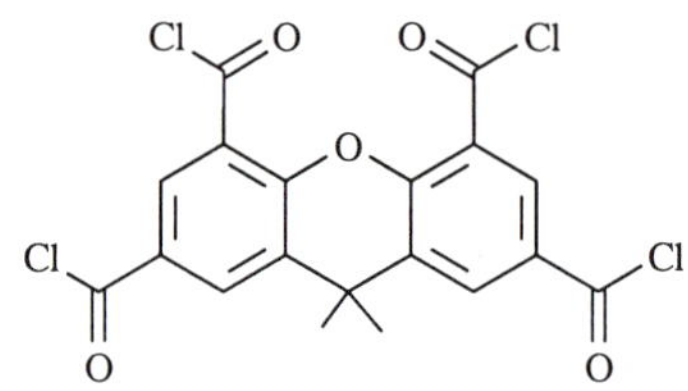

2

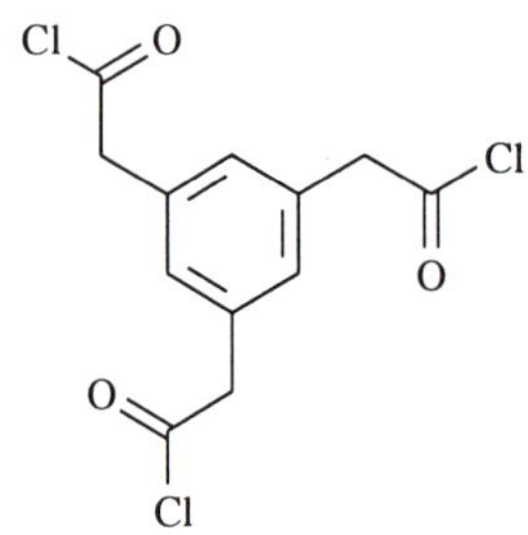

3

The first (upper left) is a cubane derivative, the second one a xanthene, and the third one is a benzene triacid chlorine. The reaction scheme consists of the substructures

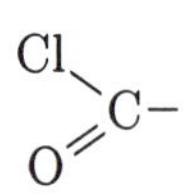

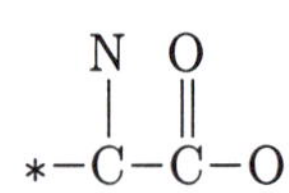

and the product substructure resulting from the reaction is of the form

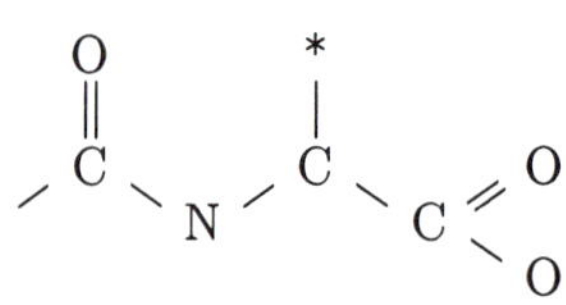

The situation just described can easily be reformulated in terms of Pólya's approach. The four active sites of the cubane or of the xanthene, as well as the three active sites of the benzene triacid chlorine, respectively, form the set X while the set Y consists of a subset of the set of the twenty natural amino acids. Attaching amino acids at the active sites is the same as forming a mapping $f \in Y^X$, and the equivalence classes of attachments are the orbits of Y^X under the symmetry group of the core. The general formula for the number of such orbits with respect to the symmetry group G is obtained from the well known Lemma of Cauchy-Frobenius.

If applied to the Pólya situation, where a group G acts on a set X and therefore also on Y^X, we obtain the following formula for the number of orbits of G on Y^X :

$$\frac{1}{|G|}\sum_{g\in G}|Y|^{c(g)},$$

where $c(g)$ means the number of cyclic factors of g on X which is in fact also the number of orbits of the group $\langle g\rangle$ generated by the single element $g\in G$. For example in the case of the cubane and if we use as symmetry group the group of proper rotations of the central molecule, which is, in mathematical terms, isomorphic to the alternating group A_4, then we obtain for the number of orbits the expression [**16**]

$$\frac{1}{12}\left(|Y|^4+11\cdot|Y|^2\right).$$

If we take Y to be the set of 20 amino acids, then we obtain 13 700 orbits of A_4. A table with all the values, for the alternating group, is given here[2]

n	cubane	xanthene	triacid
1	1	1	1
2	5	10	4
3	15	45	10
4	36	136	20
5	75	325	35
6	141	666	56
7	245	1225	84
8	400	2080	120
9	621	3321	165
10	925	5050	220
11	1331	7381	286
12	1860	10440	364
13	2535	14365	455
14	3381	19306	560
15	4425	25425	680
16	5696	32896	816
17	7225	41905	969
18	9045	52650	1140
19	11191	65341	1330
20	13700	80200	1540
21	16611	97461	1771

In the xanthene case, the symmetry group is of order 2, and so the formula for the size of the library turns out to be

$$\frac{1}{2}(|Y|^4+|Y|^2).$$

The full library, i. e. $|Y|=20$ gives the number 80 200.

[2]its entries can be evaluated online via our internet page
http://www.mathe2.uni-bayreuth.de/axel/perm_iso.html

In the case of the benzene triacid chlorine, the symmetry group is the dihedral group of order 6 on the set of three active sites, and so the size of the library is

$$\frac{1}{6}(|Y|^3 + 3 \cdot |Y|^2 + 2 \cdot |Y|).$$

For $|Y| = 20$, we obtain the number 1540.

3. The Elements of the Library by Weight

The evaluation of the size of the library is, of course, only the very first step in order to get an idea how big the library in principle might be. Much more important is the *construction of the elements of it.* A first step towards this refinement can also be done using Pólya's theory of enumeration under finite group actions.

Pólya's Theorem gives the numbers of orbits of G on Y^X by weight (=multiplicities of amino acids) in terms of the so-called *cycle index polynomial:*

$$Cyc(G,X) := \frac{1}{|G|}\sum_{g\in G}\prod_{i=1}^{|X|} x_i^{a_i(g)},$$

if $a_i(g)$ denotes the number of orbits of the subgroup $\langle g\rangle$ generated by $g \in G$ on X which are of length i, i.e. the number of i-cycles of g on X. The number of orbits of G on Y^X which consist of mappings f that take the value $y \in Y$ with multiplicity b_y is then the coefficient of the monomial $\prod_{y\in Y} y^{b_y}$ in the polynomial

$$\frac{1}{|G|}\sum_{g\in G}\prod_{i=1}^{|X|}(\sum_{y\in Y} y_i)^{a_i(g)}$$

which arises from the cycle index by replacing the indeterminate x_i by the polynomial $\sum_y y^i$.

Using SYMMETRICA[3], we obtain, for example, in the dioxin case the following generating function for the permutational isomers:

$$Cl_8 + 2H_1Cl_7 + 10H_2Cl_6 + 14H_3Cl_5 + 22H_4Cl_4 + 14H_5Cl_3 + 10H_6Cl_2 + 2H_7Cl_1 + H_8.$$

The summand $22H_4Cl_4$ means that there are exactly 22 permutational isomers containing both 4 hydrogen and 4 chlorine atoms. We should like to explain now in detail how this generating function can be obtained.

There exists an interesting description of the generating functions which Pólya obtained in terms of irreducible representations $[\alpha]$ of the symmetric group S_X which correspond (see e.g. [**5**] or [**6**]) to Young diagrams $[\alpha]$ with $|X|$ boxes. This description is in fact due to H. O.Foulkes (see [**4**]). If we denote by f^α the dimension of the representation $[\alpha]$ and by ζ^α its character, then we can define the Schur functions $\{\alpha\}$ as follows:

$$\{\alpha\} := \frac{f^\alpha}{|X|!}\sum_{\pi\in S_X}\zeta^\alpha(\pi)\prod_{i=1}^{|X|}(\sum_y y^i)^{a_i(\pi)}.$$

In terms of the multiplicities $(IG \uparrow S_X, [\alpha])$ of these $[\alpha]$ in the representation $IG \uparrow S_X$ of S_X which is induced from the identity representation of the symmetry group G, we obtain that

[3]SYMMETRICA can be obtained via internet from http://www.mathe2.uni-bayreuth.de

$$\frac{1}{|G|}\sum_{g\in G}\prod_{i=1}^{|X|}(\sum_{y\in Y} y_i)^{a_i(g)} = \sum_{\alpha\vdash|X|}(IG\uparrow S_X,[\alpha])\cdot\{\alpha\}.$$

The cubane case is trivial as well as the triacid case, since the symmetry group is the symmetric group in both cases, and so the resulting generating function is simply the Schur polynomial $\{4\}$ in the cubane case, while in the triacid case it is $\{3\}$. They both correspond to the one rowed Young diagram.

Less trivial is the situation of the xanthene, where the symmetry group G is a cyclic group of order 2, and the resulting generating function for the elements of the library by weight is

$$\{4\}+\{3,1\}+2\cdot\{2^2\}+\{2,1^2\}.$$

In the case when Y consists of 4 amino acids, say of types which we denote by a, b, c and d, then we obtain, using the computer algebra package SYMMETRICA:

$$c^4+2bc^3+4b^2c^2+2b^3c+b^4+2ac^3+6abc^2+6ab^2c$$

$$+2ab^3+4a^2c^2+6a^2bc+4a^2b^2+2a^3c+2a^3b+a^4$$

The summand $6ab^2c$, for example, shows that there are six essentially different attachments of an amino acid of type a, two amino acid of type b and one amino acid of type c.

4. The Generation of the Library

Here we can use — at least for our particular examples taken from [**1**] — the very same method as in the case of the construction of permutational isomers: *there exists a canonical bijection onto a set of double cosets.* The reason is that the symmetric group S_X on X acts transitively on the set of mappings $f \in Y^X$ which have a prescribed weight (which means prescribed multiplicities $f^{-1}(y)$ for the values $y \in Y$, for our chemical examples this means: prescribed multiplicities of amino acids). Hence the orbit of a fixed mapping (= labeled molecule) under that symmetric group is essentially the same as the set $S_X/(S_X)_{f_0}$ of left cosets of the stabilizer $(S_X)_{f_0}$. Therefore the set of orbits $G(f)$ of the symmetry group G (of the cubane derivative, say) on this orbit $S_X(f_0)$ can be mapped bijectively onto the corresponding set of double cosets

$$\varphi\colon \mathrm{Orb}(G, S_X(f_0)) \to G\backslash S_X/(S_X)_{f_0} : G(gf_0)\mapsto Gg(S_X)_{f_0}.$$

Thus, once we have obtained a complete system of representatives of these double cosets we obtain, by an application of the inverse mapping φ^{-1}, a complete system of molecules with prescribed multiplicities of the admissible amino acids. This can be carried out, as it is described in several papers in detail, by going up and down along a leporello of Young subgroups, from S_X to $(S_X)_{f_0}$, a method that applies in the same way to the construction of full sets of permutational isomers. This method was invented by B. Schmalz (see [**12**]), applied to the construction of combinatorial designs and called the *ladder game.* Here is a selfexplanatory example of such a folder which can be used in the dioxin case, in order to construct all the permutational isomers successively:

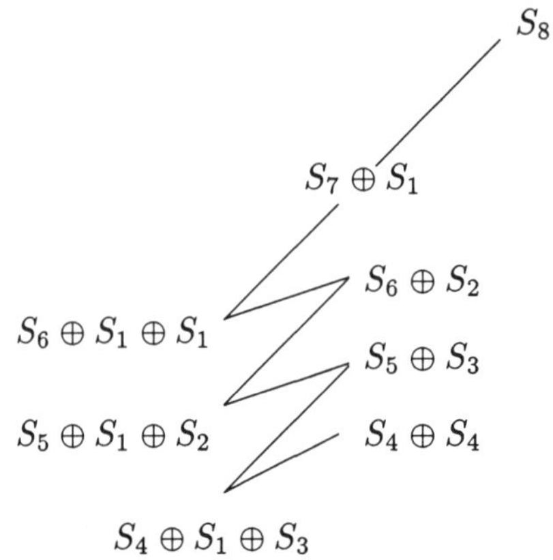

The method used is to start from above with the subgroup S_8 of S_8. It has exactly one left coset and so there is exactly one orbit of V_4 on this set of left cosets, and hence exactly one isomer containing 8 hydrogen atoms and no chlorine. In the second step this one-element set of left cosets of S_8 is split up into the 8 left cosets of $S_7 \oplus S_1$. It turns out that V_4 has exactly two orbits on this set, representatives of which yield the two different permutational isomers containing 7 hydrogen atoms and 1 chlorine, ... , until we reach $S_4 \oplus S_4$, obtaining from the orbits of V_4 on the set of left cosets the corresponding permutational isomers with 4 hydrogen and 4 chlorine atoms. The "Homomorphism Principle", see [**6**],[**7**], shows how we can obtain a transversal of this bigger set of orbits from the smaller set we had before. This principle is combined with orderly generation which is popular in constructive combinatorics (see [**9**],[**3**]).

This rather general and efficient approach allows a very rapid generation of the combinatorial libraries. In the three cases from our example (taking the natural amino acids as building blocks), we examined a computing speed of 40 structures per second on a Pentium 90 MHz PC, writing all solutions to the hard disk. Figure 1 shows six molecules from each of the three libraries. More details are given in [**15, 16, 17**].

5. Summary

We have shown that enumeration under finite group action can cover famous examples of combinatorial chemistry. The methods used are Pólya's Ansatz to choose suitable sets X, Y and a finite group G acting on X. The examination of the induced action of G on the set of mappings Y^X allows to evaluate the total number of elements in a combinatorial library arising from a core with the set X as set of active sites by attaching elements of a set Y of building blocks, and with respect to the symmetry group G of X. This number is obtained by an easy application of the Cauchy-Frobenius Lemma. Moreover we can obtain (by an application of Pólya's Theorem) a generating function for the elements of the library, which enumerates these elements by weight. Finally we can even construct the library using the double coset reformulation of the orbits, the Homomorphism Principle and Orderly Generation. The corresponding software is under development.

Core 1:

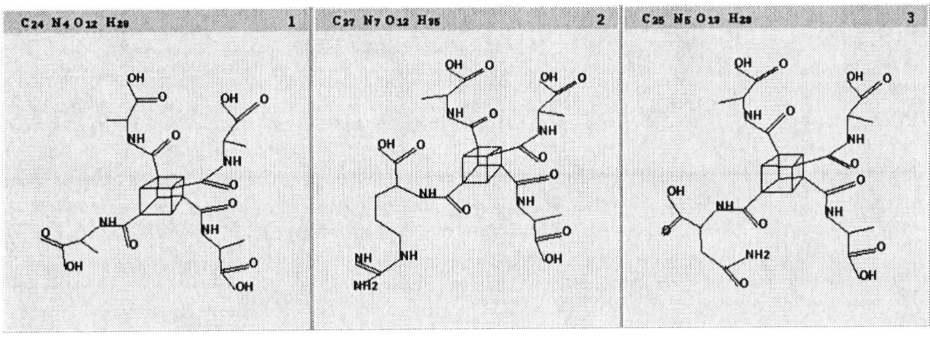

Core 2:

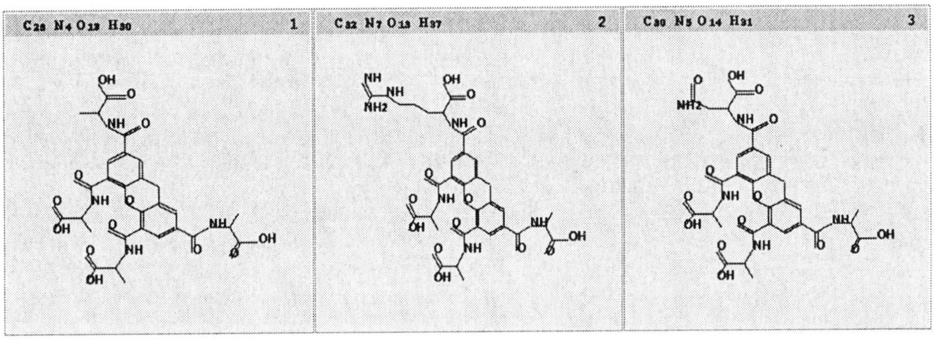

Core 3:

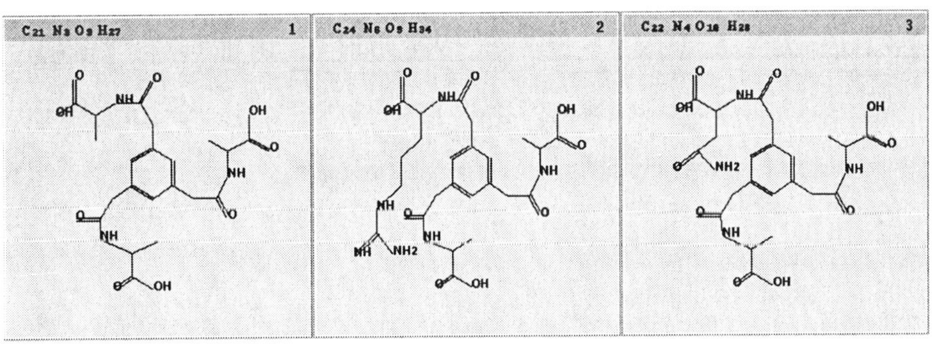

FIGURE 1. Extracts from the combinatorial libraries produced from the structures **1**, **2** and **3** and the natural amino acides

References

[1] T. Carell, E.A. Wintner, A.J. Sutherland, J. Rebek, Jr., Y.M. Dunayevskiy, and P. Vouros, New Promise in Combinatorial Chemistry: Synthesis, Characterization, and Screening of Small-molecule Libraries in Solution, *Chem. & Biol.*, **2** (1995) 171–183.

[2] J. Dugundji, and I.K. Ugi, An Algebraic Model of Constitutional Chemistry as a Basis for Chemical Computer Programs, *Top. Curr. Chem.*, **39** (1973) 19–64.

[3] I.A. Faradžev, Constructive Enumeration of Combinatorial Objects, *Problèmes combinatoires et théorie des graphes (Colloq. Internat. CNRS, Univ. Orsay, Orsay, 1976)*, **260** (1978) 131–135.
[4] H.O. Foulkes, On Redfield's Group Reduction Functions, *Can. J. Math.*, **15** (1963) 272–284.
[5] G.D. James, A. Kerber, *The Representation Theory of the Symmetric Group*, Encyclopedia of Math., vol. 16, Addison-Wesley, 1981.
[6] A. Kerber, *Algebraic Combinatorics Via Finite Group Actions*, BI-Wissenschaftsverlag, Mannheim, Wien, Zürich, 1991.
[7] R. Laue, Construction of Combinatorial Objects – a Tutorial, *Bayreuther Mathematische Schriften*, **43** (1993) 53–96.
[8] G. Pólya, Kombinatorische Anzahlbestimmungen für Gruppen, Graphen und chemische Verbindungen, *Acta mathematica*, **68** (1937) 145–253.
[9] R.C. Read, Every One a Winner, *Ann. Discr. Math.*, **2** (1978) 107–120.
[10] E. Ruch, W. Hässelbarth, and B. Richter, Doppelnebenklassen als Klassenbegriff und Nomenklaturprinzip für Isomere und ihre Abzählung, *Theoretica Chimica Acta*, **19** (1970) 288–300.
[11] E. Ruch, and D.J. Klein, Double Cosets in Chemistry and Physics, *Theoretica Chimica Acta*, **63** (1983) 447–472.
[12] B. Schmalz, Verwendung von Untergruppenleitern zur Bestimmung von Doppelnebenklassen, *Bayreuther Mathematische Schriften*, **31** (1990) 109–143.
[13] I. Ugi, A. Dömling, B. Gruber, M. Heilingbrunner, C. Heiss, and W. Hörl, Formale Unterstützung bei Multikomponentenreaktionen – Automatisierung der Synthesechemie, In: *R. Moll (Hrsg.): Software-Entwicklung in der Chemie 9*, S. 114-128, GDCh, Frankfurt am Main, 1995.
[14] I. Ugi, Fast and Permanent Changes in Preparative and Pharmaceutical Chemistry Through Multicomponent Reactions and their 'Libraries', In: *Proc. Estonian Acad. Sci. Chem.*, **44**(4) (1995) 237–273.
[15] T. Wieland, Mathematical Simulations in Combinatorial Chemistry, *MATCH*, **34** (1996) 179–206.
[16] T. Wieland, Combinatorics of Combinatorial Chemistry, *J. Math. Chem.*, **21** (1997) 141–157.
[17] T. Wieland, Konstruktionsalgorithmen bei molekularen Graphen und deren Anwendung, *MATCH*, **36** (1997) 5–155.

University of Bayreuth, Department of Mathematics, D-95440 Bayreuth, Germany
E-mail address: kerber@uni-bayreuth.de, http://www.mathe2.uni-bayreuth.de

University of Bayreuth, Department of Mathematics, D-95440 Bayreuth, Germany
E-mail address: laue@uni-bayreuth.de, http://www.mathe2.uni-bayreuth.de

Deutsches Zentrum für Luft- und Raumfahrt e.V., Deutsches Fernerkundungsdatenzentrum, Oberpfaffenhofen, D-82234 Wessling, Germany
E-mail address: thomas.wieland@dlr.de, http://www.dfd.dlr.de

DIMACS Series in Discrete Mathematics
and Theoretical Computer Science
Volume 51, 2000

Carbon Networks on Cubic Infinite Periodic Minimal Surfaces

R. Bruce King

ABSTRACT. Polymeric low density allotropes of carbon are predicted to arise from the decoration of negative curvature cubic infinite periodic minimal surfaces (cubic IPMS's) with networks of trigonal carbon atoms containing hexagons and one type of polygon with more than six edges. The cubic IPMS's of interest are the P surface and D surface, respectively, which are adjoint surfaces related by a Bonnet transformation with an association parameter of $\pi/2$. Such cubic IPMS's have genus 3 and are thus required by the generalized Euler's theorem to have 24 heptagons, 12 octagons, or 8 nonagons in their unit cells. Leapfrog transformations from trigonal carbon networks on cubic IPMS's containing only non-hexagonal rings (e.g., the Klein surface in the case of heptagons and the Dyck surface in the case of octagons) provide the minimum number of carbon hexagons required for dilution of the carbon non-hexagons so that the isolated non-hexagon rule (INHR) is satisfied, i.e., no edges are shared by pairs of non-hexagonal faces. The structures of such cubic IPMS's decorated by trigonal carbon atoms are conveniently described by the octants of their unit cells, which are required to have three-fold symmetry. However, if the carbon atom decoration of each octant is chiral (i.e., with only the three-fold axis and no plane of symmetry), then a D surface corresponding to the face-centered cubic lattice is required.

1. Introduction

Two allotropes of elemental carbon have been recognized since antiquity, namely diamond and graphite. The structure of diamond [**1**] (Figure 1a) consists of an infinite three-dimensional network of tetrahedral (degree 4) carbon vertices which form interlocking non-planar six-membered rings. The structure of graphite [**2, 3**] (Figure 1b) consists of a stack of parallel planar hexagonal layers of trigonal (degree 3) carbon vertices. The strong chemical bonding of the diamond structure in all three directions is responsible for the extreme hardness of diamond whereas the relatively slight forces between the hexagonal layers in the graphite structure are consistent with its softness and lubricity.

Until the 1980's diamond and graphite were the only well-characterized allotropes of elemental carbon. However, during the 1980's new allotropes of carbon were discovered exhibiting finite molecular cage structures rather than the infinite

1991 *Mathematics Subject Classification.* Primary 49Q05; Secondary 30F50.

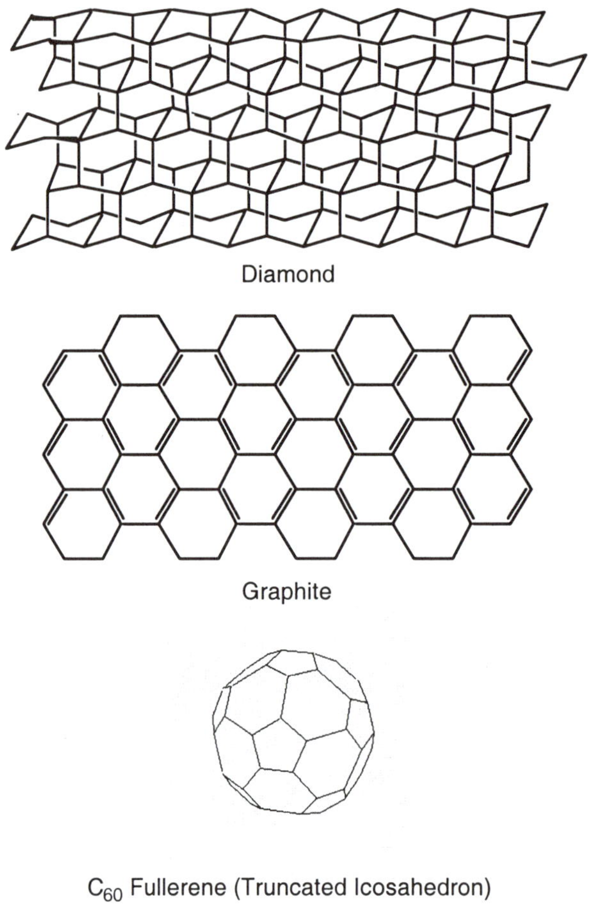

FIGURE 1. (a) A portion of the diamond lattice; (b) A portion of a planar hexagonal sheet of the graphite structure; (c) A front view of the truncated icosahedral structure of the C_{60} fullerene.

polymeric structures found in diamond and graphite. The first such molecular carbon allotrope was C_{60}, which was postulated and subsequently shown to have a truncated icosahedral structure with all 60 equivalent trigonal carbon vertices, 20 hexagonal faces, and 12 pentagonal faces (Figure 1c) [**4**]. The discovery of C_{60} was followed almost immediately by the discovery of other molecular C_n allotropes (e.g., $n = 70$, 76, 78, 80, 82, 84, 86, 88, 90, and 96) exhibiting other polyhedral cage structures albeit with much lower symmetry. Such molecular carbon cages, generally known as *fullerenes* in view of their resemblance to the architectural creations of R. Buckminster Fuller, were all found to have exactly 12 pentagonal faces; the remaining faces of all known fullerenes are all hexagons. All realizable fullerene structures satisfy the so-called *isolated pentagon rule* (IPR), i.e., the pentagonal faces must be sufficiently isolated from one another so that no edges of the polyhedron are shared by two pentagonal faces.

All fullerenes may be regarded as topologically homeomorphic to a sphere, the prototypical surface exhibiting positive curvature and having genus zero. Thus the introduction of carbon pentagons into the flat graphite hexagonal lattice (Figure 1b) generates sites of positive curvature. Euler's theorem can be used to show that a closed polyhedron with all trigonal vertices and only pentagonal and hexagonal faces, such as all known fullerene polyhedra, must have *exactly* 12 pentagonal faces.

A topic of interest is the favored shapes of networks of trigonal carbon vertices in which only hexagonal and heptagonal or octagonal carbon rings are present. Such networks cannot form closed polyhedra of positive curvature homeomorphic to the sphere but instead must form infinite structures of negative curvature known as infinite periodic minimal surfaces (IPMS's). The simplest and most symmetrical IPMS's exhibit cubic symmetry and have unit cells of genus three. The two fundamental IPMS's of this type are the P surface related to a simple cubic lattice and the D (or F) surface related to a face-centered cubic lattice. Negative curvature allotropes of carbon based on IPMS's are conveniently called *schwarzites*, in honor of the mathematician H.A. Schwarz, who was the first to study IPMS's in detail [**5**].

Mackay and Terrones [**6**] appear to have been the first to recognize the possibility of negative curvature allotropes of carbon. Thus in 1991 they proposed the $P192$ structure on the simple cubic P surface with a cubic unit cell of 192 carbon atoms containing 12 octagons and 80 hexagons. Shortly thereafter followed the first proposals of a negative curvature allotrope of carbon with only hexagons and heptagons, namely the $P216$ and $D216$ proposals of Lenosky, Gonze, Teter, and Elser [**7**] with 216 carbon atoms containing 24 heptagons and 80 hexagons in the cubic unit cell of the P or D surfaces, respectively, and the $D168$ proposal of Vanderbilt and Tersoff [**8**] with 168 carbon atoms containing 24 heptagons and 56 hexagons in the cubic unit cell of the D surface. Note that in these carbon allotropes, the heptagonal and octagonal carbon rings are not planar but bent. Furthermore in all of these proposed negative curvature carbon structures the IPR rule of fullerene structures can be generalized to the *isolated non-hexagonal rule* (INHR), i.e., the non-hexagonal faces (e.g., the heptagons in $P216$, $D216$, and $D168$ and the octagons in $P192$) must be sufficiently isolated from each other so that no pair of non-hexagons shares any edges. The negative curvature carbon allotrope proposed by O'Keeffe, Adams, and Sankey [**9**] with a much smaller unit cell containing 24 carbon atoms, 12 octagons, and 8 hexagons violates the INHR since many of its edges are shared by two octagons. None of these negative curvature allotropes of carbon have yet been obtained experimentally. They are predicted to have cross-linked insoluble polymeric structures with significant lower densities that the other allotropes of carbon because of their relatively porous structures.

The truncated icosahedron of the C_{60} structure can be generated from the regular dodecahedron by omnicapping followed by dualization (Figure 2). Such a process is called a *leapfrog transformation*, and the resulting network is called the *leapfrog* of the original network. A leapfrog transformation of a network with all degree 3 vertices and only non-hexagonal faces triples the number of vertices and assures that the non-hexagonal faces will be "diluted" by enough hexagonal faces so that the INHR is satisfied. Figure 3 shows the effects of a leapfrog transformation on pentagonal, hexagonal, and heptagonal faces as well as edges and trivalent vertices. In 1996 the author [**10**] showed how the $D168$ structure [**8**] can be generated by an analogous leapfrog transformation starting with the genus 3 figure containing 24 heptagons and 56 vertices first described by Klein [**11, 12**] in 1879. This paper

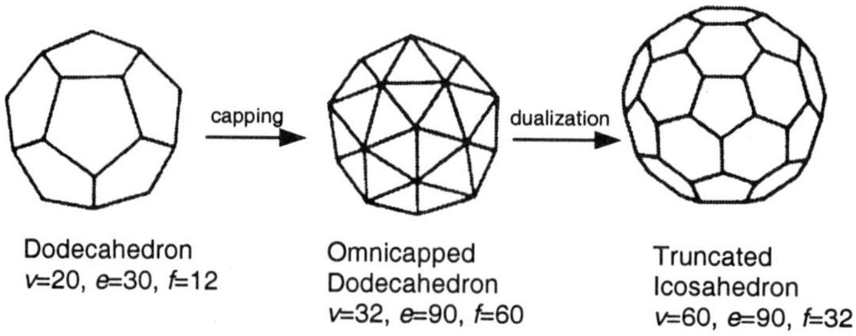

FIGURE 2. Application of the leapfrog transformation to the regular dodecahedron to give the truncated icosahedron of C_{60} through an omnicapped dodecahedron intermediate.

examines further the possible structures of carbon allotropes based on cubic IPMS's with particular reference to the following questions:

(1) What determines whether the P or D surface is used?
(2) Why do the number of hexagons in the unit cells of cubic IPMS's appear to follow a "$24k + 8$ rule," i.e., 56 hexagons (i.e., $k = 2$) in the $D168$ structure and 80 hexagons (i.e., $k = 3$) in the $P192$, $P216$, and $D216$ structures?

2. The Cubic Infinite Periodic Minimal Surfaces

Consider a curved surface, such as one formed by a network of trigonal carbon atoms. At each point on the surface there are two principal curvatures k_1 and k_2. The mean curvature H and Gaussian curvature K are defined as follows:

$$H = \frac{1}{2}(k_1 + k_2) \tag{2.1a}$$

$$K = k_1 k_2 \tag{2.1b}$$

A spherical or ellipsoidal shell has positive Gaussian curvature (i.e., it is "convex"), a hyperbolic sheet has negative Gaussian curvature (i.e., it is "concave"), and a cylinder, cone, or plane has zero Gaussian curvature. Minimal surfaces are surfaces where the mean curvature H at each point is zero so that $k_1 = -k_2$ by equation (2.1a) and $K \leq 0$ by equation (2.1b). They are thus saddle-shaped everywhere except at certain "flat points" which are higher order saddles.

It is not possible to construct an infinite surface with a *constant* negative Gaussian curvature. However, before 1865 Schwarz found that patches of *varying* negative Gaussian curvature, K, and constant zero mean curvature, H, could be smoothly joined to give an infinite surface with zero mean curvature which is periodic in all three directions. Such surfaces are called *infinite periodic minimal surfaces* (IPMS's) [**13**]. These IPMS's can be defined by their unit cells as the minimum repeating unit in a similar manner that unit cells are used to describe crystal structures. The five classical IPMS's known by 1880 have octahedral, tetrahedral, cubic, trigonal prismatic, and triangular unit cells and are conventionally known as the$P$, D, T, CLP, and H surfaces, respectively

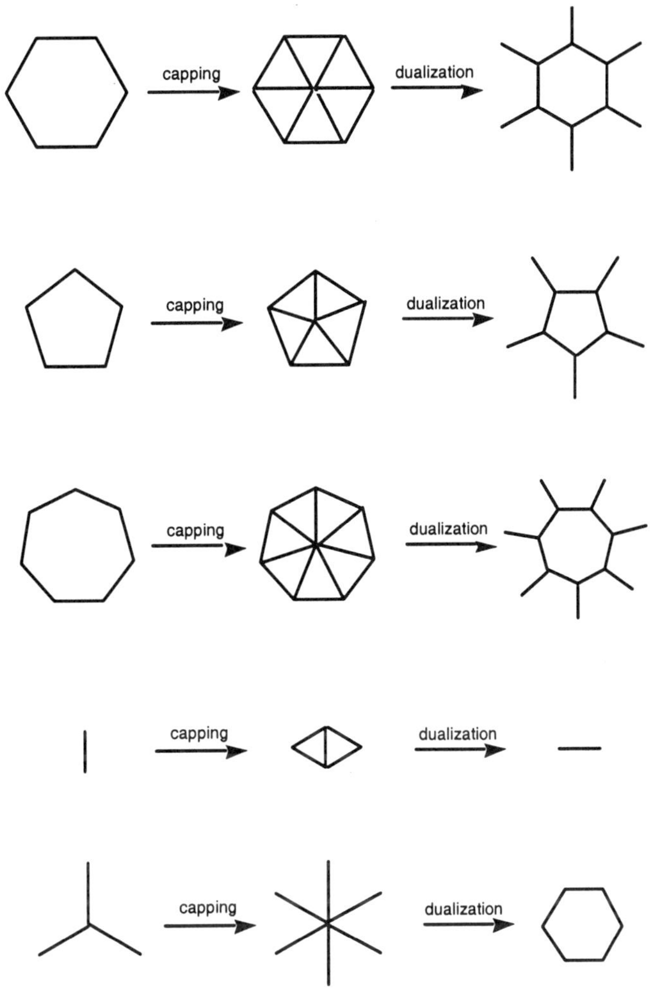

FIGURE 3. A schematic description of the effect of the leapfrog transformation on a hexagonal face, a pentagonal face, a heptagonal face, an edge, and a trivalent vertex. The second column shows elements of the intermediate omnicapped deltahedron and the third column their counterparts in the final leapfrog polyhedron.

These IPMS's cannot be defined by analytic functions in the usual three-dimensional space but instead require elliptic or hyperelliptic integrals. This description is facilitated by the fact that minimal surfaces are the only surfaces other than the sphere whose spherical representation is conformal. This means that every complex analytic function can be used to describe a minimal surface. In order to generate such a description, each point $P(x, y, z)$ of the surface is mapped onto a point $P'(x', y', z')$ on the unit sphere given by

$$P'(x', y', z') = \frac{\left(\frac{dF}{dx}, \frac{dF}{dy}, \frac{dF}{dz}\right)}{\sqrt{\left(\frac{dF}{dx}\right)^2 + \left(\frac{dF}{dy}\right)^2 + \left(\frac{dF}{dz}\right)^2}} \tag{2.2}$$

The same sphere, taken as a special unit sphere called the *Riemann sphere*, can also be used to represent the complex numbers $\omega = \sigma + \tau i$ by points on the (σ, τ) complex plane (Figure 4a). The complex plane can correspond to the equatorial plane of the Riemann sphere in which the north pole is ∞ and the south pole is 0; the $0 - \infty$ axis can be called the *polar axis* [**14**]. In this way the point $P'(x', y', z')$ on the Riemann sphere can be projected to the point $P''(\sigma, \tau)$ on the complex plane (Figure 4). Weierstrass showed in the 19th century that the Cartesian coordinates of a point $P(x, y, z)$ on an IPMS are related to the coordinates $P''(\sigma, \tau)$ on the complex plane by the integrals

$$x = Re \int (1 - \omega^2) R(\omega) d\omega \tag{2.3a}$$

$$y = -Im \int (1 + \omega^2) R(\omega) d\omega \tag{2.3b}$$

$$z = Re \int 2\omega R(\omega) d\omega \tag{2.3c}$$

in which Re and Im denote the real and imaginary parts of the complex integral and $\omega = \sigma + i\tau$ as above.

The Weierstrass function $R(\omega)$ determines the intrinsic parameters of the surface and can be expressed in the general form

$$R(\omega) = \frac{1}{\left(F(\omega)\right)^{1/b}} = \frac{1}{\prod_{i=1}^{k} (\omega - \omega_i)^{1/b}} \tag{2.4}$$

The D and P surfaces, have the same Weierstrass function namely

$$R(\omega) = \frac{1}{\sqrt{1 - 14\omega^4 + \omega^8}} \tag{2.5}$$

The locations on the Riemann sphere of the eight roots of the degree 8 polynomial in the denominator of equation (2.5) correspond to the vertices of a cube inscribed in the Riemann sphere in accord with the cubic symmetry of the unit cell of the D and P surfaces. Polynomials of this type, which vanish at the vertices of a regular polyhedron inscribed in the Riemann sphere, are called *polyhedral polynomials* and have been studied for the regular polyhedra, particularly in connection with the solution of the general quintic equation using elliptic functions [**14**]–[**17**].

The transformation defined by the coordinates

$$x = Re \int e^{i\theta} (1 - \omega^2) R(\omega) d\omega \tag{2.6a}$$

$$y = -Im \int e^{i\theta} (1 + \omega^2) R(\omega) d\omega \tag{2.6b}$$

$$z = Re \int e^{i\theta} 2\omega R(\omega) d\omega \tag{2.6c}$$

is used to relate IPMS's which have the same Weierstrass function $R(\omega)$. This transformation is called a *Bonnet transformation* and the angle θ is called the *association*

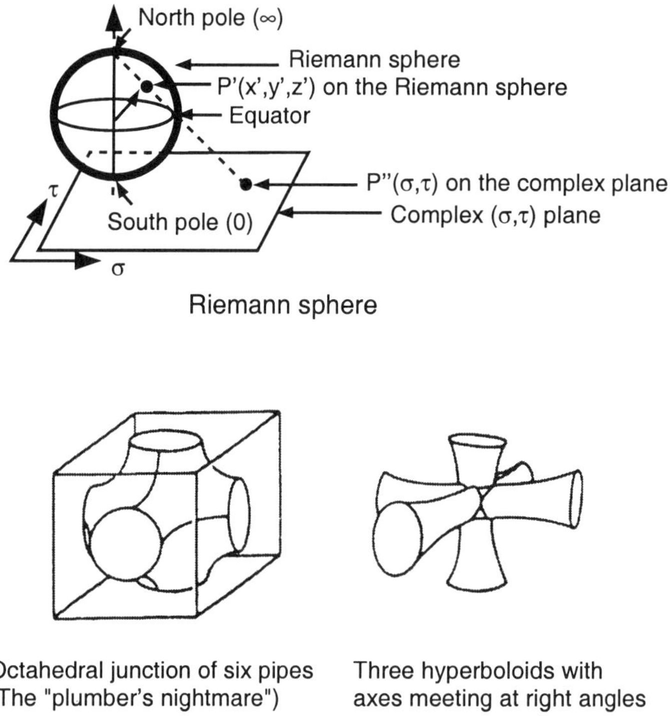

FIGURE 4. (a) The Riemann sphere and the projection of a point $P'(x', y', z')$ on the Riemann sphere to a point $P''(\sigma, \tau)$ in the complex plane; (b) Descriptions of the unit cell of the P surface as a "plumber's nightmare" intersection of six pipes coming through the faces of a cube to meet at an octahedral junction and as three hyperboloids whose axes intersect at right angles.

parameter. Surfaces related by a Bonnet transformation are called *associate surfaces.* All associate surfaces have the same metrics so that lengths are preserved during any Bonnet transformation. Thus a Bonnet transformation only bends the surface without stretching. The P and D surfaces are special cases of associate surfaces since the Bonnet transformations converting either the P to the D surface or vice versa both have $\theta = \pi/2$. Surfaces related by Bonnet transformations having $\theta = \pi/2$ are called *adjoint* surfaces.

The use of IPMS's to construct possible carbon allotropes by decorating them with networks of trigonal carbon atoms requires consideration of their genus. In this connection the P and D surfaces suggested for schwarzite structures are found to have unit cells with genus 3. The unit cell of the P surface is depicted in two different ways in Figure 4b. Thus it can be viewed as an octahedral junction of six pipes or tubes that has been called a "plumber's nightmare" or equivalently as three hyperboloids whose axes meet at right angles. Connecting the open pipes emerging from each of the three pairs of adjacent faces of the plumber's nightmare thereby generating a closed surface leads to a sphere with three handles thereby indicating that a unit cell of this surface has genus 3.

The description of the unit cell of a P surface by the plumber's nightmare (Figure 4b) suggests a natural division of the unit cell into eight equivalent octants, each corresponding to a face of the underlying octahedron so that the numbers of vertices, edges, and faces in a unit cell of a P or D surface are eight times those of an individual octant. In the case of the P surface all octants must be equivalent and contain the symmetry elements of the O_h point group describing the symmetry of the underlying octahedron or its dual cube. The local symmetry of the octant of the P surface is that of an equilateral triangle, namely D_{3h}. Thus in order to be compatible with the symmetry of the P surface, the structure within an individual octant must contain a C_3 axis passing through the center of the octant and also must contain a reflection plane (σ_v in O_h) through each altitude of the triangle representing the octant corresponding to D_{3h} local octant symmetry. This symmetry clearly corresponds to that of a simple cubic lattice. The requirement of reflection planes in an octant of the P surface means that only achiral octants are suitable for P surfaces.

The D surface is adjoint to the P surface as noted above and thus must have the same Weierstrass function with singular points at the vertices of a cube. However, the local symmetry of a D surface is only tetrahedral (T_d) which means that the eight vertices are partitioned into two equivalent sets of four vertices each corresponding to the vertices of the two subtetrahedra making up the cube. In other words, the Bonnet transformation of a P surface into a D surface introduces a distortion such that the image of a reflection plane cutting the P surface (i.e., a mirror line on the surface) is no longer a reflection plane cutting the D surface. Three-fold symmetry is still required in the individual octants of the D surface but the reflection plane is no longer necessary. The D surface can thus accommodate chiral octants with the two sets of four vertices each corresponding to enantiomeric pairs of octants.

The relationship between the unit cells of P and D surfaces is depicted in Figure 5 as well as the face-centered cubic symmetry of a $2 \times 2 \times 2$ block of eight D surface unit cells. Thus the descent in unit cell symmetry from O_h for the P surface to T_d for the D surface can be depicted either by elongation of the "plumber's nightmare" octahedral junction (Figure 4b) in the P surface to two tetrahedral junctions in the D surface [**18**] or by the partitions of eight equivalent octants in the P surface unit cell (represented schematically as A in Figure 5) into two complementary sets of tetrahedrally disposed octants in the D surface unit cell (represented schematically as A and B in Figure 5).

3. Decorating Cubic Infinite Periodic Minimal Surfaces with Trigonal Carbon Atoms

3.1. Generalization of Euler's Theorem to Surfaces of Non-zero Genus. Euler's theorem can be generalized to polyhedra (or other polygonal networks) embedded in a surface of genus g by using

$$v - e + f = 2(1 - g) \tag{3.1}$$

Note that if $g = 0$ (i.e., for polyhedra homeomorphic to a sphere) equation (3.1) reduces to the familiar version of Euler's theorem. In the case of the cubic IPMS's with genus 3 equation (3.1) becomes

$$v - e + f = -4 \tag{3.2}$$

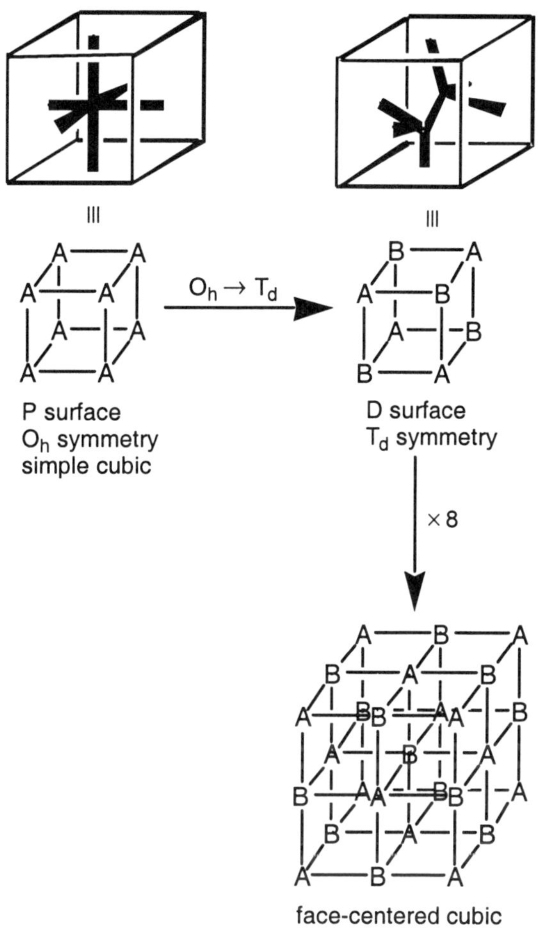

FIGURE 5. Relationships between the P and D surfaces and cubic structures. The reduction of the O_h symmetry of the unit cell of the P surface to the T_d symmetry of the D surface is depicted schematically either by elongation of the "plumber's nightmare" octahedral junction in the P surface to two tetrahedral junctions in the D surface [**18**] or by the partitions of eight equivalent octants in the P surface unit cell (represented schematically as A) into two complementary sets of tetrahedrally disposed octants in the D surface unit cell (represented schematically as A and B).

Now consider a cubic IPMS decorated with only trigonal carbon atoms so that

$$2e = 3v \tag{3.3}$$

since each edge connects exactly two vertices and each vertex is an endpoint of exactly three edges. In addition each edge is shared by exactly two faces leading to the relationship

$$\sum_n n f_n = 2e \tag{3.4}$$

Substitution of equations (3.3) and (3.4) into equation (3.2) gives

$$\sum (6-k) f_k = -24 \tag{3.5}$$

Now apply equation (3.5) to trigonal carbon networks on genus 3 surfaces containing only carbon hexagons and a single type of larger carbon polygon. The following solutions are of particular interest:

(a) Hexagons and heptagons $\Rightarrow f_{\neq 7} = 0 \Rightarrow f_7 = 24$ (i.e., 3 heptagons per octant)
(b) Hexagons and octagons $\Rightarrow f_{\neq 8} = 0 \Rightarrow f_8 = 12$ (i.e., 1 1/2 octagons per octant)
(c) Hexagons and nonagons $\Rightarrow f_{\neq 9} = 0 \Rightarrow f_9 = 8$ (i.e, 1 nonagon per octant).

These three cases will be discussed later in this paper with reference to cubic IPMS's.

3.2. Conditions Imposed by Symmetry on Trigonal Carbon Networks on Cubic Infinite Periodic Minimal Surfaces: the "$24k+8$ Hexagon Theorem". The unit cell of a cubic IPMS, i.e., either a P or D surface, can be decomposed into eight equivalent octants, each of which must have a three-fold symmetry axis. Now consider such a surface decorated with carbon hexagons and one type of carbon non-hexagon with h vertices or edges. The requirement of a three-fold symmetry axis in the center of each octant and the INHR means that each octant must contain a hexagon surrounding its center except in the following cases: (1) h is divisible by three; (2) three hexagons share a vertex located in the center of the octant at a C_3 axis; (3) three hexagons related by a C_3 axis are shared equally between two adjacent octants so that their midpoints are located at C_2 axes. The possibility that three of the non-hexagons share a vertex at the center of each octant is excluded by the INHR and thus will not be considered further. Any non-central hexagons must come in triplets because of the required three-fold symmetry. Thus in an octant of a cubic IPMS decorated with hexagons and one type of non-hexagon with h vertices or edges where h is not divisible by three, the INHR is satisfied, and cases (2) and (3) above are excluded, there must be exactly $3k+1$ hexagons in each octant where k is an integer. Since a unit cell of a cubic IPMS consists of exactly eight equivalent octants the following theorem is proved:

THE $24k+8$ HEXAGON THEOREM: *Consider a cubic IPMS decorated with trigonal vertices forming hexagons and one type of non-hexagon containing h vertices satisfying the following conditions: (1) h is not divisible by 3; (2) the vertex located in the center of each octant through which the C_3 axis passes is not shared by three hexagons; (3) there are no C_3 symmetry-related triplets of hexagons shared equally between two adjacent octants so that C_2 axes pass through their midpoints. Then a structure satisfying the INHR must contain $24k+8$ hexagons in the unit cell.*

Thus for a network of trigonal carbon atoms embedded on a cubic IPMS containing carbon hexagons and one type of carbon non-hexagon, the number of non-hexagons per unit cell is determined by the generalized Euler's theorem discussed above (i.e., equation 3.5). The number of hexagons per unit cell is independent of Euler's theorem since f_6 does not appear in equation (3.5). However, the number

TABLE 1. Comparison of several leapfrog transformations

Polygon combination	Pentagons + Hexagons		Heptagons + Hexagons		Octagons + Hexagons	
leapfrog process	C_{20} $\xrightarrow{\text{leapfrog}}$	C_{60}	$D56$ $\xrightarrow{\text{leapfrog}}$	$D168$	$D32$ $\xrightarrow{\text{leapfrog}}$	$D96$
original figure	I_h dodecahedron		Klein surface		Dyck surface	
vertices	20	60	56	168	32	96
edges	30	90	84	252	48	144
faces	12	32	24	80	12	44
$f_{\neq 6}$	12	12	24	24	12	12
f_6	0	20	0	56	0	32
$f_{\neq 6} + f_{\neq 6}$ edges	30	0	84	0	48	0
$f_{\neq 6} + f_6$ edges	0	60	0	168	0	96
$f_6 + f_6$ edges	0	30	0	84	0	48

of hexagons must satisfy the $24k + 8$ rule indicated by the above theorem. For this reason the promising candidates for negative curvature carbon allotropes based on cubic IPMS's have 56 ($k = 2$) or 80 ($k = 3$) hexagons in their unit cells.

3.3. Networks Containing Only Hexagons and Heptagons. The generalized Euler's theorem indicates that a genus 3 structure of only hexagons and heptagons of trigonal carbon atoms has exactly 24 heptagons in the unit cell corresponding to 3 heptagons per octant. The simplest version of such a structure satisfying the INHR is generated by a leapfrog transformation on a genus 3 network of 24 heptagons with 56 vertices in the unit cell described by Klein in 1879 [**11, 12**] and now conveniently called the $D56$ structure. Table 1 compares this leapfrog transformation with the more familiar leapfrog transformation of a regular dodecahedron to a truncated icosahedron (Figure 2). Figure 6 shows an octant of the original Klein surface and an octant of its leapfrog, which has 168 vertices in the unit cell and which was proposed by Vanderbilt and Tersoff [**8**] as a possible allotrope of carbon designated as $D168$. Note that the octants of both the Klein surface and its leapfrog are chiral (Figure 6) and thus do not satisfy the symmetry requirements for a P surface but require the D surface of lower symmetry. The number of hexagons in the unit cell of $D168$ is 56, which satisfies the $24k + 8$ theorem for $k = 2$. Because of the generation of the $D168$ structure by a leapfrog transformation from the prototypical network of trigonal carbon heptagons with 24 heptagons in the unit cell, the $D168$ structure has the minimum number of hexagons required to generate a structure satisfying the INHR. Thus a structure with 24 heptagons and 32 hexagons ($= 24k + 8$ for $k = 1$) would violate the INHR even though it satisfies the $24k + 8$ theorem.

A different carbon allotrope structure containing only hexagons and heptagons is the $P216$ structure proposed by Lenosky, Gonze, Teter, and Elser [**7**] with 216 vertices, 24 heptagons, and 80 hexagons in the unit cell (Figure 6). An octant of this structure is achiral and thus meets the symmetry requirements for the P surface as well as the D surface. The 80 hexagons in the unit cell of the $P216$ or $D216$ structures satisfy the $24k + 8$ theorem for $k = 3$.

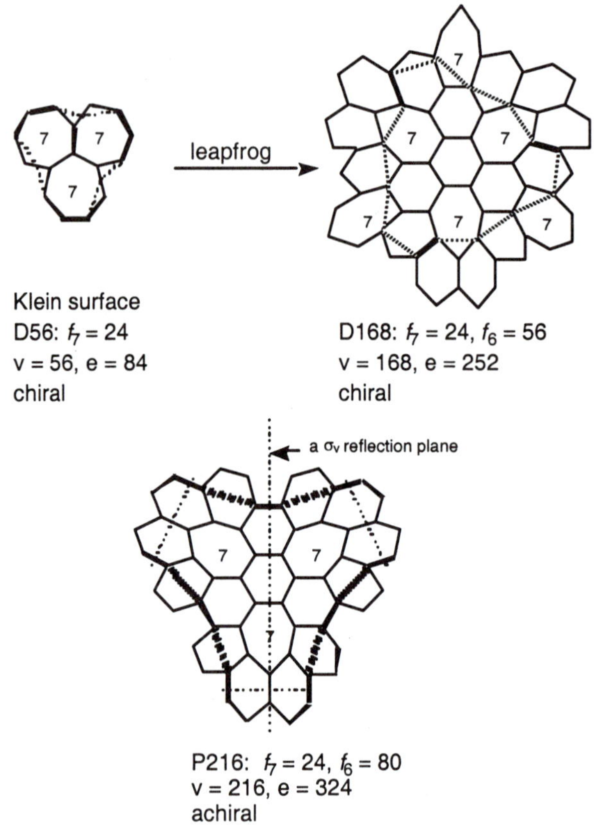

FIGURE 6. Octants of cubic IPMS's decorated with only hexagons and heptagons including the Klein surface ($D56$), its leapfrog ($D168$) and the $P216$ carbon structure proposed by Lenosky, Gonze, Teter and Elser [**7**].

3.4. Networks Containing Only Hexagons and Octagons. The generalized Euler's theorem indicates that a genus 3 structure of only hexagons and octagons of trigonal carbon atoms has exactly 12 octagons in the unit cell or 1 1/2 octagons per octant. The simple configuration of 12 octagons in the unit cell of a genus 3 surface leading to 32 vertices and 48 edges was studied by Dyck [**19**] in 1880 and is conveniently called the $D32$ structure. The leapfrog of this configuration (Table 1) has 96 vertices, 144 edges, and 32 hexagons in its unit cell and is conveniently called the $D96$ structure. Note again that the octants of both the $D32$ and $D96$ structures are chiral (Figure 7a) and thus do not satisfy the symmetry requirements for a P surface but require the D surface of lower symmetry (Figure 5). The number of hexagons is seen to satisfy the $24k + 8$ theorem for $k = 1$.

Mackay and Terrones [**6**] have proposed negative curvature carbon allotropes containing hexagons and octagons of trigonal carbon atoms and satisfying the IPMS. The simplest such allotrope is $P192$ in which the cubic unit cell contains

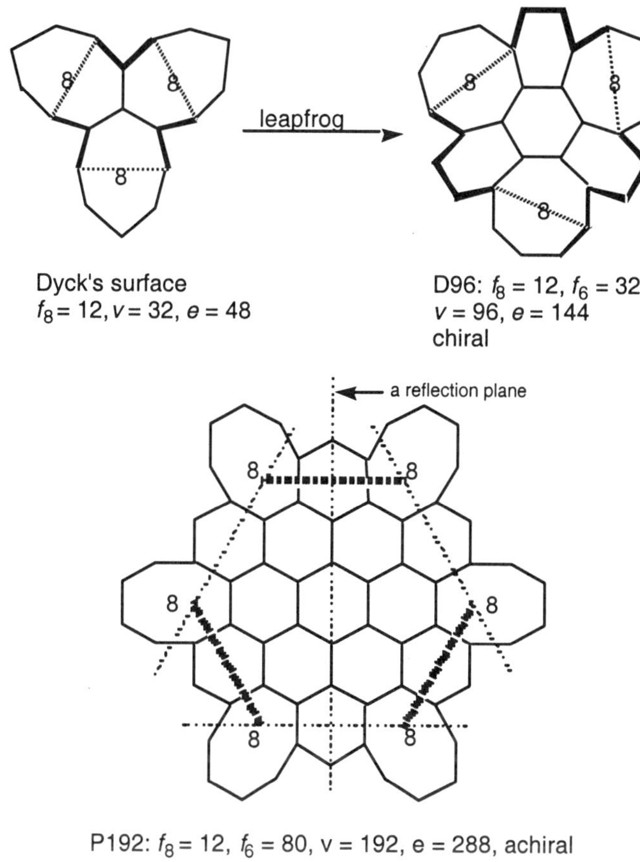

FIGURE 7. Octants of cubic IPMS's decorated with only hexagons and octagons including the Dyck surface, its leapfrog, and the $P192$ carbon structure proposed by Mackay and Terrones [**6**].

the required 12 octagons as well as 80 hexagons. Mackay and Terrones [**6**] also mention another carbon allotrope, conveniently designated as $P432$, with a large unit cell containing 432 carbon atoms in 12 octagons and 200 hexagons. Note that the number of hexagons in both $P192$ and $P432$ satisfy the $24k + 8$ theorem for $k = 3$ and $k = 8$, respectively.

3.5. Networks Containing Only Hexagons and Nonagons. The generalized Euler's theorem indicates that a genus 3 structure of only hexagons and nonagons of trigonal carbon atoms has exactly 8 nonagons in the unit cell. A structure with only the 8 nonagons and no hexagons based on a cubic IPMS would have one necessarily non-planar nonagon in each octant leading to 24 vertices and 36 edges per unit cell. A leapfrog transformation on this structure would then lead to a INHR structure containing the 8 nonagons as well as 72 vertices, 108 edges, and 24 hexagons in the unit cell. Since the number of vertices of a nonagon is divisible by three, the required three-fold symmetry for a P or D surface can be preserved with a nonagon in the center of each octant so that the number of hexagons in the unit cell of a cubic IPMS decorated with only hexagons and nonagons does not satisfy the $24k + 8$ theorem but satisfies a "$24k$ rule."

References

[1] E. Bruton, *Diamonds*, NAG Press, London, 1970.

[2] C.L. Mantell, *Carbon and Graphite Handbook*, Interscience, New York, 1968.

[3] W.N. Reynolds, *Physical Properties of Graphite*, Elsevier, Amsterdam, 1968.

[4] H.W. Kroto, A.W. Allaf, and S.P. Balm, C_{60}: Buckminsterfullerene, *Chem. Rev.*, **91** (1991) 1212.

[5] H.A. Schwarz, *Gesammelte, Mathematische Abhandlungen*, Springer, Berlin, 1890.

[6] A.L. Mackay, and H. Terrones, Diamond from Graphite, *Nature*, **352** (1991) 762.

[7] T. Lenosky, X. Gonze, M. Teter, and V. Elser, Energetics of Negatively Curved Graphitic Carbon, *Nature*, **355** (1992) 333.

[8] D. Vanderbilt, and J. Tersoff, Negative Curvature Fullerene Analog of C_{60}, *Phys. Rev. Lett.*, **68** (1992) 511.

[9] M. O'Keeffe, G.B. Adams, and O.F. Sankey, Predicted New Low Energy Forms of Carbon, *Phys. Rev. Lett.*, **68** (1992) 2325.

[10] R.B. King, Chemical Applications of Topology and Group Theory. 29. Low Density Polymeric Carbon Allotropes Based on Negative Curvature Structures, *J. Phys. Chem.*, **100** (1996) 15096.

[11] F. Klein, Über die Transformation siebenter Ordnung der elliptischen Funktionen, *Math. Ann.*, **14** (1879) 428.

[12] F. Klein, *Gesammelte Mathematischen Abhandlungen*, Springer Verlag, Berlin, 1923, Vol. 3, pp. 90–136.

[13] S. Andersson, S.T. Hyde, K. Larsson, and S. Lidin, Minimal Surfaces and Structures: From Inorganic and Metal Crystals to Cell Membranes and Biopolymers, *Chem. Rev.*, **88** (1988) 221.

[14] F. Klein, *Vorlesungen über das Ikosaeder*, Teubner, Leipzig, 1884, Part I, Chapter II.

[15] L.E. Dickson, *Modern Algebraic Theories*, Sanborn, Chicago, 1930, Chapter XIII.

[16] R.B. King, and E.R. Canfield, Icosahedral Symmetry and the Quintic Equation, *Comp. and Math. with Appl.*, **24** (1992) 13.

[17] R.B. King, *Beyond the Quartic Equation*, Birkhäuser, Boston, 1996.

[18] J.F. Sadoc, and J. Charvolin, Infinite Periodic Minimal Surfaces and their Crystallography in the Hyperbolic Plane, *Acta Cryst.*, **A45** (1989) 10.

[19] W. Dyck, Notiz über eine reguläre Riemann'sche Fläche vom Geschlechte Drei und die zugehörige "Normalcurve" vierter Ordnung, *Math. Ann.*, **17** (1880) 510.

Department of Chemistry, University of Georgia, Athens, Georgia 30602
E-mail address: rbking@sunchem.chem.uga.edu

DIMACS Series in Discrete Mathematics
and Theoretical Computer Science
Volume **51**, 2000

Applications of Isometric Embeddings to Chemical Graphs

Sandi Klavžar

ABSTRACT. Applications of isometric embeddings of benzenoid graphs are surveyed. Their embeddings into hypercubes provide methods for computing the Wiener index and the Szeged index, while embeddings into the Cartesian product of trees lead to fast algorithms. A new method for computing the hyper-Wiener index of partial cubes in general, and of benzenoid graphs and trees in particular, is also presented.

1. Introduction

Graphs that can be isometrically embedded into a hypercube are called partial cubes. The structure of these graphs is well understood via characterizations of Djoković [**8**], Winkler [**31**] and Chepoi [**3**]. In addition, there are two algorithms of complexity $O(n^2 \log n)$ for recognizing these graphs, where n denotes the number of vertices of the graph considered. The algorithms are due to Aurenhammer, Hagauer [**2**] and Imrich, Klavžar [**18**], see also [**19**] for a survey on related algorithmic results.

Partial cubes have found several applications, in particular in the chemical graph theory. It was observed in [**22**] that benzenoid graphs are partial cubes. In this paper we wish to present the developments arising from this point of departure. We proceed as follows. In the rest of this section we recall the concepts needed later. In Sections 2 and 3 we survey results which were obtained using isometric embeddings of benzenoid graphs. Section 2 contains (theoretical) methods for computing the Wiener and the Szeged index of benzenoid graphs which in particular enable us to obtain expressions for the two indices of several families of benzenoid graphs. Section 3 contains linear algorithms for computing the two indices on benzenoid graphs. In Section 4 we propose a new method for computing the hyper-Wiener index of benzenoid graphs. At the end of the paper we also briefly mention the so-called scale embeddings, a generalization of isometric embeddings.

For a graph $G = (V(G), E(G))$ let $d_G(u,v)$ be the usual shortest path distance between vertices u and v of G. Given two connected graphs H and G, we say that H admits an *isometric embedding* into G if there exists a mapping

$$\iota : V(H) \to V(G)$$

1991 *Mathematics Subject Classification.* Primary 92E10, 05C12; Secondary 05C78, 68R10.

Supported by the Ministry of Science and Technology of Slovenia under the grant J1-0498-0101.

such that for all vertices $u, v \in V(H)$ we have

$$d_G(\iota(u), \iota(v)) = d_H(u, v)\,.$$

Clearly, ι is injective and maps edges to edges, thus H can be considered as an induced subgraph of G. We can also say that H is an *isometric subgraph* of G. An example of an isometric embedding is given in Fig. 1. We refer to [**9, 10, 30**] for more information on isometric embeddings of graphs.

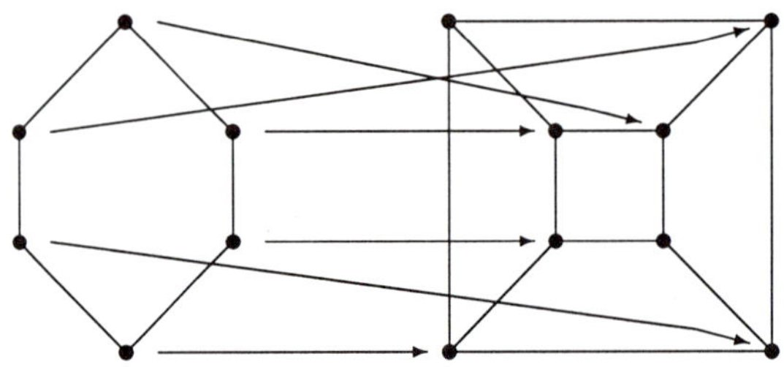

FIGURE 1. C_6 is an isometric subgraph of the 3-cube Q_3

The *Cartesian product* $G \Box H$ of graphs G and H is the graph with vertex set $V(G) \times V(H)$ and $(a, x)(b, y) \in E(G \Box H)$ whenever $ab \in E(G)$ and $x = y$, or $a = b$ and $xy \in E(H)$. The Cartesian product is commutative and associative. Hence we may write $G = G_1 \Box G_2 \Box \cdots \Box G_k$ for the Cartesian product of graphs $G_1, G_2, \ldots, G_k$. In this case, the vertex set of G is $V(G_1) \times V(G_2) \times \cdots \times V(G_k)$ and two vertices $(u_1, u_2, \ldots, u_k)$ and $(v_1, v_2, \ldots, v_k)$ are adjacent if they differ in exactly one position, say in ith, for which $u_i v_i$ is an edge of G_i. An example of the Cartesian product is given in Fig. 2 where we can see the Cartesian product of the path on four vertices P_4 with two copies of K_2.

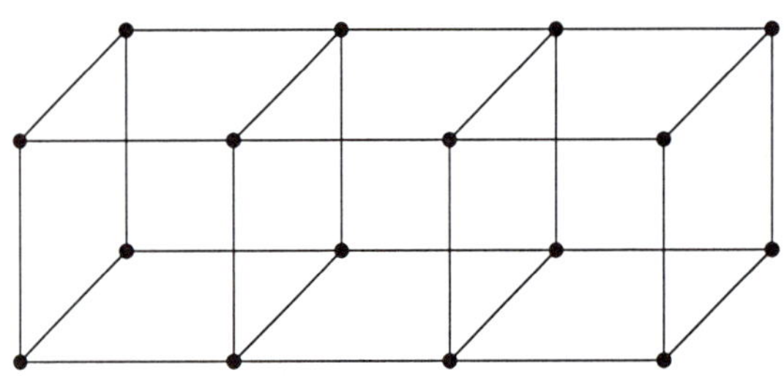

FIGURE 2. The Cartesian product $P_4 \Box K_2 \Box K_2$

The *n-cube* Q_n is the Cartesian product of n copies of the complete graph on two vertices K_2. In other words, if we set $V(K_2) = \{0, 1\}$, then the vertex set of Q_n consists of all words of length n over $\{0, 1\}$ and two such words are adjacent if they differ in exactly one position. A graph is called a *hypercube* if it is isomorphic to some n-cube.

The 3-cube, i.e. the graph $Q_3 = K_2 \Box K_2 \Box K_2$ can be seen on Fig. 1. Consider now the graph from Fig. 2. Since $K_2 \Box K_2 = Q_2$, the product graph can also be interpreted as $P_4 \Box (K_2 \Box K_2) = P_4 \Box Q_2$. Observe also that if we restrict P_4 to any of its three edges we find a copy of Q_3 in the product $P_4 \Box Q_2$.

A graph G is a *partial cube*, if there is an isometric embedding $\iota : V(G) \to V(Q_n)$ for some n. In other words, partial cubes are isometric subgraphs of hypercubes. We have seen already on Fig. 1 that C_6 is a partial cube. For another example see Fig. 3.

Another way of describing partial cubes is the following. Let $\Sigma = \{0, 1\}$ and let w_1 and w_2 be words of equal length over Σ. Then the *Hamming distance* between w_1 and w_2, $H(w_1, w_2)$, is the number of positions in w_1 and w_2 in which the two words differ. Then a graph G is a partial cube if and only if each vertex $v \in V(G)$ can be labeled by a word of fixed length, $\ell(v)$, such that for all $u, v \in V(G)$ we have

$$H(\ell(u), \ell(v)) = d_G(u, v) .$$

The partial cube from Fig. 3 is equipped with such a labeling.

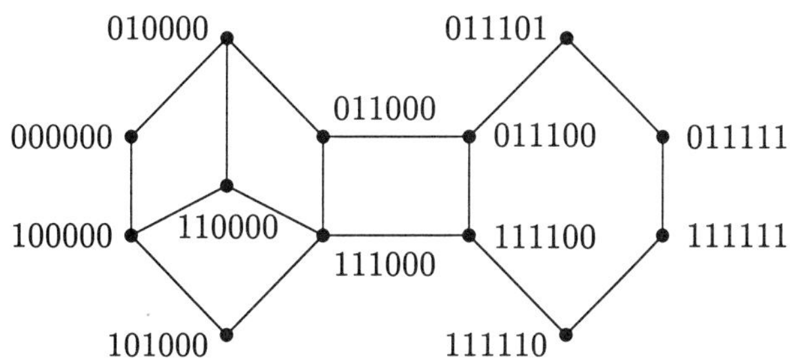

FIGURE 3. A partial cube with a corresponding labeling

Benzenoid graphs are graphs constructed as follows. Let H be the hexagonal (graphite) lattice and let Z be a circuit on it. Then a benzenoid system is formed by the vertices and edges of H, lying on Z and in the interior of Z. Note that the edge set of a benzenoid system can be partitioned into three classes of parallel edges. For more information on benzenoid graphs we refer to the book of Gutman and Cyvin [**11**].

2. Wiener and Szeged Index of Benzenoid Graphs

The *Wiener index* $W(G)$ of a (connected) graph G is defined as

$$W(G) = \frac{1}{2} \sum_{u,v \in V(G)} d_G(u, v) .$$

The Wiener index, introduced in 1947 in [**29**], is one of the most important topological indices of chemical graphs. As many papers have already been written on this topic, we only refer to two recent surveys [**17, 26**] for more information on it.

For partial cubes we have the following result from [**22**]:

THEOREM 2.1. *Let G be a partial cube on n vertices embedded into the q-cube. For $i = 1, 2, \ldots , q$ let n_i be the number of vertices of G whose ith component of the embedding is equal to 1. Then*

$$W(G) = \sum_{i=1}^{q} n_i(n - n_i) .$$

In order to make Theorem 2.1 specific to benzenoid graphs, we need to recall the following concept. Let G be a benzenoid graph. A straight line segment $\mathbf{C}$ in the plane with end points P_1 and P_2 is called a *cut segment* if $\mathbf{C}$ is orthogonal to one of the three edge directions, each P_1 and P_2 is the center of an edge and the graph obtained from G by deleting all edges intersected by $\mathbf{C}$ has exactly two connected components. An *elementary cut* is the set of all edges intersected by a cut segment. Let $\mathcal{C}(G)$ be the set of all elementary cuts of G. For $C \in \mathcal{C}(G)$ let $n_1(C)$ and $n_2(C)$ be the number of vertices in the connected components of $G \setminus C$. As shown in [**14**], Theorem 2.1 refines as follows.

THEOREM 2.2. *For a benzenoid graph G we have*

$$W(G) = \sum_{C \in \mathcal{C}(G)} n_1(C) n_2(C)\,.$$

Theorem 2.2 provides a simple method for computing the Wiener index of benzenoid graphs by hand. For a simple example consider the molecular graph of naphthalene together with its 5 elementary cuts as shown in Fig. 4. Then the contribution of each of the cuts C_1, C_2, C_3 and C_4 is $3 \cdot 7$ and the contribution of C_5 is $5 \cdot 5$. Thus, the Wiener index of naphthalene is $4 \cdot (3 \cdot 7) + 5 \cdot 5 = 109$.

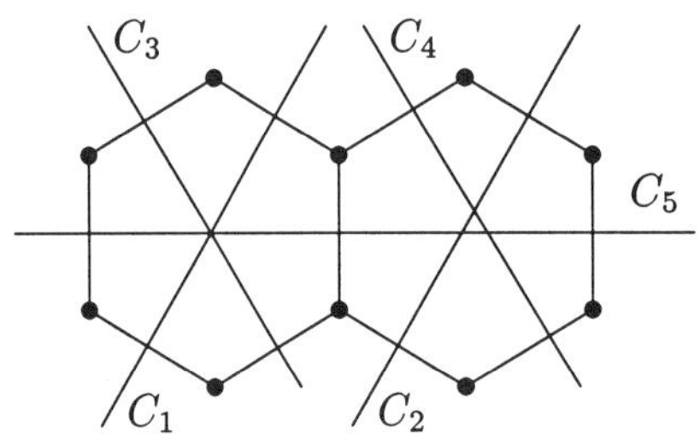

FIGURE 4. A benzenoid graph G with its elementary cuts

Theorem 2.2 can also be used to obtain expressions for the Wiener index of many classes of benzenoid graphs. For instance, it was shown in [**14**] that for the so-called coronene/circumcoronene series H_r we have:

COROLLARY 2.3. *For any $r \geq 1$ we have*

$$W(H_r) = \frac{1}{5}(164r^5 - 30r^3 + r)\,. \tag{2.1}$$

The finding of a general expression for $W(H_r)$ was, for a long time, considered as a special challenge in the theory of the Wiener index, but using Theorem 2.2 it becomes a routine matter. Corollary 2.3 was independently obtained by Shiu and Lam [**28**] with a much more complicated method. (In fact, formula (2.1) is the result of [**28**].)

Several formulas of type (2.1) are given in [**23**]. For instance, if $T(n,k)$ denotes the trapezium benzenoid graph with n columns and k rows of hexagons, then we have:

$$W(T(n,k)) = \frac{4n^3(k^2+2k+1)}{3} - \frac{2n^2(k+1)(2k^2-8k-3)}{3} + \frac{2n(k^4-4k^3+6k^2+9k+1)}{3} - \frac{k(8k^4+35k^2-45k-28)}{30}\,.$$

In particular, for $k = 1$ the above expression reduces to

$$W(P(n,1)) = \frac{1}{3}(16n^3 + 36n^2 + 26n + 3)\,,$$

which is the well-known formula for the Wiener index of polyacenes.

The Szeged index is a recently introduced topological index [**20**] and is defined as follows. For an edge $e = uv$ of a graph G let

$$W_{uv} = \{w \mid w \in V(G), d_G(w,u) < d_G(w,v)\}$$

and

$$W_{vu} = \{w \mid w \in V(G), d_G(w,v) < d_G(w,u)\}\,.$$

The *Szeged index* $Sz(G)$ of a graph G is defined as

$$Sz(G) = \sum_{uv \in E(G)} |W_{uv}|\,|W_{vu}|\,.$$

In [**13**] the following result similar to Theorem 2.2 was established:

THEOREM 2.4. *For a benzenoid graph G we have*

$$Sz(G) = \sum_{C \in \mathcal{C}(G)} |C|\, n_1(C) n_2(C)\,.$$

As in the case of the Wiener index, Theorem 2.4 can be used to obtain formulas for the Szeged index of classes of benzenoid graphs. For instance:

COROLLARY 2.5. *For any $k \geq 1$ we have*

$$Sz(H_k) = \frac{3}{2}k^2(36k^4 - k^2 + 1)\,.$$

3. Fast Algorithms for W and Sz of Benzenoid Graphs

Let G be an arbitrary connected graph on n vertices and m edges. Then Mohar and Pisanski [**25**] showed that $W(G)$ can be computed in $O(mn)$ time. For the Szeged index the same result is due to Žerovnik [**32**]. In [**25**] it is also shown how to compute $W(G)$ in $O(n)$ time provided that G is a tree. In this section we present an $O(n)$ algorithm from [**6**] for computing $W(G)$ and $Sz(G)$ of a benzenoid graph G. Even faster algorithm of complexity $O(s)$ for the Wiener index of benzenoid graphs is proposed in [**7**], where s is the length of the outer cycle of a benzenoid graph considered. However, the last algorithm is of more theoretical than practical importance.

Let G be a benzenoid graph and let E_1, E_2, and E_3 denote the edges of G of a given direction. For $i = 1, 2, 3$ let G_i be the graph which is obtained from G be deleting all the edges of E_i. Define a graph T_i whose vertices are the connected components of G_i and two such components P' and P'' are adjacent in T_i if and only if there are vertices $u \in P'$ and $v \in P''$ which are end–vertices of an edge from E_i. Note that every T_i is a tree (the existence of a cycle in T_i would imply that G contains a non–hexagonal interior face). Then define the canonical embedding

$$\alpha : V(G) \to V(T_1 \Box T_2 \Box T_3)$$

as

$$\alpha(v) = (P, Q, R)\,,$$

where P, Q, and R are the connected components of the graphs G_1, G_2, and G_3, respectively, sharing the vertex v. Chepoi [**4**] proved the following result:

THEOREM 3.1. *The canonical embedding map α provides an isometric embedding of a benzenoid graph G with n vertices into the graph $H = T_1 \Box T_2 \Box T_3$. The trees T_1, T_2, T_3 and the corresponding labels of the vertices of G can be computed in $O(n)$ time.*

In order to make use of the isometric map α we introduce the Wiener index of weighted graphs. A *weighted graph* (G, w) is a graph G together with a function $w: V(G) \to \mathbb{N}^+$. The Wiener index $W(G, w)$ of a weighted graph (G, w) is defined as

$$W(G,w) = \frac{1}{2} \sum_{u,v \in V(G)} w(u)\, w(v)\, d_G(u,v).$$

Note that if all the weights are 1 then $W(G,w) = W(G)$.

The following result of [**6**] is the key to a fast algorithm for $W(G)$. We also include the proof since it reflects most of the tools needed.

THEOREM 3.2. *Let G be a benzenoid graph and let α be the canonical embedding of G into $T_1 \Box T_2 \Box T_3$. Then*

$$W(G) = W(T_1, w_1) + W(T_2, w_2) + w(T_3, w_3).$$

PROOF. Let $V = V(G)$ and set $H = T_1 \Box T_2 \Box T_3$. For $u \in V$ let $\alpha(u) = (u_1, u_2, u_3)$. Then we have:

$$\begin{aligned}
W(G) &= \frac{1}{2} \sum_{u \in V} \sum_{v \in V} d_G(u,v) \\
&= \frac{1}{2} \sum_{u \in V} \sum_{v \in V} d_H(\alpha(u), \alpha(v)) && (3.1) \\
&= \frac{1}{2} \sum_{u \in V} \sum_{v \in V} \sum_{i=1}^{3} d_{T_i}(u_i, v_i) && (3.2) \\
&= \sum_{i=1}^{3} \left(\frac{1}{2} \sum_{u \in V} \sum_{v \in V} d_{T_i}(u_i, v_i) \right) \\
&= \sum_{i=1}^{3} \left(\frac{1}{2} \sum_{u_i \in T_i} \sum_{v_i \in T_i} w_i(u) w_i(v) d_{T_i}(u_i, v_i) \right) && (3.3) \\
&= \sum_{i=1}^{3} W(T_i, w_i). && (3.4)
\end{aligned}$$

In the above computation (3.1) holds because α is isometric. For (3.2) recall the following important and well-known property of the Cartesian product of graphs. If $u = (u_1, \ldots, u_k)$ and $v = (v_1, \ldots, v_k)$ are vertices of $G = G_1 \Box \cdots \Box G_k$ then $d_G(u,v) = \sum_{i=1}^{k} d_{G_i}(u_i, v_i)$. The equality (3.3) holds by the definition of the weighted trees (T_i, w_i). Finally, (3.4) is just the definition of the weighted Wiener index. □

By Theorem 3.2, a linear algorithm for computing $W(G)$ will be provided by a linear algorithm for computing the Wiener index of a weighted tree. This task was done implicitly by Mohar and Pisanski [**25**] and explicitly in [**6**], thus we can state:

THEOREM 3.3. *The Wiener index of a benzenoid graph on n vertices can be computed in $O(n)$ time.*

Moreover, along the same lines we obtain:

THEOREM 3.4. *The Szeged index of a benzenoid graph on n vertices can be computed in $O(n)$ time.*

4. Hyper-Wiener Index

The hyper-Wiener index was in the case of trees introduced by Randić [**27**] and extended to all graphs by Klein, Lukovits and Gutman [**24**]. The *Hyper-Wiener index* $WW(G)$ of a graph G is defined as

$$WW(G) = \frac{1}{2} \sum_{u,v \in V(G)} \binom{d_G(u,v)+1}{2}$$

which can be rewritten as

$$WW(G) = \frac{1}{4} \sum_{u,v \in V(G)} d_G(u,v)^2 + \frac{1}{4} \sum_{u,v \in V(G)} d_G(u,v)\,. \tag{4.1}$$

We are going to develop a method for computing the hyper-Wiener index of partial cubes and thus in particular of benzenoid graphs and trees.

Let G be a partial cube on n vertices embedded into the q-cube and let ℓ be a corresponding Hamming labeling. For any $1 \le i, j \le q$ set

$$n_{ij}^{11} = \sum_{u \in V(G)} \ell_i(u)\ell_j(u)\,, \qquad n_{ij}^{00} = \sum_{u \in V(G)} (1-\ell_i(u))(1-\ell_j(u))\,,$$

$$n_{ij}^{10} = \sum_{u \in V(G)} \ell_i(u)(1-\ell_j(u))\,, \qquad n_{ij}^{01} = \sum_{u \in V(G)} (1-\ell_i(u))\ell_j(u)\,,$$

For instance, n_{ij}^{00} is the number of vertices of G whose ith and jth component of ℓ is equal to 0.

THEOREM 4.1. *Let G be a partial cube on n vertices embedded into the q-cube. Then*

$$WW(G) = W(G) + \sum_{i=1}^{q} \sum_{j=i+1}^{q} (n_{ij}^{11} n_{ij}^{00} + n_{ij}^{01} n_{ij}^{10})\,. \tag{4.2}$$

PROOF. As the second term of (4.1) is one half of the Wiener index, we need to concentrate on the first term only. Set $V = V(G)$ and for vertices u, v of G let $\delta_i(u,v)$ be 0 if ith components of $\ell(u)$ and $\ell(v)$ agree, and 1 otherwise. We have:

$$
\begin{aligned}
\sum_{u\in V}\sum_{v\in V} d_G(u,v)^2 &= \sum_{u\in V}\sum_{v\in V}(H(\ell(u),\ell(v)))^2 \\
&= \sum_{u\in V}\sum_{v\in V}(\sum_{i=1}^{q}\delta_i(u,v))^2 \\
&= \sum_{u\in V}\sum_{v\in V}\sum_{i=1}^{q}\sum_{j=1}^{q}\delta_i(u,v)\delta_j(u,v) \\
&= \sum_{i=1}^{q}\sum_{j=1}^{q}(\sum_{u\in V}\sum_{v\in V}\delta_i(u,v)\delta_j(u,v)) \\
&= \sum_{i=1}^{q}\sum_{u\in V}\sum_{v\in V}\delta_i(u,v)\delta_i(u,v) + \sum_{i=1}^{q}\sum_{\substack{j=1\\ j\neq i}}^{q}\sum_{u\in V}\sum_{v\in V}\delta_i(u,v)\delta_j(u,v) \\
&= \sum_{i=1}^{q}\sum_{u\in V}\sum_{v\in V}\delta_i(u,v) + \sum_{i=1}^{q}\sum_{\substack{j=1\\ j\neq i}}^{q}\sum_{u\in V}\sum_{v\in V}\delta_i(u,v)\delta_j(u,v) \\
&= 2W(G) + 2\sum_{i=1}^{q}\sum_{\substack{j=1\\ j\neq i}}^{q}(\frac{1}{2}\sum_{u\in V}\sum_{v\in V}\delta_i(u,v)\delta_j(u,v)) \\
&= 2W(G) + 2\sum_{i=1}^{q}\sum_{\substack{j=1\\ j\neq i}}^{q}(n_{ij}^{11}n_{ij}^{00} + n_{ij}^{01}n_{ij}^{10}) \\
&= 2W(G) + 4\sum_{i=1}^{q}\sum_{j=i+1}^{q}(n_{ij}^{11}n_{ij}^{00} + n_{ij}^{01}n_{ij}^{10})
\end{aligned}
$$

Inserting this expression into (4.1) yields the result. □

In the case of benzenoid graphs, Theorem 4.1 provides a particular simple procedure for computing the hyper-Wiener index. We compute the Wiener index and for the second term of (4.2) we proceed as follows. Let C_1 and C_2 be two elementary cuts of a benzenoid graph G. There are two different cases, as shown in Fig. 5. With a, b, c and d we denote the number of vertices in the corresponding parts of G. Then the contribution of the pair C_1, C_2 to (4.2) is $ab + cd$ in the first case and ab in the second one.

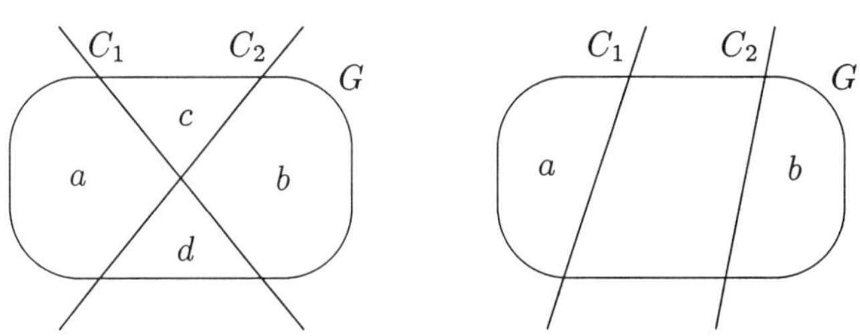

FIGURE 5. Different positions between two elementary cuts

For a concrete example consider the benzenoid graph (naphthalene) from Fig. 4 together with its elementary cuts C_1, C_2, C_3, C_4 and C_5. The contributions of pairs of cuts

$$C_1, C_2;\ C_1, C_3;\ C_1, C_4;\ C_1, C_5;\ C_2, C_3;$$
$$C_2, C_4;\ C_2, C_5;\ C_3, C_4;\ C_3, C_5;\ C_4, C_5$$

are, respectively:

$$3 \cdot 3 + (1 \cdot 1 + 2 \cdot 6) + 3 \cdot 3 + (2 \cdot 4 + 1 \cdot 3) + 3 \cdot 3 +$$
$$(1 \cdot 1 + 2 \cdot 6) + (1 \cdot 3 + 2 \cdot 4) + (1 \cdot 1 + 2 \cdot 6) + 3 \cdot 3 + (1 \cdot 3 + 2 \cdot 4) = 106.$$

Therefore $WW(G) = W(G) + 106 = 109 + 106 = 215$.

As already mentioned, trees are partial cubes as well. For trees, the formula (4.2) reduces to

$$WW(T) = W(T) + \sum_{e \in E(T)} \sum_{\substack{f \in E(T) \\ f > e}} n_1(e, f) \cdot n_2(e, f)$$

where $n_1(e, f)$ and $n_2(e, f)$ are the number of vertices in the two "extremal" connected components in the graph $T \setminus \{e, f\}$, see Fig. 6. More precisely, a pair of cuts in a tree is just a pair of edges, and its contribution to (4.2) is always the second case from Fig. 5. We don't want to go into more details in this particular case, because this was previously observed by Aringhieri, Hansen and Malucelli [**1**]. Moreover, in [**1**] implementation details are also provided to obtain a linear algorithm for computing the hyper-Wiener index of a tree.

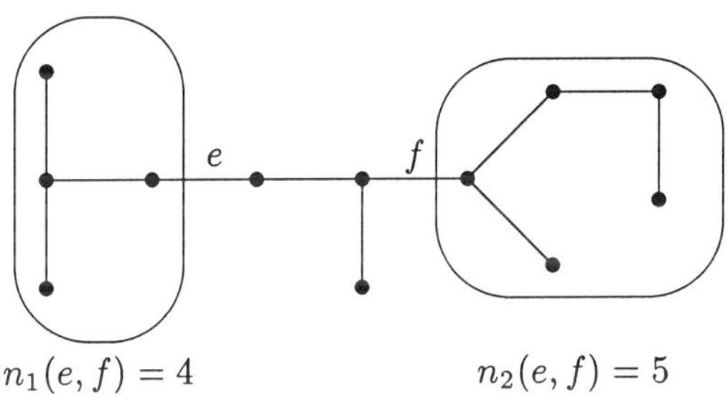

FIGURE 6. A tree and the two extremal components with respect to e, f

5. Concluding Remarks

Isometric embeddings were also applied by Gutman, Khadikar and Khaddar [**12**] in the study of benzenoid graphs containing a linear polyacene fragment as well as in [**15, 16, 21**] for the study of phenylenes.

A generalization of isometric embeddings is provided with the following concept. Let $\lambda \in \mathbb{N}$ and let G and H be two graphs. Then we say that H is *scale λ embeddable* into G if there exists a mapping

$$\iota : V(H) \to V(G)$$

such that for all vertices $u, v \in V(H)$ we have

$$d_G(\iota(u), \iota(v)) = \lambda d_H(u, v)\,.$$

Note that if $\lambda = 1$ then we are back to the isometric embeddings. A graph is an L_1-*graph* if it admits a scale embedding into a hypercube. Chepoi, Deza and Grishukhin [**5**] extended Theorem 2.1 to the class of all L_1-graphs. Their theorem is important for non-bipartite graphs, since in the bipartite case the L_1-graphs coincide with the partial cubes. In the non-bipartite case the class of L_1-graphs is a rich class of graphs containing many chemical graphs and thus a lot of work is waiting to be done.

References

[1] R. Aringhieri, P. Hansen, and F. Malucelli, A Linear Algorithm for the Hyper-Wiener Number of Chemical Trees, manuscript, 1998.

[2] F. Aurenhammer, and J. Hagauer, Recognizing Binary Hamming Graphs in $O(n^2 \log n)$ Time, in: *Proc. 16th Int. Workshop on Graph Theoretical Concepts in Computer Science, Lecture Notes in Comput. Sci.* **484** (Springer, New York, 1991) 90–98. Also: *Math. Systems Theory,* **28** (1995) 387–395.

[3] V. D. Chepoi, *d*-Convexity and Isometric Subgraphs of Hamming Graphs, *Cybernetics,* **1** (1988) 6–9.

[4] V. Chepoi, On Distances in Benzenoid Systems, *J. Chem. Inf. Comput. Sci.*, **36** (1996) 1169–1172.

[5] V. Chepoi, M. Deza, and V. Grishukhin, Clin d'oeil on L_1-embeddable planar graphs, *Discrete Appl. Math.*, **80** (1997) 3–19.

[6] V. Chepoi, and S. Klavžar, The Wiener Index and The Szeged Index of Benzenoid Systems in Linear Time, *J. Chem. Inf. Comput. Sci.*, **37** (1997) 752–755.

[7] V. Chepoi, and S. Klavžar, Distances in Benzenoid Systems: Further Developments, *Discrete Math.*, **192** (1998) 27–39.

[8] D. Djoković, Distance Preserving Subgraphs of Hypercubes, *J. Combin. Theory Ser. B*, **14** (1973) 263–267.

[9] R. L. Graham, Isometric Embeddings of Graphs, in: *Selected Topics in Graph Theory* III, (L.W. Beineke and R.J. Wilson, eds.) (Academic Press, London, 1988), 133–150.

[10] R. L. Graham, and P. M. Winkler, On Isometric Embeddings of Graphs, *Trans. Amer. Math. Soc.*, **288** (1985) 527–536.

[11] I. Gutman, and S. J. Cyvin, *Introduction to the Theory of Benzenoid Hydrocarbons,* (Springer–Verlag, Berlin, 1989).

[12] I. Gutman, P.V. Khadikar, and T. Khaddar, Wiener and Szeged Indices of Benzenoid Hydrocarbons Containing a Linear Polyacene Fragment, *Comm. in Math. and Comp. Chem. (MATCH)*, **35** (1997) 105–116.

[13] I. Gutman, and S. Klavžar, An Algorithm for the Calculation of the Szeged Index of Benzenoid Hydrocarbons, *J. Chem. Inf. Comput. Sci.*, **35** (1995) 1011–1014.

[14] I. Gutman, and S. Klavžar, A Method for Calculating Wiener Numbers of Benzenoid Hydrocarbons, *ACH Models in Chemistry,* **133** (1996) 389–399.

[15] I. Gutman, and S. Klavžar, Relations between Wiener Numbers of Benzenoid Hydrocarbons and Phenylenes, *ACH Models in Chemistry,* **135** (1998) 45–55.

[16] I. Gutman, L. Popović, and L. Pavlović, Elementary Edge-Cuts in the Theory of Benzenoid Hydrocarbons – An Application, *Comm. in Math. and Comp. Chem. (MATCH)*, **36** (1997) 217–229.

[17] I. Gutman, Y. N. Yeh, S. L. Lee, and Y. L. Luo, Some Recent Results in the Theory of the Wiener Number, *Indian J. Chem.* **32A**, (1993) 651–661.

[18] W. Imrich, and S. Klavžar, A Simple $O(mn)$ Algorithm for Recognizing Hamming Graphs, *Bull. Inst. Comb. Appl.*, **9** (1993) 45–56.

[19] W. Imrich, and S. Klavžar, On the Complexity of Recognizing Hamming Graphs and Related Classes of Graphs, *European J. Combin.*, **17** (1996) 209–221.

[20] P.V. Khadikar, N.V. Deshpande, P.P. Kale, A. Dobrynin, and I. Gutman, The Szeged Index and an Analogy with the Wiener Index, *J. Chem. Inf. Comput. Sci.*, **35** (1995) 547–550.

[21] S. Klavžar, and I. Gutman, Wiener Number of Vertex-Weighted Graphs and a Chemical Application, *Discrete Appl. Math.*, **80** (1997) 73–81.

[22] S. Klavžar, I. Gutman, and B. Mohar, Labeling of Benzenoid Systems which Reflects the Vertex-Distance Relations, *J. Chem. Inf. Comput. Sci.*, **35** (1995) 590–593.

[23] S. Klavžar, I. Gutman, and A. Rajapakse, Wiener Numbers of Pericondensed Benzenoid Systems, *Croat. Chem. Acta* **70**, (1997) 979–999.

[24] D.J. Klein, I. Lukovits, and I. Gutman, On the Definition of the Hyper-Wiener Index for Cycle-Containing Structures, *J. Chem. Inf. Comput. Sci.*, **35** (1995) 50–52.

[25] B. Mohar, and T. Pisanski, How to Compute the Wiener Index of a Graph, *J. Math. Chem.*, **2** (1988) 267–277.

[26] S. Nikolić, N. Trinajstić, and Z. Mihalić, The Wiener Index: Developments and Applications, *Croat. Chem. Acta*, **68** (1995) 105–129.

[27] M. Randić, Novel Molecular Descriptor for Structure-Property Studies, *Chem. Phys. Lett.*, **211** (1993) 478–483.

[28] W.C. Shiu, and P.C.B. Lam, The Wiener Number of a Hexagonal Net, *Discrete Appl. Math.*, **73** (1997) 101–111.

[29] H. Wiener, Structural Determination of Paraffin Boiling Points, *J. Amer. Chem. Soc.*, **69** (1947) 17–20.

[30] E. Wilkeit, Isometric Embeddings in Hamming Graphs, *J. Combin. Theory Ser. B*, **50** (1990) 179–197.

[31] P. Winkler, Isometric Embeddings in Products of Complete Graphs, *Discrete Appl. Math.*, **7** (1984) 221–225.

[32] J. Žerovnik, Computing the Szeged Index, *Croat. Chem. Acta*, **69** (1996) 837–843.

Department of Mathematics, PeF, University of Maribor, Koroška cesta 160, 2000 Maribor, Slovenia

E-mail address: `sandi.klavzar@uni-lj.si`

DIMACS Series in Discrete Mathematics
and Theoretical Computer Science
Volume **51**, 2000

Geometrical and Combinatorial Questions About Fullerenes

Joseph Malkevitch

ABSTRACT. From a mathematical point of view the fullerenes can be thought of as those convex 3-dimension polytopes which are 3-valent and have exactly 12 pentagons and some number h, not 1, of hexagons. This note surveys some of the graph theory tools that are useful in studying the fullerenes and surveys a variety of combinatorial and geometrical questions about them that are of mathematical interest.

1. Introduction

By a fullerene I will mean a 3-valent (i.e. cubic) convex 3-polytope with exactly 12 pentagons and h hexagons for some value of h (see Grünbaum [**16**] for basic definitions not explicitly defined here). Intuitively, convex 3-polytopes are 3-dimensional solids with flat (plane) surfaces and no notches or holes. The first systematic work within the mathematics community dealing with the polytopes that today are referred to as fullerenes seems to have been that of Branko Grünbaum and Theodore Motzkin. In their paper Grünbaum and Motzkin [**19**], Grünbaum and Motzkin showed that there were fullerenes for every value of h different from 1. Additional light has been shed on this result by others, including a simpler proof which appears in Malkevitch [**24**]. Much of this early work was aimed at understanding what were the admissible values of the number of hexagons for 3-valent 3-polytopes whose faces, except for the hexagons, all have five or fewer sides.

The purpose of this note is to raise a few questions related to the geometry and combinatorics of fullerenes and to provide a variety of bibliographic references for those interested in the combinatorics and geometry of fullerenes. The questions that will be raised appear to be of some mathematical interest though their importance for chemistry and physics are less clear.

2. Background

Among the many results that prove useful or are of intrinsic interest in studying 3-polytopes in general and fullerenes in particular are:

1. Euler's formula (extended to graphs by Cauchy):

$$V(\text{vertices}) + F(\text{faces}) - E(\text{edges}) = 2$$

1991 *Mathematics Subject Classification.* Primary 52; Secondary 05.
The remarks of two anonymous referees are gratefully acknowledged.

for a connected graph drawn in the plane.

2. Ernst Steinitz's Theorem (as reformulated in modern terminology by Branko Grünbaum) (see Ziegler [**34**]).

   A graph G is 3-polytopal (i.e. isomorphic to the edge-vertex graph of a convex, bounded, 3-dimensional polyhedron) if and only if G is planar and 3-connected.

   (A planar graph is one that is isomorphic to a graph drawn in the plane whose edges intersect only at vertices. A graph drawn in the plane so that its edges meet only at vertices will be called a plane graph. A graph is 3-connected if for any pair of vertices u and v there are at least three paths from u to v whose only vertices in common are u and v.)

   Steinitz's Theorem is complemented by a theorem of Hasslar Whitney (see Ziegler [**34**]) which states that planar 3-connected graphs can be embedded in the plane in essentially only one way and by K. Kuratowski's Theorem that characterizes which graphs are planar in a combinatorial way (see Bollobás [**3**]).

3. K. Wagner/I. Fáry's Theorem (see Ziegler [**34**]).

   If G is a 3-polytopal graph, then G can be drawn in the plane with straight line segments.

4. William Tutte's Theorem (see Ziegler [**34**]).

   If G is a 3-polytopal graph, then G can be drawn in the plane in such a manner that all the faces are convex polygons with the exception of the unbounded face. The circuit which bounds the unbounded face can be chosen to be a convex polygon.

5. Peter Mani's Theorem (see Ziegler [**34**]).

   Let G be a 3-polytopal graph having H as its automorphism group (i.e. its group of combinatorial symmetries). Then G can be realized by a convex 3-polytope P (i.e. P's graph is isomorphic to G) such that the (geometrical) symmetry group of P is isomorphic to H.

(Note: If K is a subgroup of H then there may be no polytope P^* with graph isomorphic to G and whose (geometrical) symmetry group is K.)

In order to formulate some of the problems and questions below it will be useful to introduce some terminology. Suppose G is a plane, 3-valent, 3-connected graph (3-polytopal by Steinitz's Theorem). Let C be a circuit in G. The circuit C can be represented by a binary code as follows: For each vertex v of C the additional edge e_v of the graph G at v is either in the interior or exterior of C. If e_v is in the interior of C, label v with a 1, while if e_v is in the exterior of C, label v with a 0. The resulting (cyclic) binary sequence listed in clockwise order will be referred to as the code of C. We can associate with a code a shell diagram which consists of a circuit associated with the code and short spikes indicating whether an edge at a vertex points into the cycle or out from the cycle. For example, the shell diagram associated with the code 001101010101 is shown in Figure 1.

When the interior of a shell diagram is completed to a 3-valent graph, this will be referred to as a filling of the shell diagram. When the exterior of a shell diagram is completed to a 3-valent graph, this will be referred to as an icing of the shell diagram. If a shell diagram has the property that interchanging the roles of zero and one for its associated code yield the same code or the same code up to a reversal of cyclic order, we say the code is self dual. Note that when a shell diagram

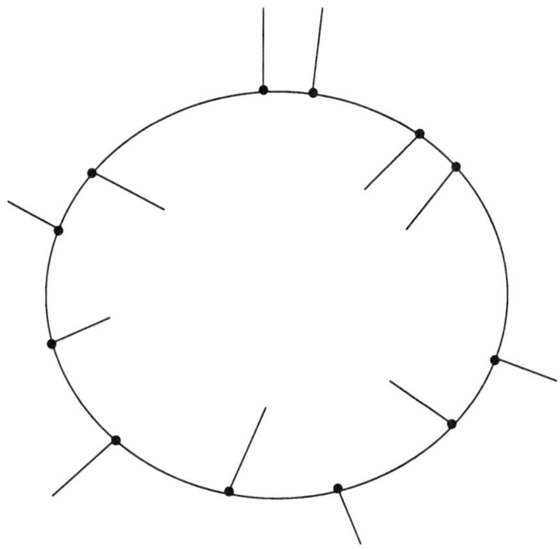

FIGURE 1. A shell diagram.

is self-dual, then a filling of it can be used as an icing and an icing of it can be used as a filling. We say that a shell diagram is replicable with k-hexagons if adjoining k hexagons to the circuit defined by the shell diagram gives rise to a bounding circuit with the same code as the original shell diagram. Of especial interest are replicable self-dual shell diagrams. The proof of Grünbaum and Motzkin that there are fullerenes for all h other than 1 is based on the use of shell diagrams which are 6-replicable. Somewhat surprisingly shell diagrams which are 1-replicable exist (Figure 2), and it is based on such a shell that the simpler proof (see Malkevitch [**24, 29**]) of the theorem of Grünbaum and Motzkin referred to above is based. Figure 2.

The proof follows from the fact that the unique fullerene with 2 hexagons admits the self-dual shell diagram that is 1-replicable: $0011(01)^4$.

One can make an analogous definition of code of a circuit C for an infinite 3-valent tiling of the plane by k-gons, Tk, where k is 6 or greater. Using Euler's formula it is possible to show that for any such circuit:

$$Z - O - 6 = (k - 6)p_k.$$

where Z denotes the number of zeros in the code, O denotes the number of ones in the code, and p_k denotes the number of k-gons in the interior of C.

A variety of other Euler's formula inspired equations that relate to fullerenes can be found in the literature of fullerenes.

3. Questions

Question 1:
Characterize those binary sequences which can arise as the boundary of some patch in the regular tiling of the plane by hexagons. (Note: Fullerenes embedded in the plane can admit large collections (patches) of faces which locally consist of only

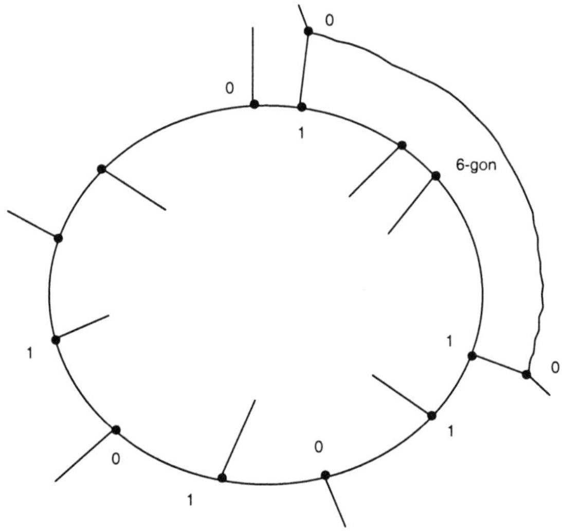

FIGURE 2. The shell diagram of Figure 1, indicating how it can be replicated by a single hexagon.

hexagons. Hence, this question is of interest for fullerenes as well as in its own right.)

If one knows that code C arises from a circuit in some T_k, can one determine what the value of k must be? (Note: what is being asked for here is a combinatorial condition that can be obtained from the code C itself. If one has available a copy of the tiling and carries out the code C starting at any edge of the tiling, one can check whether or not it closes up.)

Question 2:
Characterize those binary sequences which can arise as the boundary of some circuit in a fullerene. Sufficient conditions for a binary sequence to be present in some fullerene would also be of interest.

Question 3:
What are the automorphism groups that can arise for fullerenes with the property that they contain a code which allows augmentation by one hexagon (exactly k hexagons)?

Question 4:
For which values of h are there equilateral fullerenes? (A polyhedron is equilateral if all of its edges have the same length.)

Question 5:
To what extent can one realize a fullerene so that all or some of its pentagonal faces are regular pentagons and all or some of its hexagonal faces are regular hexagons? So that its pentagonal faces are congruent? So that its hexagonal faces are congruent?

REMARK: As one referee noted if all the faces are regular polygons the solid would have to be one of the Johnson Solids and only two of these are fullerenes, the dodecahedron and the truncated icosahedron.

Question 6:
[illegible] many fullerenes admit a selfdual shell diagram?

Question 7:
Do fullerenes which admit self-dual diagrams of many kinds have any special properties? Some fullerenes have the property that every edge is incident with a self-dual diagram. Do such fullerenes have any special properties?
Question 8:
For each self-dual shell diagram determine the values of h for which there is a fullerene with h hexagons that admit such a shell diagram. (Note that what I mean here is for, say, the family $(01)^n$, for each n, which values of h are admissible for that given value of n?)
Question 9:
If one knows that a fullerene has a certain group as the automorphism group of its graph, does this guarantee that any particular shell diagrams must be present in the graph of this fullerene?
Question 10:
Determine necessary and sufficient conditions that codes such as L,R; L,L,R,R; L,L,L,R,R,R yield simple circuits when applied to any edge of a 3-valent 3-polytope. In particular, which fullerenes admit L,R; L,L,R,R; and L,L,L,R,R,R codes starting with any edge? Here L (R) refers to making a left (right) turn when one arrives at a 3-valent vertex in the graph, moving along any edge of the graph. When one gets to the end of the code, one starts again from the beginning.
REMARK: It is known Malkevitch [**24**] that multi k-gon graphs ($k = 3, 4, 5$) (i.e. plane graphs where all faces have a multiple of k for the number of sides of every face) give rise to simple circuits when the L,R code is used. Extensions to graphs that are 4-valent and 5-valent have also been studied in Malkevitch [**24**]. (See also Grünbaum and Motzkin [**19**].)
Question 11:
A spanning tree of a graph is a (1,3)-tree if all its vertices are either 1-valent or 3-valent. Which fullerenes admit spanning trees which are (1,3)-trees? (See Grünbaum [**18**] and Malkevitch [**27**].)

Partial answers to questions 1, 2, and 10 by M. Deza and V. P. Grishukhin have been obtained in a research report entitled Hexagonal Sequences and can be viewed at: www.dmi.ens.fr/EDITION/preprints.

Another open problem of considerable interest is whether or not every fullerene has a Hamiltonian circuit.

Clearly these problems have unequal levels of difficulty and in some cases partial answers to the questions are already available. I believe that answers to these questions will help provide additional useful insights and methods to study this fascinating collection of polyhedra.

References

[1] H. Aldersey-Williams, The Most Beautiful Molecule: The Discovery of the Bucky Ball, Wiley, New York, 1995.
[2] D. Babic, D. Klein, and C. Sah, Symmetry of Fullerenes, *Chemical Physics Letters*, **211** (1993) 235–243.
[3] B. Bollobás, Modern Graph Theory, Springer-Verlag, New York, 1998.
[4] G. Brinkmann, Generating Cubic Graphs Faster than Isomorphism Checking, preprint, SFB 343 No. 92–047, Bielefeld, 1992.
[5] G. Brinkmann, Fast Generation of Cubic Graphs, *J. of Graph Theory*, **23** (1996) 139–149.
[6] G. Brinkmann, and A. Dress, A Constructive Enumeration of Fullerenes, *J. of Algorithms*, **23** (1997) 345–358.

[7] G. Brinkmann, and A. Dress, PentHex Puzzles-A Reliable and Efficient Top-Down Approach to Fullerene-Structure Enumeration, *Pro. Nat. Acad. Sci.*, (to appear).
[8] F. Budden, The Fascination of Groups, Cambridge U. Press, London, 1972.
[9] R. Burn, Groups: A Path to Geometry, Cambridge U. Press, New York, 1985.
[10] F. Chung, and S. Sternberg, Mathematics and the Buckyball, *American Scientist*, **81** (1993) 56–71.
[11] P. Cromwell, Polyhedra, Cambridge U. Press, New York, 1997.
[12] R. Curl, and R. Smalley, Fullerenes, *Scientific American*, **265** (1991) 54–63.
[13] A. Dress, and G. Brinkmann, Phantasmagorical Fulleroids, *Match*, **33** (1996) 87–100.
[14] P. Fowler, and D. Manolopoulos, An Atlas of Fullerenes, Oxford U. Press, Oxford, 1995.
[15] P. Fowler, D. Manolopoulos, D. Redmond, and R. Ryan, Possible Symmetries of Fullerene Structures, *Chemical Physics Letters*, **202** (1993) 371–378.
[16] B. Grünbaum, Convex Polytopes, Wiley-Interscience, New York, 1967.
[17] B. Grünbaum, Some Analogues of Eberhard's Theorem on Convex Polytopes, *Israel J. Math.*, **6** (1968) 398–411.
[18] B. Grünbaum, Polytopes, Graphs, and Complexes, *Bull. Amer. Math. Soc.*, **76** (1970) 1131–1201.
[19] B. Grünbaum, and T. Motzkin, The Number of Hexagons and the Simplicity of Geodesics on Certain Polyhedra, *Canadian J. Math.*, **15** (1963) 744–751.
[20] D. Klein, and X. Liu, Theorems for Carbon Cages, *J. of Math. Chem.*, **11** (1992) 199–205.
[21] W. Kratschmer, L. Lamb, K. Fostiropoulos, and D. Huffman, Solid C_60: A New Form of Carbon, *Nature*, **347** (1990) 354–358.
[22] X. Liu, D. Klein, T. Schmalz, and W. Seitz, Generation of Carbon-Cage Polyhedra, *J. of Computational Chem.*, **12** (1991) 1252–1259.
[23] X. Liu, D. Klein, W. Seitz, and T. Schmalz, Sixty-Atom Carbon Cages, *J. Computational Chem.*, **12** (1991) 1265–1269.
[24] J. Malkevitch, Planar Graphs with Uniform Vertex and Face Structure, Doctoral Thesis, University of Wisconsin (Madison), 1969.
[25] J. Malkevitch, Planar Graphs with Uniform Vertex and Face Structure, Memoir 99, American Mathematical Society, Providence, 1970.
[26] J. Malkevitch, 3-Valent 3-Polytopes with Faces having Fewer than 7 Edges, *Annals of the New York Academy of Sciences*, **75** (1970) 285–286.
[27] J. Malkevitch, Spanning Trees in Polytopal Graphs, *Annals of the New York Academy of Sciences*, **319** (1979) 362–367.
[28] J. Malkevitch, Polytopal Graphs, in Selected Topics in Graph Theory 3, ed. L. Beineke and R. Wilson, Academic Press, New York, 1988.
[29] J. Malkevitch, A Note on Fullerenes, *J. of Fullerene Science*, **2** (1994) 423–426.
[30] D. Manolopoulos, and P. Fowler, A Fullerene without a Spiral, *Chemical Physics Letters*, **204** (1993) 1–7.
[31] T. Pisanski, On Planar Graphs with 12 Vertices of Degree Five, *Glasnik Mat.*, **12** (1977) 233–235.
[32] C. Sah, Combinatorial Construction of Fullerene Structures, *Croatia Chem. Acta*, **66** (1993) 1.
[33] B. Yakobson, and R. Smalley, Fullerene Nanotubes: $C_{1,000,000}$ and Beyond, *American Scientist*, **85** (1997) 324–337.
[34] G. Ziegler, Lectures on Polytopes, Springer-Verlag, New York, 1995.

Department of Mathematics York College (CUNY) Jamaica, New York 11451
E-mail address: joeyc@cunyvm.cuny.edu

DIMACS Series in Discrete Mathematics
and Theoretical Computer Science
Volume 51, 2000

Topological Methods of Molecular Shape Analysis: Continuum Models and Discretization

Paul G. Mezey

ABSTRACT. Some of the topological methods of shape analysis of electron density clouds of molecules, and the associated topological techniques designed to evaluate molecular similarity, are based on the concept of topological resolution. Within the context of an earlier technique of topological shape analysis, based on the shape group methods, SGM, a discretization step provides a simplification that also serves as a basis for a specific, chemically motivated choice for topological resolution. The interrelations between discretization and topological resolution are the tools for the introduction of similarity/dissimilarity measures. These interrelations, as applied to the quantum chemical shape properties of fuzzy molecular electron density distributions, are studied in some detail.

1. Introduction

Recent developments in computational chemistry and in molecular modeling, originally based on novel applications of continuum mathematics to molecular problems, have set the stage for new potential advances in the applications of discrete mathematics to chemistry.

Many of the general features of these latest developments in the mathematical description of chemistry mirror the typical, earlier evolution of the interrelations between continuum and discrete mathematics in other fields. The classical, fundamental laws of nature, such as Newton's Laws, are usually formulated in terms of continuous functions, yet some of the more useful representations of these laws involve discrete mathematics. Many natural phenomena and problems appear continuous, nevertheless, most of the practically useful mathematical models we apply to describe these apparently continuous phenomena are discrete. Sound perception and vision appear to provide continuous impressions, however, if we need reliable communication, we usually discretize: for example, in writing we use discrete letters, we digitalize music in CD players, discretization of pictorial information is involved in sending and receiving fax messages using telephone lines, etc. Ultimately, all digital computers use discrete algorithms, based on integer representations.

1991 *Mathematics Subject Classification.* 68R99, 68U05, 54A10.
This study was supported by a research grant from the Natural Sciences and Engineering Research Council of Canada.

When a scientific field undergoes a significant development, it often occurs by recognizing new analogies, that leads, at least initially, to the advancement of continuum models. This advance, then becomes incorporated into the general body of knowledge, often followed by an associated development of a discrete model. Such a discrete model is often analogous, or even identical with an algebraic structure; in mathematics, the essential aspects of new mathematical advances can often be described by an algebraic structure.

In the field of molecular modeling and computational chemistry, many advances have been achieved in recent years, some of these in continuum representations of molecules and chemical reactions. Among these developments one field that has some importance in molecular engineering and pharmaceutical drug design is progress in macromolecular quantum chemistry.

Recently, it has become possible to compute high quality (*ab initio* quality) electron densities, as well as many other properties of large molecules containing hundreds, even thousand atoms; the macromolecules which can be studied by the new, *ab initio* quality quantum chemical computational techniques include proteins. For example, it is now possible to calculate the forces acting on individual atomic nuclei in any, distorted conformation of a macromolecule, such as the forces involved in the folding processes of proteins (for a recent review, see Mezey [**3**]).

These new methods are based on an additive, fuzzy electron density fragmentation (AFDF) scheme, where individual fuzzy electron density fragments are generated from accurate, "custom-made" quantum chemical representations of smaller molecules. These "parent molecules" include a "coordination shell" representing correctly all the local interactions (within any desired finite range) about each fuzzy density fragment within the actual macromolecule.

A related theoretical development, serving as the foundation of these approaches, is the proof of the "Holographic Electron Density Theorem", stating that any nonzero-volume piece of a fuzzy, boundaryless molecular electron density contains the complete information about the entire molecule. In other words, any small, positive-volume part of a molecular electron density may serve as a tool for establishing correlations with reactivity and activity, a result that serves as a firm foundation of the long-established practice of drawing conclusions concerning the global behavior of molecules based on some local features. The local electron density pieces used in the holographic electron density theorem and those used in the AFDF computational techniques are fundamentally different. Whereas the theorem establishes a relation between any finite, nonzero-volume electron density piece of some boundary surface and the complete, boundaryless, fuzzy molecular electron density, the local density pieces used in the new AFDF computational methods are fuzzy, boundaryless fragment densities. Nevertheless, the theorem gives a justification for the claim that the local interactions between fuzzy electron density fragments are properly represented by these methods within any range selected for the "parent molecules" involved in the computation of the fuzzy fragments.

The advancements in this new era of continuum models in chemistry indicates a new need for discretization approaches. As it is often the case, differential and algebraic topology can provide useful tools for discretizing continuum models, and in this contribution, some of these possibilities are discussed.

2. Some Topological Tools for the Study of Interrelations and Similarities Among Molecular Fragments

One very useful tool of discretization is the selection of a series of topologies which provide some detail but not the complete information present in the original continuum model. Whereas in the following discussion we shall have a general description of a possible topological methodology that serves as a discretization tool, we should keep in mind that the actual motivation and also the applications are related to fuzzy molecular electron densities or similar fuzzy electron density contributions of various molecular fragments. For the applied topological background tailored to the problem of molecular electron densities, the reader is referred to the literature (Mezey [**1**]).

The concept of topological resolution will be used in two roles: in the selection of the appropriate level of detail that provides a useful discretization scheme, and also for the evaluation of similarity, based on the level of topological resolution required to distinguish objects. First we recall a simple relation between generating subbases and topologies. If for two generating subbases S_1 and S_2 containing families of subsets from an underlying space X the relation

$$S_2 \supset S_1 \tag{2.1}$$

holds, then the corresponding topologies generated by these subbases are necessarily comparable, furthermore, topology $\boldsymbol{T}_2$ is finer than topology $\boldsymbol{T}_1$:

$$\boldsymbol{T}_2 \supset \boldsymbol{T}_1. \tag{2.2}$$

We shall assume that in the underlying set X a family $\mathbb{T}$ of topologies $\boldsymbol{T}_i$ is given,

$$\mathbb{T} = \{\boldsymbol{T}_1, \boldsymbol{T}_2, \ldots, \boldsymbol{T}_i, \boldsymbol{T}_{i+1}, \ldots\}, \tag{2.3}$$

where these topologies $\boldsymbol{T}_i$ are fully ordered by the finer-cruder relation, that is,

$$\boldsymbol{T}_{i+1} \supset \boldsymbol{T}_i, \tag{2.4}$$

or, as expressed for topological spaces,

$$(X, \boldsymbol{T}_{i+1}) \supset (X, \boldsymbol{T}_i) \tag{2.5}$$

for every index pair i and $i+1$, for which $\boldsymbol{T}_i$, and $\boldsymbol{T}_{i+1}$ are included in the family $\mathbb{T}$ of topologies.

In fact, one may claim that the choice of family $\mathbb{T}$ of topologies is appropriate for the given molecular problem, if from the series of topological spaces

$$\cdots \supset (X, \boldsymbol{T}_{i+1}) \supset (X, \boldsymbol{T}_i) \cdots \supset (X, \boldsymbol{T}_2) \supset (X, \boldsymbol{T}_1) \tag{2.6}$$

it is possible to select at least one that provides the desired type of discretization. Alternative choices of topologies provide the tools for selecting different levels of topological resolutions. In a later section we shall present an illustrative example.

If the relative level of topological resolution is defined in terms of the finer-cruder relations, then relation (2.5) implies that the topological space $(X, \boldsymbol{T}_{i+1})$ gives a higher topological resolution for the underlying space X than the topological space $(X, \boldsymbol{T}_i)$.

According to relation (2.5), the topological space, $(X, \boldsymbol{T}_{i+1})$ describes more or at least as much detail of the underlying space X than the topological space

$(X, \boldsymbol{T}_i)$. If the strict inclusion relation applies in relation (2.5) that is, if in addition to relation (2.5), the constraint

$$(X, \boldsymbol{T}_{i+1}) \neq (X, \boldsymbol{T}_i) \tag{2.7}$$

also applies, than it is possible that a function f that maps X onto itself is $\boldsymbol{T}_i$ - continuous but not $\boldsymbol{T}_{i+1}$ - continuous, and a topological description $(X, \boldsymbol{T}_{i+1})$ with a higher topological resolution provides more information, (in particular, it gives more contrast, using a computer graphics analogy) than a description $(X, \boldsymbol{T}_i)$ with a topological resolution of lower level.

Molecular electron densities can be regarded as three-dimensional, fuzzy sets embedded in the ordinary $3D$ Euclidean space E^3. A tool introduced for the shape analysis of these electron densities are the algebraic-topological Shape Groups: homology groups of level sets of electron densities truncated according to some curvature or other criteria. The Shape Groups and their applications for molecular shape analysis and similarity assessment have been reviewed extensively (Mezey [**1, 2**]). Here only a brief summary will be given, the reader may find more detail in the original references.

The nuclear arrangement of the molecule is denoted by K, and the molecular electron density function is denoted by $\rho(K, \boldsymbol{r})$. This three-dimensional function $\rho(K, \boldsymbol{r})$ can be represented by its level sets, that is, by an infinite family of molecular isodensity contour (MIDCO) surfaces $G(K, a)$. The density threshold a can take values from the $[0, \infty)$ interval. Each MIDCO level set is defined as

$$G(K, a) = \{\boldsymbol{r} : \rho(K, \boldsymbol{r}) = a\}. \tag{2.8}$$

In the simplest implementation of the shape group method, the local shape of each MIDCO $G(K, a)$ is compared to a range of reference curvatures b. All points $\boldsymbol{r}$ along each MIDCO $G(K, a)$ are classified according to the local curvatures. Accordingly, at a point $\boldsymbol{r}$ the contour surface $G(K, a)$ is either

(i) locally convex relative to reference curvature b (then $\boldsymbol{r}$ belongs to a domain of type $D_2(b)$),
(ii) locally of the saddle type relative to reference curvature b (then $\boldsymbol{r}$ belongs to a domain of type $D_1(b)$), or
(iii) locally concave relative to curvature b (then $\boldsymbol{r}$ belongs to a domain of type $D_0(b)$).

For each curvature type m relative to a reference curvature value b, the domains $D_\mu(b)$ of each MIDCO $G(K, a)$ generate a pattern $P(K, a, b)$ of domains on the contour surface $G(K, a)$. By analyzing the topology of these patterns $P(K, a, b)$, within each, topologically distinct pattern $P(K, a, b)$ a topological description of the interrelations among the domains $D_\mu(b)$ is obtained.

In order to visualize the process of generating these topological descriptors, one may select a curvature type μ relative to a reference curvature value b, and assume that the corresponding domains $D_\mu(b)$ are excised from the MIDCO surface $G(K, a)$ of the given pattern $P(K, a, b)$. After the domains $D_\mu(b)$ of a specified curvature type μ with reference to a curvature value b are removed from a MIDCO $G(K, a)$, a truncated object $G_\mu(K, a, b)$ with a certain set of holes is obtained.

The Shape Groups of the molecule, denoted by $H^k_\mu(K, a, b)$, are the algebraic-topological homology groups of the truncated MIDCO surfaces $G_\mu(K, a, b)$.

For molecular electron densities, the Shape Groups fall into three classes: the zero-, one- and two-dimensional Shape Groups $H^k_\mu(K, a, b)$, where the dimension is indicated by the index k. Furthermore, in this notation the truncation type μ, the nuclear configuration K, the density threshold a, and the reference curvature b are also specified; some of these specifications may be omitted when this does not lead to ambiguity.

If the molecule contains N nuclei ($N \geq 3$; for $N \leq 2$ special considerations apply), the family of chemically accessible nuclear configurations K (excluding processes involving nuclear reactions) form a nuclear configuration space M, a metric space of $3N - 6$ dimensions; those configurations which preserve the identity of the molecule form a subset (a catchment region) of the nuclear configuration space M.

The Shape Groups show some invariance properties within small ranges of the nuclear configuration space M, and within small intervals of the electron density threshold values a and small intervals of the reference curvature values b. These local invariances of the shape groups $H_\mu^k(K, a, b)$ within domains of the nuclear configuration space M and domains of the parameter plane (a, b) spanned by the density and curvature thresholds a and b imply that for each molecule there are only a finite number of shape groups (Mezey [**1, 2**]).

This fact is an important feature that can be exploited for discretization. If the Shape Groups themselves are partially characterized by a set of integers, for example, by their ranks (the Betti numbers of the respective homology groups), then a finite number of these integers provides a discretized representation of the shape of the fuzzy electron density.

Indeed, one of the simplest choices of discretization is offered by the zero-, one-, and two-dimensional Betti numbers, by $b_\mu^0(a, b)$, $b_\mu^1(a, b)$, and $b_\mu^2(a, b)$, that is, by the ranks of the zero-, one-, and two-dimensional Shape Groups, $H_\mu^0(a, b)$, $H_\mu^1(a, b)$, and $H_\mu^2(a, b)$, respectively. The sets of Betti numbers determined for the entire range of MIDCOs $G(K, a)$ of the molecule M and for the appropriate range of curvature parameters b, generate a set of numerical shape descriptors.

An additional discretization step is used in most applications: the actual curvatures b and density thresholds a are sampled only at finite number, n_a and n_b values, respectively, usually chosen at equal increments in a logarithmic scale.

As a result, the distribution of various values of Betti numbers $b_\mu^p(a, b)$ as a function of the density threshold a and curvature parameter b is described by various, discretized (a, b)-maps. These discretized (a, b)-maps are then represented numerically in terms of shape matrices $\mathbb{M}^{(a,b)}$. In fact, these shape matrices are numerical shape codes for molecules, where the total number of elements in each shape matrix $\mathbb{M}^{(a,b)}$ is

$$t = n_a n_b. \tag{2.9}$$

Based on these shape codes, a simple shape-similarity measure is obtained. A quantity $m[\mathbb{M}^{(a,b),A}, \mathbb{M}^{(a,b),B}]$ is defined as the number of matches between corresponding elements of the two shape codes $\mathbb{M}^{(a,b),A}$ and $\mathbb{M}^{(a,b),B}$ where A and B refers to either two molecules, or to two fuzzy (AFDF) molecular fragments.

Since this quantity $m[\mathbb{M}^{(a,b),A}, \mathbb{M}^{(a,b),B}]$ is based on the two shape codes, the expression

$$s(A, B) = m\left[\mathbb{M}^{(a,b),A}, \mathbb{M}^{(a,b),B}\right]/t \tag{2.10}$$

defines a shape-similarity measure $s(A, B)$, describing the degree of similarity between the global shape group representations of the two molecules or fuzzy molecular electron density fragments A and B.

The numerical comparison of shape codes is a much simpler process than any direct comparisons of the electron densities of two molecules or molecular fragments. In most direct density comparisons the mutual orientation and superposition of the two molecules or molecular fragments must be optimized before a shape comparison can be performed. By contrast, no such optimum superposition is required when the shape-similarity of two molecules or molecular fragments is determined based on their shape codes.

Note that, by a simple numerical transformation, the shape similarity measure $s(A, B)$ can also be used for the evaluation of shape complementarity (Mezey [**1**]).

3. Topological Resolution as a Tool for Discretization

The practical implementation of the Shape Group method, described above, can be regarded as a reduction of the information content of a fuzzy, molecular electron density cloud $\rho(K, \boldsymbol{r})$, a three-dimensional continuum, to a finite set of discrete Betti numbers. This discretization process retains most of the essential shape information of the molecules or molecular fragments.

The shape group approach and the associated shape codes provide a concise representation of the molecular shape information. Such a detailed shape description is often required to detect and interpret the roles of shape features of the electron density $\rho(K, \boldsymbol{r})$ in various chemical interactions and processes. On the other hand, in many instances there is no need for a complete shape analysis of the fine details of the molecular electron density $\rho(K, \boldsymbol{r})$, and the interest lies in some of the more prominent, large scale shape features.

Furthermore, it may be impractical to consider all fine details of shape, especially, in the case of macromolecules. In such cases, for example, in the case of a protein's electron density, a complete shape group analysis of the entire electron density $\rho(K, \boldsymbol{r})$ may produce a large amount of detail that may be both unnecessary for the given purpose and also computationally prohibitive.

Alternative shape characterization methods are applicable in such cases, where it is possible to adjust the level of detail described. In these techniques, the level of topological resolution serves as a suitable tool to adjust the level of detail detected by the shape analysis. Whereas the principle of using topological resolution as a tool for adjusting the detail for analysis is rather general, the concept of topological resolution of fuzzy electron density clouds $\rho(K, \boldsymbol{r})$ is particularly useful for adjusting the sensitivity of molecular similarity measures.

If our goal is simply to shift the emphasis between the local details and the large scale features of the shapes of molecular electron densities or fuzzy fragment densities $\rho(K, \boldsymbol{r})$, then a choice of different ranges for the electron density threshold parameter a and for the reference curvature b provides a natural scaling, since low electron density thresholds a and large reference curvature values b are usually associated with the large scale features of molecular electron densities. By contrast, density contours $G(K, a)$ of high a values are often highly curved and show more detail, that is, they exhibit important topological properties at large curvature values b, since these MIDCOs run near the atomic nuclei.

In particular, the role of range selection of the reference curvature parameter b appears more important than the range selection of density threshold a, since in the vicinity of different nuclei the density values can be very different, yet the local curvatures of isodensity surfaces at a given small enough distance from all nuclei are nearly uniform. If a reference curvature b of high negative value is selected, then at most points $\boldsymbol{r}$ along a MIDCO $G(K,a)$, including some MIDCOs with rather high values of the electron density threshold a, it is likely that both of the local canonical curvatures of the MIDCO are greater than b, the reference curvature. The local canonical curvatures are given as the eigenvalues of the local Hessian matrix expressing the local curvatures at the given point $\boldsymbol{r}$ of the MIDCO $G(K,a)$. This implies that with reference to this curvature b, all or nearly all points $\boldsymbol{r}$ of the MIDCO $G(K,a)$ belong to a curvature domain of type $D_0(b)$ of relative concavity. Consequently, the usual truncation of type $\mu = 2$ eliminates no domains or only few domains from the MIDCO $G(K,a)$. In the simplest case, if both of the local canonical curvatures at all points $\boldsymbol{r}$ of the MIDCO $G(K,a)$ are greater than the reference curvature b, then no truncation occurs, that is, no domain is removed, and one obtains

$$G_2(K,a,b) = G(K,a), \tag{3.1}$$

a complete coincidence. Consequently, the associated shape group $H_2^1(K,a,b)$ is the trivial group,

$$H_2^1(K,a,b) = \{0\}. \tag{3.2}$$

Consider now a gradual increase of the reference curvature b. As a function of the increasing value of b, the local canonical curvatures will fall below the value b within an increasing subset of the MIDCO $G(K,a)$. Accordingly, more and more domains become the subject to elimination in the truncation step of the shape group generation, resulting in a variety of objects $G_2(K,a,b)$ topologically different from $G(K,a)$,

$$G_2(K,a,b) \neq G(K,a). \tag{3.3}$$

Consequently, one obtains a variety of shape groups $H_2^1(K,a,b)$ which are no longer the trivial groups, that is, more detail of the shape of the molecular electron density function $\rho(K,\boldsymbol{r})$, as represented by the MIDCO $G(K,a)$, becomes registered in terms of the topological descriptors.

The analysis may focus on the topological changes of the truncated objects $G_2(K,a,b)$ as the value of the reference curvature b is increased. For example, the actual threshold values for b where the topological change occurs may be registered, that involves the actual scanning of the entire range of relevant reference curvature values. An alternative, less precise but computationally much simpler approach relies on a finite series of pre-selected, monotonically increasing b values,

$$b_1, b_2, \ldots, b_i, \ldots, b_m. \tag{3.4}$$

The truncated MIDCO $G_2(K,a,b)$ is determined for each b_i value from the series, where the condition

$$b_i < b_{i+1} \tag{3.5}$$

simplifies the computations.

Once the corresponding series of truncated MIDCOs

$$G_2(K,a,b_1), G_2(K,a,b_2), \ldots, G_2(K,a,b_i), \ldots, G_2(K,a,b_m) \tag{3.6}$$

is known, the associated truncation patterns

$$P(K,a,b_1), P(K,a,b_2), \ldots, P(K,a,b_i), \ldots, P(K,a,b_m) \tag{3.7}$$

can be determined, where the focus is not necessarily restricted to the curvature domains $D_\mu(b)$ of type $\mu = 2$.

Since each of these patterns $P(K, a, b_i)$ is generated on the same MIDCO $G(K, a)$, it is possible to take the combined pattern of several and possibly all of these patterns, for example, the pattern

$$P(K, a, b_1 \dots b_{i'}), \tag{3.8}$$

obtained by superimposing all patterns $P(K, a, b_i)$ in a subsequence of all indices from 1 to i':

$$P(K, a, b_1), P(K, a, b_2), \dots, P(K, a, b_{i'}). \tag{3.9}$$

Evidently, if not all these patterns in the subsequence are geometrically identical, then the superimposed pattern $P(K, a, b_1 \dots b_{i'})$ generates a more detailed characterization of the MIDCO $G(K, a)$ than any one of the individual patterns from the subsequence. Note that even if the patterns in the subsequence are topologically identical but show some geometrical differences, then the superposition produces a topologically richer description than any one of the individual patterns from the subsequence.

For each index $i, 1 \leq i \leq i'$, a particular topology $\boldsymbol{T}(K, a, b_i)$ and an associated topological space

$$\big(G(K, a), \boldsymbol{T}(K, a, b_i)\big) \tag{3.10}$$

can be defined on the MIDCO $G(K, a)$, where the set of $D_\mu(b_i)$ domains of the pattern $P(K, a, b_i)$ may serve as a defining subbase $S(K, a, b_i)$:

$$S(K, a, b_i) = \{D_\mu(b_i)\}. \tag{3.11}$$

On the other hand, for a topological representation of the combined pattern $P(K, a, b_1 \dots b_{i'})$, a subbase $S(K, a, b_1 \dots b_{i'})$ can be chosen as the union of subbases $S(K, a, b_1), S(K, a, b_2), \dots, S(K, a, b_{i'})$:

$$S(K, a, b_1 \dots b_{i'}) = \underset{k = 1, i'}{\cup} S(K, a, b_k). \tag{3.12}$$

This subbase $S(K, a, b_1 \dots b_{i'})$ defines a topology $\boldsymbol{T}(K, a, b_1 \dots b_{i'})$ that is in fact the same topology taken using the family of all intersections of all domains $D_\mu(b_i)$ for all indices $i = 1, 2, \dots, i'$ of patterns $P(K, a, b_1), P(K, a, b_2), \dots, P(K, a, b_{i'})$.

As a consequence of this construction, for the defining subbases $S(K, a, b_1 \dots b_{i'})$ of these topologies $\boldsymbol{T}(K, a, b_1 \dots b_{i'})$ of various upper indices i', where

$$1 \leq i' \leq m - 1, \tag{3.13}$$

the relation

$$S(K, a, b_1 \dots b_{i'+1}) \supset S(K, a, b_1 \dots b_{i'}) \tag{3.14}$$

applies. This, in turn implies that for the corresponding topologies defined by these subbases the relation

$$\boldsymbol{T}(K, a, b_1 \dots b_{i'+1}) \supset \boldsymbol{T}(K, a, b_1 \dots b_{i'}) \tag{3.15}$$

must hold. That is, the series of subbases $S(K, a, b_1 \dots b_{i'})$ and the corresponding topologies $\boldsymbol{T}(K, a, b_1 \dots b_{i'})$ are fully ordered,

(3.16)

$$S(K, a, b_1 \dots b_{m}) \supset S(K, a, b_1 \dots b_{m-1}) \supset \dots \supset S(K, a, b_1 \dots b_{i'}) \supset \dots \supset S(K, a, b_1),$$

and

(3.17)

$$\boldsymbol{T}(K,a,b_1\dots b_m)\supset\boldsymbol{T}(K,a,b_1\dots b_{m-1})\supset\cdots\supset\boldsymbol{T}(K,a,b_1\dots b_{i'})\supset\cdots\supset\boldsymbol{T}(K,a,b_1).$$

As a simple consequence of their construction, one can utilize the fact that these topologies $\boldsymbol{T}(K,a,b_1\dots b_{i'})$ are completely ordered. The corresponding monotonic series of topologies can serve as the basis for a systematic implementation of the general technique of topological resolution for the shape characterization of the MIDCO $G(K,a)$. Using index i' as an integer parameter that changes monotonically within the interval $[1,m]$, a series of gradually increasing topological resolutions is obtained.

A similarity measure based on topological resolutions can be constructed using these topologies $\boldsymbol{T}(K,a,b_1\dots b_{i'})$. This similarity measure is a special case of a resolution-based similarity measure, RBSM, where the fact that it is a *topological* resolution-based similarity measure, is reflected in the acronym TRBSM.

If two nuclear configurations K_1 and K_2, of two molecules M_1 and M_2, respectively, are considered, and the analysis is focused on two associated MIDCOs $G(K_1,a_1)$ and $G(K_2,a_2)$, with respect to two respective density thresholds a_1 and a_2, then this similarity measure is generated as follows.

First, construct the two series of topologies $\boldsymbol{T}(K_1,a_1,b_1\dots b_{i'})$ and $\boldsymbol{T}(K_2,a_2,b_1\dots b_{i'})$, respectively, for the range $1\le i'\le m$ of indices i', for both molecules M_1 and M_2.

With respect to these two series of topologies, the two MIDCOs $G(K_1,a_1)$ and $G(K_2,a_2)$ have equivalent shapes at some level i' of topological resolution if and only if there is a one to one and onto correspondence between the two defining subbases $S(K_1,a_1,b_1\dots b_{i'})$ and $S(K_2,a_2,b_1\dots b_{i'})$ where the curvature index μ assignment of each element of the subbase for each sublevel k, $1\le k\le i'$ is preserved. We may denote this equivalence of shapes of MIDCOs at level i' of topological resolution by

$$G(K_1,a_1)\ \mathrm{eq}\ [b_1\dots b_{i'}]\,G(K_2,a_2). \tag{3.18}$$

In this notation the corresponding reference curvatures $[b_1\dots b_{i'}]$ are also specified, nevertheless, if the curvature values are set by some general convention or are otherwise evident from the context, these specification can be omitted and one may write

$$G(K_1,a_1)\ \mathrm{eq}\ G(K_2,a_2). \tag{3.19}$$

The general properties of the convergence properties of molecular electron densities at large distances from the nuclei imply that if the first reference curvature b_1 is chosen as a large enough negative value, then the equivalence

$$G(K_1,a_1)\ \mathrm{eq}\ [b_1]\,G(K_2,a_2), \tag{3.20}$$

holds for any two nuclear configurations K_1 and K_2 of practically any two molecules M_1 and M_2, since for small enough density thresholds both of the MIDCOs $G(K_1,a_1)$ and $G(K_2,a_2)$ are, in most typical cases, topological spheres. This is the case if both patterns $P(K_1,a_1,b_1)$ and $P(K_2,a_2,b_1)$ are trivial patterns, each just a single domain $D_0(b_1)$. This feature is advantageous, since this implies that a common initial pattern exists for any two molecules M_1 and M_2, taken at a low level of topological resolution. At this level, their shapes appear equivalent, practically independently of the chemical nature and the actual nuclear configurations K_1 and K_2 of the two molecules.

Nevertheless, for two different molecules M_1 and M_2, a gradual increase of the topological resolution eventually reveals the shape differences. At a sufficiently high positive value of the reference curvature $b_{i'}$, even slight conformational deviations of the same molecule are detectable using topological resolution. Such a non-equivalence of the shapes of the two MIDCOs at the some level i' of topological resolution is formally stated as

$$G(K_1, a_1) \text{ noneq } [b_1 \ldots b_{i'}]\, G(K_2, a_2). \tag{3.21}$$

Clearly, a higher threshold index i' for equivalence implies a higher level of similarity, where a finer (stronger) topology provides a higher level of topological resolution.

As an example, consider the three MIDCO surfaces $G(K_1, a_1)$, $G(K_2, a_2)$, and $G(K_3, a_3)$, of three molecules, M_1, M_2, and M_3, taken at nuclear arrangements K_1, K_2, and K_3, and density thresholds a_1, a_2, and a_3, respectively. We assume that these density thresholds are comparable. For the three possible pairs the maximum indices of equivalence are denoted by as follows:

$$i_{12} = i_{12}\big[G(K_1, a_1), G(K_2, a_2)\big], \tag{3.22}$$

$$i_{13} = i_{13}\big[G(K_1, a_1), G(K_3, a_3)\big], \tag{3.23}$$

and

$$i_{23} = i_{23}\big[G(K_2, a_2), G(K_3, a_3)\big]. \tag{3.24}$$

Here in the detailed notation the actual MIDCO pairs are specified within the square brackets. Let us assume further that these indices are ordered as

$$i_{12} < i_{13} < i_{23}. \tag{3.25}$$

Accordingly, as manifested by the discretized device of topological resolution, the pair $G(K_2, a_2)$ and $G(K_3, a_3)$ of MIDCOs shows the highest degree of similarity, whereas the topological resolution indicates the lowest degree of similarity for the pair $G(K_1, a_1)$ and $G(K_2, a_2)$ of MIDCOs.

The tool of the discretized topological resolution is applicable in the role of dissimilarity measure as well. With respect to a fixed set of topologies, the quantity $m - i_{12}$ may serve as a dissimilarity measure. One should point out, however, that, in part as a consequence of the discretization step, this quantity $m - i_{12}$ is not a metric on the abstract space of all shapes, since a formal zero distance may be obtained for two geometrically different objects, that is, for two objects of different shapes at infinite resolution.

One advantage of a similarity measure or dissimilarity measure based on the level of topological resolution is the fact that such similarity/dissimilarity measures can be adjusted according to some physical or chemical shape property that is relevant to the actual problem studied. It is possible to select the appropriate level of resolution that allows one to emphasize those shape properties which appear essential for the given purpose. For the given chemical problem, one is able to select that level of topological resolution that is suitable for the actual requirements of detail in the analysis of local and global size and shape features. If the topological resolution is gradually increased, finer and finer details of shape become detectable by the method of topological resolution. Whereas such choices are available, the actual similarity/dissimilarity measures obtained are well-defined, and can be re-adjusted in a systematic, and mathematically rigorous manner, allowing only such freedoms that do not generate excessive ambiguity. The significance of any uncertainty associated with this freedom of choice of the actual topologies is clear from the construction of the corresponding subbases.

4. An Adaptation of the Discretization Tool of Topological Resolution to Local Molecular Moieties and Functional Groups

We may summarize the features discussed above: the similarity/dissimilarity measures based on the concept of topological resolution, combined with a discretization step, are tools which are suitable to focus on the essential shape features of molecular electron densities, and to re-emphasize, in a systematic manner, the actual range or the detail that is selected for analysis.

The quantum chemical treatment of functional groups and other, chemically significant local molecular moieties has been based on the concept of Density Domains (for detail the reader is referred to the recent literature, Mezey [**1, 2, 3**]). Essentially, a quantum chemical functional group has been defined as the set of nuclei and the associated fuzzy "share" of the molecular electron density that possesses some MIDCO surface $G(K,a)$ separating the selected nuclei from the rest of the nuclei of the molecule, indicating some limited "autonomy" of this fuzzy electron density contribution within the given molecule. In fact, in all such cases the corresponding MIDCO $G(K,a)$ must have at least two components, one component, $G_F(K,a)$, enclosing the selected nuclei, that is, the nuclei of the functional group F, whereas the rest, $G_R(K,a)$ of the MIDCO $G(K,a)$ contains all the other nuclei of the molecule:

$$G(K,a) = G_F(K,a) \cup G_R(K,a) \tag{4.1}$$

where these two subset of the MIDCO are disjoint,

$$G_F(K,a) \cap G_R(K,a) = \emptyset, \tag{4.2}$$

where $\emptyset$ is the empty set.

One interesting possibility arising in the application of the topological resolution approach to functional groups is the computational analysis of similarities between shape features of various functional groups, indicating potential similarities in their chemical roles.

In this context, one may consider the finest topological resolution that still does not distinguish two functional groups, F_1 and F_2, and the associated numerical similarity measure, obtained using the tool of topological resolution. If the similarity is high, that is, if even at high topological resolution the functional groups F_1 and F_2 show topological equivalence, then it is expected that the two functional groups may act in similar chemical roles in the two actual molecules considered. For example, it is likely that in some reactions the two functional groups F_1 and F_2 are interchangeable, resulting in similar chemical processes.

Using the formalism described in the previous sections, a greater index

$$i_{12} = i_{12}\big[G(K_{F_1}, a_{F_1}), G(K_{F_2}, a_{F_2})\big] \tag{4.3}$$

indicates a greater degree of similarity.

One potential application involves catalysis. If the catalytic action of a given molecule is primarily associated with a given type of functional group F_1 and if in a different molecule the local shape of the electron density of another functional group F_2 shows topological equivalence with that of the former at some level of topological resolution, than a high topological resolution indicates the possibility of potential replacement of the former catalyst with the latter molecule in a similar catalytic role.

Search for new catalysts is a major task in modern chemistry, and similarity approaches, possibly based on the concept of topological resolution and the associated discretization approach, may provide new clues.

References

[1] P.G. Mezey, *Shape in Chemistry: An Introduction to Molecular Shape and Topology*, VCH Publishers, New York, 1993.

[2] P.G. Mezey, *Shape in Quantum Chemistry*. In *Conceptual Trends in Quantum Chemistry*, Vol. 3, Calais. J.-L.; Kryachko, E.S. (eds.), Kluwer Academic Publ., Dordrecht, The Netherlands, pp. 519–550, 1997.

[3] P.G. Mezey, Quantum Chemistry of Macromolecular Shape, *Internat. Rev. Phys. Chem.*, **16** (1997) 361–388.

[4] P.G. Mezey, Generalized Chirality and Symmetry Deficiency, *J. Math. Chem.*, **23** (1998) 65–84.

Mathematical Chemistry Research Unit, Department of Chemistry and Department of Mathematics and Statistics, University of Saskatchewan, 110 Science Place, Saskatoon, SK, Canada, S7N 5C9

E-mail address: mezey@sask.usask.ca

DIMACS Series in Discrete Mathematics
and Theoretical Computer Science
Volume 51, 2000

Random Networks in Two Dimensions. Simulations and Correlations

H. M. Ohlenbusch, N. Rivier, T. Aste, and B. Dubertret

ABSTRACT. We perform extensive simulations of two-dimensional disordered cellular systems (random, planar, regular graphs of degree 3). Familiar examples are foams, soap froths, biological tissues, fullerenes, or carbon nanotubes. We construct structures with different degree of disorder by randomly performing local elementary topological transformations (bond switch, or 3-sided cell disappearance). Coalescence of cells and division are modeled by using combinations of these elementary transformations. We analyze the structures and correlations between cells, either as topological trees rooted on a central cell, or as closed shells arranged concentrically around a germ cell, thereby introducing two definitions of topological distance in discrete geometry. The relations between the parameters that statistically characterize the organization of these structures are discussed.

1. Introduction

Artificial or natural, cellular structures are widespread [**13**][**7**] (soap foams, biological tissues, carbon nanotubes, geological jointings (Fig.1)). Ordered structures, such as honeycombs, have rotation and translation symmetries. By contrast, disordered systems, have no symmetry requirements and are only constrained by the topological conditions of partitioning space. In two dimensions, these structures correspond to partitions of the plane by irregular polygons.

A two-dimensional cellular structure can be modeled topologically as a regular planar graph, with 3 edges incident on a vertex, and 2 cells incident on an edge. A vertex with more than 3 incident edges is neither structurally, nor topologically stable. A small deformation splits it into two or more vertices of degree 3. An edge is bounded by two vertices, and a cell, by n edges. n is the only local random variable of the problem. Cells with two edges only are also rejected: they only decorate edges and play no topological role. If they occur fortuitously, for example in soap froths, their lifetime is very short. The cellular structure has a dual, a triangulation (tiling by triangles), where the local random variable n is now the degree of each vertex.

Elementary, local topological transformations shuffle the random variable n and explore the set of possible configurations. In two dimensions the elementary topological transformations are exchange of neighbors (T1) and triangular cell disappearance into a single vertex (T2) [**1**] (Fig.2). The inverse T2 transformation

1991 *Mathematics Subject Classification.* Primary 54C40, 14E20; Secondary 46E25, 20C20.

FIGURE 1. The Giant's causeway. Situated in Ulster, this causeway (a basalt slab) cracked into polygonal based columns.

can be seen as the division of a n-sided cell into a triangular and an $n+1$-sided daughter. Other topological transformations, such as cell division, coalescence, or disappearance of a non-triangular cell are combinations of the elementary T1 and T2 processes. These transformations are observed in all natural foams. They are the means of simulating random cellular structures.

An infinite two-dimensional Euclidean froth (or one with periodic boundary conditions) has average number of neighbors per cell equal to 6. One introduces the topological charge ($q = 6 - n$) of a cell. The froth is topologically neutral on average, and remains neutral at all times, even though n is shuffled between neighboring cells by elementary topological transformations. No topological charge can be generated or destroyed.

We study the structural organization in terms of distances between cells. The most natural and powerful approach is to foliate the structure into successive layers around a given central cell. A layer contains all cells at the same distance from the

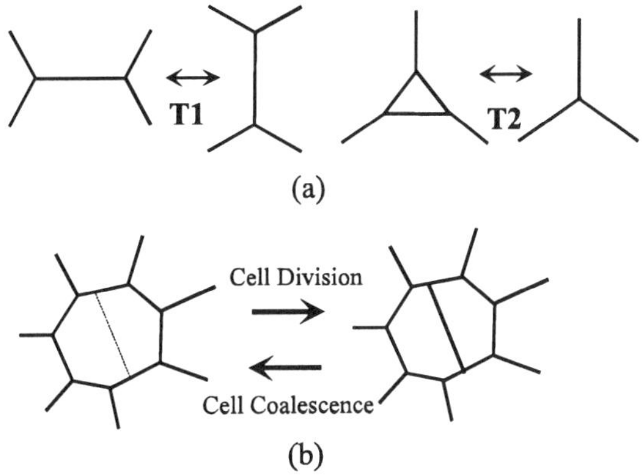

FIGURE 2. The two elementary topological transformations that we use to generate disordered cellular structures are: exchange of neighbors (T1) and triangular cell disappearance into a single vertex (T2). The division and coalescence of a cell, respectively appearance and disappearance of an edge, can be regarded as combinations of the T1 and T2 transformations.

central cell. We define two different topological distances: the shell distance [**11, 12, 3**] and the bond distance [**6**]. The shell distance between two cells is the minimum number of edges that must be crossed to join the two cells. It is the shortest path between two vertices in the dual triangulation. The bond distance (also called T1 distance), is the shortest path or minimum number of edges connecting two cells of the cellular structure. These two distances are not equal, but are strongly correlated [**9**] .

The shell distance naturally introduces a hierarchy of successive layers in the foam. The bond distance reduces the froth to a set of n trees rooted on the vertices of an arbitrary, n-sided central cell.

2. Simulation

We use computer simulations to generate different types of disordered cellular structures. (For other simulations, more specifically adapted to biological tissues, see [**4, 5, 6**]. Using a vertex base model we wrote a C program implementing the two elementary topological transformations T1 and T2.

For each vertex we have a data structure which stores the reference of the three other vertices to which it is connected (implemented using C pointers). These references to other vertices represent the connections between the vertices (i.e. the edges of the structure). This way of structuring the data very closely matches its representation as a graph and gives us the same freedom (such as generating non planar graphs). In order to keep track of the cells, we implemented a similar data structure for the cells, which are the vertices of the dual triangulation. Using this model, we were able to implement the T1 and T2 processes as being only reference changes as shown in Figure 3.

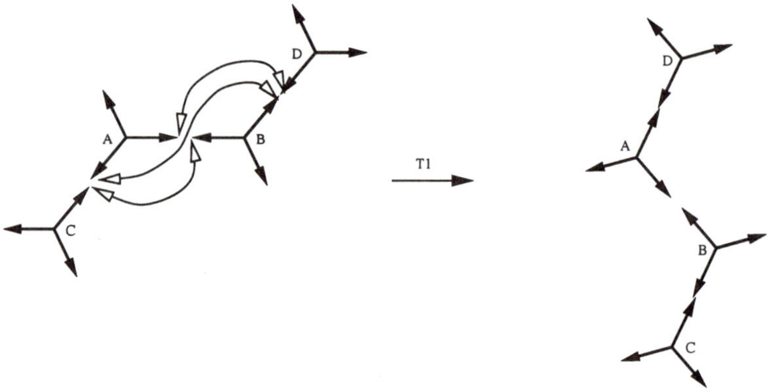

FIGURE 3. Representation of the T1 transformation as reference changes. To flip the edge AB, six references need to be modified: (C $\to$ A) is changed to (C $\to$ B), (B $\to$ A) changed to (B $\to$ C), (D $\to$ B) to (D $\to$ A), (A $\to$ B) to (A $\to$ D). The two references (A $\to$ B) and (B $\to$ A) are rearranged to prevent twists to occur in the structure.

To visualize the quality of the structures generated we use the X11 library facilities. Each vertex is assigned a location (x,y). For a clear visualization, we must compute the new position of the vertices directly affected by the topological transformations (T1 and T2). Since one local transformation also affects its surrounding cells, we recompute the new location of all the vertices before the final representation.

The algorithm that we use computes the new coordinates of each vertex from the position of its three neighbors (center of gravity). This relaxation minimizes the energy of the structure from a mechanical standpoint, and converges to a solution, different in general from that obtained by minimizing the length of all the edges (i.e. minimizing the link tension energy using the Torricelli solution of the Fermat problem, also called Steiner problem).

The algorithms used to find the layers of equidistant cells, are based on visiting flags; cell flags for the shell distance, and vertices flags for the bond distance.

We simulate three different types of disordered systems with more than 10^5 cells generated from an hexagonal lattice by applying different types of elementary topological transformations: (*i*) C-D simulation, (*ii*) T1 simulation, (*iii*) T1-T2 simulation.

2.1. Coalescence and Division (C-D Simulation). We implement the cell division and coalescence as combinations of T1 and T2 (reverse T2 for the division). We choose a symmetric mode of division: The two "daughters" of an n-sided "mother" cell have both $2+n/2$ edges for n even, and $2+(n+1)/2$ and $2+(n-1)/2$ edges if n is odd.

We start the simulation by generating an hexagonal lattice of N=144400 cells (380 cells high * 380 cells wide) with periodic boundary conditions. These boundary conditions imply that our 2D structures can be represented on a torus. We first perform 5*N T1 transformations on edges chosen at random to disorder the system. Then, a cell is chosen at random, and divided into two parts if it has more than

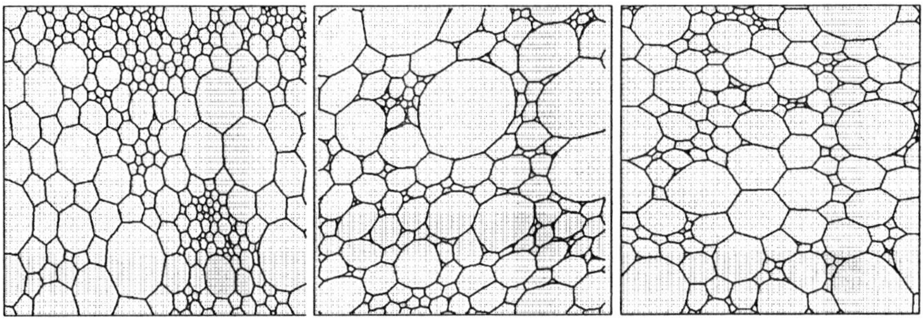

FIGURE 4. Typical cellular structures generated by: (a) Cell coalescence and cell division; (b) T1 transformations only; (c) T1 and T2 transformations.

6 sides, or fused by coalescence with one of its neighbors chosen at random, if it has less than 6 sides. These simple local rules are biologically relevant [**5**][**4**]. A cell with 6 sides is divided with probability p_d, otherwise it remains unchanged. By changing the probability p_d we are able to vary the rate of increase or decrease of the number of cells. Per move, on average, $\Delta N = \sum_{n \geq 7} N_n - \sum_{n \leq 5} N_n + p_d N_6$. The condition of stationarity $\Delta N = 0$ implies that the median $Me = 6 - (p_d/2)$ of the continuous distribution interpolating $p(n)$ lies between $n = 5.5$ ($p_d = 1$, all 6-sided cells divide) and $n = 6$ ($p_d = 0$, no 6-sided cells divides). Given that the mean of the distribution $p(n)$ is $\langle n \rangle = 6$, and that the distribution has a lower bound $n \geq 3$ but is unbounded for large n, $\sum_{n \geq 7} N_n - \sum_{n \leq 5} N_n < 0$ in general, and $5.5 < Me < \langle n \rangle = 6$ (see Fig.5.a). The number of cells remains stationary for some finite value of p_d. Only for a symmetrical distribution ($Me = 6 = \langle n \rangle$), does the number of cells remains stationary if 6-sided cells do not divide, $p_d = 0$. In our simulations, we found that for p_d=0 the number of cells decreases dramatically, and for p_d=1/2 the number of cells increases. For p_d=1/4 the number of cells is stationary with considerable fluctuations, and $Me = 5.875$.

After repeating the coalescence-division operation 10*N times, the resulting structure (Fig.4.a) reaches a stationary number of N=114475 cells. The final distribution is shown in Figure 5.a.

We varied the number of initial, disordering T1 transformations, and found that the statistical properties of the final structure are independent of the initial configuration. Depending on the initial structure, the system evolves towards order or towards disorder as illustrated by Figure 6. The former case is an example of self-organization induced by local random transformation only.

Given transition probabilities for a n-sided cell to divide or coalesce with one of its neighbors, the tissue always reaches a steady state, regardless of its initial configuration. We have indeed a Markov process. But this steady state could be death ($N = 0$) or proliferation ($N = \infty$) of all the cells. Only the topological transition kernels given above, and more finely tuned in [**4**], lead to the narrow, stationary distribution of cells, with small fluctuations in variance observed in living

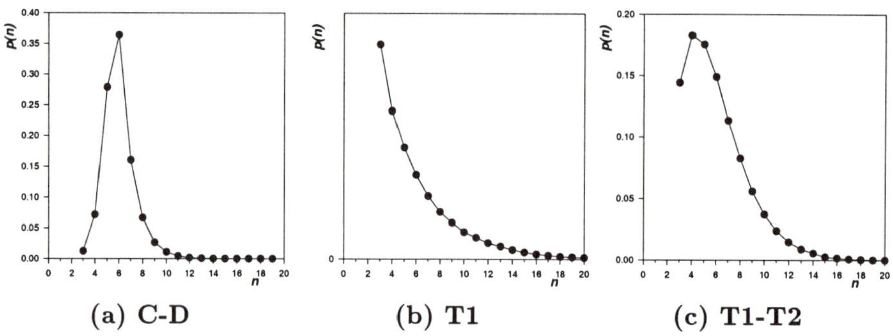

FIGURE 5. Normalized distributions $p(n)$ of the number of edges per cell in the froth generated (a) by applying cell-coalescence and division transformations (C-D simulation), (b) by T1 transformations only (T1 simulation), (c) by T1 and T2 transformations (T1-T2 simulation).

tissues. The number of cells remains finite and stationary over a long time, but it shows large fluctuations.

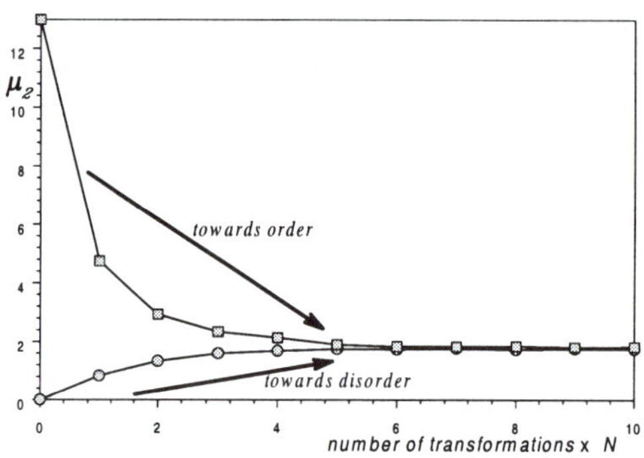

FIGURE 6. Second moment (μ_2) of the edges distribution v.s. the number of elementary topological transformations, in the system generated by cell-division and coalescence transformations (C-D simulation). N is the total number of cells of the starting structure. The system converges toward a steady state which is independent of the starting configuration. If the simulation starts from a very disordered state, the system self-organizes, reducing the disorder by random local transformations.

2.2. Random T1 (T1 Simulation). We start with an hexagonal lattice of N=100172 cells (316 cells high * 317 cells wide) with periodic boundary conditions. The T1 transformation is used for flipping edges chosen at random. We repeat this process 10*N times.

The T1 process conserves the number of cells. A T1 on an edge bounding a 3-sided cell is refused (it would create a cell with two edges). The probability for a T1 to be refused increases during the disordering process, it reaches the value 0.21

at the end of the simulation. The average number of sides per cell is still equal to 6, but most of the cells in the final structure are triangular. This accumulation of 3-sided cells is understandable since T1 on their edges are rejected.

2.3. Random T1 and T2 (T1-T2 Simulation). This is similar to the T1 simulation but, every time a T1 is refused, as in the case where the randomly chosen edge bounds a triangular cell, we make that cell disappear by a T2 transformation. Because the T2 process destroys a cell, the number of cells decreases during the simulation. We found that when 5*N transformations are applied, the number of cells is reduced by half.

By the time 10*N transformations have been performed, we always find that the froth only contains 3 hexagonal cells. This is the minimum possible number of cells tiling a torus (Fig.7). Each cell has two neighbors and 6 edges. It is also the final state of the froth, because a futher T1 would cause a cell to be its own neighbor, and must be refused. No cell disappearance can be performed for the same reason. Note that, in this configuration, there are two families of closed paths of only 4 edges, not bounding a cell, that wind around the torus. Because we are operating with local elementary topological transformations conserving $\langle n \rangle = 6$ (no topological charge created or destroyed), and conserving the torus structure, i.e. the double periodicity of the froth, the final structure must be that of a torus, with minimal number (2) of cells around at least one period. If elementary topological transformations are performed anisotropically, or even at random, the torus may thin down more along one direction than along the other, and reach a hexagonal tiling with more than 3 cells, such as two adjacent copies of Fig.7. At that stage, several T1, causing a cell to be its own neighbor, are rejected. But a few are allowed; together with T2, they will ultimately drive the froth to the final state of Fig.7. This may take a long time, because a lot of moves are rejected.

Starting from a lattice of N=250000 cells (500*500), after performing 5*N transformations, 115149 cells remain. From the distribution obtained (Fig.5.c), we see that 4-sided cells are the most frequent.

3. Discussion

The distribution of the number of edges per cell $p(n)$ characterizes the disorder of the different cellular structures. The normalized distribution $p(n)$ is the probability that a cell with n edges exists. We recall that the first moment of $p(n)$ is $\langle n \rangle = 6$. For the C-D simulation, the distribution $p(n)$ corresponding to a stationary, albeit fluctuating number of cells, has a maximum at $n = 6$ and decreases exponentially for $n \geq 6$ as $p(n) = p_6 exp[-\alpha(n-6)]$ with $p_6 = .37$ and $\alpha = .9$. Likewise, the normalized distribution for the T1 simulation has a maximum at $n = 3$ and $p(n) = p_3 exp[-\alpha(n-3)]$ with $p_3 = .27$ and $\alpha = .3$. For the T1-T2 simulation after 5*N transformations, we have a maximum at $n = 4$ and then the distribution decreases with an exponential tail $p(n) = p_4 exp[-\alpha(n-4)]$ with $p_4 = .84$ and $\alpha = .53$.

The second moment of the distribution $p(n)$, μ_2, can be used to classify the structures. The value of μ_2 increases with the disorder. Both T1 and T1-T2 simulations start with μ_2 equal to 0 (hexagonal lattice). The T1 simulation ends with $\mu_2 = 11.9$, and the T1-T2 simulation with $5*N$ transformations has $\mu_2 - 6.33$. In the case of the CD simulation we have a much lower $\mu_2 = 1.77$. The final distribution of the CD simulation does not depend on the initial one. We have an

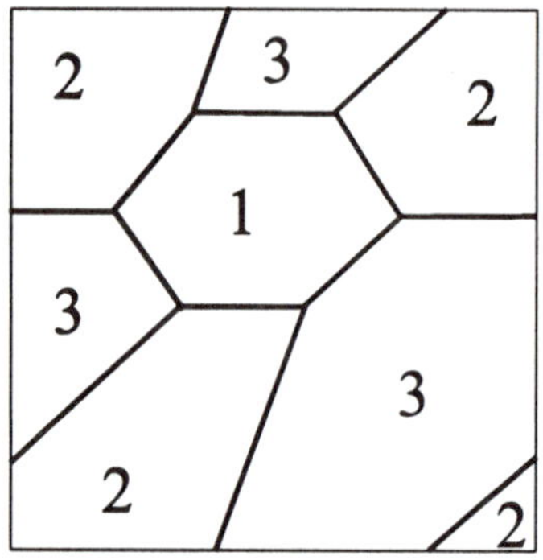

FIGURE 7. Topological representation of the final configuration for the T1-T2 simulation. The structure contains only three 6-sided cells. No further T1 or T2 transformations can be applied on this minimal configurations without generating a cell that is a neighbor of itself. Two families of closed paths of only 4 edges, not bounding a cell, wind around the torus.

increase or decrease of the value of μ_2 towards 1.77 (Fig.6). These gives us a first insight that the T1 produces the most disordered structures, and the CD the most ordered cellular structure. The epidermis of mammals is indeed one of the least disordered natural cellular structure [**6**].

We now present two different approaches to analyze these various structures. First, the shell foliation is a most natural way of looking at the structures and is probably the way a human eye will make sense of them. The second one is very similar in its essence but is based on edges connecting the cells. The approach is closer to how a signal would propagate [**2**] along the edges of the structures. Finally, we combine the two approaches to see how they are correlated.

For each simulation, the cellular systems are characterized by the number of cells in the shell structure (K_j), in the bond trees (F_t), and in the combined structures($N(j,t)$), at the respective distances j and t of the central cell.

4. Shell Foliation

The shell foliation is obtained starting from one cell. This cell is the center of the shell structure and makes up layer 0. All the cells that share an edge with the central cell (neighbors of the central cell) constitute the first layer. The other layers are obtained by repeating the same process, outwards. By definition, all the cells in layer j have at least one neighbor in the layer $j-1$. But some cells in layer j may not have a neighbor in layer $j+1$. These cells are defect inclusions. They fill up the meanders of the $j-1$ shell (the outer boundary of layer $j-1$), which are due to disorder. This additional length of a shell at distance j can be seen as a negative Gaussian curvature [**10**] induced by disorder. Since the cellular structure fills the Euclidean plane, defect inclusions are required to straighten the meanders and eliminate the curvature caused by disorder. This shows very clearly that topology (discrete geometry) has a geometrical foundation (filling Euclidean plane), encoded in Euler's formula for polyhedra.

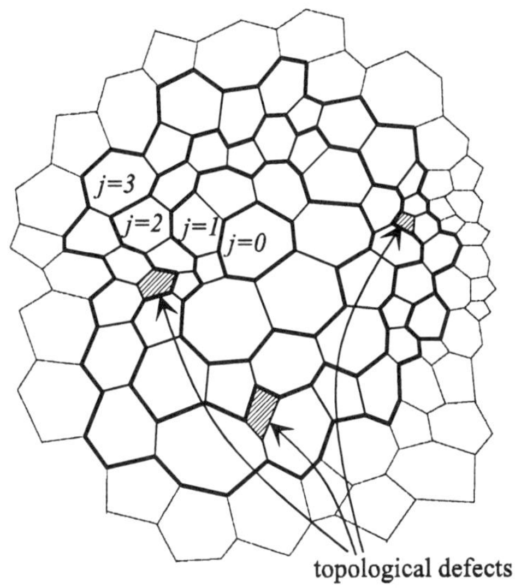

FIGURE 8. The shell foliation. Any froth can be analyzed as structured in a system of closed concentric layers of cells at the same shell distance with respect to the central cell. The cells (in gray) in layer j with neighbors in layers $j-1$ and j only are local topological defects included between the layers.

The results presented here are based on our simulations. For each type of structures obtained by simulation we apply the shell layering process, counting the number of cells in each layer, so that each cell in turn is taken as being the center. The number of cells K_j at shell distance j increases linearly with j after the first few layers, as expected for filling Euclidean space (Fig.9). The rate of increase is different for each simulation considered. We obtain a slope of 19.35 for the C-D simulation, 56.6 for the T1 simulation and 25.7 for the T1-T2 simulation. The slopes are much larger than for an hexagonal lattice (slope 6), or for an Euclidean continuum (slope 2π). This fact indicates the presence of defect inclusions filling up the meandering shells. The wiggling of the concentric layers is reduced and their thickness is increased by these defect inclusions. The proportion of the cells that are defect inclusions is 18% for the C-D simulation, 52% for the T1 simulation, and 38% for T1-T2 simulation. Note that defect inclusions are defined with respect to a given central cell. A cell may be a defect for some centers, and not for others. A triangular cell is always a defect inclusion (except when it is the center).

5. Bond Trees

A bond tree is constructed starting from a central cell, in the same way as with the shell foliation. The central cell and all the cells that share an edge with it are counted as being in layer zero. In order for a cell to be in layer 1 there must be one edge connecting this cell to the central cell. The bond distance is also called "T1 distance" because its value ascribed to the cell tells us the minimum number

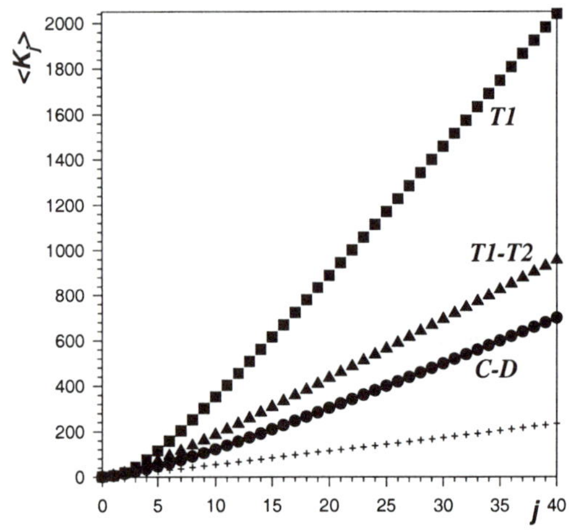

FIGURE 9. Average number of cells $\langle K_j \rangle$ at distance j in the shell layers, for the three simulations C-D, T1 and T1-T2 and for the hexagonal lattice (+ symbol). The average is made over all central cells.

of T1 one needs to perform, so that this cell becomes neighbor with the central cell. The algorithm to compute the bond tree layer is similar to that used for the shell foliation. But we consider vertices instead of cells. We begin by assigning the value 0 to all the vertices of the central cell. Vertices that are connected to one of the vertices of layer $t-1$, and do not already have a number assigned, belong to layer t, and so on recursively outwards. Once all vertices have been assigned a number, a cell belongs to the layer of its vertex with the smallest number. The process is repeated by considering each cell as the root cell. F_t denotes the total number of cells at a bond distance t.

One can compute the mean number of cells in the bond tree $\langle F_t \rangle$, where the averaging is made over the choice of the central cell, as a function of the bond distance t (Fig. 11). For the three sets of simulations, $\langle F_t \rangle$ increases linearly with the distance as t goes to infinity, as expected from filling Euclidean space. We found a slope of 3.58 for the C-D simulation, 4.68 for the T1 simulation and 3.11 for the T1-T2 simulation.

6. Combined Bond and Shell Foliation

In order to extract information on angular fluctuations, it is interesting to study the interplay between bond and shell foliations. Call $N(j,t)$ the number of couples of cells that are simultaneously at a shell distance j and at a bond distance t from each other.

Figure 12 represents the distribution $N(j,t)$ of couples of cells at shell distance j and bond distance t as a function of t for different values of j. The symbols are data taken from the T1 simulation, but the two others types of simulation exhibit a similar behavior. Beyond $j = 10$, the distribution is symmetric, peaked at the mean value $t =$ [illegible], with a variance $\sigma^2 =$ [illegible]

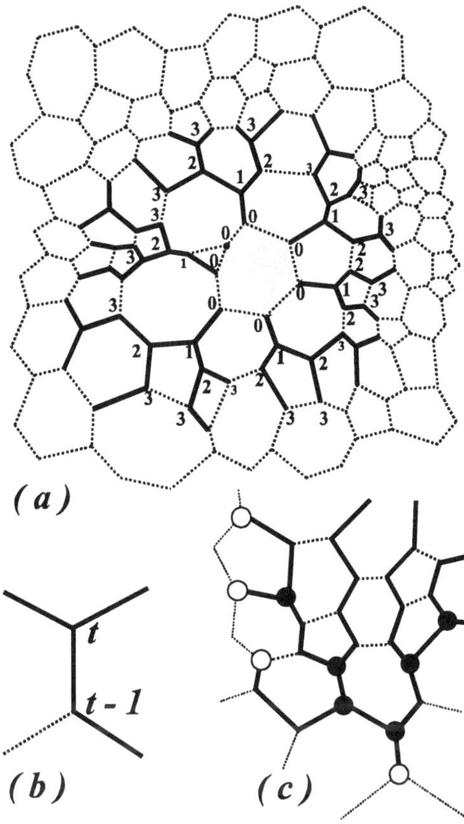

FIGURE 10. (a) The forest of n bond trees (thick lines). (b) In a tree, any vertex at distance t is connected with one bond to a vertex at distance $t-1$. Two of the three cells incident on a vertex at distance t are also incident on a vertex at distance $t-1$. The mean vertex connectivity $\langle z_l \rangle$ tends asymptotically to 2, which is typical of a tree structure spanning vertices distributed uniformly. In (c), vertices with connectivity z=1 and z=3 are labeled with white and black circles respectively.

To go from one cell layer to the next in the bond structure, one needs at least 2 bonds. Thus, the shortest bond path crosses at least $t_{min} = 2j - 3$ vertices to connect the central cell with a cell in the layer j. In a disordered structure, the bond distance t is, in general, much larger than that.

With $\langle v \rangle$, the mean number of additional bounds from one layer to the next, $\langle v \rangle (j-1)$ extra bounds are needed to go from the central cell to a cell at distance j. Therefore, the total number of bonds between layers is $t = (2 + \langle v \rangle)j - 3 - \langle v \rangle$. The number v of extra bonds between layers is a random variable with probability distribution $w(v)$ and mean $\langle v \rangle$. If one makes the reasonable assumption that

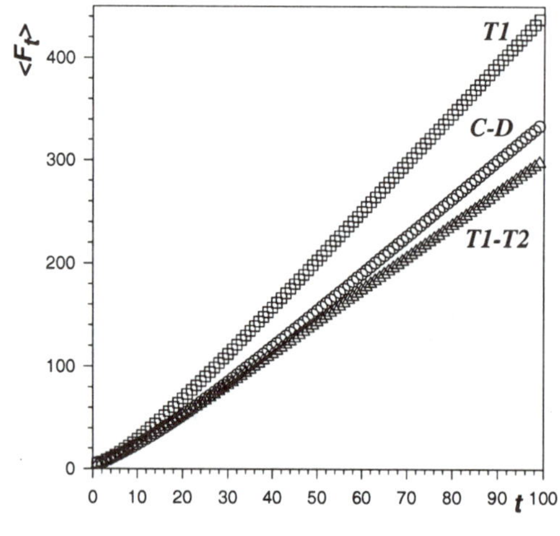

FIGURE 11. Average number of cells $\langle F_t \rangle$ at distance t in the bond trees, for the three simulations C-D, T1 and T1-T2. The average is made over all central cells.

$w(v)$ is Poissonian, the conditional probability $P(t|j)$ of finding a cell at a bond distance t, given that it lies at a shell distance j, is a shifted Poissonian with average $t = \langle v \rangle (j-1) + 2j - 3$ and variance $\sigma^2 = \langle v \rangle (j-1)$. The full lines on Fig.12 represent the theoretical distribution based on the shifted Poissonian:

$$N(j,t) = NK_j P(t|j) = NK_j \frac{e^{-\langle v \rangle (j-1)} [\langle v \rangle (j-1)]^{t-2j+3}}{(t-2j+3)!} \tag{6.1}$$

7. Conclusion

We have simulated the dynamics of 2D random cellular structures with over 10^5 cells. Different degrees of disorder were obtained and characterized as structures reach steady state. When only neighbors switches (T1) are performed, the structures obtained are very disordered, cells are mainly three-sided, but no fractal pattern was observed [8]. Combining T1 transformations with 3-sided cells disappearance (T2) yielded more ordered systems, characterized by a monomodal distribution of the cells sides centered on 4. In an attempt to simulate more realistic systems, we studied the steady state of a system evolving with cells division and cells coalescence. We obtained fairly ordered systems whose statistics are very close to natural 2D foams. For all types of dynamics, the steady states reached were found to be independent of the initial size and statistical state of the foam.

The disorder of the structure is characterized quantitatively by the correlations between arbitrary central cells and cells j layers apart. Two distances were used: the bond distance and the shell distance. When the bond distance is used, a given n-sided central cell defines a forest of n bond trees, rooted on its n vertices. Each tree has an asymptotic vertex coordination $z_\infty = 2$ and the number of vertices at a given distance increases linearly with t.

When the shell distance is used, the number of cells K_j at a shell distance j from a given central cell, increases linearly with j. The rate of increase of K_j

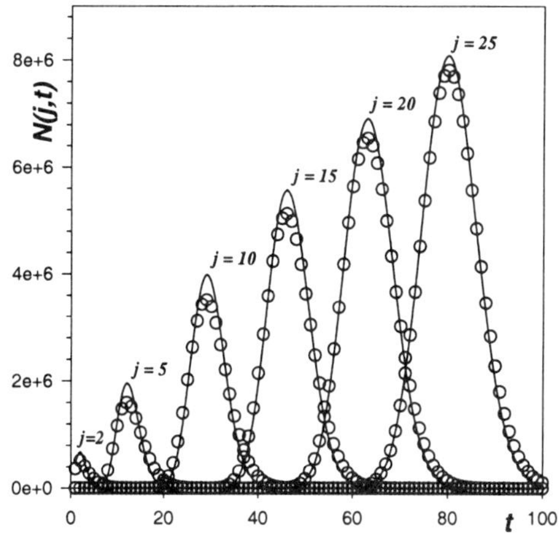

FIGURE 12. Number of couples of cells $N(j,t)$ at shell distance j and bond distance t, as a function of t, for several values of j. The symbols are data taken from the T1 simulation. The full lines are the theoretical solution Eq.6.1 with $\langle v \rangle$=1.39. There is a linear correlation between the most probable value of t, and j.

is 19.35 for the C-D simulation, 56.6 for the T1 simulation and 25.7 for the T1-T2 simulation. The slope of K_j is small for ordered systems (as low as 6 for the hexagonal tiling) and increases as the disorder of the system increases. It therefore appear to be a good measure of the degree of disorder of a system.

The two topological structures, bond trees, and shell foliation, are intimately related. The combined structure was studied in terms of the number $N(j,t)$ of cells which are simultaneously at a shell distance j and at a bond distance t from a given central cell. We find that the combined structure is well described by only one parameter $\langle v \rangle$, where $\langle v \rangle + 2$ is the mean number of bonds between adjacent layers.

We have also performed simulations of the C-D type but more finely adapted to the dynamic of the epidermis [**4**]. In particular, we have obtained the very restrictive condition on division and coalescence for the number of cells of the tissue to remain stationary.

The simulation program is available on our web site http://maxent.u-strasbg.fr/ or upon request to the authors.

References

[1] J.W. Alexander, The Combinational Theory of Complexes, *Ann. Math.*, **31** (1930), 292–320.

[2] T. Aste, Random Walks on Disordered Networks, *Phys. Rev. E*, **55** (1997), 6233.

[3] T. Aste, D. Boose, and N. Rivier, From one Bubble to the Whole Froth: A Dynamical Map, *Phys. Rev. E*, **53** (1996), 6181–6191.

[4] B. Dubertret, T. Aste, H.M. Ohlenbusch, and N. Rivier, Two-Dimensional Froths and The Dynamics of Biological Tissues, *Phys. Rev. E*, **58** (1998), 6368–6379.

[5] B. Dubertret, and N. Rivier, The Renewal of the Epidermis: A Topological Mechanism, *Biophys. J.*, **73** (1997), 38–44.

[6] B. Dubertret, *Études théorique des systèmes cellulaires désordonnés. Application à la modélisation des tissus épithéliaux*, Ph.D. thesis, Université Louis Pasteur, Strasbourg, 1998.

[7] d'A.W. Thompson, *On Growth and Form*, 2nd ed., Cambridge Univ. Press, New-York, 1942.
[8] G. Le Caër, and R. Delannay, Topological Models of 2d Fractal Cellular Structures, *J. Phys. I France*, **5** (1995), 1417–1429.
[9] H.M. Ohlenbusch, T. Aste, B. Dubertret, and N. Rivier, The Topological Structure of 2d Disordered Cellular Systems, *Euro. Phys. J. B*, **2** (1998), 211–220.
[10] N. Rivier, and T. Aste, Curvature and Frustration in Cellular Systems, *Phil. Trans. R. Soc. London A*, **354** (1996), 2055–2069.
[11] N. Rivier, *Jigsaw Crystallography*, Unpublished notes, 1985, Seminar, Imperial College, 1986.
[12] J.-P. Troadec, *Échantillon construit grain par grain (monodiamètre). Triangulation complète*, Unpublished notes, 1986.
[13] D. Weaire, and N. Rivier, Soap, Cells and Statistics – Random Patterns in Two Dimensions, *Contemp. Phys.*, **25** (1984), 59–99.

Department of Computer Science, Worcester Polytechnic Institute, Massachusetts, 01609-2280
E-mail address: helgo@cs.wpi.edu

LDFC, Universite Louis Pasteur, Institut de Physique, 3 rue de l'Universite,F 67084 Strasbourg cedex
E-mail address: nick@ldfc.u-strasbg.fr

LDFC, Universite Louis Pasteur, Institut de Physique, 3 rue de l'Universite,F 67084 Strasbourg cedex
E-mail address: aste@ldfc.u-strasbg.fr

LDFC, Universite Louis Pasteur, Institut de Physique, 3 rue de l'Universite,F 67084 Strasbourg cedex
E-mail address: benoit@ldfc.u-strasbg.fr

DIMACS Series in Discrete Mathematics
and Theoretical Computer Science
Volume **51**, 2000

Group and Graph Theoretical Perspectives on the Structures of Large Icosahedral Cages

Charles M. Quinn, David B. Redmond, and Patrick W. Fowler

ABSTRACT. The Mollweide projection, in combination with local inversion operations, is used to explore the geometrical diversity of the polyhedral orbits of the icosahedral point group. Using the example of a large icosahedral cage, the hypothetical fullerene, C_{320}, the "orbit by orbit" based procedure is shown to be a useful general technique for the analysis and construction of polyhedral cages with large numbers of vertices.

1. Introduction

The discovery of the cluster molecules C_{60} and C_{70} [**1, 2**] and the detailed characterization of the class of cage structures known as the fullerenes [**3**], in recent years, have emphasised the need for a two-dimensional mapping procedure, in which the three-dimensional details of such cage structures are evident in the plane. We proposed [**4, 5, 6**], some years ago, the use of the geographical projection, known as the Mollweide Projection [**7, 8, 9**], as a very useful mapping of three-dimensional detail onto the plane and as the "Spherical Shell Technique", this device has been applied in chemistry, with considerable success in Group Theory [**10, 11, 12**] and Qualitative Molecular Orbital Theory [**13, 14**].

In this paper, we propose, now, the use of the same device as a general procedure for the representation of polyhedral cage structures and, indeed, the construction of all such cages as sums of the projections for particular choices for the geometries of contributing orbits. Throughout, the example of the icosahedral group, I_h, is used because of its importance in the definition of the structures of many fullerene cages.

The construction of a Mollweide projection for any object that can be inscribed in a unit sphere is illustrated in Figure 1, using the example of the cube. In the diagram, on the left, the conventional construction is drawn and the inscribing sphere is identified by two great circles containing pairs of coordinate axes as particular radii. Then, in the second diagram, the two-dimensional map of this structure as a Mollweide projection is presented. Note how one great circle, *2684*, defines the limiting boundary of the front hemisphere, the middle circle of the projection. Then, the other, *1573*, defines the outer elliptical boundary, which can be considered to

1991 *Mathematics Subject Classification.* Primary 92E10; Secondary 05E20.

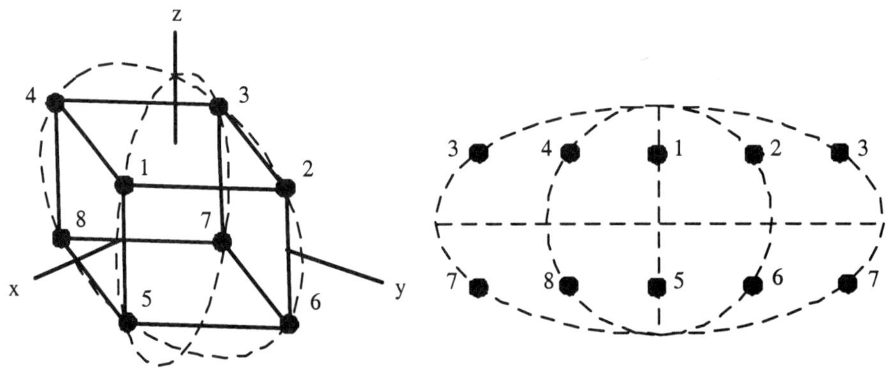

FIGURE 1. The construction of the Mollweide projection for the eight vertices of the cube, as described in the text. Note that, in this diagram, the vertex symbols for the vertices imagined to be split in two by the construction process *(vertices 3 and 7)*, for clarity, are marked with the whole symbols and so appear at both the left and right boundaries, the half-great circle from $+e_z$ through $-e_x$ to $-e_z$.

result from a splitting of the unit sphere on the arc from e_z through $-e_x$ to $-e_z$ and the smearing out of the surface of the unit sphere onto the plane to form the projection and the unsplit half of this circle is the vertical line from $+e_z$ to $-e_z$. On the projection the xy-plane is distinguished as the horizontal line running from $-e_x$ through $-e_y$, e_x, e_y and back to $-e_x$.

The traditional structural net for any polyhedron is to be found in standard textbooks on geometry [**15**] and, on any such a map, the vertices of the cage polyhedral structure can be drawn. As a graph theoretical exercise, the map of the structure can be built on the net by propagating a representative patch [**16, 17**] over the surface of the unit sphere. The procedure is shown in Figure 2, in this form, and then, in the diagrams of the second column, each step is represented as the corresponding Mollweide projection. Comparisons of the diagrams in the bottom row of Figure 2, for the 120-vertexed cage of the regular orbit of the I_h group demonstrate the superiority of the Mollweide projection device. All the structural details over the complete surface of the unit sphere are presented on an equal footing. Moreover, all the effects of group operations as co-ordinate transformations on the vertices, can be examined, over the net of the structure, and individual features emphasised at will.

2. Pathways Preserving Point Group Symmetry

In particular, it is straightforward to present pictorially all pathways on the unit sphere that preserve the point symmetry of a given structure and so to generate all possible structural isomers, or all the orbits of particular geometries corresponding to a cage of the given symmetry. The key to this analysis, for single or multi-orbit cages, is the division of the structure into local sets of vertices defining, in each orbit, the particular symmetry about the rotational axes.

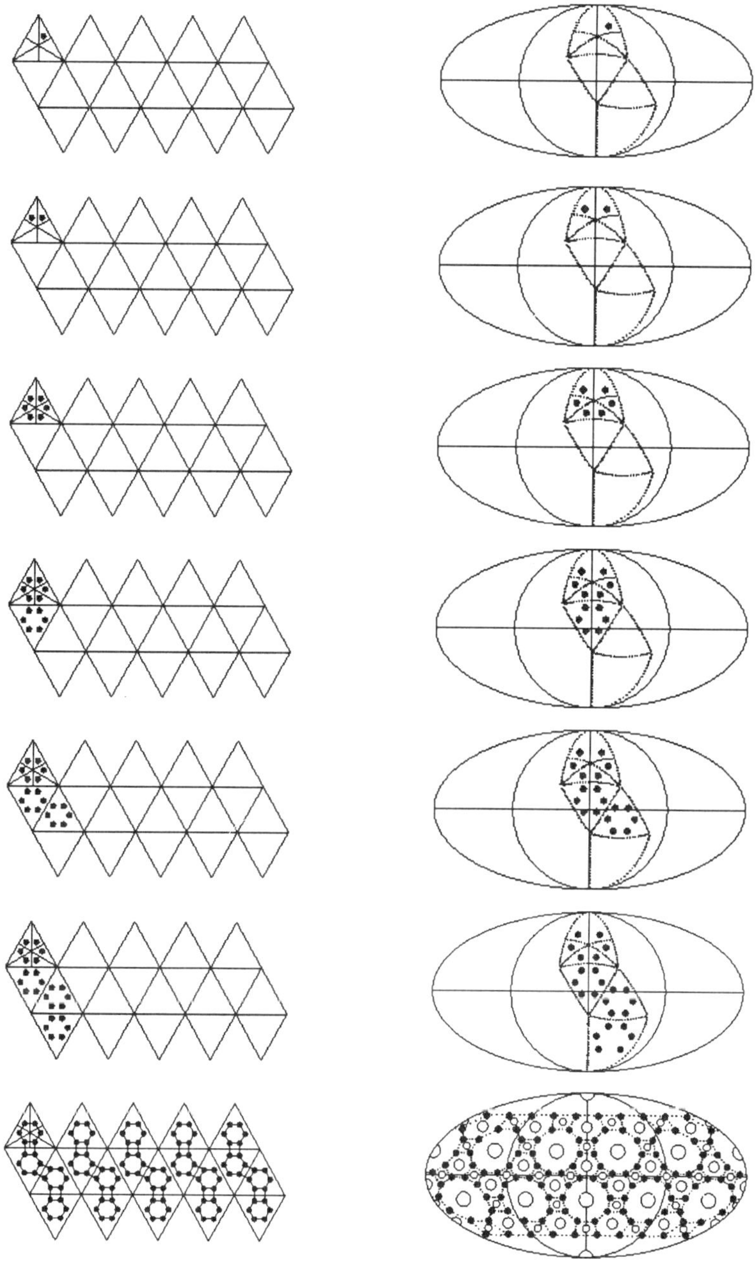

FIGURE 2. Steps in the construction of the 120-vertex decorated icosahedron by the actions of symmetry operations on the fundamental scalene spherical triangle. On the left the actions of various operations are shown on a net, on the right, the same sequence on the Mollweide map. The bottom row shows the completed great rhombicosidodecahedral structure, with the poles of C_5, C_3 and C_2 rotations indicated by open circles on the projection.

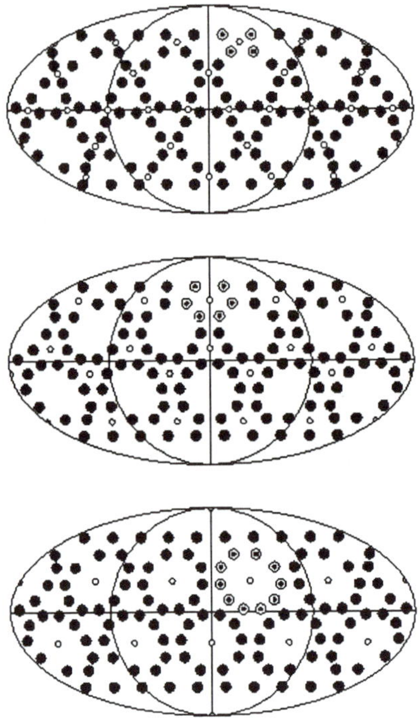

FIGURE 3. Mollweide projections of the regular 120-vertexed orbit of the I_h point group on the unit sphere. In the diagrams, the vertices are grouped into local sets about the rotational axes, C_2, C_3 and C_5 of 4, 6 and 10 points. The locations of these sets of rotational axes, on the unit sphere, are marked as the open circles on the appropriate diagrams [*compare Figure 2*].

Such a division is shown in Figure 3 for the vertices of the regular orbit cage of I_h point symmetry. One can choose 30 sets of 4 vertices about the locations of the C_2 on the unit sphere, 20 sets of 6 vertices about the C_3 axes and 12 sets of 10 vertices about the C_5 axes. Co-ordinate transformations that can be carried out on this regular orbit cage, which preserve the point symmetry, involve contractions or expansions of these local sets of vertices about the central symmetry axes. For example, Figure 4 is a simple illustration of the nature of the local transformation for, say, a triangular face of a truncated spherical polyhedron. It shows successive stages of the contraction/inversion of a triangular set onto, and finally, through the associated local C_3 axis.

Since such motions are restricted to the unit sphere, there are two degrees of freedom available and these can be described, for example, by carrying out transformations locally involving the vertices about two sets of the locations of the rotational axes.

2.1. The Lower Order Geometric Orbits of I_h. Any of the diagrams of Figure 3 can be used to identify all the orbits of lower order, by carrying out local inversions of the sets of vertices about the rotational axes of each kind.

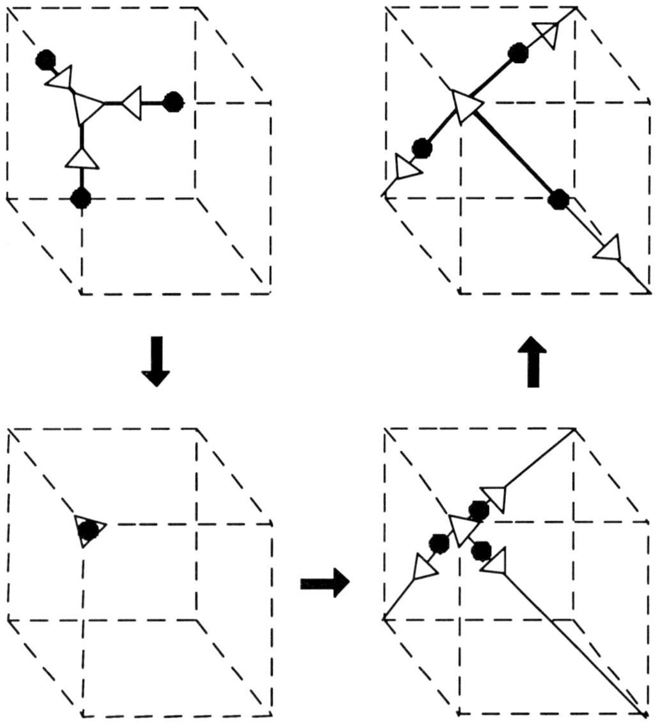

FIGURE 4. The local contraction/inversion operation illustrated for the three vertices of the O_{12} orbit of the tetrahedron or an O_{24} orbit of the cube, about a typical C_3 axis.

For example, consider the sets of ten vertices, identified about each C_5 axis of the regular orbit in the third projection of Figure 3. There are two straightforward transformations, which lead to orbit cages with fewer vertices. It is clear that contraction of each set of ten vertices, locally onto the C_5 axes leads to the lowest order orbit, O_{12}, of the I_h point group, with site group, C_{5v}, at each of the twelve C_5 axis points. Equally, too, a small local expansion of the rings of decagons leads to the coalescing of adjacent pairs of vertices and the formation of the truncated dodecahedron [**15**], which structure is representative of the O_{60} orbit of the point group, with site group C_s.

These transformations are shown in Figure 5. The second diagram is the projection of the regular orbit geometry with the locations of the rotational axes identified as before. The first diagram of the figure is the net of the O_{12} orbit obtained by local contractions onto the C_5 axes points. The third diagram of the figure is the net of the dodecahedron obtained, when the expansion about the C_5 axes leads to the coincidence of two vertices in every set of four about each C_2 axis, and again is a particular geometry of the O_{60} orbit of the point group.

Apart from the trivial orbit of I_h, the single vertex at the origin, with site group I_h, there remain to be identified, in this construction, the O_{20} orbit with site group C_{3v} and the O_{30} orbit with site group C_{2v}. Figure 6 shows that these follow directly by the carrying out of local contractions about the C_3 and C_2 axes.

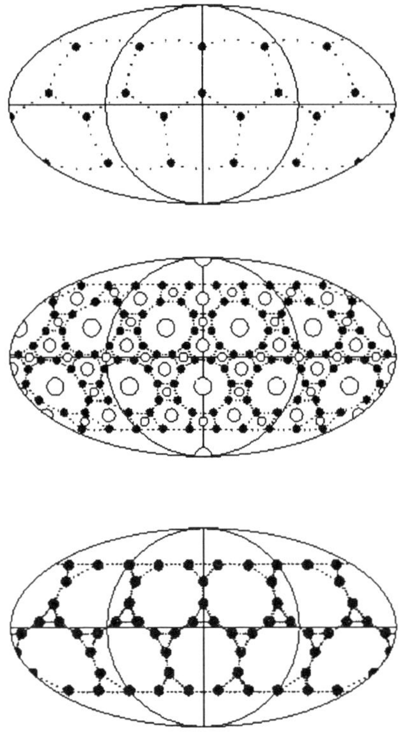

FIGURE 5. Generation of the O_{12} and O_{60} orbits of the I_h point group by contractions (O_{12}) or expansions (O_{60}) about the C_5 axes on the regular orbit in the middle diagram of the figure [*compare Figure 1*].

Contraction onto the C_3 axes yields the truncated dodecahedral net of the O_{20} orbit, the regular dodecahedron of 20 vertices, and contraction onto the C_2 axes yields the icosidodecahedron of 30 vertices, the net of the O_{30} orbit of the point group.

2.2. Fullerenes. The fullerenes are tri-valent carbon cages and a large number of fullerenes, including the archetypal one, C_{60}, exhibit the icosahedral point symmetry, I_h. The C_{60} fullerene structure is the cage of 12 pentagons and twenty hexagons arranged in I_h point symmetry on the unit sphere with no pentagon adjacencies.

The O_{60} orbit, identified in Figure 5, is the geometry associated with a hypothetical isomer of C_{60} [**18**]–[**21**] and is, in fact, only one of an infinite set of possible geometries of 60 vertices exhibiting I_h point symmetry. This is so, because the mappings, corresponding to the local inversions, described previously, represent continuously variable pathways preserving symmetry. Included in these pathways are two other particular geometries, one the C_{60} fullerene structure and, the other, the rhombicosidodecahedron [**15**], a geometry that of itself is four-valent, but which is found as a component in many other tri-valent fullerene structures.

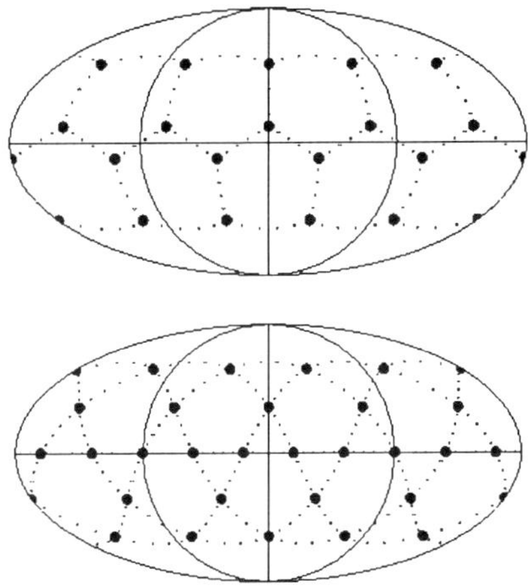

FIGURE 6. Identification of the remaining non trivial orbits, O_{20} and O_{30} of the I_h point group by local contractions of the sets of two and three vertices of the truncated dodecahedron onto the C_3 and C_2 axes of the truncated dodecahedron drawn as the second projection in Figure 3.

These statements are demonstrated in Figure 7. From the regular orbit, local inversions are carried out to generate, in turn, the tri-valent C_{20} fullerene, the four-valent geometry of the O_{60} orbit and, in the final diagram of the figure, the array of pentagons and hexagons corresponding to the cage of the C_{60} fullerene[1].

The rhombicosidodecahedral [**15**] geometry of the O_{60} orbit cage in Figure 7 occurs commonly in fullerenes when these consist of combinations of the geometrical orbits of the icosahedral point group and the simplest case of this is illustrated in Figure 8. The net of the fullerene C_{60}, is included, in the figure, to emphasise that there is only one O_{60} orbit of the point group, but, that it can be realised in different geometries as required by, in this case, the requirement that all vertices be three-valent.

The number of degrees of freedom in an orbit is less than or equal to 2 and is usually clear by inspection. As a vertex on a rotational axis is fixed, one, in a mirror plane, has, at most, freedom of movement along a great circle and a point in the regular orbit is in the general position and has two angular degrees of freedom, as mentioned earlier. In a group-theoretical description of the vertex positions, the number of degrees of freedom of a set described by the permutation representation $\Gamma_\sigma(v)$ is one less than the number of copies of the totally symmetric representation, Γ_1, occurring in the vibrational representation

$$\Gamma_{vib} = \Gamma_\sigma(v) \times \Gamma_{trans} - \Gamma_{trans} - \Gamma_{rot}$$

where Γ_{trans} and Γ_{rot} are the translational and rotational representations in the appropriate point group (the difference of *"one"* allows for the other motion, the

[1]These structures are found to repeat at various angles of local inversion about the rotational axes. Furthermore, the pathways described herein are only one set of many which can be used to generate the various cages corresponding to the different geometric orbits of the point group.

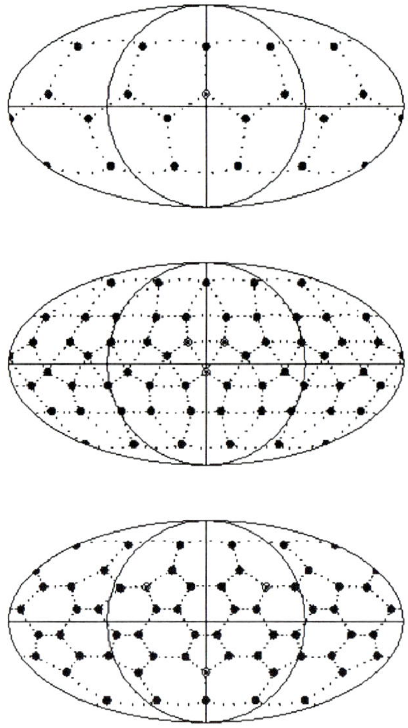

FIGURE 7. The generation of other geometrical arrangements of the O_{60} orbit of I_h, of interest in fullerene chemistry. (i) Starting from the 120-vertex regular orbit [*see Figure 2 bottom right*] contraction of local sets of vertices onto the C_3 axes, yields the O_{20} orbit; (ii) inversion generates the 60-vertexed rhombicosidodecahedron [**15**] to be found, for example, in the hypothetical fullerene, C_{80}; (iii) further inversion leads to the archetypal fullerene, C_{60}, the third diagram of the figure.

purely radial expansion or contraction of the orbit geometry, absent in the projection onto the unit sphere). The number of degrees of freedom, also, is equal to the number of independent edge lengths for a spherical polyhedral representation of the orbit, again restricted to the unit sphere.

3. Large Clusters

This freedom to choose a particular geometry for the representation of an orbit as a cage of vertices, is fundamental to the construction of multi-orbit cages, especially for clusters containing more vertices than are to be found in the regular geometric orbit. Consider the hypothetical cage, C_{320}, which is drawn in Figure 9 as a three-valent hypothetical fullerene of I_h point symmetry with a single carbon-carbon bond length. This is decomposed, straightforwardly, into the component orbits O_{20}, O_{120} and O_{60}, this last orbit occurring three times corresponding to different local expansions of the pentagons of vertices about the C_5 axes as shown in the figure.

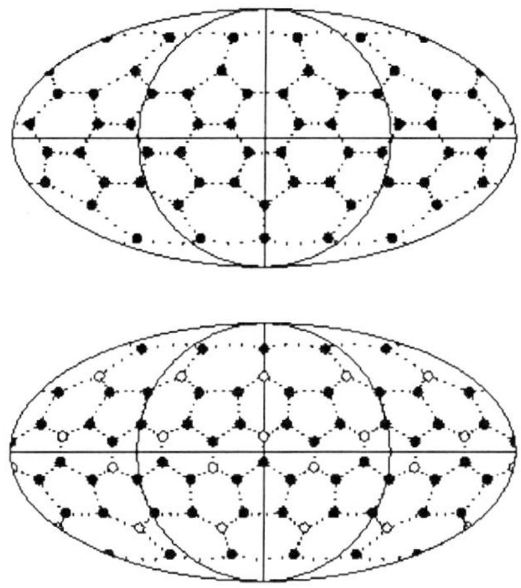

FIGURE 8. Mollweide projections of the archetypal fullerene, C_{60}, a carbon cage for a particular choice of the geometrical net of the O_{60} orbit of the point group I_h. The fullerene, C_{80}, formed by combination of the O_{20} vertices (o) with the vertices of the O_{60} orbit net obtained, as a rhombicosidodecahedron, by local inversion of the C_{60} pentagons through the C_5 axes positions on the unit sphere.

Note that the two copies of 4-valent O_{60} geometry differ only in the distances of the vertices of the pentagons from the locations of the C_5 axes, the centres of the pentagons, while the third 60-vertexed orbit is the isomer geometry of C_{60}.

Of course, restriction of any multi-orbit structure such as C_{320} to the unit sphere, often may be an approximation, since there is no general requirement that each orbit should be on a single sphere about the origin. Corrugated structure can occur and such corrugations increase the numbers of isomers because the third degree of freedom of radial position becomes available.

4. Conclusions

In conclusion, it is to be noted that the notion of continuous variation of a polyhedral orbit has several applications in theoretical chemistry. The ideas of correlation diagrams, whereon electronic energy levels of a molecular structure are tracked as the molecule is collapsed to a united-atom limit, or of Walsh diagrams [**22**], whereon orbital energies vary with a geometric parameter, have natural Mollweide counterparts for spherical cage molecules.

A rationalisation of the typical electron deficiency of fullerene molecules [**23**], for example, relies on the correlation between two extreme limits (30 isolated pairs and 12 isolated pentagonal rings) and the equilateral truncated icosahedral representation of the O_{60} orbit of I_h. In future work, with Mollweide projections, it is intended to map out the *"basins"* of parameter space corresponding to the topologically distinct polyhedral forms of O_{120} in I_h and connect this to a parametric model of π-electronic structure of model carbon cages. Parallel studies in the cubic groups, based on O_h, also, will be of interest.

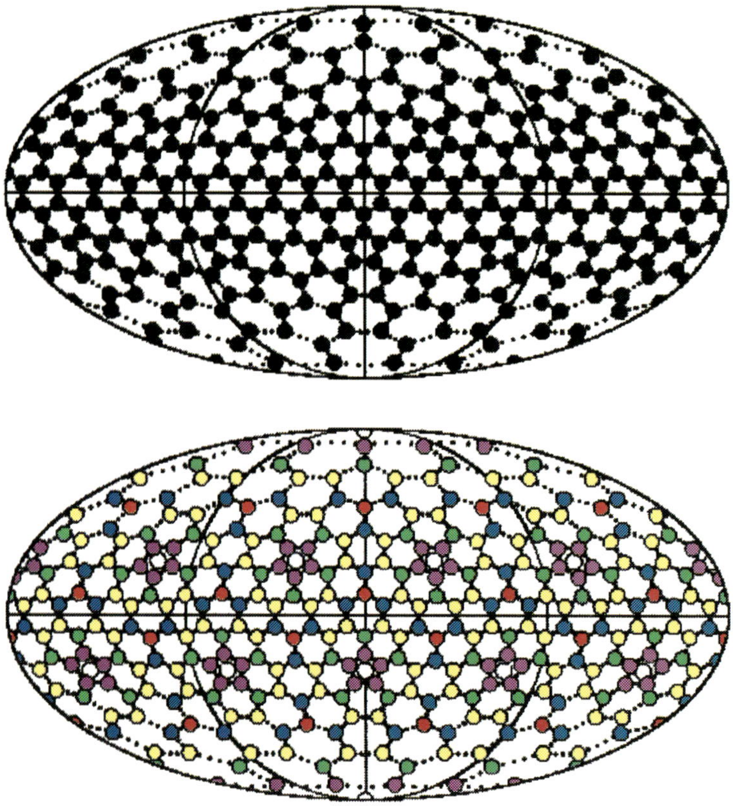

Figure 9. The net of a hypothetical fullerene, C_{320}, of I_h point symmetry and, then, in the second projection, with each contributing orbit identified. Note how the complete net can be constructed from one O_{20} orbit (●) three O_{60} orbits (●, ●, ●) and one O_{120} orbit (○). Of the O_{60} orbits, the first (●) is a geometry of the dodecahedron of Figure 3 and the other two (●,●) are geometries of the rhombicosidodecahedron of Figure 5, chosen so that the 320-point net is a single-bond length structure. The C_5 axes locations are marked as circles on the second diagram. Local expansions and contractions about these axes can obtain all the geometries shown. Note, too, that only for the case of the regular orbit can the rectangles of local vertices be seen about the locations of the C_2 axes. Because for the other cases, all lower order orbits, the site groups at the vertices are greater than the group of the identity and, so, some of the regular orbit vertices are coincident.

References

[1] H.W. Kroto, J.R. Heath, S.C. O'Brien, R.F. Curl, and R.E. Smalley, "C_{60}: Buckminsterfullerene", *Nature*, **318** (1985) 162–163.

[2] W. Krätschmer, L.D. Lamb, K. Fostiropoulos, and D.R. Huffman, "Solid C_{60} – A New Form of Carbon", *Nature*, **347** (1990) 354.

[3] P.W. Fowler, and D.E. Manolopoulos, *An Atlas of Fullerenes*, Clarendon Press, Oxford, 1995.

[4] C.M. Quinn, D.B. Redmond, and J.G. McKiernan, "Mollweide Projections: Molecular Orbital Symmetries on the Spherical Shell", *J. Chem. Education*, **61** (1984) 569–572.

[5] C.M. Quinn, D.B. Redmond, and J.G. McKiernan, "The Spherical Shell Technique for Molecular Wavefunctions: Some New Theorems in Group Theory", *J. Chem. Education*, **61** (1984) 572–579.

[6] D.B. Redmond, C.M. Quinn, and J.G. McKiernan, "Point Group Symmetries on the Spherical Shell", *J. Chem. Soc., Faraday II*, **79** (1983) 1791–1809.

[7] P.W. McDonnell, Jnr., *Introduction to Map Making*, Marcell Decker, New York, 1979.

[8] A.H. Robinson, and R.D. Sale, *Elements of Cartography*, John Wiley, New York, 1969.

[9] D.H. Maling, *Co-Ordinate Systems and Map Projections*, Phillip, London, 1973.

[10] C.M. Quinn, D.B. Redmond, and J.G. McKiernan, "Mollweide Projections: Molecular Orbital Symmetries on the Spherical Shell, Tetrahedral and oOther Symmetries, and δ Orbitals in Metal Clusters", *Inorganic Chemistry*, **22** (1983) 2310–2315.

[11] P.W. Fowler, and C.M. Quinn, "σ, π and δ-Representations of the Molecular Point Groups", *Theoretica Chimica Acta*, **70** (1986) 333–350.

[12] C.M. Quinn, "Normal Modes and Symmetry Coordinates on the Spherical Shell", *Journal of the University of Kuwait*, **15** (1988) 255–267.

[13] D.M.P. Mingos, and R.L. Johnston, in *Structure and Bonding*, **68**, Springer-Verlag, Berlin, 1987.

[14] R.L. Johnston, Mathematical cluster chemistry, in *Structure and Bonding*, **87**, Springer-Verlag, Berlin, 1997. 1–34.

[15] H.M. Cundy, and A.P. Rollet, *Mathematical Models*, OUP, Oxford, England, 1951.

[16] Au Tang, Fu Qiang Huang, Qian Shu Li, and Ruo Zhuang Liu, "Hückel Treatment of Carbon Species with Icosahedral Symmetry", *Chem. Phys. Letters*, **227** (1994) 579.

[17] Fu Qiang Huang, and Au Chin Tang, "Symmetries of Closed-Shell Icosahedral Fullerenes", *Chem. Phys. Letters*, **248** (1996) 361–365.

[18] P.W. Fowler, and J. Woolrich, "π Systems in Three Dimensions", *Chem. Phys. Letters*, **127** (1986) 78–83.

[19] A.J. Stonen, and D.J. Wales "Theoretical Studies of Icosahedral C_{60} and Some Related Species", *Chem. Phys. Letters*, **128** (1986) 501–503.

[20] T.G. Schmalz, W.A. Seitz, D.J. Klein, and G.E. Hite, "C_{60} Carbon Cages", *Chem. Phys. Letters*, **130** (1986) 203–206.

[21] T.I. Shibuya, and M. Yoshitani, "Two Icosahedral Structures for the C_{60} Cluster", *Chem. Phys. Letters*, **137** (1987) 13–16.

[22] A.D. Walsh, "The Electronic Orbitals, Shapes and Spectra of Polyatomic Molecules, Parts 1-10", *J. Chem. Soc.*, (1953) 2260–2331.

[23] P.W. Fowler, and A. Ceulemans, "Electron Deficiency of the Fullerenes", *J. Chem. Phys.*, **99** (1995) 508–510.

Department of Chemistry, The National University of Ireland Maynooth, Maynooth, Co. Kildare, Ireland

E-mail address: cm.quinn@may.ie

Department of Mathematics, The National University of Ireland Maynooth, Maynooth, Co. Kildare, Ireland

E-mail address: dredmond@maths.may.ie

Department of Chemistry, The University of Exeter, Stocker Road, Exeter, England, U.K.

E-mail address: P.W.Fowler@exeter.ac.uk

DIMACS Series in Discrete Mathematics
and Theoretical Computer Science
Volume **51**, 2000

Use of Path Matrices for a Characterization of Molecular Structures

Milan Randić, Xiaofeng Guo, and Sol Bobst

ABSTRACT. Is there a simple structural invariant, a single number with an apparent structural interpretation, that can characterize molecular graphs uniquely? We review briefly several attempts to derive a unique characterization of trees by a single number. A recently introduced branching topological index derived from path matrix of a tree appears as a promising candidate for the single number characterization of trees sought. A selection of trees examined that could offer a counterexample to uniqueness of the index failed to produce a duplicate characterization.

1. Introduction

Topological indices introduced for application in structure-property and structure-activity studies show rather limited discrimination among molecular graphs. This is manifested already for the well known and widely used topological indices like the Wiener number W [**1**], the Hosoya topological index Z [**2**] and the connectivity index χ [**3**]. One finds the smallest trees having the same Wiener index among heptane isomers ($n = 7$). Similarly the smallest trees having the same Hosoya Index Z have $n = 7$ vertices, and the smallest trees having the same connectivity index are found among octane isomers ($n = 8$).

One of the first indices that has shown greater discrimination was Balaban's J index [**4**], which can be derived from the row sums of the graph distance matrix analogous to the construction of the connectivity index from the row sums of the adjacency matrix. Among graphs having $n = 12$ vertices there are several pairs of trees having the same Balaban J index (Fig. 1) [**5, 6**]. Indices with much higher discrimination power were subsequently developed. The first among these indices of quite impressive discriminatory power was the molecular ID number, which represents the sum of weighted paths in a graph [**7**]. Szymanski and coworkers [**8**] have tested all trees having $n = 20$ vertices or less. They found the smallest trees having identical ID number to have $n = 15$ vertices (Fig. 2). That means that among 4347 isomers of $C_{15}H_{32}$ there was one pair of trees having the same ID

1991 *Mathematics Subject Classification.* 05C05, 05C38, 05C50, 05C60.

We thank Professor D. J. Klein (Galveston, TX) and Dr. M. Vracko (Ljubljana, Slovenia) for reviewing the manuscript and suggesting improvements in the presentation. We also thank NSFC (National Natural Science Foundation of China) and CSC (China Scholarship Council) for supporting this research and the visit of X. Guo to Drake University. This work was supported in part by the grant J1-8901-0104-97 from the Ministry of Science and Technology of Slovenia.

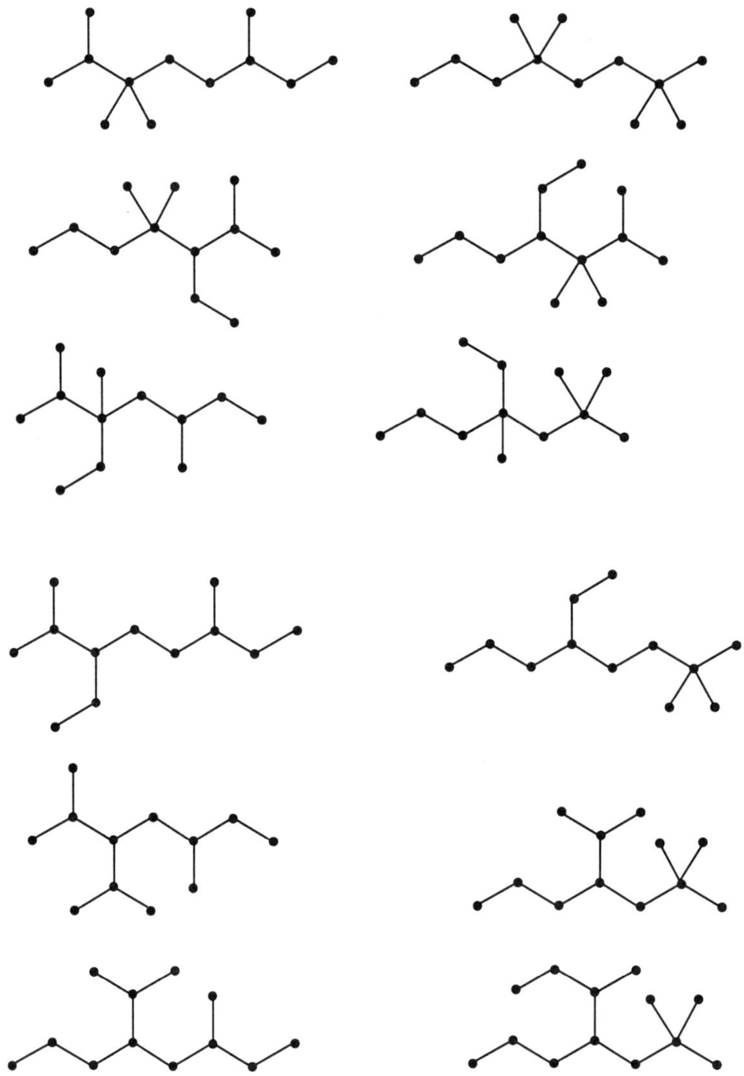

FIGURE 1. The smallest trees having the same Balaban J index

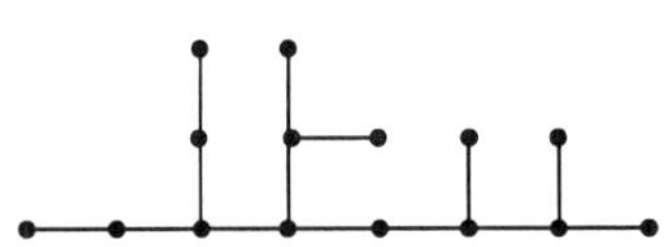

FIGURE 2. The smallest pair of trees (having $n = 15$ vertices) which have identical ID number

number. A similarly constructed quantity which use different prime numbers for the weights of different bond (m, n), the so called prime ID number [**9**], shows even greater discrimination. The smallest trees having identical prime-based ID number as shown by Szymanski and coworkers have $n = 20$ vertices [**10**] (Fig. 3). In Table 1 we list path numbers for the symmetry inequivalent vertices for the two graphs of

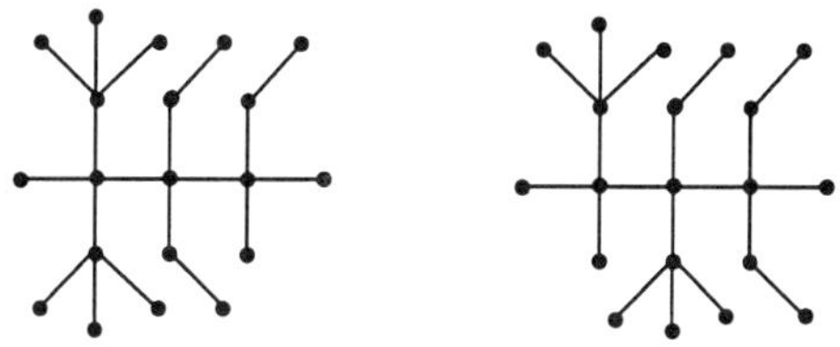

FIGURE 3. The smallest pair of trees (having $n = 20$ vertices) which have identical prime number ID characterization

Fig. 3. When all paths are counted the two graphs produce the same molecular path sequence: 19, 33, 45, 52, 35, 6. Path numbers of length k are obtained by counting all paths of length k in a molecule. Hence, p_0 gives the number of atoms, p_1 gives the number of bonds, p_2 gives the number of pairs of consecutive bonds, etc. The path sequence is defined as the list of path counts: $p_0, p_1, p_2, p_3, p_4, p_5, \ldots$.

TABLE 1. Path numbers for the two trees of Fig. 3 having $n = 20$ vertices and the same prime number ID. Only path numbers for symmetry inequivalent vertices are listed.

Vertex	Paths	Vertex	Paths
1	1, 3, 3, 6, 5, 1	1	1, 3, 3, 3, 7, 2
2	4, 3, 6, 5, 1	2	4, 3, 3, 7, 2
3	1, 3, 9, 5, 1	3	1, 3, 6, 7, 2
4	4, 9, 5, 1	4	4, 6, 7, 2
5	1, 1, 3, 7, 7	5	1, 3, 3, 7, 5
6	2, 3, 7, 7	6	4, 3, 7, 5
7	4, 8, 7	7	4, 10, 5
8	1, 3, 4, 5, 6	8	2, 3, 9, 5
9	4, 4, 5, 6	9	1, 1, 3, 9, 5
10	2, 3, 3, 5, 6	10	1, 1, 3, 4, 7, 3
11	1, 1, 3, 3, 5, 6	11	2, 3, 4, 7, 3
		12	4, 5, 7, 3
		13	1, 3, 5, 7, 3
Molecule		**Molecule**	
19, 33, 45, 52, 35, 6		19, 33, 45, 52, 35, 6	

Another topological index of high discriminating power was reported by Kier and Hall [**11**]. The index, referred to as the total topological index τ, can be viewed as the Wiener sum (i.e., the sum of all matrix elements above the main diagonal) of a matrix whose entries are the components used in the construction of the connectivity indices [**12**] and valence connectivity indices [**13**]. Hence, the components used for construction of the index have a structural interpretation, e.g., in the case of alkanes the index τ is analogous to the molecular ID number. Hu and Xu [**14**] investigated the total topological index τ and confirmed its great discriminatory power. They found the smallest pair of alkane trees having the same τ to have $n = 19$ vertices (Fig. 4). Additional pairs of alkane trees having $n = 20$ vertices were also reported, one of which is shown in the lower part of Fig. 4. Each pair of graphs of Fig. 4 has also the same molecular path sequences.

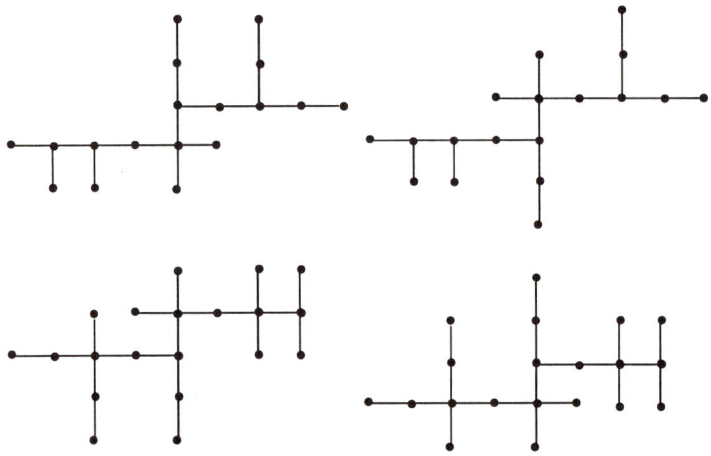

FIGURE 4. The smallest trees (having $n = 19$ and $n = 20$ vertices) which have the same total topological state τ

Hu and Xu [**14**] also introduced an extended adjacency matrix (EA matrix) which is then used for the construction of another matrix (EA^* matrix in their notation) obtained from the sum of the first $n - 1$ powers of the corresponding matrix elements. Here n is the number of vertices in a graph. The resulting ID number is defined as the trace of the EA^* matrix. They checked all alkanes having $n = 22$ or fewer carbon atoms (almost 4 million trees) and no single duplicate was found.

2. Invariants of High Discrimination Power

As we have seen graph-theoretical invariants can be found that show an impressive power of discrimination among the trees of the same size. Such invariants include the ID number (the first duplicate appears for $n = 15$), Kier and Hall's total topological state (the smallest graphs having the same ID when $n = 19$), the prime ID number (the smallest pair of graphs having the same ID when $n = 20$), and finally the trace of EA^* matrix, as reported by Hu and Xu [**15**] show no duplicate for $n \leq 22$. There is no doubt that additional indices of very high resolution power will be proposed in the future, either because they may have a simpler construction, or because they may have simpler structural interpretation.

In view of the exponential growth of the number of trees with increase in the number n of vertices an exhaustive checking of all trees beyond a certain size becomes prohibitive. In Table 2 we show the number of isomers $N(n)$ for given n as reported in the literature [**16**], and also show the quotient $Q = N(n+1)/N(n)$. As we see not only $N(n)$ increases but also the quotient Q, which means that $N(n)$ grows faster than a geometrical sequence. Exhaustive testing of all trees having more than twenty vertices appears not to be practical, even though today such exhaustive testing represents the only reliable route to check for uniqueness of an invariant. In view of no available proofs for uniqueness of topological indices, search for counterexamples remains an attractive route in this field. However, a failure of an index to discriminate non isomorphic graphs can not be guaranteed outside the sets of graphs thus examined. Because of the vastness of the pool of graphs to be tested exhaustively one has to select set of graphs that are more likely to produce such counterexamples. Often such select set of graphs includes graphs that already have been reported to have several identical topological indices.

TABLE 2. The exponential growth of the number of trees with their size.

n	$N(n)$	**Quotient**
12	355	2.2592
13	802	2.3180
14	1859	2.3383
15	4347	2.3830
16	10,359	2.4031
17	24,894	2.4312
18	60,523	2.4500
19	148,284	2.4704
20	366,319	2.4862
21	910,726	2.5020
22	2,278,658	

3. Challenge

It is generally believed in mathematical circles that no single *simple* invariant suffices to discriminate all graphs, i.e., for any selection of graph invariants it is believed that there may exist a pair of graphs that will have the same set of the considered invariants. Hence, no finite set of simple invariants may guarantee that two graphs having common values of all the selected invariants are isomorphic or non-isomorphic. On the other hand one can expect that any two non isomorphic graphs will differ at least in one invariant. The problem, of course, is that we do not know in advance what that invariant is. The premise that simple invariants will not suffice to discriminate among graphs leaves a possibility that an invariant that is not simple may lead to a unique characterization of graphs. However, probably deliberately and prudently, no one has defined what a "simple" invariant is. The question therefore is open whether there are structural invariants that are unique even if not "simple". The topological index of Hu and Xu appears so far to be unique; at least it is unique for the domain of over four million structures upon which it was tested. Is there another such index? Is there an index which is also structural but apparently simpler? We outline here one such candidate index. We have to emphasize the attribute "candidate", because, before the approach is tested on a much larger set of structures, a claim to uniqueness would be premature.

Before outlining the approach let us point to the distinction between molecular invariants and molecular codes. Strictly speaking an invariant is the quantity (expression) that does not change with a change of the coordinate system (e.g., translation or rotation of the coordinate system). For example, the moment of inertia of a molecule is an invariant as it does not depend on the orientation chosen when computing this quantity. By extension, an invariant is a quantity that is independent of the choice of numbering of atoms. This means that computation of an invariant quantity gives the same result if two different choices of labeling of atoms are assumed. Illustrations are the connectivity index, Hosoya Z topological index or Wiener number.

Quantities that are based on a prescribed labeling of atoms do not qualify as invariants. An illustration of such quantities are Read's "walk around valence code", [**17**], the n-tuple code of Knop and coworkers [**18**], and the minimal binary number derived by reading the rows of the adjacency matrix from the left to the right and form the top to the bottom as a single number [**19**]. All these cases lead to a faithful representation, i.e., graph can be reconstructed from such codes, and all these are uniquely defined. However, they either require canonical labels or imply such labeling of vertices, and hence do not qualify as invariants in the strict sense of the word. For more discussion on this topic, consult Ref. [**20**].

4. Molecular Branching

Lovasz and Pelikan [**21**] suggested the leading eigenvalue λ_1 of the adjacency matrix as a branching index. This index of branching, however, shows serious limitations. It is not uncommon to find trees having different branching patterns that have the same leading eigenvalue. An example is given by the molecular graphs of 2,2-dimethylhexane and 2,5-dimethylhexane (Fig. 5). This should not be surprising because the leading eigenvalue of graphs, λ_1, does not depend on molecular size. The Frobenius theorem states that the bounds for the leading eigenvalue are given by the largest and the smallest row sums of the adjacency matrix, (i.e., by the largest and the smallest valence of the graph).

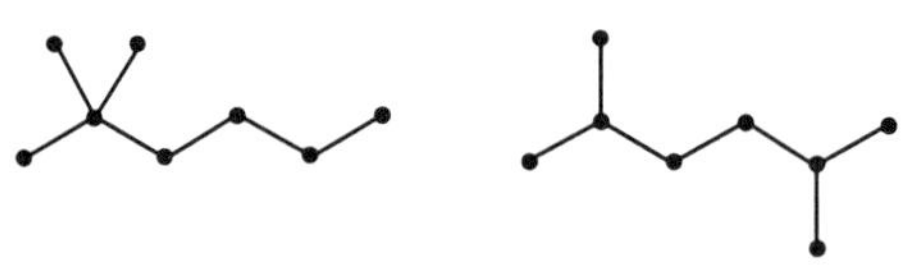

FIGURE 5. Two octane isomers showing distinct branching and having the same leading eigenvalue λ_1

Is there a way to quantify molecular branching that would assign different branching index to graphs of apparently distinct branching pattern? Can the adjacency matrix of a tree be suitably modified so that the limitations of the first eigenvalue of the adjacency matrix are removed?

The use of matrix invariants for a characterization of the molecular branching appears attractive for several reasons. First, such an approach eliminates a need for canonical rules to resolve ambiguities when different molecules have the same λ_1. For example, Bonchev and Trinajstić [**22**] proposed a set of rules for the characterization of branching which were based on the Wiener number, which is an invariant of the distance matrix. However, the Wiener number for many graphs (even graphs of relatively small size) is the same, even though graphs show different branching. To resolve such ambiguities *ad hoc* rules have to be introduced. An alternative would be to consider other graph matrices which may show greater diversity in their properties [**23, 24**]. The first eigenvalue of such matrices, λ_1 may offer an alternative characterization of the connectivity and branching. The leading eigenvalue of some of these matrices have already shown interesting structural properties. For example, the leading eigenvalue of the D/D matrix [**20**], a structural matrix whose elements are given as the quotient of the Euclidean and the graph theoretical distance between vertices i and j, represents a measure of the folding of a chain [**25**].

5. Path Matrices

Can we construct a novel matrix which can better reflect molecular branching than the adjacency matrix? Recently such a matrix has been introduced [**26**]. This is the Path matrix, P, the (i, j) element of which is defined as the subgraph of the shortest paths connecting vertices i and j [**26**]–[**28**]. Observe that here the matrix elements are not numerical entries but subgraph of the graph considered (including occasionally the graph itself). In the case of trees all the shortest distance subgraphs are paths, p_k, of length k. Hence, for trees the elements of the path matrix bears some relationship to the corresponding elements of the distance matrix. The difference is that now the entry k (a number) of a distance matrix which gives the shortest distance between vertices (i, j) is replaced by a subgraph p_k (path of length k).

Below we illustrate path matrix for the first graph of Fig. 5 (2,2-dimethylhexane). By assuming the standard labeling of vertices we obtain for the path matrix P the following:

$$
\begin{array}{cccccccc}
0 & p_1 & p_2 & p_3 & p_4 & p_5 & p_2 & p_2 \\
p_1 & 0 & p_1 & p_2 & p_3 & p_4 & p_1 & p_1 \\
p_2 & p_1 & 0 & p_1 & p_2 & p_3 & p_2 & p_2 \\
p_3 & p_2 & p_1 & 0 & p_1 & p_2 & p_3 & p_3 \\
p_4 & p_3 & p_2 & p_1 & 0 & p_1 & p_4 & p_4 \\
p_5 & p_4 & p_3 & p_2 & p_1 & 0 & p_5 & p_5 \\
p_2 & p_1 & p_2 & p_3 & p_4 & p_5 & 0 & p_2 \\
p_2 & p_1 & p_2 & p_3 & p_4 & p_5 & p_2 & 0
\end{array}
$$

In order to arrive at the numerical form of such matrices one has to select a particular graph invariant and replace each path by the numerical value for the invariant selected. Different choices will produces different matrices for the graph considered. In this way one can derive a variety of novel matrices for graphs, and from each such matrix one can select a variety of different matrix invariants. Hence, the label "double invariants" was suggested for the invariants obtained from this kind of graph matrices [**24**], because twice we make a choice in calculating an invariant of a graph.

If we select the leading eigenvalue of paths as the invariant to replace path subgraphs the path matrix P for 2,2-dimethylhexane becomes:

0	1	1.41421	1.61803	1.73205	1.80193	1.41421	1.41421
	0	1	1.41421	1.61803	1.73205	1	1
1.41421	1	0	1	1.41421	1.61803	1.41421	1.41421
1.61803	1.41421	1	0	1	1.41421	1.61803	1.61803
1.73205	1.61803	1.41421	1	0	1	1.73205	1.73205
1.80193	1.73205	1.61803	1.41421	1	0	1.80193	1.80193
1.41421	1	1.41421	1.61803	1.73205	1.80193	0	1.41421
1.41421	1	1.41421	1.61803	1.73205	1.80193	1.41421	0

The entries have been truncated to five decimal places, but the actual computations were made using 15 digits (listed in Table 3). The leading eigenvalue of paths of length 1, 2, 3 and 4 can be easily recognized (as shown in the last column of Table 3). Additional four entries for p_6, p_7, p_8 and p_{10} have the interpretation shown, as can found in a dictionary of real numbers [**29**].

The path matrix P should not be confused with the matrix of the same name of Diudea and Balaban [**30**]. They considered matrices constructed so that the elements (i, j) above the main diagonal are given by the length of the shortest

TABLE 3. The leading eigenvalues of the chains having increasing path length.

path	path eigenvalue	
1	1	
2	1.41421356237210	$\sqrt{2}$
3	1.61803398874990	$(1+\sqrt{5})/2$
4	1.73205080756888	$\sqrt{3}$
5	1.80193773580484	
6	1.84775906502257	$\sqrt{(2+\sqrt{2})}$
7	1.87938524157182	root of: $3x^3 + 9x^2 - 3 = 0$
8	1.90211303259031	$2\sin(2\pi/5)$
9	1.91898594722900	
10	1.93185165257814	$\sqrt{(2+\sqrt{3})}$
11	1.94188363485210	
12	1.94985582436365	
13	1.95629520146761	
14	1.96157056080646	
15	1.96594619936781	

path between vertices i, j, while the symmetrically positioned element (j, i) is given by the length of the longest path between the two vertices. In the case of trees, the shortest and the longest distances are the same (because there is only one path connecting any two vertices). Hence for trees this matrix reduces to the distance matrix. The matrix in which elements (i, j) and (j, i) represent the length of the longest paths between vertices i and j, called the detour matrix, has recently received attention in mathematical chemistry [**31**].

6. Branching Index $\lambda\lambda_1$

A new index for branching index of trees was suggested in Ref. [**26**]. It may be viewed as a generalization of the branching index proposed some time ago by Lovasz and Pelikan [**14**]. It is defined through the leading eigenvalue of the path matrix in which paths are replaced by the leading eigenvalue of their adjacency matrix. In the case of 2,2-dimethylhexane and 2,5-dimethylhexane, both of which have the same leading eigenvalue of the adjacency matrix, instead of $\lambda_1 = 2.0000$ we obtain $\lambda\lambda_1 = 10.0816$ and $\lambda\lambda_1 = 10.1712$ respectively for the leading eigenvalues of the path matrices. The new branching index is obtained by normalizing the leading eigenvalue of the path matrices: $\phi = \lambda\lambda_1/n$.

In what follows we will replace the elements p_k in path matrices of trees with the corresponding leading eigenvalues λ_1 belonging to the path of the same length (listed in Table 3 for paths up to length 15). The leading eigenvalue of such path matrices is denoted as $\lambda\lambda_1$. The new invariant $\lambda\lambda_1$ parallels the leading eigenvalue of the adjacency matrix λ_1 to a considerable extent (Fig. 6), hence it qualifies as an alternative branching index to that proposed earlier by Lovasz and Pelikan. The index $\lambda\lambda_1$ has been computed for all trees having $n = 10$ and less vertices and no coincidental $\lambda\lambda_1$ values have been observed.

The smallest n for which degeneracy occurs can be taken as a measure of deficiency of the descriptor for characterization of molecules. The reason that $\lambda\lambda_1$

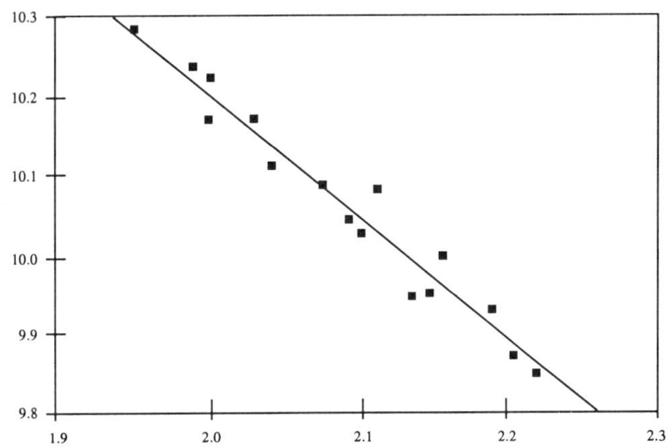

FIGURE 6. The plot of the leading eigenvalue of the path matrix ($\lambda\lambda_1$) against the leading eigenvalue of the adjacency matrix(λ_1) for octane isomers

appears to be a promising descriptor of high resolving power follows from the fact that the characteristic polynomial of a path matrix no longer has integers, or even rationals, as the coefficients. This tremendously reduces the chance for duplicate leading eigenvalues, although one cannot discard chance that two non isomorphic graphs have identical leading eigenvalue $\lambda\lambda_1$.

7. Is the Leading Eigenvalue $\lambda\lambda_1$ Unique?

The leading eigenvalue $\lambda\lambda_1$ is an attractive molecular descriptor. It was already found to lead to good regression of motor octane numbers [**26**] and other applications may follow. One of its attractive features is its conceptual simplicity, but here lies also a danger that hangs over simple descriptors as a sword of Damocles: If they are simple are they unique? The answer will be known for sure if a pair of graphs having the same $\lambda\lambda_1$ is found. Meanwhile we can examine graphs that may be suspected to produce the same value for $\lambda\lambda_1$. The following types of graphs have to be examined in particular:

(a) Graphs having the same Wiener number;
(b) Graphs having identical hyper-Wiener number;
(c) Graphs having the same Balaban J number;
(d) Graphs having the same hyper Balaban JJ number;
(e) Isospectral graph (with respect to the Adjacency matrix);
(f) Endospectral graphs;
(g) Isospectral graphs with respect to the Distance matrix;
(h) Topological twin graphs;
(i) Graphs having the same molecular ID number;
(j) Graphs having the same prime ID number;
(k) Graphs having the same atomic path sequences.

Graphs considered may have different molecular path sequences or may have identical sequences of path numbers. The distinction between these two groups is rather important. Graphs of the first group are less likely to produce a counterexample because in such cases the path matrices are built from different irrational number as the matrix entries. The graphs of the second group have as matrix elements the same p_k subgraphs, though in different relative arrangement. Hence, necessarily such matrices will be built from the same irrational numerical entries. It follows that in such instances the occurrence of the same leading eigenvalue is more likely.

We will examine separately graphs which have different path sequences and separately potentially more "threatening" graphs that have the same frequency of p_k entries. We have restricted our interest to acyclic graphs only.

8. Graphs Having Different Molecular Paths

8.1. Graphs with the Same Hyper-Wiener Number. The hyper-Wiener number WW is defined for trees in a manner analogous to the definition of the Wiener number (which has originally been also defined only for trees) as the sum of the elements above the main diagonal of a matrix. The distinction is that now instead of using the distance matrix one uses the Wiener matrix as a source. The (i, j) element of the Wiener matrix is defined as the product of the numbers n_i and n_j which signify the number of vertices on each side of the path i, j (including i and j, respectively) [**32, 33**]. If the count is restricted only to paths of length 1 (i.e., bonds) one obtains a sparse matrix with entries only for adjacent entries. The sum of entries above the main diagonal of such matrix gives the Wiener number.

The smallest trees with identical WW are two pairs of nonane isomers: 2,2,5-trimethylhexane and 3,3-dimethylheptane, with $WW = 217$, and the pair 3-methyl-3-ethylhexane and 2,3,4-trimethylhexane with $WW = 188$ (Fig. 7). The first pair has different path sequence, 8, 11, 6, 5, 6, and 8, 10, 8, 5, 4, 1 respectively, and consequently different higher order Wiener numbers: 98, 62, 34, 17, 6 and 98, 62, 34, 16, 6, 1, respectively [**34**]. The other pair of nonane isomers has the same path sequence (8, 10, 10, 6, 2) and the same higher order Wiener numbers (92, 56, 28, 10, 2). Hence, graphs with the same WW can have different path sequences or can have the same path sequence.

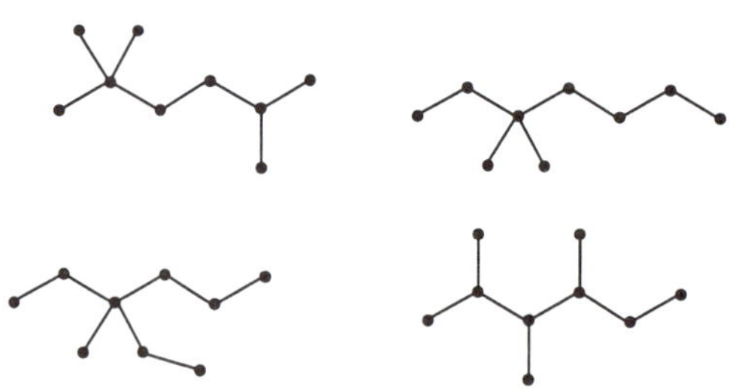

FIGURE 7. Trees having the same hyper-Wiener number WW and different path numbers

Here we have only considered trees. A generalization of the hyper-Wiener number to cyclic graphs has been considered by Klein *et al.* [**35**].

8.2. Trees Having the Same Balaban J Index. Already in Fig. 1 we illustrated trees having $n = 12$ vertices and having the same Balaban index J ($J = 4.2525095$ [**6**]). If the rows of the distance matrix for a pair of such graphs have the same sums then the graphs necessarily have the same index J. However, although the row sums of such graphs are the same the entries in the matrix have different distribution. Thus, for example, the first graph in Fig. 1 has two paths of length 7, while the second graph has three paths of length 7. This results in different molecular path counts and suggests that the two graphs are less likely to give the same leading eigenvalue $\lambda\lambda_1$.

Not all the graphs having identical index J are equally likely to lead to a counterexample. The last pair of graphs of Fig. 1 having the same J not only having a different distribution of smaller paths but also differ in the length of the longest path. The longest path in the first structure is p_7, but in the second structure it is p_6. This results in different characteristic polynomials since the leading eigenvalue of p_7 (1.87938524157182, see Table 3) will not appear in the expression for the characteristic polynomial of the first graph. Although this does not guarantee that the corresponding $\lambda\lambda_1$ will be different, it appears even more unlikely that they will be the same, than in the case of the first pair of trees of Fig. 1. That indeed all the leading eigenvalues for graphs having the same Balaban J index are different can be seen from Table 4.

TABLE 4. Normalized leading eigenvalue for selection of trees, those having different path sequences.

Graphs		**I**	**II**	**Ref.**
Same Hyper-Wiener index WW	$\lambda\lambda/18$	1.54985519135714	1.54988438019354	36
Same Balaban J index	$\lambda\lambda/12$	1.39941578732967	1.39967996258425	37
Topological twins	$\lambda\lambda/17$	1.54584073403678	1.54584213440155	38

8.3. Graphs with the Same Hyper-Balaban Number JJ. The hyper-Balaban number JJ is defined analogously to the Balaban number J however instead of using the row sums of the distance matrix one uses the row sums of the Wiener matrix. Recall that Balaban J index is defined analogous to the connectivity index χ which can be obtained by using the row sums of the adjacency matrix. The (i, j) element of the Wiener matrix is defined as the product of the numbers n_i and n_j which signify the number of vertices on each side of the path i, j (including i and j, respectively) [**32**].

In Fig. 8 we illustrate a pair of trees having the same JJ index. The pair have

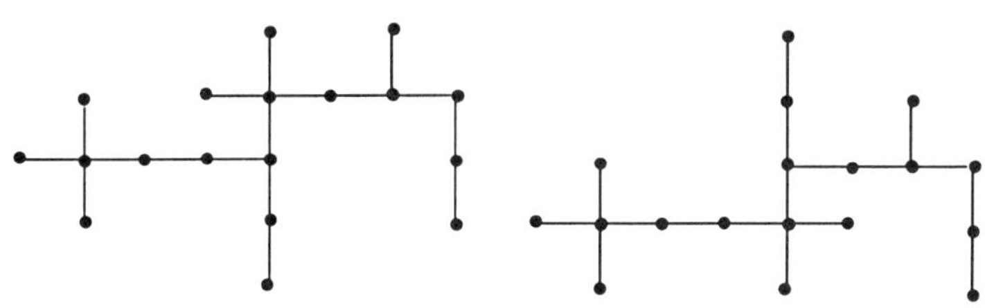

FIGURE 8. Trees having identical hyper-Balaban index JJ and different molecular path sequences

different path sequences, 17, 24, 22, 22, 24, 16, 13, 8, 4, 3 and 17, 24, 22, 23, 20, 22, 9, 9, 4, 3 respectively. However, as we will see later graphs having the same hyper-Balaban index JJ can also have the same path sequence. This is the case of trees shown in Fig. 9. For example, the trees at the top of Fig. 9 have the path count: 17, 23, 25, 24, 24, 18, 10, 8, 4. The graphs of Fig. 9 will be reviewed in the next section. The construction of graphs having the same JJ index, whether they have the same path counts or different, is outlined elsewhere [**36**].

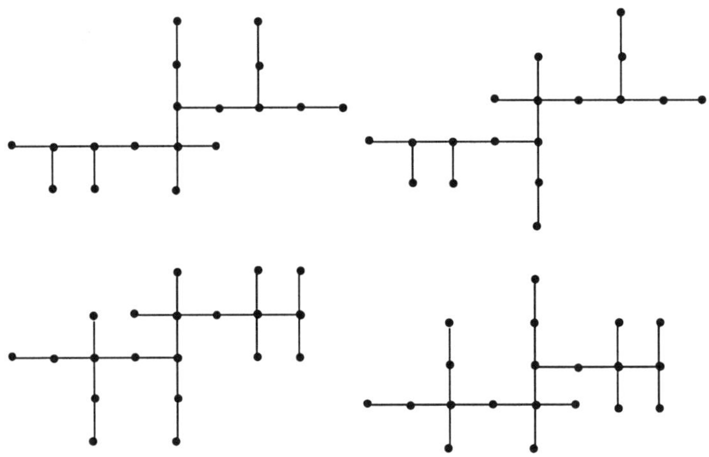

FIGURE 9. Trees having identical hyper-Balaban index JJ and the same molecular path counts

8.4. Distance Isospectral Trees. Graphs which are isospectral relative to the distance matrix are of interest in view of the fact that the path matrix and the distance matrix are closely related. The example considered is the pair of trees due to McKay [**37**] which are also isospectral respective to the adjacency matrix. This pair of graphs has also identical vertex valence distributions, as well as identical Z and W indices, where Z is Hosoya's topological index and W is Wiener number [**38**]. Hosoya referred to such graphs as "topological twin graphs". The two graphs have also the same (m, n) bond types, where m, n are the end vertices: $3(1,2) + 3(1,3) + 2(2,2) + 7(2,3) + (3,3)$, hence they will also have the same connectivity index. They differ in the *distribution* of the bond types only. However, despite all this structural similarity the vertex path numbers for the two trees are quite different, except for the terminal vertices 10, and 17. The molecular path numbers are also different, though only for paths of length four and paths of length seven. This apparently is enough to produce different leading eigenvalues for the path matrices of two graphs. Normalized $\lambda\lambda_1/n$ eigenvalues for the pair of trees are listed in Table 4.

9. Graphs Having the Same Molecular Path Counts

9.1. All Trees Having $n = 11$ Vertices and Duplicate Path Counts. Already in the late 1940's Platt [**39**] suggested path numbers as useful molecular descriptors, However, the use of paths for characterization of molecules has been overlooked for a while. The topic has been revived in the late 1970's and expanded ever since [**40**]. Path counts for smaller alkanes can be found in Ref. [**41**]. Examination of paths counts for smaller alkanes shows that there is only one pair of nonane and one pair of decane isomers with the same molecular path sequence. In case of $n = 11$ there are already 11 pairs of isomers having identical paths counts and one triple [**40**]. They are all listed in Table 5, together with the computed leading eigenvalue $\lambda\lambda_1$ (in decreasing order of the leading eigenvalue). Hence, 25 structures out of 159 isomers of undecane have no unique path count but nevertheless all have a distinct leading eigenvalue.

TABLE 5. Trees on $n = 11$ vertices having the same path numbers.

Path sequence / Isomers	$\lambda\lambda_1$	$\lambda\lambda_1/11$
10, 11, 11, 10, 8, 4, 1		
4-isopropyloctane	15.44847658485545	1.40440696225959
3-methyl-5-ethylheptane	15.38962486754609	1.39905680614055
10, 13, 12, 8, 8, 4		
2, 5-dimethyl-5-ethylheptane	15.22897443514759	1.38445222137705
2, 3, 5, 6-tetramethylheptane	15.22447642044540	1.38404331094958
10, 13, 12, 10, 6, 4		
2, 2-dimethyl-3-ethylheptane	15.20770596405252	1.38251872400477
2, 3, 4, 6-tetramethylheptane	15.20217147158091	1.38201558832554
10, 12, 14, 12, 6, 1		
4, 4-diethylheptane	15. 17090513699170	1.37917319427197
3, 5-dimethyl-4-ethylheptane	15.16587894199982	1.37871626745453
10, 13, 14, 11, 6, 1		
4-methyl-4-isopropylheptane	15.11193638699520	1.37381239881775
3, 3-dimethyl-4-ethylheptane	15.11061245627287	1.37369204147935
10, 14, 13, 9, 7, 2		
2, 2, 2, 5-tetramethylheptane	15.09975621280718	1.37270511025520
2, 4, 5, 5-tetramethylheptane	15.09963810720957	1.37269437338269
10, 13, 14, 12, 6		
2-methyl-4, 4-diethylhexane	15.08993741056539	1.3718149186958
2, 3, 5-trimethyl-4-ethylhexane	15.08609705187403	1.37146336835218
10, 13, 16, 12, 4		
3-methyl-3, 4-diethylhexane	15.04117022065084	1.36737911096826
2-methyl-3, 3-diethylhxane	15.02874234389348	1.36624930399032
10, 14, 14, 11, 6		
2, 3, 5-trimethyl-3-ethylhexane	15.03047452365549	1.36640677487777
2, 2, 3-trimethyl-4-ethylhexane	15.02939439177423	1.36630858107038
10, 14, 16, 11, 4		
2, 3-dimethyl-3-isopropylhexane	14.96931503202586	1.36084682109 326
2, 3, 3-trimethyl-4-ethylhexane	14.96802435574828	1.36072948688621
2, 3, 4-trimethyl-4-ethylhexane	14.96734717755215	1.36066792523201
10, 15, 16, 10, 4		
2, 2, 3-trimethyl-3-ethylhexane	14.91258514798757	1.35568955890796
2, 3, 3, 4, 5-pentamethylhexane	14.90711970712323	1.35519270064757
10, 15, 18, 12		
2, 2-dimethyl-3, 3-diethylpentane	14.82472939541925	1.34770267231084
2, 3, 4-trimethyl-3-isopropylpentane	14.82162969543441	1.34742088140313

9.2. Graphs with the Same Hyper-Balaban Number JJ. Graphs having the same hyper-Balaban index (shown in Fig. 1) can be constructed by considering fragments that have the same path count for a root vertex but different distributions of more distant neighbors. For example, the vertex path sequence 1, 2, 2 describing the two local situations shown at the top of Fig. 10. If we now join the two vertices

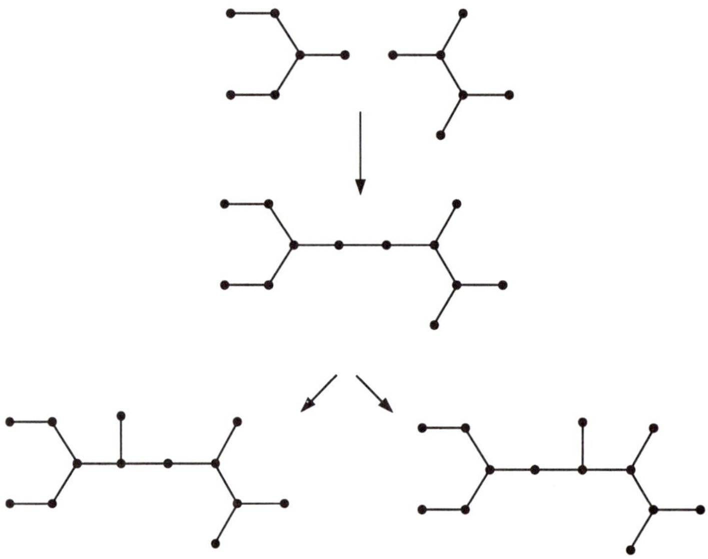

FIGURE 10. Construction of trees having the same hyper-Balaban index JJ and the same molecular path count

having the same count of paths but different distribution of the second neighbors we obtain the graph in the middle of Fig. 10. Adding asymmetrically a substituent, or adding two different substitutions at the rooted vertices, will produce a pair of graphs that will have the same hyper-Balaban number (because both graphs have the same row sums) and maintain the same path count.

9.3. Graphs Having the Same Prime-Based ID. The graphs shown in Fig. 3 are the smallest graphs having identical ID numbers when the weights for different bond types (m, n) were based on prime numbers [**9**]. Using consecutive prime numbers for weights eliminates not only the accidental degeneracies that are caused by bonds $(2,2)$ and $(1,4)$ having the same weights, but degeneracies caused by combinations of $\sqrt{2}, \sqrt{3}, \sqrt{6}, \sqrt{8}, 3$ and $\sqrt{12}$ that arise from bond types $(1,2), (1,3), (2,3), (2,4), (3,3)$ and $(3,4)$ respectively. The path numbers for the two graphs were already listed in Table 1. Although all the vertices in two graphs have different atomic path sequences, the path sequences for the two molecules are the same. The leading eigenvalues of the path matrices for the two graphs however remain different. The normalized leading eigenvalues for the path matrices for graph of Fig. 3 are shown in Table 6.

9.4. Graphs with the Same Total Topological Index τ. In Fig. 4 we illustrated three pairs of trees having $n = 19$ or $n = 20$ vertices and having identical total topological index τ of Hall and Kier [**14**]. These cases have been reported by Hu and Xu [**15**, **42**] who systematically tested trees to find those having the

TABLE 6. Normalized leading eigenvalues for selection of trees, those having the same path sequence.

Graphs		I	II
Same *JJ*	$\lambda\lambda_1/18$	1.54783765707002	1.54861921325200
		17, 23, 25, 24, 24, 18, 10, 8, 4	
	$\lambda\lambda_1/20$	1.56289185791922	1.56289295457248
		19, 29, 31, 29, 30, 21, 15, 12, 4	
Same Hall & Kier τ	$\lambda\lambda_1/19$	1.55386065115372	1.55386190437726
		18, 23, 27, 33, 32, 23, 13, 2	
	$\lambda\lambda_1/20$	1.56640775811016	1.56640535378175
		19, 24, 32, 36, 33, 24, 16, 16, 2	
Same Prime ID	$\lambda\lambda_1/20$	1.51810508858122	1.51812023632592
Slater's trees	$\lambda\lambda_1/18$	1.47888103830098	1.47888122990892

same total topological index τ. These trees represent potential counterexamples to the uniqueness of the leading eigenvalue $\lambda\lambda_1$ because they all have the same distribution of paths of different length in their path matrix. For example, the first pair has distribution: $p_1^{18}p_2^{23}p_3^{27}p_4^{33}p_5^{32}p_6^{23}p_7^{13}p_8^{2}$, that is, the molecular path sequence is 18, 23, 27, 33, 32, 23, 13, 2 for both graphs. Similarly the second pair has the distribution: $p_1^{19}p_2^{24}p_3^{32}p_4^{36}p_5^{33}p_6^{24}p_7^{14}p_8^{6}, p_9^{2}$ and the molecular path: 19, 24, 32, 36, 33, 24, 14, 6, 2. Hence the graphs of Fig. 4 satisfy the necessary condition of having their path matrix built from the same entries, only in a different arrangement. Only by calculating $\lambda\lambda_1$ for such cases one can be assured that no counterexample is found. As we see from Table 6 all the graphs examined produce different $\lambda\lambda_1$.

9.5. Slater's Pair of Graphs. In Fig. 11 we illustrate the pair of trees constructed by Slater [**43**], the path numbers for which are listed in Table 7. This

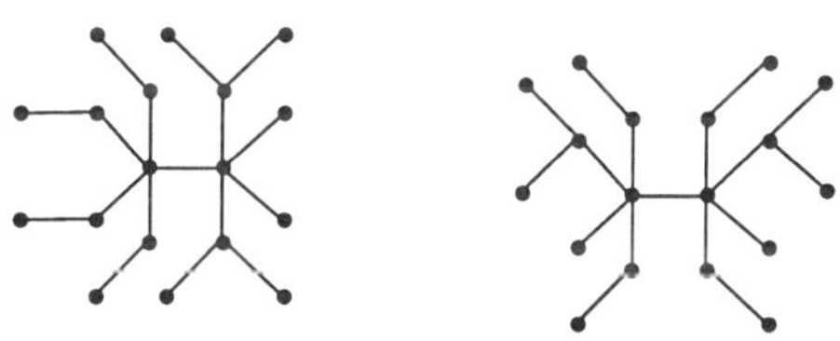

FIGURE 11. Slater's pair of graphs (having the same list of atomic path numbers)

pair of graphs has an interesting property: both graphs have the same list of the vertex path numbers. Ivanciuc and Balaban [**44**] have constructed a number of similar pair of trees for which both graphs have the same list of vertex path numbers. It was conjectured that even though individual vertex path numbers may be the same in different molecules the totality of path numbers for all atoms in two molecules may be different [**40**]. If that would be the case one could use path numbers (i.e., invariants) to uniquely encode structures, a dream that has not been realized hitherto. Because the same path numbers appear in both graphs their characteristic polynomials will be built from the same components, which makes this pair of trees a candidate for a counterexample to the conjecture on uniqueness of the leading eigenvalue of the path matrix (the novel branching index). However, the calculated leading eigenvalue are found to be different: 26.61985868941762 and

26.61986213836063 for the two trees respectively. In Table 6 we show the corresponding normalized values $\lambda\lambda_1/n$.

TABLE 7. The path numbers for the two Slater's graphs.

Vertex	**Paths**	**Vertex**	**Paths**
1	1, 2, 4, 6, 4	1	1, 2, 4, 6, 4
2	1, 2, 4, 6, 4	2	1, 2, 4, 6, 4
3	1, 4, 8, 4	3	1, 4, 8, 4
4	1, 4, 8, 4	4	1, 4, 8, 4
5	1, 2, 4, 6, 4	5	1, 2, 4, 6, 4
6	1, 2, 4, 6, 4	6	1, 2, 4, 6, 4
7	3, 4, 6, 4	7	3, 4, 6, 4
8	3, 4, 6, 4	8	3, 4, 6, 4
9	5, 8, 4	9	5, 8, 4
10	5, 8, 4	10	5, 8, 4
11	2, 4, 7, 4	11	2, 4, 7, 4
12	1, 1, 4, 7, 4	12	1, 1, 4, 7, 4
13	2, 4, 7, 4	13	2, 4, 7, 4
14	1, 1, 4, 7, 4	14	1, 1, 4, 7, 4
15	2, 4, 7, 4	15	2, 4, 7, 4
16	1, 1, 4, 7, 4	16	1, 1, 4, 7, 4
17	2, 4, 7, 4	17	2, 4, 7, 4
18	1, 1, 4, 7, 4	18	1, 1, 4, 7, 4
Molecule	17, 30, 48, 42, 16	Molecule	17, 30, 48, 42, 16

10. Concluding Remarks

The main purpose of this paper has been to draw attention of readers to the possibility of the construction of simple graph invariants of unusual discrimination power. Although at this point we can not claim the uniqueness for the here discussed leading eigenvalue $\lambda\lambda_1$ of the path matrices based on the leading eigenvalues λ_1 of paths at least it appears that in the case of several graphs that could produce a counterexample, none did. Of all graphs considered most likely a counterexample may come from graphs having the same path distributions, and among such in particular from graphs having the same atomic path sequences (such as Slater's graphs). We hope in the near future to extend the present work by (i) systematically search for counterexamples among larger trees; (ii) extend the consideration to cyclic graphs; and finally (iii) extend the similar consideration to graphs having multiple bonds and heteroatoms.

References

[1] H. Wiener, Structural Determination of Paraffin Boiling Points, *J. Am. Chem. Soc.*, **69** (1947) 17–20.

[2] H. Hosoya, Topological Index. A Newly Proposed Quantity Characterizing the Topological Nature of Structural Isomers of Saturated Hydrocarbons, *Bull. Chem. Soc. Japan*, **44** (1971) 2332–2339.

[3] M. Randić, On Characterization of Molecular Branching, *J. Am. Chem. Soc.*, **97** (1975) 6609–6615.

[4] A.T. Balaban, Highly Discriminating Distance-Based Topological Index, *Chem. Phys. Lett.*, **89** (1982) 399–404.

A.T. Balaban, Topological Index Based on Topological Distances in Molecular Graphs, *Pure Appl. Chem.*, **55** (1983) 199–206.
[5] A.T. Balaban, and L.V. Quintas, The Smallest Graphs, Trees, and 4-Trees with Degenerate Topological Index J, *MATCH*, **14** (1983) 213–233.
[6] M. Razinger, On Highly Discriminating Distance Based Topological Index, *Vestn. Slov. Kem. Drus.*, **31** (1984) 333–337.
[7] M. Randić, On Molecular Identification Numbers, *J. Chem. Inf. Comput. Sci.*, **24** (1984) 164–175.
[8] K. Szymanski, W.R. Müller, J.V. Knop and N. Trinajstić, On Randić's Molecular Identification Number, *J. Chem. Inf. Comput. Sci.*, **25** (1985) 413–415.
[9] M. Randić, Molecular ID-Numbers - by Design, *J. Chem. Inf. Comput. Sci.*, **26** (1986) 134–136.
[10] K. Szymanski, W.R. Müller, J.V. Knop, and N. Trinajstić, Molecular ID Numbers, *Croat. Chem. Acta*, **59** (1986) 719–723.
[11] L.H. Hall and L.B. Kier, Determination of Topological Equivalence in Molecular Graphs from the Topological State, *Quant. Struct.-Act. Relat.*, **9** (1990) 115.
[12] M. Randić, On Characterization of Molecular Branching, *J. Am. Chem. Soc.*, **976** (1975) 6609.
[13] L.B. Kier, W.J. Murray, M. Randić, and L.H. Hall, Molecular Connectivity V. Connectivity Series Concept Applied to Density, *J. Pharm. Sci.*, **65** (1976) 1226–1230.
[14] L.B. Kier and L.H. Hall, Molecular Connectivity in Chemistry and Drug Research, Academic Press: New York, 1996.
[15] C.-Y. Hu and L. Xu, On Highly Discriminating Molecular Topological Index, *J. Chem. Inf. Comput. Sci.*, **36** (1996) 82–90.
[16] N. Trinajstić, Chemical Graph Theory, CRC Press, Boca Raton, 1992.
[17] R.C. Read, A New System for the Designation of Chemical Compounds. 1. Theoretical Preliminaries and Coding of Acyclic Compounds. *J. Chem. Inf. Comput. Sci.*, **23** (1983) 135–149.
[18] J.V. Knop, W.R. Müller, K. Szymanski and N. Trinajstić, Computer Generation of Certain Classes of Molecules, (SKHT/Kemija u Industriji, Zagreb, 1985).
[19] M. Randić, A Graph Theoretical Basis for Structural Chemistry I. Structures Based on Trivalent Graphs with $n = 10$ Vertices, *Acta Crystallogr.*, **A 34** 275–282.
[20] M. Randić, Generalized Molecular Descriptors, *J. Math. Chem.*, **7** (1991) 155–168.
[21] L. Lovasz and J. Pelikan, On the Eigenvalues of Trees, *Periodica Math. Hung.*, **3** (1973) 175–182.
[22] D. Bonchev and N. Trinajstić, Information Theory, Distance Matrix and Molecular Branching, *J. Chem. Phys.*, **67** (1977) 4517–4533.
D. Bonchev and N. Trinajstić, On Topological Characterization of Molecular Branching, *Int. J. Quant. Chem: Quant. Chem. Symp.*, **12** (1978) 293–303.
[23] S.S. Tratch, M.I. Stankevich, and N.S. Zefirov, Combinatorial Models And Algorithms in Chemistry. The Expanded Wiener Number — A Novel Topological Index, *J. Comput. Chem.*, **11** (1990) 899–908.
M. Randić, Hosoya matrix — A Source of New Molecular Descriptors, *Croat. Chem. Acta*, **67** (1994) 415–429.
M. Randić, Restricted Random Walks on Graphs, *Theor. Chim. Acta*, **92** (1995) 97–106.
[24] M. Randić, A.F. Kleiner and L.M. DeAlba, Distance/Distance Matrices, *J. Chem. Inf. Comput. Sci.*, **34** (1994) 277–286.
[25] L. Bytautas, D.J. Klein, M. Randić, and T. Pisanski, Foldedness in Linear Polymers: A Difference Between Graphical and Euclidean Distances, *Proceedings DIMACS Workshop on Discrete Mathematical Chemistry*, Dimacs Center, Piscataway, NJ (March 23-25, 1998).
M. Randić, Topological Indices, Encyclopedia of Computational Chemistry (P. von Rague Schleyer, editor-in-chief), Wiley & Sons, London (1998).
[26] M. Randić, On Molecular Branching, *Acta Chim. Sloven.*, **44** (1997) 57–77.
[27] M. Randić, On Structural Ordering and Branching of Acyclic Saturated Hydrocarbons, *J. Math. Chem.*, **24** (1998) 345–358.
[28] M. Randić, D. Plavšić and M. Razinger, Double Invariants, *MATCH*, **35** (1997) 243–259.
[29] J. Borwein and P. Borwein, A Dictionary of Real Numbers, Wadsworth & Brooks / Cole Advanced Books & Software, Pacific Grove, California 1990.

[30] M. Diudea, Cluj Matrix Invariants, *J. Chem. Inf. Comput. Sci.*, **33** (1997) 300–305.
A.T. Balaban, Numerical Modeling of Chemical Structures: Local Graph Invariants and Topological Indices. In: Graph theory and Topology, (R.B. King and D.H. Rouvray, Eds.) Elsevier, Amsterdam 1987, pp. 159–176.

[31] D. Amić and N. Trinajstić, On the Detour Matrix, *Croat. Chem. Acta*, **68** (1995) 53–62.
S. Nikolić, N. Trinajstić, A. Jurić and Z. Mihalić, The Detour Matrix and the Detour Index of Weighted Paths, *Croat. Chem. Acta*, **69** (1996) 1577–1591.
I. Lukovits, The Detour Index, *Croat. Chem. Acta*, **69** (1996) 873–882.
I. Lukovits and M. Razinger, On Calculation of the Detour Index, *J. Chem. Inf. Comput. Sci.*, **37** (1997) 283–286.
N. Trinajstić, S. Nikolić, and Z. Mihalić, On Computing the Molecular Detour Matrix, *Int. J. Quant. Chem.*, **65** (1997) 415–419.
N. Trinajstić, S. Nikolić, B. Lučić, D. Amić and Z. Mihalić, The Detour Matrix in Chemistry, *J. Chem. Inf. Comput. Sci.*, **37** (1997) 631–638.
M. Randić, L.M. DeAlba and F. Harris, Graphs with the Same Detour Matrix, *Croat. Chem. Acta*, **71** (1998) 53–68.

[32] M. Randić, Novel Molecular Descriptor for Structure-Property Studies, *Chem. Phys. Lett.*, **211** (1993) 478–483.

[33] M. Randić, X. Guo, T. Oxley, and H. Krishnapriyan, Wiener Matrix: Source of Novel Graph Invariants, *J. Chem. Inf. Comput. Sci.*, **33** (1993) 709–716.
M. Randić, X. Guo, T. Oxley, H. Krishnapriyan and L. Naylor, Wiener Matrix Invariants, *J. Chem. Inf. Comput. Sci.*, **34** (1994) 361–367.

[34] I. Gutman, W. Linert, I. Lukovits and A. Dobrynin, Trees with Extremal Hyper-Wiener Index: Mathematical Basis and Chemical Applications, *J. Chem. Inf. Comput. Sci.*, **37** (1997) 349–354.

[35] D. J. Klein, I. Lukovits, and I. Gutman, On the Definition of the Hyper-Wiener Index for Cycle-Containing Structures, *J. Chem. Inf. Comput. Sci.*, **35** (1995) 50–52.

[36] X. Guo and M. Randić, Trees with the Same Topological Index *JJ*, *SAR and QSAR* (in press)

[37] B. D. McKay, On the Spectral Characterization of Trees, *Ars Combinatorica*, **3** (1977) 219–232.

[38] H. Hosoya, U. Nagashima and S. Hyugaji, Topological Twin Graphs. Smallest Pair of Isospectral Polyhedral Graphs with Eight Vertices, *J. Chem. Inf. Comput. Sci.*, **34** (1994) 428–431.

[39] J. R. Platt, Influence of Neighbor Bonds on Additive Bond Properties in Paraffins, *J. Chem. Phys.*, **15** (1947) 419–420.

[40] M. Randić, Characterization of Atoms, Molecules and Classes of Molecules Based on Path Enumerations, *MATCH*, **7** (1979) 5–64.

[41] M. Randić and C. L. Wilkins, Graph Theoretical Ordering of Structures as a Basis for Systematic Searches for Regularities in Molecular Data, *J. Phys. Chem.*, **83** (1979) 1525–1540.

[42] C.-Y. Hu and L Xu, On Hall and Kier's Topological State and Total Topological Index, *J. Chem. Inf. Comput. Sci.*, **34** (1994) 1251–1258.

[43] P. J. Slater, Counterexample to Randić's Conjecture on Distance Degree Sequence for Trees, *J. Graph Theory*, **6** (1982) 89–91.

[44] O. Ivanciuc and A. T. Balaban, Non Isomorphic Graphs with Identical Atomic Counts of Self-Returning Walks: Isocodal Graphs, *J. Math. Chem.*, **11** (1992) 155–167.

Department of Mathematics and Computer Science, Drake University, Des Moines, Iowa 50311, USA
E-mail address: `milan.randic@drake.edu`

Department of Mathematics and Computer Science, Drake University, Des Moines, Iowa 50311, USA; and Institute of Mathematics and Physics, Xinjiang University, Wulumuqi Xinjiang 830046, People's Republic of China
E-mail address: `xfguo@xju.edu.cn`

Undergraduate student of chemistry, Department of Mathematics and Computer Science, Drake University, Des Moines, Iowa 50311, USA
E-mail address: `smb009@drake.edu`

DIMACS Series in Discrete Mathematics
and Theoretical Computer Science
Volume 51, 2000

Star Sets and Star Complements in Finite Graphs: A Spectral Construction Technique

Peter Rowlinson

ABSTRACT. Let μ be an eigenvalue of the graph G with multiplicity k. A *star set* for μ in G is a set X of k vertices of G such that μ is not an eigenvalue of the *star complement* $G-X$. The paper discusses the construction of connected graphs with multiple eigenvalues as extensions of connected star complements.

1. Background

This paper is an update of Chapter 7 of the research monograph [**5**]; proofs given there will not be repeated here. The general context is the relation between the structure of a graph G and an individual eigenvalue μ of G. Here we are concerned in particular with the construction of connected graphs which have a multiple eigenvalue. We take G to be an undirected graph without loops or multiple edges, with vertex set $V(G) = \{1, 2, \ldots, n\}$, and with $(0,1)$-adjacency matrix A. Let P denote the orthogonal projection of $\mathbb{R}^n$ onto the eigenspace $\mathcal{E}(\mu)$ of A, and let $\{\mathbf{e}_1, \ldots, \mathbf{e}_n\}$ be the standard orthonormal basis of $\mathbb{R}^n$. Since $\mathcal{E}(\mu)$ is spanned by the vectors $P\mathbf{e}_j$ $(j = 1, \ldots, n)$ there exists $X \subseteq V(G)$ such that the vectors $P\mathbf{e}_j$ $(j \in X)$ form a basis for $\mathcal{E}(\mu)$. Such a subset X of $V(G)$ is called a *star set* for μ in G. (The terminology reflects the fact that the vectors $P\mathbf{e}_1, \ldots, P\mathbf{e}_n$ form a eutactic star in the sense of Seidel [**13**]. In the context of star partitions [**5**, Section 7.1], star sets are called *star cells.*) An equivalent definition which is useful in a chemical or computational context is the following: if μ has multiplicity k then a star set for μ in G is a set X of k vertices of G such that μ is not an eigenvalue of $G - X$ [**5**, Theorem 7.2.9]. Here $G - X$ is the subgraph of G induced by $\overline{X}$, the complement of X in $V(G)$. Accordingly, the graph $G - X$ is called the *star complement* for μ corresponding to X. (Such graphs are called *μ-basic* subgraphs in [**6**].) If a single vertex is deleted from a graph then by the Interlacing Theorem [**3**, Theorem 0.10] the multiplicity of an eigenvalue changes by 1 at most. Accordingly the deletion of any r vertices from X $(0 < r < k)$ results in a graph for which μ is an eigenvalue of multiplicity $k - r$. We can also make the following observation.

1991 *Mathematics Subject Classification.* 05C50.

The author is grateful for financial support from the Royal Society (Grant number 18630/974/A1B).

PROPOSITION 1.1. *Let μ be an eigenvalue of the graph G. If Y is a subset of $V(G)$ such that μ is not an eigenvalue of $G-Y$ then there exists a star set X for μ such that $X \subseteq Y$.*

PROOF. It suffices to show that the vectors $P\mathbf{e}_j$ ($j \in Y$) span $\mathcal{E}(\mu)$. Suppose by way of contradiction that they span a proper subspace W of $\mathcal{E}(\mu)$. Then $\mathcal{E}(\mu) \cap W^{\perp}$ contains a non-zero vector $\mathbf{x}$, and we have $\mathbf{x}^T\mathbf{e}_j = (P\mathbf{x})^T(\mathbf{e}_j) = \mathbf{x}^T(P\mathbf{e}_j) = 0$ for all $j \in Y$. If we write A_Y for the principal submatrix of A determined by Y we obtain

$$\begin{pmatrix} A_Y & \star \\ \star & A_{\overline{Y}} \end{pmatrix}\begin{pmatrix} \mathbf{0} \\ \mathbf{x}' \end{pmatrix} = \mu\begin{pmatrix} \mathbf{0} \\ \mathbf{x}' \end{pmatrix}, \text{ where } \begin{pmatrix} \mathbf{0} \\ \mathbf{x}' \end{pmatrix} = \mathbf{x}.$$

It follows that $\mathbf{x}'$ is a μ-eigenvector of $G-Y$, a contradiction. □

If X is a star set for the non-zero eigenvalue μ of G then $\overline{X}$ is a dominating set in G [**5**, Theorem 7.3.1]; moreover if $\mu \notin \{-1, 0\}$ then $\overline{X}$ is a location-dominating set, meaning that the $\overline{X}$-neighbourhoods of vertices in X are distinct and non-empty [**5**, Corollary 7.3.6]. It follows that if $\mu \notin \{-1, 0\}$ and $|\overline{X}| = t$ then $|X| < 2^t$; thus there are only finitely many graphs which have an eigenspace $\mathcal{E}(\mu)$ ($\mu \neq -1, 0$) with prescribed codimension t. In particular there are only finitely many graphs with a prescribed star complement $H = G - X$ corresponding to an eigenvalue μ other than -1 or 0. If we prescribe not only μ and $G-X$ but also the set $E(X, \overline{X})$ of edges between X and $\overline{X}$ then G is determined uniquely: this is the Reconstruction Theorem [**5**, Theorem 7.4.1] (see Theorem 3.1 below). Taken together these observations provide the basis for determining the graphs with a prescribed star complement H (cf. [**7**], [**8**]) and characterizing certain graphs by properties of H which have implications for $E(X, \overline{X})$ (cf. [**9**], [**10**]). The eigenvalues -1 and 0 are essential exceptions: if $\mu = -1$, X may contain arbitrarily many pairwise adjacent vertices with the same closed neighbourhood (cf. the proof of Proposition 2.1(ii)(c) below), while if $\mu = 0$, X may contain arbitrarily many independent vertices with the same open neighbourhood. This is because the rank of the adjacency matrix A (when $\mu = 0$) or $A + I$ (when $\mu = -1$) is independent of the number of these so-called *duplicate* vertices (cf. [**6**, Section 1]). If duplicate vertices are excluded then again only finitely many graphs arise for given $t = |\overline{X}|$: these are the graphs described by Ellingham [**6**] as *reduced* ($\mu = 0$) or *co-reduced* ($\mu = -1$).

We note that when $\mu \notin \{-1, 0\}$ the exponential bound $|X| < 2^t$ may be improved to the quadratic bound $|X| \leq \frac{1}{2}(t-1)(t+4)$ when $t > 1$ [**11**]. In Section 3 we shall establish in passing the slightly weaker inequality $|X| \leq \frac{1}{2}t(t+3)$. Consideration of the line graph $L(K_t)$, for which $\mathcal{E}(-2)$ has codimension t, shows that an upper bound of order $\frac{1}{2}t^2$ for $|X|$ is asymptotically best possible as $t \to \infty$.

We shall use the notation '$u \sim v$' to indicate that vertices u, v of a graph are adjacent.

2. Connectedness

Suppose that the graph G has the star complement $H = G - X$ corresponding to μ. If $\mu \neq 0$ and H is connected then G is connected because $\overline{X}$ is a dominating set. We can say a little more:

PROPOSITION 2.1. *Let μ be a non-zero eigenvalue of the graph G, and let X be a star set for μ in G with corresponding star complement H.*

(i) *If H is connected then G is connected.*

(ii) *If H is 2-connected then one of the following holds:*

(a) *G is 2-connected;*

(b) *$\mu \neq -1$ and G has a pendant edge at a vertex of $\overline{X}$;*

(c) *$\mu = -1$ and G has a cutvertex v in $\overline{X}$ whose neighbours in X induce a complete subgraph which is a component of $G - v$.*

PROOF. *Proof of (ii).* Suppose that H is 2-connected, while G has a cutvertex v. Necessarily $v \in \overline{X}$ because $\overline{X}$ is a dominating set. By hypothesis $H - v$ is connected. Let u be a neighbour of v which lies in a component of $G - v$ other than that containing $\overline{X} \setminus \{v\}$. Then $\{v\}$ is the $\overline{X}$-neighbourhood of u; moreover if $\Gamma(v) = \{u \in X : u \sim v\}$ then the subgraph of G induced by $\Gamma(v)$ is a union of components of $G - v$.

If $\mu \neq -1$ then $\Gamma(v) = \{u\}$ because $\overline{X}$ is a location-dominating set, and we have case (*b*) of the Proposition. Accordingly, suppose that $\mu = -1$. In the notation of Section 1 we have:

$$-P\mathbf{e}_u = AP\mathbf{e}_u = PA\mathbf{e}_u = \sum_{j\sim u} P\mathbf{e}_j \text{ for each } u \in X.$$

In particular, if u_1, u_2 are vertices in $\Gamma(v)$ with X-neighbourhoods Γ_1, Γ_2 respectively then

$$-P\mathbf{e}_{u_1} = P\mathbf{e}_v + \sum_{h\in\Gamma_1} P\mathbf{e}_h \text{ and } -P\mathbf{e}_{u_2} = P\mathbf{e}_v + \sum_{h\in\Gamma_2} P\mathbf{e}_h.$$

If we now eliminate $P\mathbf{e}_v$ and use the fact that the vectors $P\mathbf{e}_j$ $(j \in X)$ are linearly independent, we find that $u_1 \sim u_2$ and $\Gamma_1 \setminus \{u_2\} = \Gamma_2 \setminus \{u_1\}$. It follows that $\Gamma(v)$ induces a complete subgraph, and we have case (c) of the Proposition. □

The following examples show that each of the possibilities (a), (b), (c) of Proposition 2.1(ii) does arise.

EXAMPLE 2.2. (i) The Petersen graph is a 2-connected graph with any 5-cycle as a 2-connected star complement for the eigenvalue 1.

(ii) If in the preceding example we delete any four vertices in a star set for the eigenvalue 1, we obtain a 5-cycle with a pendant edge as an illustration of case (b) of Proposition 2.1(ii).

(iii) Fig. 1 shows the coalescence of a 6-vertex graph H with a complete graph K_m $(m \geq 2)$. The subgraph H (with vertices circled) is a star complement for the eigenvalue -1, and the corresponding star set induces a complete subgraph of size $m - 1$, illustrating case (c) of Proposition 2.1(ii).

We shall prove a converse of Proposition 2.1(i): for any eigenvalue of a connected graph there is a connected star complement. This can be proved by induction on the number of vertices, but the proof we give here is shorter and depends on the following lemma (part of Theorem 1.1 of [**6**]).

LEMMA 2.3. *If the column space of the symmetric matrix* $\begin{pmatrix} C & D^T \\ D & E \end{pmatrix}$ *has the columns of* $\begin{pmatrix} C \\ D \end{pmatrix}$ *as a basis, then the columns of C are linearly independent.*

PROOF. Since each column of $\begin{pmatrix} D^T \\ E \end{pmatrix}$ is a linear combination of the columns of $\begin{pmatrix} C \\ D \end{pmatrix}$, there exists a matrix L such that $D^T = CL$, equivalently $D = L^TC$. Thus if $C\mathbf{x} = \mathbf{0}$ then $\begin{pmatrix} C \\ D \end{pmatrix}\mathbf{x} = \mathbf{0}$, whence $\mathbf{x} = \mathbf{0}$ as required. □

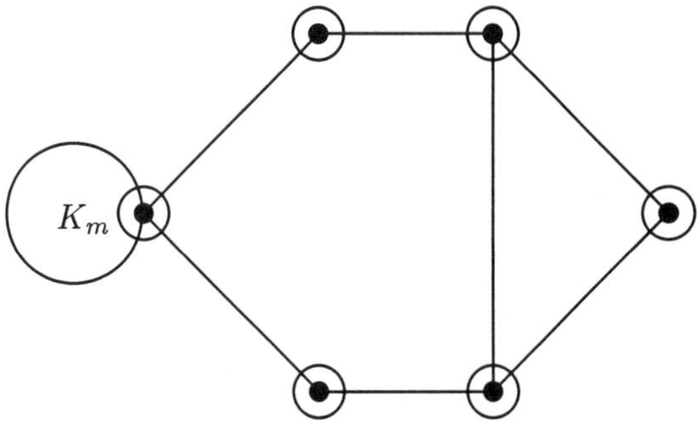

FIGURE 1. The graph of Example 2.2 (iii).

THEOREM 2.4. *Let μ be an eigenvalue of the graph G. If G is connected then there exists a connected star complement for μ in G.*

PROOF. (Communicated by F. K. Bell and ascribed to S. Penrice in [**6**].) Since G is connected we may order its vertices so that each vertex after the first is adjacent to a predecessor. Let A be the adjacency matrix of G and let $\{\mathbf{c}_j : j \in Y\}$ be the basis of the column space of $\mu I - A$ obtained by deleting each column which is a linear combination of its predecessors. By Lemma 2.3 the principal submatrix of $\mu I - A$ determined by Y is invertible. Since $|Y| = \text{codim}\mathcal{E}(\mu)$, $\overline{Y}$ is a star set for μ and the subgraph H induced by Y is a star complement.

We prove that H is connected by showing that each vertex y of Y with $y > 1$ is adjacent to a previous vertex j of Y. We take j to be least such that $j \sim y$ in G. Then $j < y$ and the y-th entry of $\mathbf{c}_j$ is -1. On the other hand the y-th entry of all preceding columns is 0, and so $\mathbf{c}_j$ is not a linear combination of its predecessors. Accordingly $j \in Y$ as required. □

Theorem 2.4 can be used to construct recursively all the connected graphs with an eigenspace $\mathcal{E}(\mu)$ ($\mu \neq -1, 0$) of prescribed codimension t. Since all connected graphs have a simple eigenvalue (namely the largest [**3**, Theorem 0.3]), the objective is to determine for $k = 2, 3, 4, \ldots$ the family $\mathcal{G}(t,k)$ of all connected graphs on $t+k$ vertices with an eigenvalue $\mu \notin \{-1, 0\}$ of multiplicity k. If $G' \in \mathcal{G}(t, k+1)$ then (by Theorem 2.4) G' has a connected star complement $H = G' - X'$ with t vertices. For each $u \in X'$, $G' - u$ has H as a star complement, and so $G' - u \in \mathcal{G}(t,k)$ by Proposition 2.1(i). In particular, each graph G' in $\mathcal{G}(t, k+1)$ can be found by adding a vertex u to a graph G in $\mathcal{G}(t,k)$. The graph G' is determined by the neighbourhood Δ of μ. If $\mathbf{b}$ is the characteristic vector of Δ and μ is an eigenvalue of G of multiplicity k then μ is an eigenvalue of G' of multiplicity $k+1$ if and only if $\mathbf{b} \in \mathcal{E}(\mu)^{\perp}$ and G' has a μ-eigenvector whose u-component is non-zero [**12**, Lemma 1.2]. Accordingly one can construct $\mathcal{G}(t, k+1)$ from $\mathcal{G}(t,k)$ by inspecting the graphs G' for which $\mathbf{b} \in \mathcal{E}(\mu)^{\perp}$. In view of an earlier remark we know that $\mathcal{G}(t,k)$ is empty for all $k > \frac{1}{2}(t-1)(t+4)$. For $t \leq 4$ and $k \in \mathbb{N}$ the graphs in $\mathcal{G}(t,k)$ are known from the published tables of spectra of graphs with at most 7 vertices (see [**2**],[**3**] and [**4**]). In particular, $\mathcal{G}(4,k) = \emptyset$ for all $k \geq 3$ because no 7-vertex graph has an eigenvalue $\mu \notin \{-1, 0\}$ of multiplicity 3. The foregoing considerations have been used in [**12**] to find the graphs in each $\mathcal{G}(5,k)$; in particular, it is shown that $\mathcal{G}(5,k)$ is empty if and only if $k > 5$.

A different approach to constructing the finite family $\cup_{k\in I\!N}\mathcal{G}(t,k)$ is to consider each graph therein as an extension of a (connected) star complement by k vertices rather than an extension of a (connected) vertex-deleted subgraph by a single vertex. The problem of extending star complements is addressed in the next section. The crucial property is that if H is a star complement for the eigenvalue μ then an extension of H by a star set X is determined by μ and the H-neighbourhoods of vertices in X.

3. Extending Star Complements

Our starting point is the Reconstruction Theorem [**5**, Theorem 7.4.1] and its converse [**5**, Theorem 7.4.4], which we combine as follows.

THEOREM 3.1. *Let G be a graph with adjacency matrix* $\begin{pmatrix} A_X & B^T \\ B & C \end{pmatrix}$. *Then X is a star set for μ in G if and only if μ is not an eigenvalue of C and*

$$\mu I - A_X = B^T(\mu I - C)^{-1}B \tag{3.1}$$

Note that if X is a star set for μ then the corresponding star complement $H(= G - X)$ has adjacency matrix C, and equation (3.1) tells us that G is determined by μ, H and the H-neighbourhoods of vertices in X. In this situation, let $|\overline{X}| = t$ and define a bilinear form on $I\!R^t$ by $\langle \mathbf{x}, \mathbf{y}\rangle = \mathbf{x}^T(\mu I - C)^{-1}\mathbf{y}$. Denote the columns of B by $\mathbf{b}_u$ $(u \in X)$. We note in passing that $|X| \le \frac{1}{2}t(t+3)$ when $\mu \notin \{-1, 0\}$: this is proved in [**11**] as a preliminary to establishing the slightly stronger upper bound $\frac{1}{2}(t-1)(t+4)$ by adapting the 'absolute bound' argument of [**14**, Section 6]. For each $u \in X$ define $f_u : I\!R^t \to I\!R$ by

$$f_u(\mathbf{x}) = \frac{\langle \mathbf{b}_u, \mathbf{x}\rangle^2 + \langle \mathbf{b}_u, \mathbf{x}\rangle}{\mu^2 + \mu}.$$

This is the sum of quadratic and linear homogeneous functions in the t components of $\mathbf{x}$. The space of such functions has dimension $\binom{t}{2} + 2t = \frac{1}{2}t(t+3)$. On the other hand the functions $f_u(u \in X)$ are linearly independent because $f_u(\mathbf{b}_v) = \delta_{uv}$, and so $|X| \le \frac{1}{2}t(t+3)$.

To discuss graphs with a prescribed star complement we shall use the following consequence of Theorem 3.1.

COROLLARY 3.2. *Suppose that μ is not an eigenvalue of the graph H. There exists a graph G with a star set X for μ such that $G - X = H$ if and only if the characteristic vectors $\mathbf{b}_u$ $(u \in X)$ satisfy*

(i) $\langle \mathbf{b}_u, \mathbf{b}_u\rangle = \mu$ *for all* $u \in X$, *and*

(ii) $\langle \mathbf{b}_u, \mathbf{b}_v\rangle \in \{-1, 0\}$ *for all pairs u, v of distinct vertices in X.*

REMARK 3.3. By Theorem 3.1, conditions (i) and (ii) of Corollary 3.2 are equivalent to the condition that, for each pair u, v of vertices in X, μ is a double eigenvalue of $H + u + v$ (the subgraph induced by $\overline{X} \cup \{u, v\}$). This in turn is equivalent to the condition that μ is an eigenvalue of $H+u$, $H+v$ and $H+u+v$; for if μ is not a double eigenvalue of $H + u + v$ in this situation then by Proposition 1.1 (with $G = H + u + v$, $Y = \{u, v\}$), one of $\{v\}, \{u\}$ is a star set, contradicting the fact that μ is an eigenvalue of $H + u$ and $H + v$.

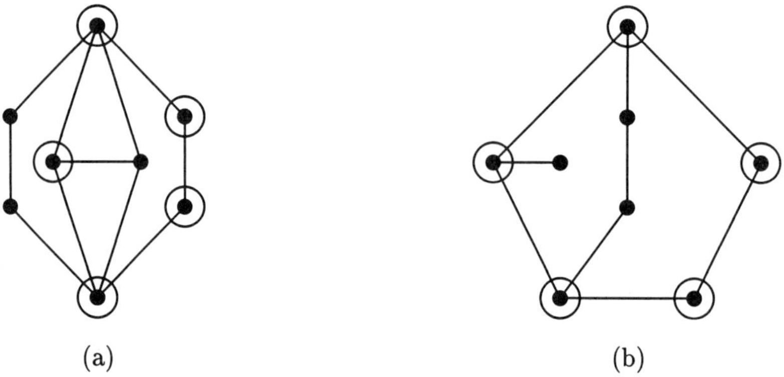

(a) (b)

FIGURE 2. The graphs of Example 3.4.

If G has H as a star complement for μ with corresponding star set X then each induced subgraph $G - Y$ $(Y \subseteq X)$ also has H as a star complement for μ. Moreover any graph with H as a star complement for μ is an induced subgraph of such a graph G for which X is maximal, because H-neighbourhoods determine adjacencies among vertices in a star set. Accordingly in determining all the graphs with H as a star complement for μ, it suffices to describe those for which a star set X is maximal.

EXAMPLE 3.4. Fig. 2 shows two 8-vertex graphs with a 5-cycle H which is a star complement for the eigenvalue 1. (The vertices of H are circled.) For the graph in Fig. 2(a) the corresponding star set is maximal: the graph cannot be extended to a 9-vertex graph in which 1 is an eigenvalue of multiplicity 4. The graph in Fig. 2(b) however can be extended to such a graph, and then to the Petersen graph (with 1 as an eigenvalue of multiplicity 5).

Let H be a graph with t vertices, and for $\Delta \subseteq V(H)$ let H_Δ denote the graph obtained from H by adding a vertex with Δ as its neighbourhood. For $\mu \in \mathbb{R}$ let $\mathcal{S}(H,\mu)$ be the set of those Δ for which μ is an eigenvalue of H_Δ but not of H. As a convenient descriptive device when $\mathcal{S}(H,\mu)$ is not empty we can define the *extendability graph* $\Gamma(H,\mu)$ as the graph on $\mathcal{S}(H,\mu)$ with Δ_1 adjacent to Δ_2 in $\Gamma(H,\mu)$ if and only if Δ_1, Δ_2 feature as H-neighbourhoods in a $(t+2)$-vertex graph for which H is a star complement. If we identify a set Δ with its characteristic vector, this graph is the 'compatibility graph' of [**6**, Algorithm 2.4]: the vertices of $\Gamma(H,\mu)$ are the $(0,1)$-vectors $\mathbf{b}$ in $\mathbb{R}^t$ such that $\langle \mathbf{b}, \mathbf{b} \rangle = \mu$, and $\mathbf{b}_1 \sim \mathbf{b}_2$ if and only if $\langle \mathbf{b}_1, \mathbf{b}_2 \rangle \in \{-1, 0\}$. A graph G with a maximal star set X for μ such that $G - X = H$ now corresponds to a clique in $\Gamma(H,\mu)$.

EXAMPLE 3.5. Fig. 3 shows the graph $\Gamma(H,1)$ in the case that H is the 5-cycle 123451. A clique of size 5 corresponds to the Petersen graph, one of size 3 to the graph illustrated in Fig. 2(a), and one of size 2 to the graph numbered 12-542 in [**2**].

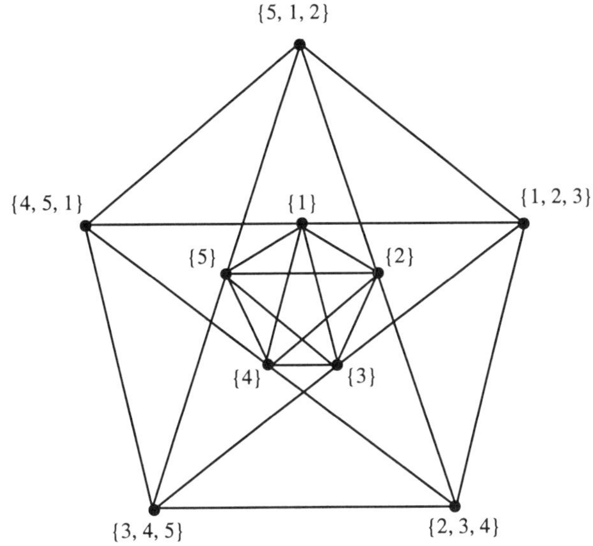

FIGURE 3. The graph $\Gamma(C_5, 1)$.

In general, finding all the graphs with a prescribed star complement H is equivalent to finding all the solutions μ, A_X, B of equation (3.1) for given C. We illustrate the procedure in the case that $H \cong K_{1,5}$, one of the smallest cases not treated in full generality elsewhere. We begin with some remarks from [**7**] on the general case $H \cong K_{r,s}$ $(r + s > 2)$. If H is the complete bipartite graph on $R\dot{\cup}S$, where $|R| = r$ and $|S| = s$, then we say that a vertex u added to H is of *type* (a, b) if the H-neighbourhood $\overline{\Gamma}(u)$ of u consists of a vertices in R and b vertices in S. If H has adjacency matrix C then C has minimal polynomial $x(x^2 - rs)$, and $\mu(\mu^2 - rs)(\mu I - C)^{-1} = (\mu^2 - rs)I + \mu C + C^2$. We may write equation (3.1) in the form

$$\mu(\mu^2 - rs)(\mu I - A_X) = B^T\{(\mu^2 - rs)I + \mu C + C^2\}B. \tag{3.2}$$

Now suppose that u, v are distinct vertices in X of types $(a, b), (c, d)$ respectively. If we let $A_X = (a_{ij})$ and equate (u, v)-entries in equation (3.2) we obtain

$$-\mu(\mu^2 - rs)a_{uv} = (\mu^2 - rs)\rho_{uv} + \mu(ad + bc) + acs + bdr, \tag{3.3}$$

where $\rho_{uv} = |\overline{\Gamma}(u) \cap \overline{\Gamma}(v)|$.

Next suppose that $r = 1, s = 5$ and write $H + u$ for the subgraph induced by $\overline{X}\cup\{u\}$. Since H has 0 as an eigenvalue and H is a star complement for μ in $H + u$ we have $\mu \neq 0$. Hence $H + u$ is connected by Proposition 2.1(i). From the spectra [**2**] of the 11 connected graphs $H + u$ (corresponding to the 11 possible types $(a, b) \neq (0, 0)$) we find that the only non-zero eigenvalue common to different graphs is -1, which arises when u is of type $(0, 2), (1, 1)$ or $(1, 5)$. Thus there are 16 possible H-neighbourhoods in this case. By Corollary 3.2, vertices u, v with different H-neighbourhoods may be added simultaneously if and only if equation (3.3) is satisfied. The feasible solutions of (3.3) when $r = 1, s = 5$ and $\mu = -1$ are shown in Table 1.

TABLE 1. the feasible solutions when $\mu = -1$.

(a, b)	(c, d)	a_{uv}	ρ_{uv}
(0, 2)	(1, 1)	0	0
		1	1
(1, 1)	(1, 5)	1	2
(0, 2)	(1, 5)	0	2
(0, 2)	(0, 2)	0	1
		1	2
(1, 1)	(1, 1)	0	1
		1	2
(1, 5)	(1, 5)	1	6

It follows that the only pairs of vertices which cannot be added simultaneously are pairs of type $(0,2)$ with no common neighbour in S. Thus $\Gamma(K_{1,5}, -1)$ is the complement in K_{16} of a Petersen graph P whose edges correspond to disjoint pairs in S. Now $\overline{P} \cong L(K_5)$, in which a clique has size 3 or 4. Thus in $\Gamma(K_{1,5}, -1)$ a clique has size 9 or 10. By symmetry each clique of size 9 gives rise to the same 15-vertex graph (G_{15}, say), and each clique of size 4 gives rise to the same 16-vertex graph (G_{16}, say). Both G_{15} and G_{16} have a star set for -1 which is maximal subject to the condition that no vertices are duplicated.

When $\mu \neq -1, \mu$ can arise as a multiple eigenvalue only when all vertices in X are of the same type, (a, b) say. In this situation, if we equate diagonal and non-diagonal entries in equation (3.2) we obtain

$$\mu^2(\mu^2 - 5) = (\mu^2 - 5)(a + b) + 2\mu ab + 5a^2 + b^2 \tag{3.4}$$

and

$$-\mu(\mu^2 - 5)a_{uv} = (\mu^2 - 5)\rho_{uv} + 2\mu ab + 5a^2 + b^2. \tag{3.5}$$

On subtracting equation (3.5) from equation (3.4), and dividing by $\mu^2 - 5$, we obtain $\mu^2 + \mu a_{uv} = a + b - \rho_{uv}$. Note that $a + b - \rho_{uv} \in \{1, 2\}$. Suppose by way of contradiction that μ is not an integer. If $\mu^2 \notin \mathbb{Z}$ then $a_{uv} = 1, a + b - \rho_{uv} = 1$ and $\mu = \frac{1}{2}(-1 \pm \sqrt{5})$; if $\mu^2 \in \mathbb{Z}$ then $a_{uv} = 0$ and $\mu = \pm\sqrt{2}$. But none of the 11 graphs $H + u$ has $\pm\sqrt{2}$ or $\frac{1}{2}(-1 \pm \sqrt{5})$ as an eigenvalue. Accordingly $\mu \in \mathbb{Z}$. The only integer other than -1 or 0 to be found among the eleven spectra is 1, which arises as an eigenvalue of $H+u$ precisely when u is of type $(0, 2)$. Now the solutions of equation (3.5) are given by $(a_{uv}, \rho_{uv}) \in \{(0,1), (1,0)\}$. Thus *all* $\binom{1}{0}\binom{5}{2} = 10$ possible vertices may be included in X: in this case X induces a Petersen graph because $u \sim v$ if and only if the 2-element subsets $\overline{\Gamma}(u), \overline{\Gamma}(v)$ of the 5-element set of endvertices in H are disjoint. The 16-vertex graph so obtained is the Clebsch graph [**1**, p.35]. Note that foreknowledge of this example shows that $\Gamma(K_{1,5}, 1) \cong K_{10}$ and obviates the need to solve equation (3.5). We may summarize our conclusions as follows.

THEOREM 3.6. *Let G be a graph with H as a star complement for the multiple eigenvalue μ. If $H \cong K_{1,5}$ then one of the following holds:*

(a) *$\mu = -1$ and the co-reduced subgraph of G is an induced subgraph of G_{15} or G_{16},*

(b) *$\mu = 1$ and G is an induced subgraph of the Clebsch graph.*

The general case $G - X = H \cong K_{1,k}$ has been studied under the additional assumption that no vertex in X is adjacent to the central vertex of H. This condition obtains when G is k-regular and it is shown in [**8**] that then the following hold unless G is a 4-cycle or 5-cycle: μ is a positive integer; $G - \overline{X}$ is regular of degree $\mu^2(\mu+2)$; $k = \mu(\mu^2+3\mu+1)$; $|X| = (\mu^2+3\mu+1)(\mu^2+2\mu-1)$; $|V(G)| = (\mu^2+3\mu)^2$. Moreover G is the Clebsch graph when $\mu = 1$ and the Higman-Sims graph [**1**, p.107] when $\mu = 2$.

More generally the solutions for equation (3.3) with $r = 1, s = k, a = c = 0$ are described in [**11**]. Again the corresponding vertices of the extendability graph form an intersecting family of subsets of $\overline{X}$, with adjacency in $\Gamma(H, \mu)$ contingent upon the size of intersections. For example, when $k = 29$, $H + u$ has -2 as an eigenvalue if u is of type $(0,5)$ or $(0,20)$. Let Δ_i $(i = 1, \dots, 118755)$, Δ'_j $(j = 1, \dots, 10015005)$ denote the corresponding subsets of $\overline{X}$ of size 5, 20 respectively. It turns out that in $\Gamma(K_{1,29}, 2)$ we have $\Delta_i \sim \Delta_j$ if and only if $|\Delta_i \cap \Delta_j| \in \{1, 3\}$, $\Delta'_i \sim \Delta'_j$ if and only if $|\Delta'_i \cap \Delta'_j| \in \{16, 18\}$, and $\Delta_i \sim \Delta'_j$ if and only if $|\Delta_i \cap \Delta'_j| = 4$. Corresponding vertices u, v of X are adjacent precisely when the intersection number here is 3 or 18. We give an example of a 36-vertex graph for which all five intersection numbers arise, thereby correcting the example in [**11**, p.84]. Suppose that the central vertex of H is labelled 36: then the graph in question is specified by taking

$$B^T = \begin{pmatrix} 0&0&0&0&0&0&0&0&0&0&0&0&0&0&0&0&1&1&1&1&0&0&0&0&0&0&0&0&1&0 \\ 0&0&0&0&0&0&0&0&0&0&0&0&0&0&0&0&1&1&0&0&1&1&0&0&0&0&0&0&1&0 \\ 0&0&0&0&0&0&0&0&0&0&0&0&0&0&0&0&0&0&1&1&0&0&1&1&0&0&0&0&1&0 \\ 0&0&0&0&1&1&1&1&1&1&1&1&1&1&1&1&1&1&1&1&1&1&1&1&0&0&0&0&0&0 \\ 0&0&1&1&0&0&1&1&1&1&1&1&1&1&1&1&1&1&1&1&1&1&1&1&0&0&0&0&0&0 \\ 1&1&0&0&0&0&0&0&1&1&1&1&1&1&1&1&1&1&1&1&1&1&1&1&1&1&0&0&0&0 \end{pmatrix}$$

in Theorem 3.1. Further, an example of a 51-vertex graph with 2 as an eigenvalue of multiplicity 21 is obtained by taking X to consist of 21 independent vertices whose H-neighbourhoods form the design $PG(2,4)$.

References

[1] P.J. Cameron, and J.H. van Lint, *Designs, Graphs, Codes and their Links*, Cambridge University Press, 1991.

[2] D. Cvetković, M. Doob, I. Gutman, and A. Torgašev, Recent Results in the Theory of Graph Spectra, North-Holland (Amsterdam), 1988.

[3] D. Cvetković, M. Doob, and H. Sachs, *Spectra of Graphs*, 3rd edition, Johann Ambrosius Barth Verlag (Heidelberg), 1995.

[4] D. Cvetković, and M. Petrić, A Table of Connected Graphs on Six Vertices, *Discrete Math*, **50** (1984) 37–49.

[5] D. Cvetković, P. Rowlinson, and S. Simić, *Eigenspaces of Graphs*, Cambridge University Press, 1997.

[6] M.N. Ellingham, Basic Subgraphs and Graph Spectra, *Australasian J. Combinatorics*, **8** (1993) 247–265.

[7] P.S. Jackson, and P. Rowlinson, On Graphs with Complete Bipartite Star Complements, *Linear Algebra and Appl.*, to appear.

[8] P. Rowlinson, Star Sets in Regular Graphs, *J. Comb. Math. & Comb. Computing*, to appear.

[9] P. Rowlinson, Dominating Sets and Eigenvalues of Graphs, *Bull. London Math. Soc.*, **26** (1994) 248–254.

[10] P. Rowlinson, Star Sets and Regularity in Graphs, *Linear Algebra and Appl.*, **226–228** (1995) 247–265.

[11] P. Rowlinson, On Graphs with Multiple Eigenvalues, *Linear Algebra and Appl.*, **283** (1998) 75–85.
[12] P. Rowlinson, and F.K. Bell, Graph Eigenspaces of Small Codimension, to appear.
[13] J.J. Seidel, Eutactic Stars, in: *Combinatorics* (eds. A. Hajnal and V. T. Sós), North-Holland (Amsterdam, 1978), 983–999.
[14] J.J. Seidel, Strongly Regular Graphs, in: *Surveys in Combinatorics* (ed. B. Bollobás), Cambridge University Press (1979), 157–180.

Mathematics and Statistics Group, Department of Computing Science and Mathematics, University of Stirling, Scotland FK9 4LA, UK
E-mail address: p.rowlinson@stir.ac.uk

DIMACS Series in Discrete Mathematics
and Theoretical Computer Science
Volume **51**, 2000

Geometry and Energetics of High Genus Fullerenes and Nanotubes

Humberto Terrones and Mauricio Terrones

ABSTRACT. The discoveries of fullerenes and carbon nanotubes have shown that graphite is very flexible and can acquire different curvatures. In the case of fullerenes, the presence of pentagonal rings of carbon produce the positive Gaussian curvature required for the closure with genus zero (no holes). If heptagonal or bigger carbon rings are introduced in a graphitic hexagonal mesh, periodic and quasiperiodic graphitic architectures can be proposed; these heptagons produce the negative Gaussian curvature necessary for the periodicity. In this work, we propose stable high genus fullerenes (up to genus 11) and nanotubes (up to genus 21) without dangling bonds and no pentagonal rings of carbon (no positive Gaussian curvature), possessing only hexagons and heptagons. These high genus structures have holes and can be regarded as finite zeolites. In addition, the stability of high genus vesicles is presented as an analogy with graphitic structures.

1. Introduction

It is well known that carbon can form diamond and graphite: in diamond every atom possess four neighbours, arranged in a tetrahedral way, separated by a distance of 1.56 Å (sp^3 bonding). In graphite each atom is surrounded by three first neighbours separated by 1.42 Å forming hexagonal sheets (sp^2 bonding) with interlayer spacing of 3.35Å . In 1985 a new structure of carbon composed of 60 atoms was discovered [**1**]. This new arrangement had a spherical shape with the symmetry of the icosahedron (I_h) and the same shape of a soccer ball. The structure received the name of Buckminsterfullerene (in honor of Richard Buckminster Fuller, American architect who designed the first geodesic domes). In C_{60}, as in graphite, every atom has three first neighbours, however, in C_{60} there are 12 pentagonal rings of carbon which produce the bending and the spherical shape (see Figure 1). In 1990, the first crystals formed with C_{60} were obtained opening a new area in physics, chemistry and materials science [**2**]. A year later, Sumio Iijima reported that graphite also could be bent to produce tubes composed of concentric sheets of graphite [**3**] (see

1991 *Mathematics Subject Classification.* 51M20, 53A05, 53A10, 49Q05.
We thank Prof. Alan Mackay, Prof. Harold Kroto, Prof. Patrick Fowler and Dr. D.R.M. Walton for fruitful discussions. This work was supported by Consejo Nacional de Ciencia y Tecnología (México) through Grant 25237-E. The support of DGAPA-UNAM IN107-296 and TWAS 97-178 RG/Phys/LA are also gratefully acknowledged.

Figure 2). In 1992, Daniel Ugarte found that by high irradiation of amorphous carbon, spherical concentric structures (graphitic onions) were possible [**4**]. C_{60} is not the only closed structure that has been found, there are other fullerenes such as C_{70}, C_{76}, C_{84}, etc., All these discoveries show that graphite is very flexible and can form a wide range of different structures with important properties.

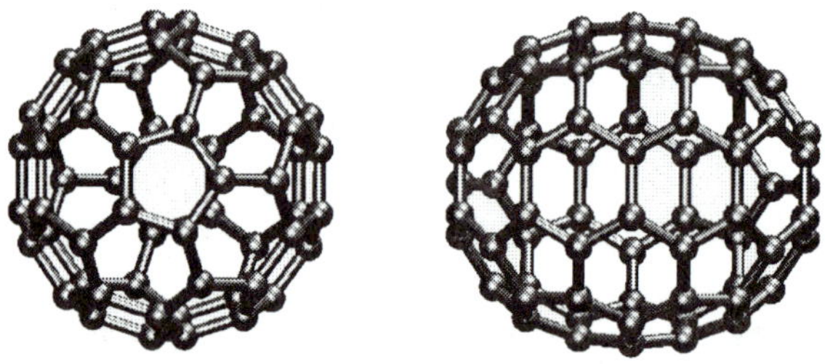

FIGURE 1. C_{60} and C_{70} fullerenes.

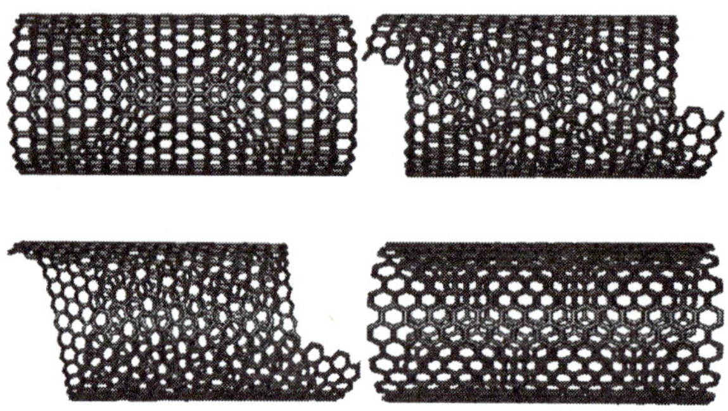

FIGURE 2. Carbon nanotubes with different helicities.

2. High Genus Graphitic Structures

In order to make a systematic study of all possible graphitic structures we have to use some concepts of differential geometry and topology, such as curvature and genus. The curvature of a curve at a point is defined as the inverse of the radius of a circle which best fits the curve at that point. In a surface two principal curvatures k_1 and k_2 can be obtained. The product of the two principal curvatures is called Gaussian curvature K ($K = k_1 k_2$) and the average value H is called mean curvature ($H = (k_1 + k_2)/2$). For a sphere the two principal curvatures are the same and have the same sign $k_1 = k_2 = 1/R$ (positive Gaussian curvature), where R is the radius of the sphere. In a cylinder $k_1 = 1/R$ and $k_2 = 0$ where R is the radius of the cylinder (zero Gaussian curvature), and in a flat sheet $k_1 = k_2 = 0$ (zero Gaussian curvature). There are surfaces which present negative Gaussian curvature, so the surface is composed of saddle points. Examples of surfaces with negative Gaussian curvature are the torus, the catenoid, the helicoid and the triply periodic minimal surfaces which also have $H = 0$ everywhere (see Figure 3).

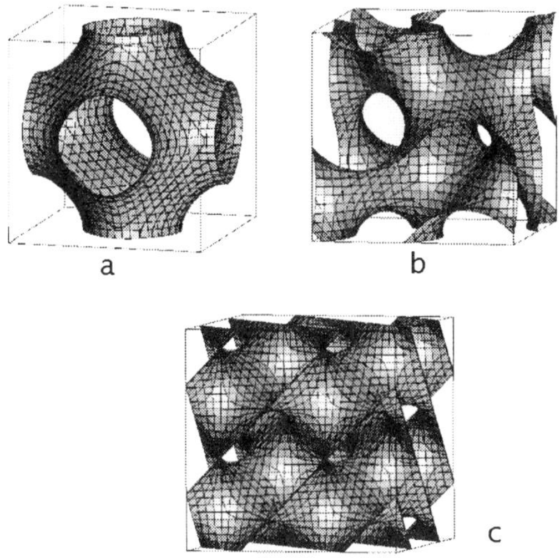

FIGURE 3. Cubic cells of Periodic Minimal Surfaces: a) P (primitive) b) G (gyroid) and c) D (diamond).

Therefore, carbon can form structures with different curvatures: zero Gaussian curvature for graphite and nanotubes, and positive Gaussian curvature for C_{60} and other fullerenes, but what about negative curvature? It has been proposed that by the introduction of heptagonal or octagonal rings of carbon in a graphitic mesh, periodic and quasiperiodic structures with negative Gaussian curvature are possible [**5**].

Considering a closed graphitic sheet in which each vertex (atom) is shared by three polygons (carbon rings) and each edge (bond) is shared by two polygonal faces, the following equation can be obtained:

$$2N_4 + N_5 - N_7 - 2\,N_8 = 6\chi = 12(1-g) \tag{2.1}$$

where N_n is the number of polygons with 4, 5, 7 and 8 sides. As we can see from Equation 2.1, N_4, N_5, N_7 and N_8 are defects in a graphitic structure which contribute to the curvature and topology of the arrangement. N_6 is not present since the hexagons do no contribute to the Gaussian curvature of the structure (hexagons produce zero Gaussian curvature). Although, four membered rings are not very stable in the case of graphite, they can be present in boron nitride (BN). χ is the Euler characteristic which is a geometric invariant related to the topology of the structure and g is the genus or number of handles of the arrangement (a sphere or a fullerene have $g = 0$, a torus has $g = 1$).

As we can see, graphite can produce architectures that belong to the three geometries of surfaces in 3-D space: Euclidean geometry (graphite), spherical geometry (fullerenes), hyperbolic geometry (negative curvature surfaces decorated with graphite).

Besides periodic negatively curved graphite, closed structures such as quasi-spherical onions –with heptagons and additional pentagons– and toroidal arrangements with genus 1 (one handle) have been studied [**6**], [**7**] (see Figure 4). Moreover, higher genus structures are possible. We have been able to generate closed higher genus fullerenes (higher than 1) without having pentagonal rings (no positive Gaussian curvature). Graphitic fullerenes up to genus 11 (holey-balls) and nanotubes up to genus 21 (holey-tubes) have been simulated in the computer [**8**]. These new hypothetical arrangements do not have dangling bonds and are composed just of hexagons and heptagons (negative Gaussian curvature) (see Figures 5 and 6). The topology of holey-balls corresponds to two concentric fullerenes in which all the pentagonal rings have been removed and replaced by heptagons, so the two structures can be connected by necks (see Figure 5c). For constructing icosahedral genus 11 holey-balls, the pentagonal cusps from two concentric icosahedral fullerenes (e.g. C_{240} and C_{500}) should be removed with the subsequent addition of heptagons in order to interconnect the two spheres by 12 necks. For our smallest holey arrangements C_{380} and C_{720}, 20 nonagons are necessary, nevertheless, the structures are still very stable. A family of genus 5 holey-balls (with cubic symmetry, O_h) has also been built using the complete cubic cells of the Schwarzites type P (see Figure 5d and Table 1). There are different ways to build and understand the topology of the holey-balls and one of them, that can be generalized, consists in joining by necks (holes) the faces of two (can be more) concentric Platonic solids.

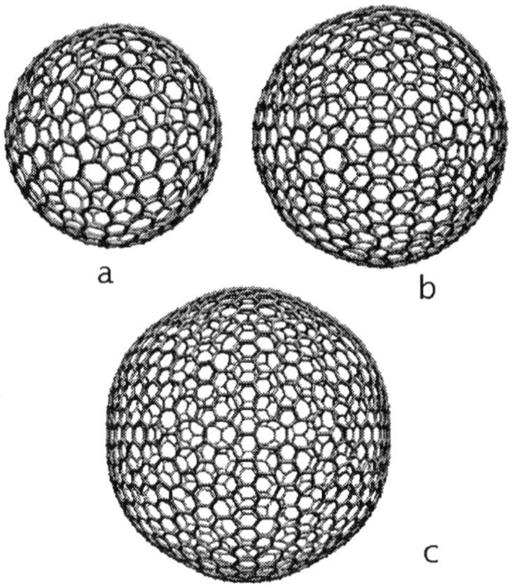

FIGURE 4. Quasispherical fullerenes. a) C_{540} with 76 pentagons and 64 heptagons. b) C_{888} with 60 pentagons and 72 heptagons. c) C_{1500} with 132 pentagons and 120 heptagons.

The holey-tubes possess genus greater than 5 and have been built by growing the 216-atom cubic cell of the Schwarzite P in one direction (the direction of the axis of the tube) closing both ends so as to avoid dangling bonds by the addition of heptagonal and hexagonal rings, in the same way as we have closed the holey-balls. These tubes present a four-fold axis of symmetry along the tube. When a cubic cell (a = 17.66 Å) is added to the tube, the tube grows in 688 atoms and the genus increases in 4. The holey-tubes can be seen as finite Schwarzites that have grown in one direction (containing the full cubic cells of the Schwazites type P in the interior of the tubes).

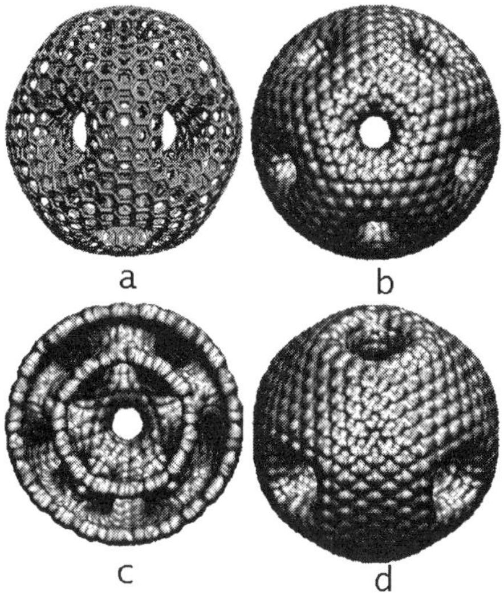

FIGURE 5. Graphitic holey Balls. a) Twofold axis of a 2040-atom holey-ball with icosahedral symmetry, genus 11. b) Five-fold axis. c) Cut perpendicular to the fivefold axis showing two concentric fullerene-like structures joined by necks. d) 1896-atom holey-ball, with genus 5 in the three fold orientation.

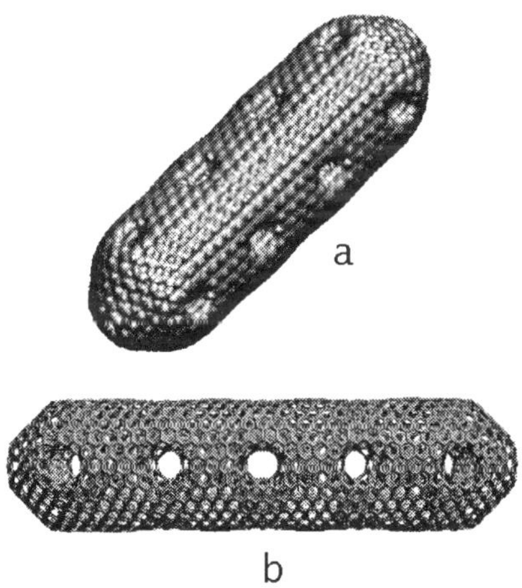

FIGURE 6. Holey-tubes. a) 2880-atom holey-tube with genus 17. b) 3568-atom holey-tube with genus 21.

The genus can be related to the number of holes in the structure: in the holey-balls the holes are located at the faces of the structural units (cube and dodecahedron), therefore, the symmetry makes these holes to meet at the center of the arrangement, so the holes are not isolated and for this reason the genus is equal to the number of holes (faces of the polyhedron) less one. Also, This is the case for the holey-tubes. In the holey-balls and holey-tubes that we have presented

TABLE 1. Holey-balls with non-positive Gaussian curvature and no dangling bonds. Energy/atom relative to graphite, approximated radius of the structure, genus and number of hexagons. C_{380} and C_{720} have 60 heptagons and 20 nonagons. C_{744} has 24 octagons and no heptagons, a suitable model for a BN higher genus molecule.

Structure	Energy/Atom eV	Appox. radius Å	genus	No. of hex.
$C_{60}(I_h)$	0.6600	3.64	0	20
C_{380} (I_h)	0.5078	8.32	11	90
$C_{720}(I_h)$	0.3110	10.74	11	260
C_{740} (I_h)	0.4432	10.58	11	230
C_{744} (O_h)	0.4864	11.54	5	340
C_{1320} (I_h)	0.3123	14.55	11	520
C_{2040} (I_h)	0.2202	18.09	11	880
C_{816} (O_h)	0.4397	11.73	5	352
C_{1536} (O_h)	0.2733	15.66	5	712
C_{1896} (O_h)	0.1982	17.74	5	892

TABLE 2. Holey-tubes with radius of 11.25Å composed of 2, 3, 4 and 5 P Schwarzite cells. Energy per atom relative to graphite, length, genus and number of hexagons.

Structure	Energy/Atom eV	length Å	genus	No. of hex.
C_{1504}	0.3515	39.59	9	640
C_{2192}	0.3156	57.25	13	928
C_{2880}	0.2963	74.91	17	1216
C_{3568}	0.2850	92.57	21	1504

the necks (holes) contain non-isolated heptagons, however, the heptagons can be isolated by hexagons in order to increase the diameter of the holes making the structures less curved, therefore, releasing strain.

In order to have an idea on the elastic strain, the Tersoff potential [**9**] has been implemented to relax the configurations; energies per atom relative to graphite and geometric properties are shown in Tables 1 and 2. The energetics of high genus graphite is comparable to those of Schwarzites, therefore, possessing a high stability. The electronic properties of these new hypothetical structures have been calculated using tight binding; the interesting result is that they behave locally as metals around the holes (were the heptagons and curved regions are) and possess a semimetalic behaviour in the flat parts (parts with just hexagons) [**10**]. Therefore, holey-balls and holey-tubes have active sites (holes) were atoms or molecules can be put, so can be regarded as finite zeolites.

3. High Genus Vesicles

We also have studied the stability of continuous closed surfaces with high genus. In particular, surfaces with the same topologies as the holey-balls and holey-tubes are considered. The motivation for this analysis is that closed surfaces also appear in other systems such as in water-lipid-surfactant mixtures, membranes and vesicles [**11**].

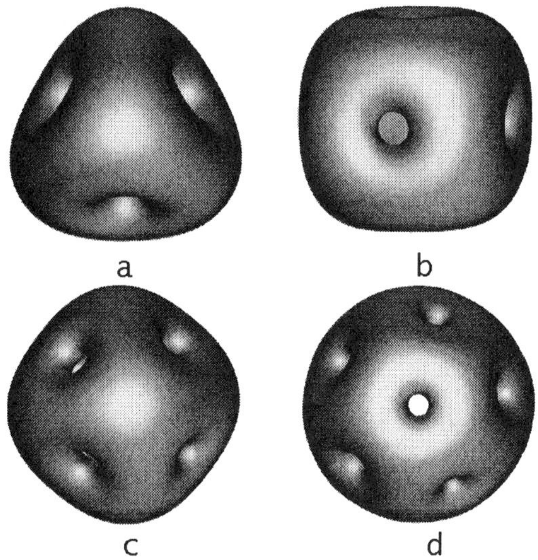

FIGURE 7. high genus vesicles. a) Tetrahedral, genus 3 (td). b) Cubic, genus 5 (Oh). c) Octahedral, genus 7 (O). d) Icosahedral, genus 11 (Ih).

The elastic energy of a surface can be written as:

$$E = \int \alpha \, dS + \int \beta \, H^2 \, ds + \int \gamma \, K \, ds \tag{3.1}$$

where H and K are the mean and Gaussian curvatures respectively. α is the surface tension, β is the bending rigidity and γ is the Gaussian bending modulus [**12**]. The first term corresponds to the stretching energy and the second and third terms refer to the bending energy. For different closed surfaces up to genus 21, we have minimized the energy functional E using the program Surface Evolver written by Keneth Brakke [**13**]. The value of the third term is a topological constant related to the genus of the structure, so it does not change during the minimization ($\int K \, ds = 4\pi(1-g)$). Minimization of just $W = \int H^2 ds$ is being carried out and the results will be published elsewhere. The minimum of W for surfaces with genus < 5 have been reported by Hsu *et al.* [**14**]. In Tables 3 and 4 we show the values of W that minimize E, the reduced volume per surface ratio v_r ($v_r = 6\sqrt{\pi} V/A^{3/2}$, where V is the volume and A is the area), the surface per volume ratio S_r ($S_r = A/V^{2/3}$), the number of vertices and faces ($Faces - Edges + Vertices = \chi$). Surface Evolver starts from a polyhedral structure with a determined topology and gradually the surface is triangulated several times, at a fixed volume, in order to have smaller triangles; at all times the surface is minimized accordingly. As the number of vertices grow, the structure approximates to the continuous surface. In order to test the program, the minimum of W for different surfaces such as a sphere, a torus and the cases studied by Hsu *et al.* [**14**] were calculated obtaining consistent results. We can say that the shapes produced are the optimal for the energy E See Figures 7, 8 and 9).

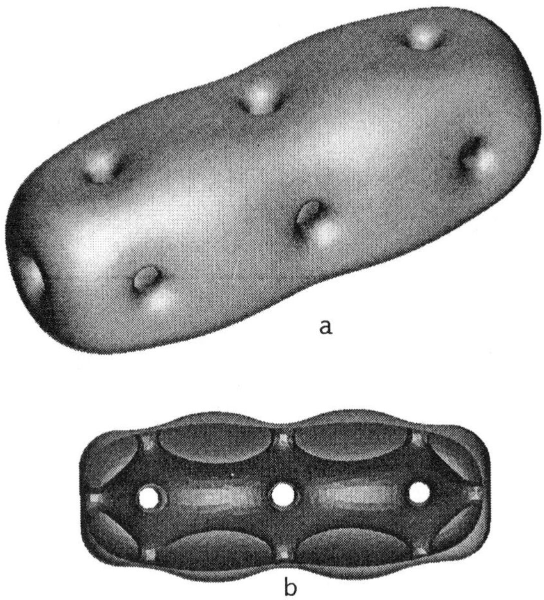

FIGURE 8. a) Genus 13 cylindrical vesicle. b) Cut along the tube to show the interior channels.

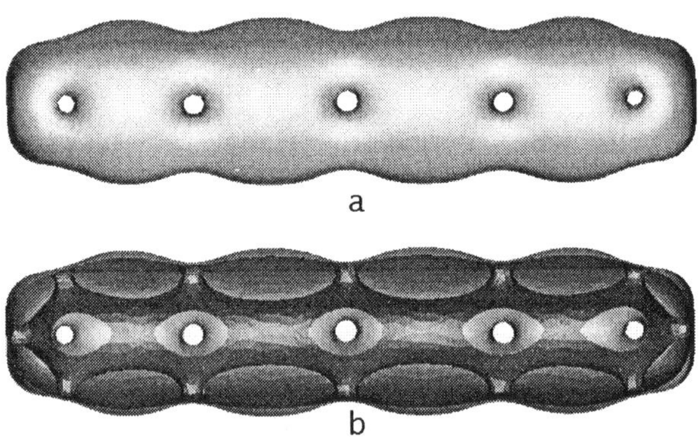

FIGURE 9. a) Genus 21 cylindrical vesicle. b) Cut along the tube to show the interior channels.

4. Conclusion

New graphitic structures with high genus (holey-balls and holey-tubes) have been proposed. An interesting feature of these new hypothetical arrangements is that they do not have dangling bonds and that pentagonal rings of carbon are not present as in fullerenes. In fact, there are just hexagons and heptagons and the

TABLE 3. High genus vesicles

genus	W	v_r	S_r	vertices	faces
3	23.4098	0.6088	6.7317	9526	19062
5	24.7272	0.5175	7.5022	9485	18986
7	25.2641	0.4507	8.2266	9260	18544
11	28.0751	0.3910	9.0440	9316	18672

TABLE 4. High genus cylindrical vesicles

genus	W	v_r	S_r	vertices	faces
9	29.4730	0.4289	8.5023	9321	18674
13	39.0488	0.3486	9.7627	9420	18888
17	46.9734	0.3047	10.6781	9971	20006
21	56.1260	0.2712	11.5409	9620	19320

curvature depends only on the heptagonal rings. Regarding the genus, holey-balls (up to genus 11) and holey-tubes (up to genus 21) are more complex than fullerenes (genus 0) being the most complicated closed graphitic architectures proposed so far. The Tersoff three-body potential has been applied to relax the holey-fullerenes finding that are quite stable. In addition, interesting electronic properties at the channels (where the heptagons are located) are predicted. We believe that one way of synthesizing holey-structures can be by annealing polyhedral carbon particles (from soot) previously treated by nitric acid and controlled oxidation. Also, the stability of vesicles with the same topologies as the holey-balls and holey-tubes has been studied by minimizing an energy functional which depends on the area and principal curvatures. In summary, what we have shown is that mathematics proves being very useful to propose structures with the same topological features in different systems such as in graphite and phospholipids.

References

[1] H.W. Kroto, J.R. Heath, S.C. O'Brien, R.F. Curl, and R.E. Smalley, *Nature*, **318** (1985) 162.
[2] W. Krätschmer, L.D. Lamb, K. Fostiropoulos, and D.R. Huffman, *Nature*, **347** (1990) 354.
[3] S. Iijima, *Nature*, **354** (1991) 56.
[4] D. Ugarte, *Nature*, **359** (1992) 707.
[5] A.L. Mackay, H. Terrones, *Nature*, **352** (1991) 761; T. Lenosky, X. Gonze, M. Teter and V. Elser, *Nature*, **355** (1992) 333;
H. Terrones and A.L. Mackay, *Chem. Phys. Lett.*, **207** (1993) 45;
D. Vanderbilt and J. Tersoff, *Phys. Rev. Lett.*, **68** (1992) 511;
M. O'Keeffe, G.B. Adams and O.F. Sankey, *Phys. Rev. Lett.*, **68** (1992) 2325;
A.L. Mackay, and H. Terrones, *Phil. Trans. Roy. Soc. A*, **343** (1993) 113.
[6] M. Terrones, and H. Terrones, *Full. Sci. Tech.*, **4** (1996) 517;
H. Terrones and M. Terrones, J. *Phys. Chem. Sol.*, **58** (1997) 1789.
[7] L.A. Chernozatonskii, *Physics Letters A*, **170** (1992) 37;
S. Ihara, S. Itoh, S. and J. Kitakami, *Phys. Rev. B*, **48** (1993) 5643.
[8] H. Terrones and M. Terrones, *Phys. Rev. B*, **55** (1997) 9969.
[9] J. Tersoff, *Phys. Rev. Lett.*, **61** (1988) 2879.
[10] J.L. Ricardo-Chávez, J. Dorantes-Dávila, M. Terrones and H. Terrones, *Phys. Rev. B*, **56** (1997) 12143.
[11] X. Michalet, D. Bensimon and B. Fourcade, *Phys. Rev. Lett.*, **72** (1994) 168;
X. Michalet, and Bensimon, *Science*, **269** (1995) 666.

[12] F. Jülicher, *J. Phys. II France*, **6** (1996) 1797.
[13] K.A. Brakke, Experimental Mathematics, Vol. 1, 142 (1992);
K.A. Brakke, *Phil. Trans. R. Soc. LOnd. A.*, **354** (1996) 2143.
[14] L. Hsu, R. Kusner and J. Sullivan, *Experimental Math.*, **1** (1991) 191.

Instituto de Física, UNAM, Apartado Postal 20-364, 01000 México, D.F., México
E-mail address: terrones@fenix.ifisicacu.unam.mx

School of Chemistry, Physics and Environmental Science, University of Sussex, Brighton BN1 9QJ, England, U.K.
E-mail address: kapa4@sussex.ac.uk

DIMACS Series in Discrete Mathematics
and Theoretical Computer Science
Volume **51**, 2000

An Abstract Representation for Molecular Graphs

Philippe Vismara and Claude Laurenço

ABSTRACT. The design of a synthesis strategy in Organic Chemistry rests on the perception of the target molecule. This one has to be perceived from several viewpoints: topology, stereochemistry, functionality... We propose an abstract representation of molecules which describes the topological viewpoint (cycles, carbon chains and their structural relationships). This representation is based on a polynomial algorithm that computes the set of relevant cycles in the molecular graph. This set is equal to the union of all the minimum cycle bases of the graph. The abstract representation is integrated in RESYN, a system for computer-aided organic synthesis planning.

Introduction

The synthesis of organic molecules is a problem which is solved in finding an efficient way to construct a given target molecule, for instance a natural product, from available simple starting materials, by applying chemical reactions in a certain sequence. These successive steps define a synthetic pathway. Retrosynthetic analysis [**4**], a reasoning backward from the target structure, is often used to design a synthetic pathway and to find a suitable set of starting materials. Because thousands of transformations are known and thousands of potential starting materials are available, it is impossible to perform a systematic analysis. To combat the combinatorial explosion, the synthetic chemist needs control strategies which reduce the size of the search space. Formalization of strategies used in synthesis planning is still a great challenge.

Since the early sixties, continuous efforts have been made to develop several classes of programs for computer-assisted organic synthesis [**29**]. In contrast to the widespread acceptance of database systems for molecule and reaction retrieval, the efficiency of computer-assisted synthesis planning programs remains to be proved [**21**]. Recently, authors of LHASA, one of the most representative programs of this class [**5**], confirmed this opinion. They wrote that entering its third decade, this program "is not yet so powerful as to be an indispensable tool to the synthetic chemists" [**27**].

Many factors have limited the development of computer-assisted synthesis planning. The combinatorics complexity of synthetic problems is not the only obstacle.

1991 *Mathematics Subject Classification.* Primary 05c85, 92e99; Secondary 05c38.

The authors are associate researchers at the LIRMM (CNRS / Université Montpellier II).

Most of the existing programs have been developed for several decades and they generally do not integrate new advances in computer science. These programs often produce too many results without explaining them, specially when heuristics have been applied to reduce the size of the search space. Moreover, creating and updating the knowledge bases of such systems remain a major difficulty and are hard tasks for experts in organic synthesis.

The RESYN project have been developed taking the weakness of current programs into consideration. The aim of the project is to mix research into modeling of organic synthesis planning [**24, 18, 8**] and computer science results in Graph Theory, Constraint Satisfaction Problems, Distributed Artificial Intelligence, Classification, Analogy and Case-based reasoning, Knowledge Acquisition and Representation [**42, 34, 26, 32, 22**]. A system for computer-assisted synthesis planning is currently under development.

The aim of this paper is not to present the whole RESYN project. We focus on the topological perception of the target molecule. This analysis is a crucial step in the definition of retrosynthetic strategies. These strategies are also based on the perception of stereochemistry and functional groups which are not presented in this paper.

The perception of a molecule involves recognizing it as a member of conceptual categories and provides diverse types of information such as properties of that molecule, its shape and parts which make it up. All this information is a base for action. If we recognize that a molecule belongs to a chemical family, we may interpret that in terms of known synthetic methods or starting materials for its construction. The process starts with the target structural formula whose information is interpreted to construct a new representation. This one describes at several abstract levels, in a structured way, multiple aspects of the molecule and gives a view, both global and detailed, of the problem which is to be solved. The topological perception of a molecular graph consists in the recognition of cyclic parts and the description of non-cyclic patterns as carbon chains, heteroatomic links,.... The main part of this paper aims at the computation of the set of relevant cycles in the molecule. The second part presents an abstract representation of the molecular graph which describes the structural relationships between patterns (cycles, carbon chains,...).

Relevant cycles

Study of cycles in molecules is a very long and rich story. It deals with finding the best set of cycles to analyze the cyclic structure of a molecule. A quite complete review of papers dealing with cycles in molecule has been done by [**9**]. Therefore, the aim of this part is to perform a more precise analysis of some works from a graph theoretic point of view.

This part is organized as follows: Section 2 presents some algorithms that search for a cycle basis of the cycle vector space associated with a molecular graph. Two major kinds of cycles basis are studied: *fundamental* basis (associated to a *spanning tree*) and *minimum* basis (often named SSSR). We show how successive papers have tried to optimize the search for a minimum basis, until [**20**] that presents a polynomial algorithm.

Generally, cycle basis notion is not satisfactory to perform a good analysis of the cyclic structure of the molecule. Hence, we present in Section 3 a few papers that deal with extended minimum cycle basis. We discuss on their interests and limits before trying to define a general notion of *Relevant Cycles*. This notion is described in Section 4. We propose a polynomial time algorithm to compute the *Set of Relevant Cycles* (denoted by $\mathcal{C}_R$) as a polynomial number of cycles families which define a partition of $\mathcal{C}_R$. Each family is represented by a single prototype from which all the other cycles of the family can be generated without backtracking.

1. Preliminaries

1.1. Definitions. A molecule is generally represented by an *undirected* graph $G = (V, E)$ without *loops* (where V denotes the set of *vertices* associated with the atoms and E defines the set of *edges* which correspond to bonds between atoms). We denote by n and m the numbers of vertices and edges respectively.

In an undirected graph, an edge is an unordered pair of vertices $\{v_1, v_2\}$.

A *path* (or a *walk*) is a route between two vertices passing along edges[1]. A *simple* path must contain each edge at most one time. A path is *elementary* if no vertex appears more than once.

For any simple path μ, we define the *length* of the μ (denoted by $|\mu|$) the number of edges in μ. Define $d(x, y)$ to be the length of any shortest path from x to y.

For any vertex v in V, we denote by $d(v)$ the degree of v in G and by $\Gamma(v)$ the set of vertices adjacent to v (so $d(v) = |\Gamma(v)|$).

A *cycle* (or a *ring*) is a closed path (i.e. a path that starts and ends at the same vertex). Simple and elementary cycles are defined according to path definitions. In this paper, we consider a cycle as either a set of edges $\{\{v_1, v_2\}, \{v_2, v_3\}, ...\}$ or a closed path $(v_1, v_2, v_3, ..., v_1)$. A simple cycle C can also be represented by a vector of $\{0, 1\}^m$ such that $C[i] = 1$ if and only if edge i belongs to C. Then, the sum of two cycles C_1 and C_2 is defined by the boolean addition of their associated vectors (regarding cycles as sets of edges, this sum is equal to $(C_1 \cup C_2) \setminus (C_1 \cap C_2)$). One can define the *vector space of cycles* on the set of simple cycles closed by addition. All the elements of this vector space (denoted by $\mathcal{C}$) are simple cycles or sum of simple cycles. Hence, $\mathcal{C}$ is the set of all the subgraphs[2] of G in which the degree of any vertex is even.

1.2. The Cyclomatic Number. The main interest of the cycle vector space is that it introduces the notion of cycle basis. A cycle basis is a set of cycles such that any element of $\mathcal{C}$ can be obtained by a sum of cycles of this basis. The dimension of the vector space defines the minimum number of cycles needed to describe all cycles of the graph. This dimension is generally called the *cyclomatic number* and denoted by ν. For a graph with n vertices and m edges, ν is equal to $m - n + 1$.

This formula has a very long history. It was introduced by **[12]** to determine the number of faces in a polyhedron. This result was generalized by **[2]** and described on a different way by **[14]** (which is generally cited as *Frerejacques number*). Ever since, some other formulas have been introduced that give the same result, like **[11]**.

---

[1]more formally, there is a path between vertices v_0 and v_k, passing through vertices v_1, v_2 ... and v_{k-1}, if each pair $\{v_i, v_{i+1}\}$ is an edge of the graph (where $i = 0, 1, ..., k-1$).

[2]a subgraph of $G = (V, E)$ is a graph $G' = (V', U')$ such that $V' \subseteq V$ and $E' \subseteq E$

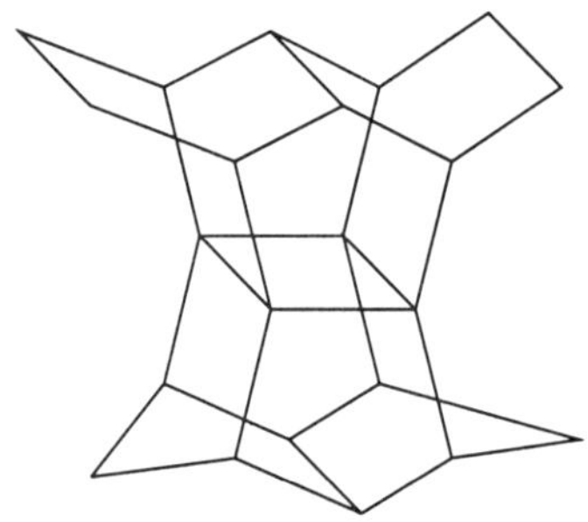

FIGURE 1. The 685 elementary cycles of this "pagodane" molecule can be generated from a cycle basis including only 11 cycles.

1.3. 2-Connected Components. A graph is *k-connected* if any vertices v_1 and v_2 are linked by at least k distinct paths. So, a molecular graph is generally *connected* (i.e. 1-connected). A *2-connected* component of a graph G (sometimes called a *block* of G) is a 2-connected maximal subgraph of G. An important property is that all the vertices of any elementary cycle belong to the same 2-connected component. Hence, studying the cycle vector space can be independently performed on each 2-connected component. This decomposition of the graph generally provides a great improvement. Firstly because it eliminates all the vertices which do not belong to any cycle (such vertices correspond to 2-connected components reduced to a single vertex). Secondly, because it divides the initial large problem into a few smallest ones.

Notion of 2-connectivity is seldom used by papers dealing with cycles in chemistry. Some of them, like [**25**], suggest to recursively eliminate terminal atoms (i.e. vertices whose degree is 1). This reduction does not only preserve vertices belonging to cycles. It also conserves the vertices of any elementary path linking two independent cycles. To solve this problem, [**1**] has proposed to eliminate vertices whose degree is 2. This method constructs a "crunched graph" on which the cycle search is performed. Some precautions must be taken to detect cycles that include vertices whose degree is 2. These reductions are not as efficient as 2-connectivity components. Hence, [**13**] has proposed to decompose the graph in *blocks* which correspond to 2-connected components. Unfortunately, the algorithm they presented is more complicated and has a greater time complexity than the one introduced by [**40**] which can be performed in linear time.

2. Cycle Basis

The set of all elementary cycles is not always relevant to analyze the cyclic structure of a graph. For example, the molecule presented in Figure 1 includes 685 elementary cycles. Hence, it is often important to be able to describe the cycle vector space without enumerating all cycles. Of course, a cycle basis provides such an optimal representation of the vector space. For a given graph, numerous cycle bases can exist. Cycle bases are compared according to their lengths (the length of a cycle basis is defined by the sum of the lengths of all cycles in the basis). This notion defines a *minimum cycle basis*.

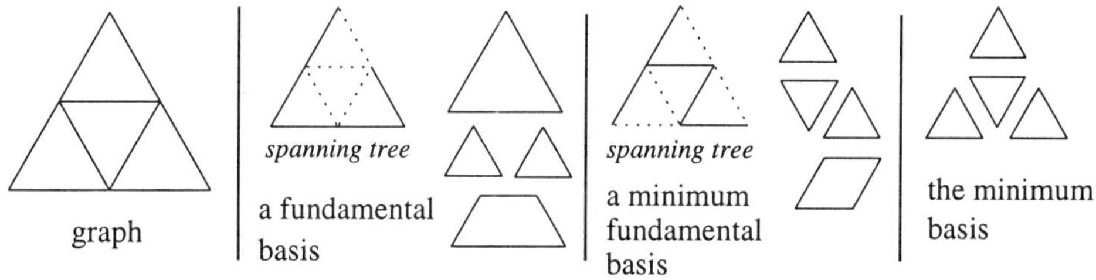

FIGURE 2. The minimum cycle basis of this graph consists of four 3-edges cycles. But no fundamental basis should include all of them.

2.1. Fundamental Basis. The first and simplest way used to construct a cycle basis consists in finding a *fundamental basis associated with a spanning tree*. This method has been introduced by [**23**]. In a *spanning tree*[3] $T = (V, E')$ of a graph $G = (V, E)$, adding any edge from $E \setminus E'$ creates a single cycle. This cycle is called a *fundamental* cycle according to the spanning tree T. Because $|E \setminus E'| = \nu$ and all these cycles are independent[4], they form a basis of the cycle space. Many papers deal with fundamental basis search [**44, 30, 17**]. Most of them use this notion to enumerate all elementary cycles of the graph. Unfortunately, [**28**] has shown that such a method can generate 2^{ν} vectors a few of which actually correspond to elementary cycles. This difficulty can be surmounted using a search algorithm, as [**33**], which needs $O(n + m + nc)$ operations (where c is the number of elementary cycles). Another search algorithm, quite easy to implement, has been proposed by [**36**]. For *planar graphs*[5], [**38**] has shown that a vector space approach can be as efficient as a search algorithm.

For a given graph, fundamental cycle bases can be numerous. [**7**] showed that finding a *minimum fundamental cycle basis* is a NP-complete problem[6]. Note that the *minimum cycle basis* is not necessarily fundamental (see Figure 2). We will see that finding a minimum cycle basis can be performed in polynomial time.

2.2. Minimum Basis: SSSR. Since a few elementary cycles can be considered as "chemically relevant", the cycle vector space is generally described by a *Smallest Set of Smallest Rings*. This notion called SSSR is equivalent to a *minimum cycle basis* (not associated with a spanning tree). Hence, an SSSR contains ν independent cycles such that the sum of their lengths is minimum.

2.2.1. *SSSR computed from a fundamental basis.* The oldest method used to find an SSSR starts by searching a fundamental cycle basis. Then, this basis $\mathcal{B}$ is optimized using the following criterion: *for each cycles $C_1, C_2 \in \mathcal{B}$, if $C_1 + C_2$ is smaller than either C_1 or C_2, replace this fundamental cycle by $C_1 + C_2$ in $\mathcal{B}$.* The new set of cycles is a basis smaller than $\mathcal{B}$. This method, used by [**6**] and [**45**], does not necessarily find a minimum basis even if some other heuristics are employed.

[3]A spanning tree is a connected subgraph without cycles which covers all the vertices of the graph.

[4]We only add in T one edge of $E \setminus E'$ at the same time. So, each fundamental cycle includes an edge which cannot belong to any other cycle.

[5]a graph is *planar* if it can be represented on a plane in such a way that no two edges cross one another.

[6]So, no polynomial time algorithm would probably be found to solve that problem

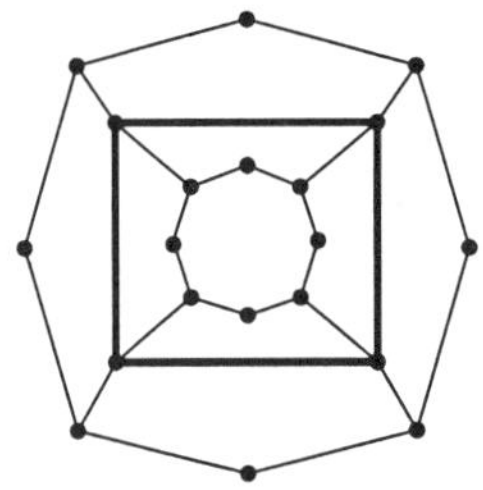

FIGURE 3. Any minimum cycle basis (SSSR) must include the 4-edges cycle. But this cycle never appears as a face of a planar representation of the graph.

Another algorithm using fundamental basis notion has been proposed by [**19**]. It directly finds a short fundamental basis that is not necessarily minimal.

2.2.2. *SSSR directly computed.* Since it is not easy to optimize a fundamental basis, many algorithms try to directly obtain a minimum basis. For "fundamental" algorithms, the main problem lies in the optimization of an initial basis. On the opposite, "direct" algorithms easily find a minimum set of independent cycles but have difficulties to obtain a complete basis. To minimize the basis length, most of the algorithms associate to any edge (or vertex) the shortest cycles that include it. But these cycles are not necessarily different or independent. The set finally obtained often includes less than ν cycles. Hence, a general method for "direct" algorithms consists in applying a heuristic search that minimizes lengths until ν independent cycles are obtained.

For example, [**46**] has proposed to use spanning notions to find a cycle basis. The algorithm chooses a vertex and searches a smallest cycle that covers it. All vertices in that cycle are marked and the process goes on until every vertex is marked. If the cyclomatic number is not reached, a similar method is performed with edges and then with faces. However, the final set of cycles is not always an SSSR. The problem lies in the absence of direct relations between minimum cycles bases and faces of a planar representation of the graph. For such a representation of the molecule, the set of faces defines a cycle basis. Like fundamental ones, this basis is not necessarily minimum. It depends on the representation of the graph. And for some particular graphs, no planar representation defines a minimum basis (see Figure 3).

Another interesting algorithm has been presented by [**16**]. Like "fundamental" algorithms, it uses a spanning tree to obtain a set of "ring-closure edges". For any ring-closure edge, a shortest cycle is searched instead of a fundamental one. The problem is that two different closure edges can be associated with the same shortest cycle (see edges a and b in Figure 4.I). Hence, Gasteiger has proposed to order the edges of the molecular graph according to the breadth first search that computes the spanning tree. Each cycle closure edge e is associated with the shortest cycle that only includes edges preceding e in this order. However, that simple but very powerful method can fail when a shortest cycle is only composed by ring-closure edges. For instance, in Figure 4.II, the 3-edges cycle would not be found respecting edge orientation. Some heuristics has been proposed to solve that problem. But none of them guarantees that the final cycle basis is minimum.

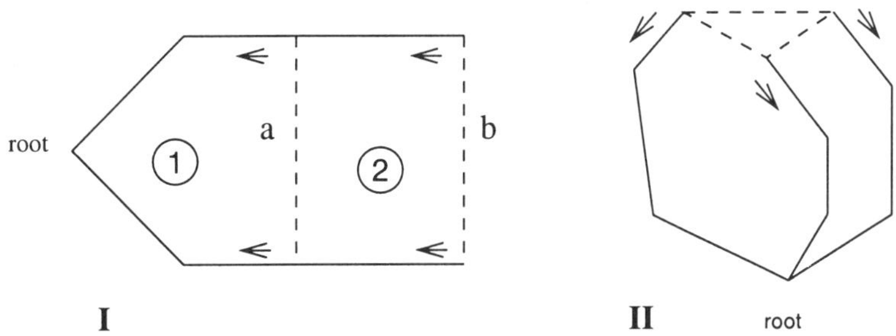

FIGURE 4. **I.** Edges a and b are associated with the same shortest cycle: Cycle 2. Using Gasteiger's heuristic, the edge a is associated with Cycle 1. **II.** A counter-example for Gasteiger's heuristic.

All the algorithms we just have studied are failing in trying to directly find an SSSR. Whatever criterion is used, it is not possible to directly recognize a cycle which is member of a minimum basis. Hence, the solution consists in finding a set including more than ν cycles from which an SSSR can be extracted.

2.2.3. *A polynomial algorithm to find an SSSR.* The weakness of the previous algorithms is due to the use of a unique spanning tree of the graph. Therefore, the solution consists in searching as many spanning trees as there are vertices in the graph. This idea has been introduced by [**37**]. But he did not prove his algorithm or study its complexity. So, we will focus on the method presented by [**20**]. It is considered as the first polynomial time algorithm that finds a minimal cycle basis.

Horton's algorithm is based on the following theorem:

THEOREM 2.1. *Let x be any vertex of any cycle C in a minimum cycle basis. There is an edge $\{y, z\}$ in C such that C consists of a shortest path from x to y, a shortest path from x to z and the edge (y, z).*

Note that any cycle satisfying this property does not necessarily belong to a minimum basis. Therefore, Horton's algorithm extracts a cycle basis from an initial set of cycles (denoted by $\mathcal{C}_I$) verifying Theorem 2.1:

1) $\forall a, b \in V$ find a shortest path $P(a, b)$ between a and b.
2) For all $v \in V$ do:
    For all $\{x, y\} \in E$ do:
        If $P(v, x) \cap P(v, y) = \{v\}$
        Then add in $\mathcal{C}_I$ the cycle $C = P(v, x) + P(v, y) + \{x, y\}$
3) Order by length the initial set of cycles $\mathcal{C}_I$
4) Use a greedy algorithm to extract a minimum cycle basis from $\mathcal{C}_I$.

First step of the algorithm chooses one path for every pair of vertices. Horton has proved that any choice can lead to an initial set $\mathcal{C}_I$ including a minimal cycle basis. This proof consists in replacing the cycles of any minimum basis by cycles belonging to $\mathcal{C}_I$. In the second stage, a cycle C can be created as many times as it contains vertices. To avoid this duplication, Horton uses the method suggested by [**41**]. Considering an order π on vertices, all cycles generated from a vertex v must only contain vertices that precede v in order π.

The two first stages of the algorithm can be implemented as follows. From any vertex r in X, we perform a breadth first search through the graph. This search generates a spanning tree which includes the edges used to reach a vertex for the first time. For any vertex y, the subpath of this tree that links y to r is necessarily a shortest path. During the breadth first search, when a cycle closing edge $\{x, y\}$ is found, a cycle is created if the paths linking x and y to r are disjoint.

Last step of the algorithm processes $\mathcal{C}_I$ according to the length of the cycles. A new cycle is added to the k cycles of the current partial basis when the new induced set is independent. This test uses a gaussian elimination on the $m \times (k+1)$ boolean matrix corresponding to the set of cycle. This elimination can be performed in $O(m \times (k+1))$. Since $k \leq \nu$ and $\mathcal{C}_I$ contains at most νn cycles[7], the minimum cycle basis can be found in $O(m\nu^2 n)$ operations.

A quite similar method as Horton's one has been introduced by [**25**]. Their algorithm can be summarized as follows: *from each cycle closure edge, search the shortest cycles of which it is a member. If all these cycles do not constitute a linearly independent set, find the next smallest cycles for each cycle closure edge. The process goes on until a minimum basis can be extracted from the current set of cycles.* The main difference between this method and Horton's algorithm is that Horton directly finds an unique and useful initial set of cycles. On the opposite, the method presented by Leach finds a sequence of ever growing sets whose biggest one is generally larger than the unique set used by Horton. Hence, Horton's algorithm is generally more efficient than the method presented by [**25**].

3. Extended Bases: Chemically Relevant Cycles

Although minimum cycles basis provides the smallest representation for the cyclic structure of the molecule, this representation is rarely the better one for chemical investigations. The problem lies in the following dilemma: on one hand, a cycle basis provides a too small set of cycles to perform a satisfactory analysis; on the other hand, the enumeration of all cycles does not allow evaluation in a reasonable time. Hence, an optimal set of cycles may be defined between these two extremes. Of course, such a definition depends on the kind of analysis performed.

1) Including cycles of a given length On a synthetic analysis point of view, [**6**] suggests to include all rings containing 6 or fewer vertices. In the same way, [**45**] adds to the basis all cycles including at most 8 vertices (note that many papers in graph theory deal with algorithms which enumerate all the cycles of a given length [**3, 35**]). The choice of the better upper bound for cycle selection is quite subjective and depends on many factors. We can say that the number of irrelevant cycles added to the set grows with the value of the upper bound.

2) ESER Another extended basis has been defined by [**15**] as the *Essential Set of Essential Rings*. This definition is based on the notion of "tied" cycles. A tied cycle is a cycle which has a *chord*[8]. The algorithm used to build ESER starts by searching all the cycles of the molecular graph. Then, it extracts the set of "tied" and "multi-tied" cycles. A cycle C is *dependent* if it is the sum of tied cycles such that: **1.** each tied cycle is the same size or smaller than C **2.** the intersection

[7] In step 2, a cycle is added to $\mathcal{C}_I$ when it consists of two distinct shortest paths. Hence, the number of cycles created from a vertex v is at most equal to the number of cycle closure edges.

[8] A chord is an edge that links two vertices of a cycle and that does not belong to this cycle.

of all these tied cycles involves no less than half of the edges of C **3.** each tied cycle includes no more *heteroatoms* (noncarbon and nonhydrogen atoms) than C.

This notion of ESER gives good results for a few examples of molecules. But it presents two inconveniences: firstly, ESER may include a great number of cycles which are not chemically relevant (particularly when the molecule has no tied cycle); secondly, the algorithm that computes ESER is not efficient because it searches all the cycles and uses a high complexity criterion to detect dependent cycles.

3) ESSR The extended SSSR introduced by [**10**] is based on faces of a planar representation of the molecular graph. It defines the *Extended Set of Smallest Rings*. This set first includes all the *simple faces* corresponding to cycles without chords which are faces of a planar embedding of the graph. But some relevant cycles are not simple faces (see Figure 3). Therefore, ESSR also includes some *cut faces* which correspond to the other cycles without chords. A cut face C is named *primary* if it includes a least one edge e such that the simple face that contains e is the same size or smaller than C. So, ESSR includes all the simple faces and primary cut faces.

ESSR notion provides a complex but original definition of cycle relevance. Unfortunately, the algorithm that builds this set also has a great time complexity. But the main weakness lies in the use of planar embedding. Although most of the simple or primary cut faces are relevant, a large part of them have no real interest. This is due to the fact that a planar embedding of the molecular graph is not a real projection of the molecule. Faces of an embedding are not equivalent to the real faces of the 3-dimensional structure of the molecule. That is the reason why using planar embeddings of the molecular graph do not lead to satisfactory definition of cycle relevance.

4) SER The last extended SSSR we will study has been presented by [**39**]. He defines the *Set of Elementary Rings* (where the word "elementary" does not correspond to the graph theoretic notion that we have mentioned in Section 1.1). To build this set of cycles, the algorithm starts including all the cycles of a SSSR in SER. Then, for every pair of cycles C_1 and C_2 in the current SER, a new cycle $C_1 + C_2$ is added if C_1 and C_2 share at least 2 contiguous edges[9]. This process goes on until SER is stable.

According to its definition, the SER may include a lot of cycles a few of which are really relevant. For instance, Figure 5 shows two irrelevant cycles added to the SER of the molecule presented in Figure 1.

4. Set of Relevant Cycles

4.1. Definition.

4.1.1. *What is cycle relevance ?* Study of the previous definition of extended bases leads us to the conclusion: whatever criterion is used, no satisfactory definition of cycle relevance has yet been found. Of course, all these works are interesting since they propose a formal definition of an abstract notion. According to particular applications, they should provide a better set of cycles than the minimum cycle basis one. But the problem of finding an universal definition of cycle relevance is still open. And it would probably never be solved.

---

[9] more precisely, Takahashi uses the notion of θ-graph. It designates a connected graph such that every vertex has 2 adjacent edges, except for two unadjacent vertices whose valence is 3. Then, cycle $C_1 + C_2$ is added to SER if the subgraph induced by $C_1 \cup C_2$ is a θ-graph.

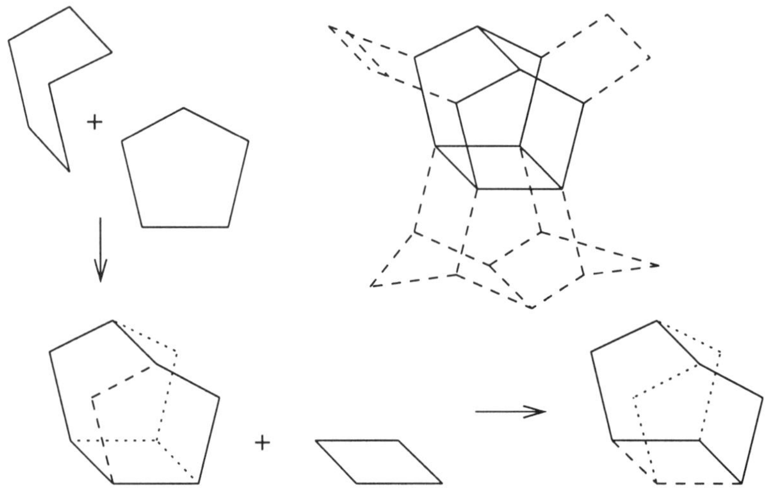

FIGURE 5. Some examples of cycles added to the SER corresponding to molecule in Figure 1.

A given cycle may be relevant according to a particular point of view, e.g. organic synthesis, but absolutely insignificant for another, e.g. structural analysis.

The other difficulty, and probably the most important, concerns the use of graphs to represent molecules. A molecular graph is not a molecule but a useful abstraction of a very complicated 3-dimensional object. Trying to find, in a molecular graph, properties that depend on geometric distance between atoms or spatial arrangement is doomed to failure. To be optimal, definition of cycle relevance probably needs to go beyond the scope of graph theory. It would require using spatial coordinates of atoms, atom interactions, chemical properties and so on. But such a definition of chemically relevant cycles would be very complex.

When we started the RESYN project, we had to choose a definition of relevant cycles. Because we are developing a system for computer-aided synthesis, this definition must satisfy some constrains. First, the definition of relevant cycles must be canonical, that is independent from the way it is computed, to allow a unambiguous representation of molecules. Then, it must be simple in order to be easy to understand by the users of the program. Finally, we need an efficient algorithm to compute the set of relevant cycles.

4.1.2. *A canonical definition.* In the scope of graph theory, notion of minimum cycle basis is absolutely unsatisfactory since different SSSR can be associated to the same molecule. For example, the graph presented in Figure 6 has three different minimum bases.

Instead of trying to elaborate another definition of relevance, we have tried to find the simplest canonical definition which is closed to the SSSR notion. Unifying previous definitions of extended bases leads us to that observation: *a cycle which is not the sum of smaller cycles is chemically relevant*. This definition has been suggested by Plotkin [**31**]. Hence, we will characterize *relevant cycles* according to that definition which can be formalized as follows (where $\mathcal{C}$ denotes the set of elementary cycles):

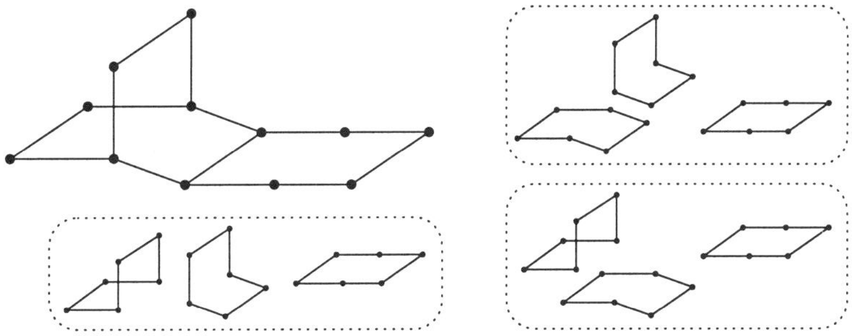

FIGURE 6. An example of a graph having several minimum bases.

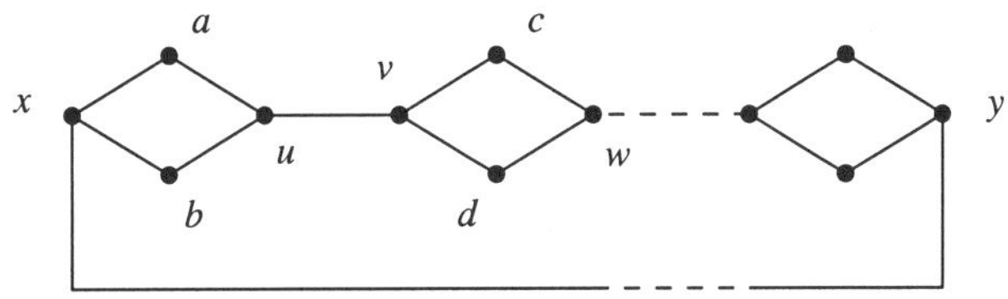

FIGURE 7. *According to the definition of relevant cycles, set $\mathcal{C}_R$ contains $2^{\frac{n}{4}}$ cycles in this graph ($(x, b, u, v, d, w, ..., y, ..., x)$, $(x, a, u, v, d, w, ..., y, ..., x)$, etc).*

DEFINITION 4.1. $C \in \mathcal{C}$ is *relevant* **iff** $\nexists C_1, ..., C_k \in \mathcal{C}$ *s.t.* $\begin{cases} \forall i, \ |C_i| < |C| \\ \text{and} \\ C = C_1 + ... + C_k \end{cases}$

This characterization defines the *Set of Relevant Cycles* which we denote by $\mathcal{C}_R$. This definition is very close to minimum basis notion. For example, in Figure 6, $\mathcal{C}_R$ contains all the 6-edges cycles whereas each SSSR only includes three of them. In fact, we should easily prove that:

LEMMA 4.2. *$\mathcal{C}_R$ is equal to the union of all the minimum cycle bases of G.*

Hence, a cycle is *relevant* if it belongs to at least one SSSR. Note that $\mathcal{C}_R$ is unique for a given molecule. This set is generally quite small for a molecular graph. For instance, in Figure 1, $\mathcal{C}_R$ contains 15 cycles whereas any SSSR would include 11 of them.

From a graph theoretic point of view, the number of *relevant* cycles is not necessarily polynomial. If G is a complete graph [10], the size of $\mathcal{C}_R$ is equal to the polynomial number of 3-edges cycles. But there exist graphs, as the one of Figure 7, which contain an exponential[11] number of *relevant* cycles.

The existence of such graphs must not hide the interest of this definition. Note that the same problem occurs with any canonical definition based on SSSR. In organic chemistry, graphs with an exponential number of *relevant* cycles are hardly possible.

Till now, no algorithm has been proposed to compute $\mathcal{C}_R$. In this paper, we present a polynomial time algorithm to compute $\mathcal{C}_R$ as a polynomial number of parts. Each part corresponds to a *family* of cycles which can be directly listed from

[10] in a *complete* graph, any vertex is adjacent to all the other ones.

[11] according to the size of the graph, i.e. number of vertices.

a particular relevant cycle. This cycle defines a *prototype* of its family. This compact representation of $\mathcal{C}_R$ gives a more precise description of the cycle vector space than a simple minimum basis. For instance, in the graph of Figure 7, apart from the $\frac{n}{4}$ 4-edges cycles that must be known, the other *relevant* cycles can be described by a single family. The prototype of this family is one of the $2^{\frac{n}{4}}$ elementary cycles with $\frac{3n}{4}$ vertices. Listing all these *relevant* cycles is not very meaningful for such a graph. But it can be very useful for a computer system to be able to perceive them and designate a single prototype.

For molecular graphs, the compact representation can be replaced by the complete enumeration of the relevant cycles.

The method we present to find the Set of Relevant Cycles is analogous to Horton's algorithm. We start by searching for an initial set of cycles (denotes by $\mathcal{C}'_I$) from which a subset of relevant prototypes is extracted. Then, all the others relevant cycles are directly generated from the set of relevant prototypes.

4.2. Partition of the Set of Relevant Cycles. This section defines a partition of the set of relevant cycles such that the number of parts is polynomial[11].

LEMMA 4.3. *If μ is a subpath of a relevant cycle C such that $|\mu| \leq \frac{1}{2}|C|$ then μ is a shortest path.*

PROOF. Assume that μ is not a shortest path. Then μ includes a subpath $\rho = (a \ldots b)$ such that $|\rho| > d(a, b)$ and there is a shortest path ρ' from a to b such that ρ and ρ' are disjoint. Replacing ρ with ρ' in C will create a new cycle C' such that $|C'| < |C|$. Define C'' to be the cycle that consists of ρ and ρ'. Since $|\rho'| < |\rho| \leq \frac{1}{2}|C|$ we have $|C''| < |C|$. Because $C = C' + C''$, C cannot be relevant. □

In the rest of this paper, we assume that the vertices are ordered by a numbering π that respects their degrees, i.e. $\forall x, y \in V,\ \pi(x) < \pi(y) \Rightarrow \deg(x) \leq \deg(y)$.

For any cycle C being considered, we denote by r_C the highest vertex in C according to the numbering π.

LEMMA 4.4. *Any relevant cycle C consists of two disjoint shortest paths $(r_C \ldots p)$ and $(r_C \ldots q)$ having the same length and linked by the edge $\{p, q\}$, if $|C|$* odd *or by the path (p, x, q), if $|C|$ is* even.

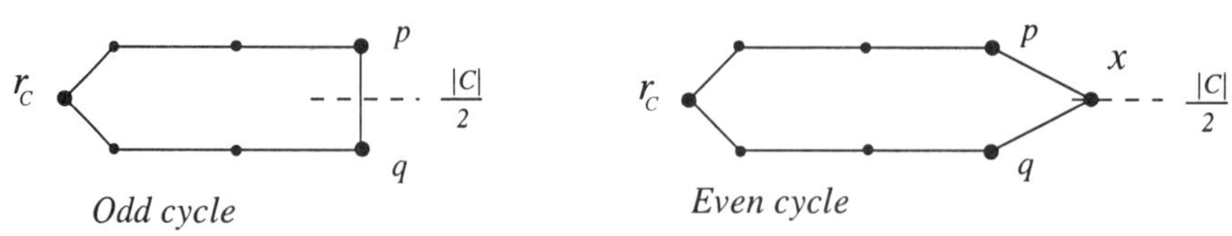

PROOF. For a given cycle C, we can define $(r_C \ldots p)$ and $(r_C \ldots q)$ to be the two longest subpaths of C whose length is strictly lower than $\frac{1}{2}|C|$. By Lemma 4.3, these paths are shortest paths in the whole graph. □

The partition of the set of relevant cycles is based on Lemma 4.4.

DEFINITION 4.5. The *cycle family* associated with a relevant cycle C, including the vertices r_C, p, q and eventually x as defined in Lemma 4.4, is:

$$\mathcal{F}(C) = \left\{ C' \in \mathcal{C}_R \;\middle|\; \begin{array}{l} |C'| = |C| \textit{ and } C' \textit{ consists of:} \\ \bullet \text{ the vertex } r_C \text{ and the edge } \{p,q\} \text{ (or the path } (p,x,q)), \\ \bullet \text{ two shortest paths } (r_C,\dots,p) \text{ and } (r_C,\dots,q) \text{ passing} \\ \text{only through vertices lower than } r. \end{array} \right\}$$

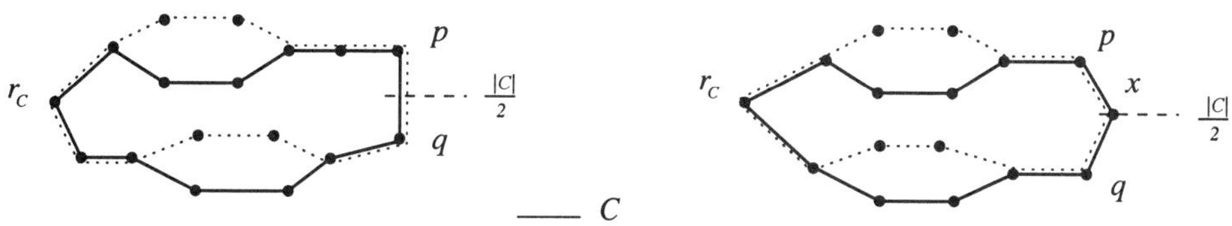

Odd cycle C' .... *Even Cycle*

Hence, two cycles C and C' belonging to $\mathcal{F}(C)$ only differ in the shortest paths from r_C to p and from r_C to q they include.

THEOREM 4.6. *The set of all the relevant cycle families defines a partition of* $\mathcal{C}_R$.

PROOF. By Lemma 4.4 and Definition 4.5, the cycle family associated with a relevant cycle C is unique for a given ordering π. So any relevant cycle belongs to exactly one family. Then, we define an equivalence relation such that two relevant cycles are equivalent if they belong to the same cycle family □

To describe the family $\mathcal{F}(C)$, we need a single cycle C which is a *prototype* of this family. All the other cycles in $\mathcal{F}(C)$ can be generated from C by replacing paths $(r_C, \dots p)$ and $(r_C, \dots q)$ with paths of the same length passing only through vertices smaller than r_C.

Hence, a cycle family is properly defined by a triple r_C, p, q for odd cycles or a quadruple r_C, p, q, x for even cycles.

The number of relevant cycle families depends on the order π used to define the families. However, we have the following result:

THEOREM 4.7. *The number of relevant cycle families is always polynomial.*

PROOF. The number of families whose prototype is an odd cycle is smaller than $n\,\nu$, since an odd cycle prototype is defined by a vertex r_C and a cycle closing edge $\{p,q\}$. As for even cycle families, their prototype is defined by 4 vertices r_C, x, p and q such that p and q are adjacent to x. Hence, the number of even cycle families is smaller than:
$\sum_{r \in V} \sum_{x \in V,\, \pi(x) \leq \pi(r)} \frac{\deg(x)^2}{2} \leq \sum_{r \in V} \left(\deg(r) \times \sum_{x \in V,\, \pi(x) \leq \pi(r)} \frac{\deg(x)}{2}\right) \leq 2m^2$,
since order π respects vertex degrees ($\forall x \in V,\ \pi(x) \leq \pi(v) \Rightarrow \deg(x) \leq \deg(r)$) □

To compute the union of all the minimum cycle bases, we may find one prototype for each relevant cycle family.

4.3. A Polynomial Algorithm that Finds the Set of Relevant Cycles. The algorithm we propose to compute cycle prototypes is based on the converse of Lemma 4.4. This converse is not necessarily true but it gives a strong condition on cycle relevance.

The outline of the algorithm can be given as follows:

- firstly, we compute the set C'_I including one cycle for each triple r_C, p, q (or quadruple r_C, p, q, x) satisfying the condition of Lemma 4.4,
- secondly, we use a greedy algorithm to extract relevant prototypes from C'_I.

4.3.1. *Computation of C'_I.* This set is generated using breadth first searches through the molecular graph G from every vertex r whose degree is higher than two[12].

Denote by V_r the set of vertices z such that $\pi(z) < \pi(r)$ and there is a shortest path in G from r to z that passes only through vertices which precede r in the ordering π.

DEFINITION 4.8. $\forall x \in V_r$, the set of *ancestors* of x according to r is defined by: $anc_r(x) = \{a \in \Gamma(r)$ *such that* $\exists$ *shortest path from* r *to* x *including* $a\}$

Now consider a cycle C that consists of the two shortest paths $(r, a_p, ..., p)$ and $(r, a_q, ..., q)$ linked by the edge $\{p, q\}$ if $|C|$ is even or the path (p, x, q) if $|C|$ is odd.

LEMMA 4.9. *If $anc_r(p) \cap anc_r(q) \neq \emptyset$ Then C is not relevant*

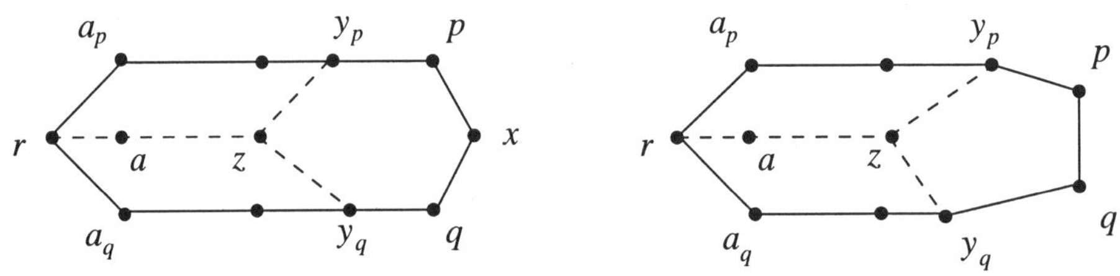

PROOF. If there exists a vertex a such that $a \in anc_r(p) \cap anc_r(q)$, cycle C is the sum of three cycles[13] smaller than C. Note that the proof holds for degenerate cases when $a = a_p$ and/or $a = a_q$. □

Lemma 4.9 gives a strong condition to detect irrelevant cycles. For a given triple r_C, p, q (or quadruple r_C, p, q, x) we add a cycle in C'_I only if $anc_r(p) \cap anc_r(q) = \emptyset$.

To generate C'_I, we also need to compute the following values, for any x in V_r:

- *fathers*$(x) = \{z \in \Gamma(x) \cap V_r$ *such that* $d(r,z) = d(r,x) - 1\}$
- *brothers*$(x) = \{z \in \Gamma(x) \cap V_r$ *such that* $d(r,z) = d(r,x)\}$
- *Path*(r,x) = a shortest path from r to x passing only through vertices in V_r

We can now consider the algorithm for the generation of C'_I as presented in Figure 8.

In the main While loop, lines 9-20 perform the breadth first search and compute the sets *fathers*(y), *brothers*(y) and $anc_r(y)$.

Note that we have the following property: *fathers*$(x) \neq \emptyset \Longrightarrow x \in V_r$.

Even and odd cycles are computed at line 21 and line 23 respectively, using a test on ancestors.

Line 22 chooses a shortest path form r to x, when $x \in V_r$.

LEMMA 4.10. *For molecular graphs, the set of cycles C'_I can be computed with $O(m^2)$ operations and the cardinality of C'_I is lower than $(2m^2 + \nu m)$.*

[12] If $\nu > 1$, every cycle includes at least one vertex r such that $d(r) \geq 3$

[13] In the example: (r, a_p, y_p, z, a, r), (r, a_q, y_q, z, a, r) and (z, y_p, p, q, y_q, z)

[Algorithm 1]

1. For each $r \in V$ such that $d(r) \geq 3$ do:
2.   For each $x \in V$ do:
3. $level(x) \leftarrow +\infty$; $fathers(x) \leftarrow \emptyset$; $brothers(x) \leftarrow \emptyset$;
4.   **endfor**
5. $fathers(r) \leftarrow \emptyset$; $anc_r(r) \leftarrow \emptyset$; $Path(r,r) \leftarrow (r)$;
6. $FIFO \leftarrow \emptyset$; AddLast($r$, $FIFO$);
7. $level(r) \leftarrow 0$; $explored \leftarrow \emptyset$;
8.   While $FIFO \neq \emptyset$ do:
9. $x \leftarrow$ ExtractFirst($FIFO$):
10.   For each $y \in \Gamma(x)$ do:
11.   If $level(x) < level(y)$ Then:
12.   If $\pi(y) < \pi(r)$ and ($fathers(x) \neq \emptyset$ or $x = r$) Then
                $fathers(y) \leftarrow fathers(y) \cup \{x\}$;
13.   If $level(y) = \infty$ Then
14. $level(y) \leftarrow level(x) + 1$; AddLast($y$, $FIFO$);
15.   If $x = r$ Then $anc_r(y) \leftarrow \{y\}$ Else $anc_r(y) \leftarrow anc_r(x)$;
16.   Else $anc_r(y) \leftarrow anc_r(y) \cup anc_r(x)$;
17.   Else If $fathers(y) \neq \emptyset$ and $fathers(x) \neq \emptyset$ Then
18. $brothers(y) \leftarrow brothers(y) \cup \{x\}$
19.   **endif**
20.   **endfor**
21. $\forall p, q \in fathers(x)$ such that $anc_r(p) \cap anc_r(q) = \emptyset$ and $\{p, q\} \notin E$
            add even cycle "$Path(r,p) + x + Path^{-1}(r,q)$" to $\mathcal{C}_I'$
22.   If $fathers(x) \neq \emptyset$ Then $Path(r,x) \leftarrow Path(r,f) + x$, where $f \in fathers(x)$;
23. $\forall b \in brothers(x) \cap explored$ such that $anc_r(b) \cap anc_r(x) = \emptyset$
            add odd cycle "$Path(r,x) + Path^{-1}(r,b)$" to $\mathcal{C}_I'$
24. $explored \leftarrow explored \cup \{x\}$;
25.   **endwhile**
26. **endfor**

FIGURE 8. Algorithm that computes $\mathcal{C}_I'$

PROOF. Except for the computation of sets anc_r, the time complexity of lines 9-20 is the standard complexity of a breadth first search, that is $O(n+m)$ for each search from a vertex r.

Computation of $anc_r(y)$ only occurs for ring-closure edges whose extremities have different breadth levels (according to the breadth first search from r). For such edges, the modification is performed in $O(d(r))$. For molecular graphs, $d(r)$ is always bounded (by 5 or 6 in organic chemistry). So we can assume that each modification is performed in $O(1)$ if we use bit vectors to represent the sets anc_r. Hence, time complexity of ancestor computation is $O(\nu n)$ for all the breadth first searches.

Line 21 is the critical step of the algorithm. The number of unordered pairs p, q checked by the program is lower than $\sum_{r \in V} \sum_{x \in V_r} \frac{1}{2} \deg(x)(\deg(x) - 1) < 2m^2$, since $\forall x \in V_r$, $\deg(r) > deg(x)$. If the sets anc_r are represented by bit vectors, each check is performed in O(1). So the complexity of Line 21 for all the breadth first searches is $O(m^2)$.

For the computation of odd cycles, the number of checks performed for each breadth first search is $O(\nu)$ since the number of "brothers" is lower than the number of ring-closure edges. So there are at most $O(\nu n)$ odd cycles.

Finally, the whole complexity is $O(m^2)$ for a molecular graph. □

LEMMA 4.11. *$\mathcal{C}_I'$ includes one and only one prototype of each relevant cycle family.*

PROOF. Let $C = (r, \ldots, p, x, q, \ldots, r)$ be a relevant even cycle (if C is odd, the proof is quite similar). Assume that r is the highest vertex in C according to π and x is the unique vertex in C such that $d(r,x) = \frac{1}{2}|C|$.
Now consider the exploration of vertex x in algorithm 1. By Lemma 4.4, $(r, \ldots, p)$ and $(r, \ldots, q)$ are shortest paths. So p and q belong to *fathers*(x). Define C' to be the cycle added to $\mathcal{C}_I'$ at line 21 for the pair of vertices p, q. By definition of the algorithm, C' is unique. Cycles C and C' only differ in the shortest paths $(r, \ldots, p)$ and $(r, \ldots, q)$ they include. The sum $C + C'$ defines a set of cycles which are smaller than C. Hence, cycle C' must be relevant since C is relevant. So, C' is the unique prototype of $\mathcal{F}(C)$ which belongs to $\mathcal{C}_I'$ □

4.3.2. *Computation of a set of relevant prototypes.* By Lemma 4.11, $\mathcal{C}_R \cap \mathcal{C}_I'$ is a set of prototypes. So, we have to extract all the relevant cycles in $\mathcal{C}_I'$.

As in Horton's algorithm, a minimum cycle basis $\mathcal{B}$ is calculated by processing the cycles of $\mathcal{C}_I'$ in order of weight. During the processing of a cycle C, we denote by $\mathcal{B}_<$ and $\mathcal{B}_=$ the subsets of cycles in the current sub-basis whose lengths are less than $|C|$ and equal to $|C|$, respectively. By Definition 4.1, C is relevant when $\mathcal{B}_< \cup \{C\}$ is independent. When this check succeeds, cycle C is added to $\mathcal{B}_=$ if $\mathcal{B}_< \cup \{C\} \cup \mathcal{B}_=$ is also independent. The processing stops when $\mathcal{B}$ is complete and all the cycles having the same length as the greatest one in $\mathcal{B}$ have been tested.

The current minimal sub-basis of $\mathcal{B}$ is represented by a $(|\mathcal{B}_<| + |\mathcal{B}_=|) \times m$ boolean matrix. To test if $\mathcal{B}_< \cup \{C\}$ is independent, we perform a Gaussian elimination on the first rows of the matrix which are associated with $\mathcal{B}_<$. When this check succeeds, the Gaussian elimination follows on the other rows to test if $\mathcal{B}_< \cup \mathcal{B}_= \cup \{C\}$ is also independent.

Since $\mathcal{B}$ contains at most ν cycles, each check of a cycle is performed in $O(\nu m)$. By Lemma 4.10, the cardinality of $\mathcal{C}_I'$ is in $O(m^2)$. So, the set of prototypes $\mathcal{C}_I' \cap \mathcal{C}_R$ can be generated in $O(\nu\, m^3)$ operations.

4.3.3. *Enumeration of $\mathcal{C}_R$.* $\mathcal{C}_R$ is completely described by a set of prototypes of the families defined in Section 4.2. From this set of prototypes, the union of all the minimum cycle bases can easily be listed.

Consider a cycle prototype $C \in \mathcal{C}_R \cap \mathcal{C}_I'$. To generate the cycle family $\mathcal{F}(C)$, we define a directed graph $D_r = (V_r, U_r)$ associated with the vertex r so that: $U_r = \{\ (y, z),\ \textit{where } z \in \textit{fathers}(y)\}$. Figure 9 gives an example of such a graph.

The computation of the digraph D_r can be directly performed in Algorithm 1.

The digraph D_r has two major properties:

- It has no directed cycle,
- $\forall x \in V_r$, any directed path $(x, \ldots, r)$ in D_r corresponds to a shortest path from r to x in G.

To list all the paths from x to r in D_r, we use a backtracking function which is based on a depth first search from x. This recursive function can be summarized as follows:

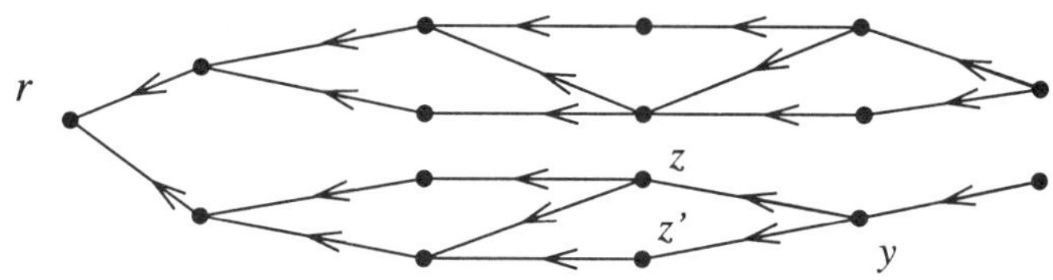

FIGURE 9. Digraph describing the shortest paths associated to the breadth first search from r

```
Function List_Paths( x; current_path )
  add x at the head-end of current_path;
  If x = r Then Return({current_path})
  Else:
    Result ← ∅;
    For any z such that (x, z) ∈ U_r Do
      Result ← Result ∪ List_Paths(z, current_path);
    Return(Result)
  endif
```

To compute $\mathcal{F}(C)$, we first replace the path $(p, \ldots, r)$ in C by each path returned by the call List_Paths(p, $\emptyset$). Then, in any cycle generated this way, we replace the path $(q, \ldots, r)$ by each path resulting from List_Paths(q, $\emptyset$).

Each cycle in $\mathcal{F}(C)$ corresponds to a pair of paths $(p, \ldots, r)$, $(q, \ldots, r)$. Since the computation of each path takes a number of operations in the order of the path cardinality, the family $\mathcal{F}(C)$ can be listed in $O(n\,|\mathcal{F}(C)|)$.

So, the generation of $\mathcal{C}_R$ from subset $\mathcal{C}_R \cap \mathcal{C}_I'$ is performed in $O(n\,|\mathcal{C}_R|)$ operations (including cycle enumeration).

Note that the use of prototypes substantially optimizes the generation of $\mathcal{C}_R$: if a cycle prototype is relevant, this implies that all cycles of the corresponding family are relevant. Therefore, the number of cycle relevance checks is polynomial since it is in $O(|\mathcal{C}_I'|)$ instead of $O(|\mathcal{C}_R|)$.

Another interest of the prototypes lies in the possibility of computing the cardinality of $\mathcal{C}_R$ without enumerating all the cycles. This method is based on the digraph D_r from which we can easily evaluate the number of shortest paths from r to any vertex x, which passe only through vertices in V_r. This method is described in [**43**]. It also provides a way of calculating, in polynomial time, the number of relevant cycles of a given length which include a given vertex.

ABSTRACT REPRESENTATION OF A MOLECULAR GRAPH

This part presents an abstract representation of the molecular graph which describes the structural relationships between topological patterns as cycles, carbon chains,... We first define the "graph of relevant cycles" which aims at the description of the cyclic systems in the molecular graph. Then we give a definition of the major non-cyclic patterns as *carbon chains, cyclic links*, and *heteroatomic links.* Finally, we define the abstract representation.

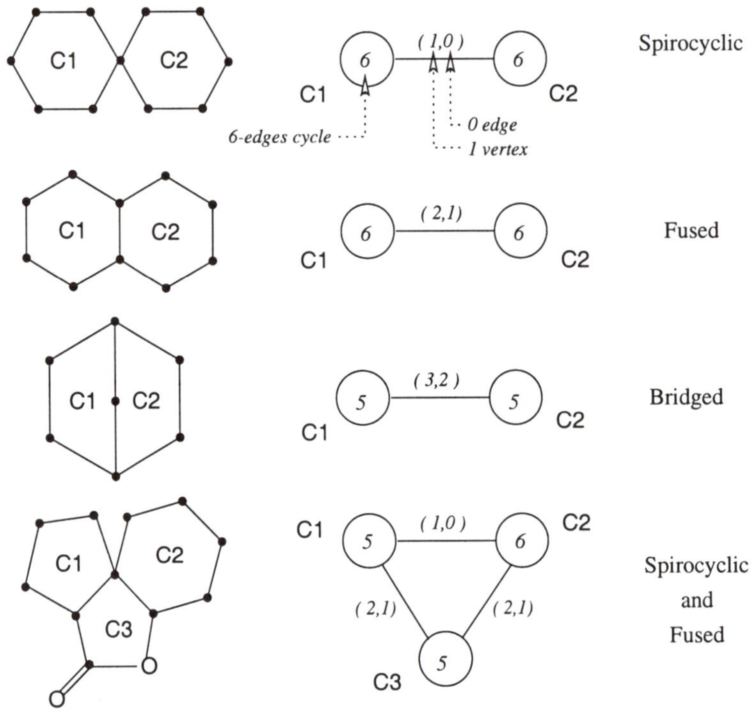

FIGURE 10. Graphs of the relevant cycles to describe cyclic systems

5. Graph of the Relevant Cycles

To analyze the cyclic system of the molecule, we define a new graph in which each relevant cycle is associated to a vertex. This vertex is labeled by the size of the cycle. Two cycles are linked in the new graph if they share at least one atom. This link is labeled by a couple of values which correspond to the numbers of vertices and edges shared by the two cycles, respectively.

Figure 10 illustrates this *graph of the relevant cycles*.

With the label of the link joining two cycles, we can analyze the relationship between these cycles and characterize the corresponding cyclic system, according to the chemical nomenclature:

- a *spirocyclic* system corresponds to the label $(1, 0)$
- a *fused* system is labeled by $(2, 1)$
- other values represent *bridged* systems

6. Non-Cyclic Topological Patterns

The analysis of the non-cyclic parts of the molecule is quite easy to implement.

In the scope of organic synthesis, it is important to distinguish the *heteroatomic* parts from the *carbon* ones. The first step of the analysis consists in finding the "*skeleton*" of the molecule (Figure 11). The *skeleton* is obtained by recursively removing terminal heteroatoms (i.e. heteroatoms whose vertex degree is 1). Hence, the *skeleton* only includes heteroatoms which are cyclic or belong to a path linking two carbons. Note that the terminal heteroatoms are suppressed because they

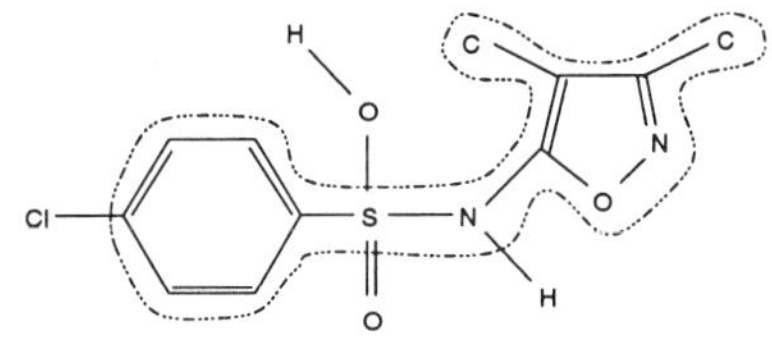

FIGURE 11. Skeleton of a molecular graph: terminal heteroatoms are recursively removed

do not affect the topological analysis of a molecular graph. For a retrosynthesis analysis, the *skeleton* represents the main structural part of the target molecule on which topological strategies can be worked out.

The analysis of the non-cyclic parts is performed on the *skeleton* of the molecule. In order to define retrosynthetic topological strategies, the RESYN system detects the following components (see Figure 12):

- a *cyclic link* is a non-cyclic edge whose at least one extremity belongs to a cycle. The perception of *cyclic link* is used to detect the boundary between cyclic and non-cyclic parts. For instance, some retrosynthesis strategies consist in cutting *cyclic links* to isolate complex cyclic systems.
- an *heteroatomic link* is a maximal non-cyclic subgraph which consists of heteroatoms. The existence of *heteroatomic links* makes the deconnection of the molecule easier.
- a *carbon chain* is a maximal non-cyclic subgraph which consists of carbons. A *carbon chain* is *linear* if all its vertex degrees are not greater than 2, otherwise the chain is *branched*.

The perception of these patterns can be performed in linear time using an algorithm based on a depth first search [**42**].

7. Abstract Representation

The abstract representation extends the graph of the relevant cycles. It also defines the relationship between the non-cyclic topological patterns. Figure 12 gives an example of abstract representation. This is an internal representation on which the system is reasoning. In the case of simple molecules it can be visually used by the chemist but for complex structures it is to dense to be useful (see Figure 14).

Abstract or reduced graph representations have been used for different purposes elsewhere. For example,[**25**] describes such a representation used in automated conformational analysis.

Our abstract representation allows to define and describe strategic goals to be reached by the system. It can also be used to perform analogical reasoning. For instance, suppose we want to find a retrosynthetic pathway for the molecule of ibogamine described in Figure 13. Computing the abstract representation of ibogamine shows that it is very close to the representation of patchouli alcohol, whereas the molecular graphs are rather different. Then, if we know a retrosynthetic pathway for patchouli alcohol, based on the Diels-Adler reaction, we can try to use the same reaction in order to build the cyclic system of ibogamine. Without the abstract representation, the analogy would have been more difficult to find.

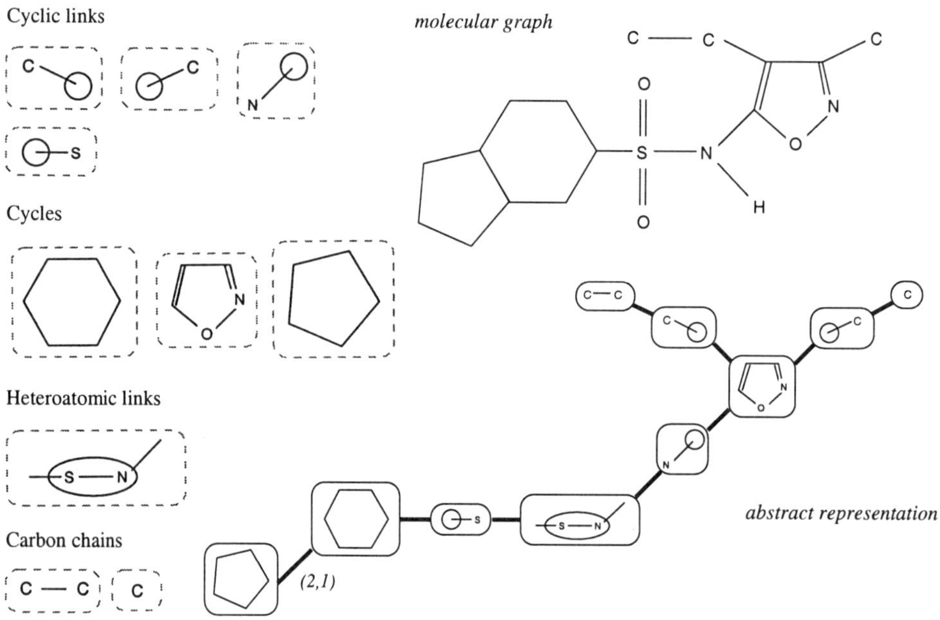

FIGURE 12. An example of abstract representation

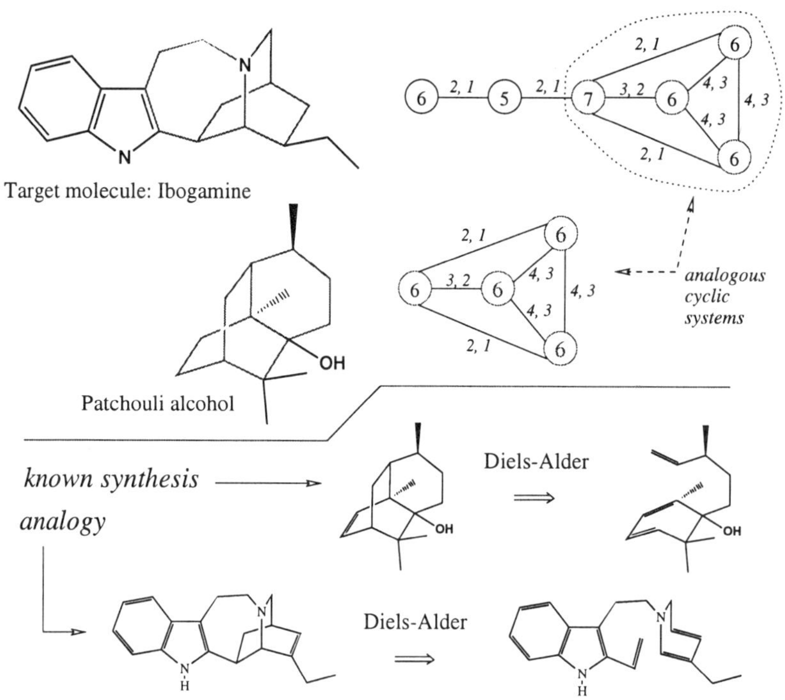

FIGURE 13. Using the abstract representation for analogical reasoning

8. Implementation

The algorithms which we presented were initially implemented in LISP. To increase the portability, a Java version of these programs has been developed. Even for complex molecules computing the abstract representation is instantaneous, i.e. a few milliseconds on a standard PC computer. Some examples are given in Figure 14.

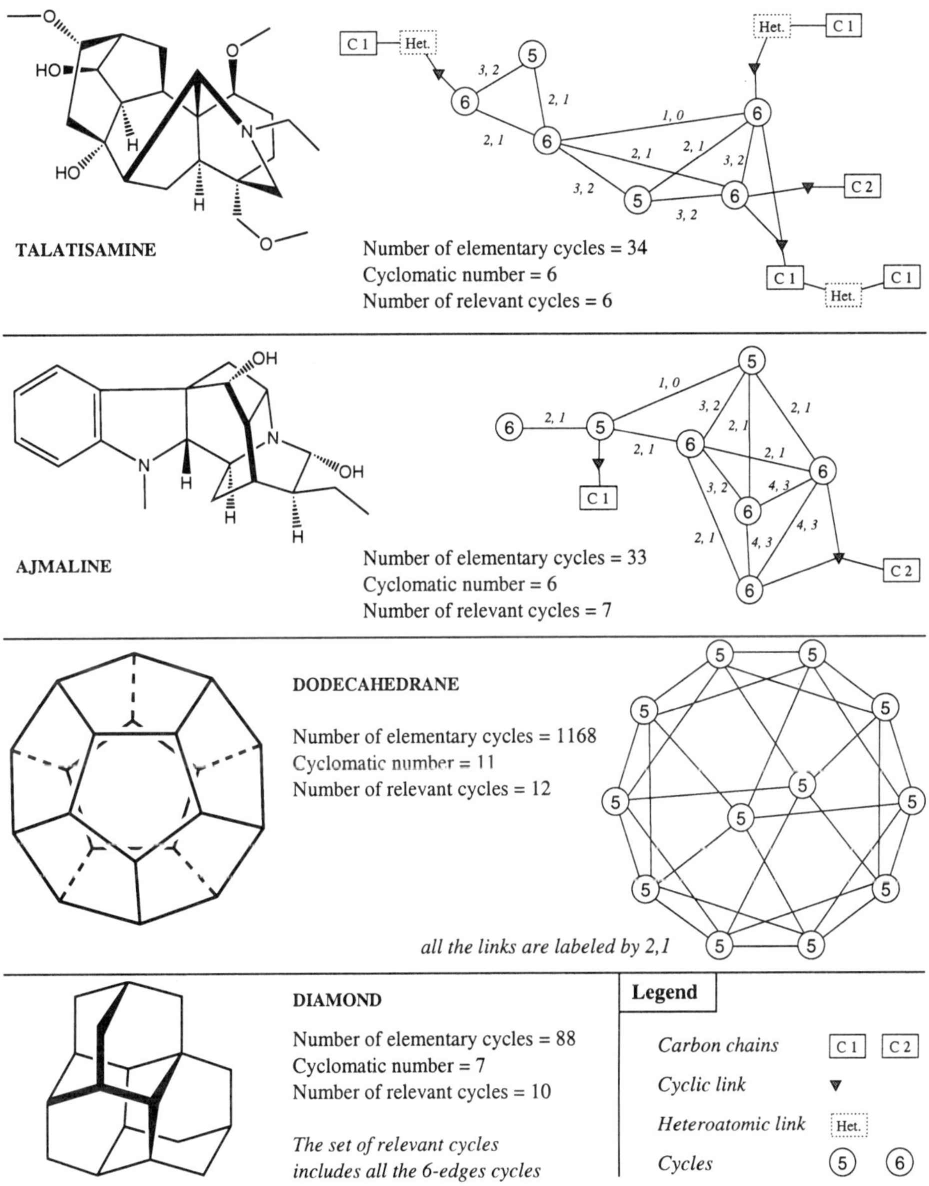

FIGURE 14. Some examples of molecules analysed by RESYN

9. Conclusion

Perception of the target molecule is a main phase in solving an organic synthesis problem. From a structural formula, this process leads to a new representation which describes the molecule components and their relationships according to different viewpoints (topology, stereochemistry, functionality). This structured representation is used for defining strategies which will guide the retrosynthetic analysis of the target. To automate this process in a synthesis planning system, it is necessary to precisely and formally define the chemical concepts used (cycles, functions,...) and to have efficient algorithms for recognizing them. The topological perception concerns the recognition of cycles, chains, links,... This perception is the subject of the work presented in this paper. We have particularly studied the problem of finding a set of relevant cycles.

We have seen, in the short review of papers dealing with SSSR, how difficult the search of an efficient algorithm is, even when the set of cycles to find is very well formalized. Minimum cycle basis does not provide a satisfactory notion to analyze the cyclic structure of a molecule. Hence, we have discussed a few papers which define extended SSSR. None of them gives a perfect notion of cycle relevance. Since these definitions try to include all the relevant cycles, they also add, in their extended basis, cycles which have no real interest in chemistry. Naturally, the criterion used to select cycles may depend on the kind of analysis which is performed. But cycle relevance should not be correctly defined in the sole scope of graph theory. It may resort to various other theories in chemistry which can explain why a given cycle is chemically relevant. Therefore, we chose the definition that corresponds to the smallest canonical set. According to this definition, the *Set of Relevant Cycles* (denoted by $\mathcal{C}_R$) is the *union of all the minimum cycles basis* of the graph. We have proposed a polynomial time algorithm that computes $\mathcal{C}_R$ in $O(\nu m^3)$ as a set of cycle families. From these families, all the relevant cycles can be generated in $O(n\,|\mathcal{C}_R|)$ (including output).

We have presented an abstract representation of molecular graphs for the definition of high level strategies allowing various kinds of reasoning (classification, analogy, induction, DAI,...). This work have been integrated in a prototype, the RESYN system, to experiment knowledge level modeling and reasoning through a constructive interaction with experts in chemistry.

References

[1] L. Baumer, G. Sala, and G. Sello, Ring Perception in Organic Structures: A New Algorithm for Finding SSSR, *Computers Chem.*, **15**(4) (1991) 293–299.

[2] A.L. Cauchy, *Recherche sur les polyèdres*, J. École Polytechnique **9**(16) (1813) 68–86.

[3] N. Chiba, T. Nishizeki, S. Abe, and T. Ozawa, A Linear Algorithm for Embedding Planar Graphs using PQ-trees, *Journal of Computer and System Sciences*, **30** (1985) 54–76.

[4] E.J. Corey, and X.-M. Cheng, *The Logic of Chemical Synthesis*, Wiley-InterScience, New York, 1989.

[5] E.J. Corey, K.A. Long, and S.D. Rubenstein, Computer-Assisted Analysis in Organic Synthesis, *Science*, **228** (1985) 408–418.

[6] E.J. Corey, and G.A. Petersson, An Algorithm for Machine Perception of Synthetically Significant Rings in Complex Cyclic Organic Structures, *J. Am. Chem. Soc.*, **92**(2) (1972) 460–465.

[7] N. Deo, G.M. Prabhu, and M.S. Krishnamoorthy, Algorithms for Generating Fundamental Cycles in a Graph, *ACM Trans. Math. Software*, **8** (1982) 26–42.

[8] A. Dietz, C. Fiorio, M. Habib, and C. Laurenço, *Representation of Stereochemistry using Combinatorial Maps*, this issue (P. Hansen et al, ed.), DIMACS workshop, AMS, 1998.

[9] G.M. Downs, V.J. Gillet, J.D. Holliday, and M.F. Lynch, Review of Ring Perception Algorithms for Chemical Graphs, *J. Chem. Inf. Comput. Sci.*, **29**(3) (1989) 172–187.

[10] G.M. Downs, V.J. Gillet, J.D. Holliday, and M.F. Lynch, Theoretical Aspects of Ring Perception and Development of the Extended Set of Smallest Rings Concept, *J. Chem. Inf. Comput. Sci.*, **29**(3) (1989) 187–206.

[11] S. B. Elk, Derivation of the Principle of Smallest Set of Smallest Rings from Euler's Polyhedron Equation and a Simplified Technique for Finding this Set, *J. Chem. Inf. Comput. Sci.*, **24**(4) (1984) 203–206.

[12] L. Euler, Elementa Doctrinae Solidorum, *Novi. Comm. Acad. Sci. Imp. Petropol.*, **4** (1752) 109–140.

[13] B.T. Fan, A. Panaye, J-P. Doucet, and A. Barbu, Ring Perception. A New Algorithm for Directly Finding the Smallest Set of Smallest Rings from a Connection Table, *J. Chem. Inf. Comput. Sci.*, **33**(5) (1993) 657–662.

[14] M. Frèrejacque, Condensation d'une molécule organique, *Bull. Soc. Chim. Fr.*, **6** (1939) 1008–1011.

[15] S. Fujita, A New Algorithm for Selection of Synthetically Important Rings: The Essential Set of Essential Rings for Organic Structures, *J. Chem. Inf. Comput. Sci.*, **28**(2) (1987) 78–82.

[16] J. Gasteiger, and C. Jochum, An Algorithm for the Perception of Synthetically Important Rings, *J. Chem. Inf. Comput. Sci.*, **19**(1) (1979) 43–48.

[17] N. Gibbs, A Cycle Generation Algorithm for Finite Undirected Linear Graphs, *J. ACM*, **16** (1969) 564–568.

[18] O. Gien, *Modĺisation des stratǵies et des tactiques en synthèse organique*, Thèse de doctorat, Université Montpellier II, 1998.

[19] J.B. Hendrickson, D.L. Grier, and A.G. Toczko, Condensed Structure Identification and Ring Perception, *J. Chem. Inf. Comput. Sci.*, **24**(4) (1984) 195–203.

[20] J.D. Horton, A Polynomial-Time Algorithm to Find the Shortest Cycle Basis of a Graph, *SIAM J. Comput.*, **16**(2) (1987) 358–366.

[21] W.-D. Ihlenfeldt, and J. Gasteiger, Computer-Assisted Planning of Organic Syntheses: The Second Generation of Programs, *Angew. Chem. Int. Ed. Engl.*, **34** (1995) 2613–2633.

[22] P. Jambaud, *Le dialogue comme processus de résolution de problème. une application en chimie organique.*, Thèse de doctorat, Université Montpellier II, 1996.

[23] G. Kirchhoff, Über die Auflösung der Gleichungen, auf welche man bei der Untersuchung der linearen Verteilung galvanischer Ströme geführt wird, *Poggendorf Ann. Physik*, **72** (1847) 497–508, English translation in *Trans. Inst. Radio Engrs.*, CT-5:4–7, 1958.

[24] C. Laurenço, *Synthèse organique assistée par ordinateur*, Thèse de doctorat d'État, Université Louis Pasteur (Strasbourg I), Septembre 1985.

[25] A.R. Leach, D.P. Dolata, and K. Prout, Automated Conformational Analysis and Structure Generation: Algorithms for Molecular Perception, *J. Chem. Inf. Comput. Sci.*, **30**(3) (1990) 316–324.

[26] J. Lieber, *Raisonnement à partir de cas et classification hiérarchique. Application à la planification de synthèses en chimie organique*, Thèse de doctorat, Université Nancy 1, octobre 1997.

[27] A.K. Long, and J.C. Kappos, Computer-Assisted Synthesis Analysis. Performance of Tactical Combinations of Transforms, *J. Chem. Inf. Comput. Sci.*, **34** (1994) 915–921.

[28] P. Mateti, and N. Deo, On Algorithms for Enumerating All Circuits of a Graph, *SIAM J. Comput.*, **5** (1975) 90–101.

[29] M.A. Ott, and J.H. Noordik, Computer Tools for Reaction Retrieval and Synthesis Planning in Organic Chemistry. A Brief Review of their History, Methods, and Programs, *Recl. Trav. Chim., Pays-Bas*, **111** (1992) 239–246.

[30] K. Paton, An Algorithm for Finding a Fundamental Set of Cycles of a Graph, *Comm. ACM*, **12** (1969) 514–518.

[31] M. Plotkin, Mathematical Basis of Ring-Finding Algorithms In CIDS, *J. Chem. Doc.*, **11** (1971) 60–63.

[32] M. Py, *Un agent rationnel pour raisonner par analogie*, Thèse de doctorat, Université Montpellier II, novembre 1992.

[33] R.C. Read, and R.E. Tarjan, Bounds on Backtrack Algorithms for Listing Cycles, Paths and Spanning Trees, *Networks*, **5** (1975) 237–252.

[34] J.-C. Régin, *Développement d'outils algorithmiques pour l'Intelligence Artificielle et application à la chimie organique*, Thèse de doctorat, Université Montpellier II, Décembre 1995.
[35] D. Richards, Finding Short Cycles in Planar Graphs using Separators, *Journal of Algorithms*, **7** (1986) 382–394.
[36] C. Shelley, Heuristic Approach for Displaying Chemical Structures, *J. Chem. Inf. Comput. Sci.*, **23** (1983) 61–65.
[37] E. Sorkau, Ringerkennung in chemischen Strukturen mit dem Computer, *Wiss. Z. Tech. Hochsch. Leuna-Meuseburg*, **27** (1985) 765–770.
[38] A.A. Syslo, An Efficient Cycle Vector Space Algorithm for Listing All Cycles of a Planar Graph, *SIAM Journal on Computing*, **10** (1981) 797–808.
[39] Y. Takahashi, Automatic Extraction of Ring Substructures from a Chemical Structure, *J. Chem. Inf. Comput. Sci.*, **34**(1) (1994) 167–170.
[40] R. E. Tarjan, Depth-First Search and Linear Graph Algorithms, *SIAM J. Comput.*, **1** (1972) 146–160.
[41] J. Tierman, An Efficient Search Algorithm to Find the Elementary Circuits of a Graph, *Comm. ACM*, **13** (1970) 722–726.
[42] P. Vismara, *Reconnaissance et représentation d'éléments structuraux pour la description d'objets complexes. Application à l'élaboration de stratégies de synthèse en chimie organique*, Thèse de doctorat, Université Montpellier II, décembre 1995.
[43] P. Vismara, Union of All the Minimum Cycle Bases of a Graph, *Electronic Journal of Combinatorics*, **4**(R9) (1997) 15.
[44] J. Welch, A Mechanical Analysis of the Cyclic Structure of Undirected Linear Graphs, *J. ACM*, **13** (1966) 205–210.
[45] W.T. Wipke, and T.M. Dyott, Use of Ring Assemblies in a Ring Perception Algorithm, *J. Chem. Inf. Comput. Sci.*, **15**(3) (1975) 140–147.
[46] A. Zamora, An Algorithm for Finding the Smallest Set of Smallest Rings, *J. Chem. Inf. Comput. Sci.*, **16** (1976) 40–43.

ENSA-M, 2 PLACE PIERRE VIALA, 34060 MONTPELLIER CEDEX 1, FRANCE
E-mail address: `vismara@ensam.inra.fr`

CCIPE, 141 RUE DE LA CARDONILLE, 34094 MONTPELLIER CEDEX 5, FRANCE
E-mail address: `cl@lirmm.fr`

DIMACS Series in Discrete Mathematics
and Theoretical Computer Science
Volume **51**, 2000

The Wulff–Shape of Large Periodic Sphere Packings

J.M. Wills

ABSTRACT. We determine the shape of large densest periodic packings of spheres with different radii. The density is measured by parametric density and the density deviation, which measures the average difference between the density of finite and infinite packings. The asymptotic shapes are the Wulff–shapes, which are polytopes and good models for crystal shapes.

1. Introduction

We determine the shape of large densest periodic packings of spheres with different radii. The special case of lattice packings was first investigated in [**13, 14**]. In [**1**] the analogue problem was investigated for quasiperiodic circle packings and in [**2**] for quasiperiodic sphere packings.

General periodic sphere packings are investigated in [**9, 10**]. It turns out that in all cases the problem of maximal packing density is equivalent to a minimum problem for an averaging function, the density deviation, and this is formally equivalent to the Gibbs–Curie problem for minimal energy. As in the Gibbs–Curie problem the optimal shapes are polytopes, the Wulff–shapes, which are good models for crystals or quasicrystals. So besides the mathematical motivation there is a strong motivation from crystallography. The general periodic case is of particular interest, because all ideal crystals can be modeled by periodic sphere packings. But the general case is much more delicate than the lattice case. The proof in [**9**] requires slight restrictions. So a proof without these restrictions is desirable and given in this paper.

The crucial point is the introduction of an appropriate generalization of lattice polytopes and the estimation of their lattice point number, which leads in a rather simple way to the Wulff shape. The tools are from discrete geometry and convex geometry. For the first we refer to [**3, 6, 8**], for the second to [**12**]. The definitions are introduced in Sections 2 and 3. The main results are in Section 4 and its proofs (as far as they are new) are in Section 5. Section 6 surveys all applications to crystals and quasicrystals until now, which are scattered in several papers.

1991 *Mathematics Subject Classification.* Primary 52C17; Secondary 82D25.
Partially supported by Deutsche Forschungsgemeinschaft.
I want to thank V. Arhelger and U. Schnell for valuable comments.

2. Basic Definitions

Let B^d denote the unit ball in Euclidean d–space $E^d, d \geq 2$. Let V denote the volume and $V(B^d) = \kappa_d$. For $x, y \in E^d$ let $\langle x, y\rangle$ denote their scalar product and $\langle x, x\rangle = \|x\|^2$.

Let $L \subset E^d$ be a lattice with $\det L > 0$ and L^* its dual or polar lattice. For $t^j \in E^d$ let $L_j = L + t^j$, $j = 1, \dots, n$ translates of L or briefly grids. Let $C_j \subset E^d$, $j = 1, \dots, n$ be convex bodies such that $(L_1 + C_1) \cup \dots \cup (L_n + C_n)$ is a packing. Then $M = L_1 \cup \dots \cup L_n$ is a periodic set and also called a periodic packing for $C_1, \dots, C_n$ (cf. [**8**]).

With $V(C_j) = v_j$, $j = 1, \dots, n$, $\mathfrak{V} = (v_1, \dots, v_n)$ and $v = \sum_{j=1}^{n} v_j$ the classical density of an infinite periodic packing is (cf. [**8**])

$$\delta_M(\mathfrak{V}) = v / \det L \tag{2.1}$$

For the applications we always assume $C_j = r_j B^d$, $j = 1, \dots, n$, i.e. the C_j are balls of radius r_j. This is no restriction, because (2.1) and (2.2) depend on the volume, not on the shape.

Concerning M we need some simple notations: An $x \in M$ is called M–point, a finite subset $M' \subset M$ is called a finite M–set. The C_j are called the decoration of M or M'. Any finite M' with its decoration is called a finite M–packing. We are interested in dense finite M–packings. So we consider only finite $M' \subset M$, which are saturated with respect to M, i.e. with

$$M' = M \cap (\operatorname{conv} M').$$

For a finite saturated M–set M' we call $P = \operatorname{conv} M'$ its M–polytope.

For any convex body $K \subset E^d$ we define its lattice point enumerator $L(K) = \operatorname{card}(K \cap L)$ and for the grids:

$$L_j(K) = \operatorname{card}(K \cap L_j) \qquad j = 1, \dots, n.$$

We are now ready for the appropriate density definition:

Definition 2.1. Let $M, \mathfrak{V}$ and $\varrho > 0$ be given. Then for an M–polytope P its parametric density is

$$\delta_M(\mathfrak{V}, P, \varrho) = \frac{\sum_{j=1}^{n} v_j L_j(P)}{V(P + \varrho B^d)}. \tag{2.2}$$

For any polytope P let $F_i(P)$, $i = 1, \dots, k$ denote its facets. For simplicity F_i denotes the point set as well as its $(d-1)$–volume. Hence $\sum_{i=1}^{k} F_i(P) = F(P)$, where F is the surface area. Further

$$f_i(P) = F_i(P)/(V(P))^{1-1/d}, \; i = 1, \dots, k$$

denotes the relative surface area of a facet. With $r(P)$ and $R(P)$ we denote the inradius and the circumradius of P.

3. Facet Densities

Any $u \in L^*(u \neq 0)$ determines an array of lattice–hyperplanes in L with normal vector u. If $M = L$ is a lattice, then there is only one type of lattice–hyperplane and so the packing density of a facet with (outer) normal u is easy to determine. In the general case, if M is the superposition of translates L_i of L equipped with spheres of different size, the determination of the facet density is more complicated.

One has to investigate the packing density for each of the L_i and has to take into consideration the influence of the packing density of the parallel layers of balls coming from the $L_j, j \neq i$. This needs a careful (but elementary) analysis which is given in this section.

For any M–polytope P its parametric density can easily be calculated. Clearly one cannot expect general results for densest packings. But for suitable 'large' M–polytopes one obtains good asymptotic results. For this we need a rather technical but useful definition: For any $u^i \in L^*$ and $L_j = L + t^j$, $j = 1, \dots, n$ let $\epsilon_{ij} = \{\langle u^i, t^j \rangle\}$, where $\{x\} = x - [x]$ and $[x]$ denotes as usual the largest integer $\leq x$. Now we define the appropriate class of polytopes:

DEFINITION 3.1. Let L and M be given and $U = (u^1, \dots, u^k) \subset L^*$ with pos $U = E^d$ (positive hull). Further let $z_i \in \mathbb{N}$, $i = 1, \dots, k$, $j_i \in \{1, \dots, n\}$ and $P' = \{x \in E^d | \langle x, u^i \rangle \leq z_i + \varepsilon_{ij_i}, i = 1, \dots, k\}$. Then $P = \operatorname{conv}(P' \cap M)$ is called an (M, U)–polytope.

REMARKS. (M, U)–polytopes are M–polytopes.

By Definition 3.1 a facet with normal u^i intersects an L_{j_i} in a $(d-1)$–grid with determinant $\|u^i\| \det L$.

If the z_i are 'large', the other facets with normals $\notin U$ are 'small' and lie 'near' the $(d-2)$–faces of P'. In general P' is not the convex hull of M–points, hence no M–polytope. In Lemma 5.2 it is shown that $V(P) - V(P') < c(V(P))^{1-2/d}$, where c depends only on M and U. For any i one can choose $j_i \in [1, \dots, n]$. It turns out that the choice of j_i is essential for the local packing density at the surface of P.

The following local density functions were first introduced in [**9**]:

DEFINITION 3.2. For $m \in \{1, \dots, n\}$ let $\tau(i, m) = \frac{1}{2} - \sum_{j=1}^{n} \frac{v_j}{v}\{\varepsilon_{im} - \varepsilon_{ij}\}$ and $\tau_i = \max_m \tau(i, m)$. $\tau(i, m)$ is called the local density distribution. $\tau(i, m)/\|u^i\|$ is called the facet density and $\tau_i/\|u^i\|$ the optimal facet density in direction of u^i.

REMARKS. Clearly $\tau(i, m) \in (-1/2, 1/2]$. If $M = L$ is a lattice, then $\tau_i = 1/2$, $i = 1, \dots, k$. The facet density measures the volume of the balls per unit of the surface area. So it has homogeneity degree 1 rather than 0 as densities usually have. We use this notation, because it gives the appropriate density description.

4. Results

Our main result is

THEOREM 4.1. *Let M, U and $\mathfrak{V}$ be given, let P be an (M, U)-polytope with $r(P)/R(P) > \varepsilon_0$, $\mu = (V(P))^{1/d}$, $j_i \in [1, \dots, n]$ and $\varrho > \max(\tau(i, j_i)/\|u^i\|)$. Then (for $\mu \to \infty$)*

$$\delta_M(\mathfrak{V}, P, \varrho) = \delta_M - \mu^{-1}\Delta_M(P, \varrho) + O(\mu^{-2}) \tag{4.1}$$

and

$$(4.2) \qquad \Delta_M(P,\varrho) = \delta_M \sum_{i=j}^{k} f_i(P)(\varrho - \tau(i,j_i)/\|u^i\|)$$

DEFINITION 4.2. $\Delta_M(P,\varrho)$ is called the Density Deviation of P and ϱ.

REMARKS. The first error term in (4.1) $\Delta_M(P,\varrho)$ measures the average difference between δ_M and parametric density. The second error term (or edge term) $O(\mu^{-2})$ depends on $M, U, \mathfrak{V}, \varepsilon_0$ and ϱ, but not on the size of P or of the f_i.

If $V(P)$ and hence μ is large, then $\Delta_M(P,\varrho)$ is the essential error term. The choice of ϱ guarantees that all summands are > 0 and hence $\Delta_M(P,\varrho) > 0$.

So among all (M,U)–polytopes of same volume, i.e. of same μ, those with minimal $\Delta_M(P,\varrho)$ are best packings and it is the central problem to find the shapes which minimize $\Delta_M(P,\varrho)$. We minimize $\Delta_M(P,\varrho)$ in two steps:

a) First we minimize the coefficients of the $f_i(P)$ in $\Delta_M(P,\varrho)$, i.e. we replace $\tau(i,m)/\|u^i\|$ by $\tau_i \,/\, \|u^i\|$, for which we have:

PROPOSITION 4.3. *$\frac{1}{2n} \le \tau_i \le \frac{1}{2}$, and both bounds are tight.*

In Section 5 we give a short and elegant proof by U. Schnell [**9**].

b) Second we observe that the density deviation can be written as

$$\Delta_M(P,\varrho) = \sum_{i=1}^{k} \sigma_i f_i(P)$$

with some $\sigma_i > 0$. If the σ_i are energies, then the problem of minimizing this sum for all polytopes with given unit normals $U = \{u^1,\dots,u^k\}$ and fixed volume $V(P)$, hence for fixed μ, is the famous Gibbs–Curie energy problem and solved in Wulff's theorem (1901): The optimal polytope is the Wulff–shape

$$\{x \in E^d | \langle x, u^i\rangle \le \sigma_i, i = 1,\dots,k\}.$$

From the mathematical point of view the meaning of the numbers $\sigma_i > 0$ is irrelevant. In fact, if we consider the facet density (Definition3.2 and (4.2)) as weighted sum with weights v_j/v, the v_j can be volumes as well as weights, potentials or energies. So we get, independent of the physical meaning of the v_j, in our case with the unit vectors $u^i/\|u^i\|$

THEOREM 4.4. (Wulff) *Let $M, U, \mathfrak{V}$ and $\varrho > \max_i(\tau_i/\|u^i\|)$ be given. Then $\Delta_M(P,\varrho)$ is minimal for the Wulff–shape*

$$W_M(U,\varrho) = \{x \in E^d | \langle x, u^i\rangle \,/\, \|u^i\| \;\le\; \varrho - \tau_i \,/\, \|u^i\|\,, i = 1,\dots,k\}.$$

REMARKS. The Wulff–shape is a normalized asymptotic shape; hence no (M,U)–polytope. The choice of ϱ guarantees that $W_M(U,\varrho)$ is nonempty.

There are many proofs of Wulff's theorem so that we do not give one. The most famous one, based on energetic arguments, is due to M. v. Laue [**7**]. The first geometric proof is by Dinghas [**4**], the shortest one by P.M. Gruber [**5**]. In fact we got the first hint on the relation between our results and Wulff–shape by P.M. Gruber in 1994.

If M and $\mathfrak{V}$ and ϱ are prescribed, the Wulff–shape of course depends strongly on U. So there is the natural question, if the artificial restriction of prescribing $U \subset L^*$ can be replaced by admitting all $u^i \in L^*$. It is one of the most interesting and surprising results, that even then $W_M(L^*, \varrho)$ is a polytope, i.e. only finitely many u^i contribute to W_M.

THEOREM 4.5. *The Wulff–shape $W_M(L^*, \varrho)$ is a polytope.*

REMARKS. The first proof of Theorem 4.5 for the special case $M = L$ was given in [**14**] and it can easily be generalized to M. In [**9**] the general case is proved with a different method, and moreover an upper bound for the number of facets of $W_M(L^*, \varrho)$, if M is rational with respect to L. So for a proof of Theorem 3 we refer to [**14**] or [**9**].

5. Proofs

For the proof of Theorem 4.1 we need 3 lemmas. The first and second lemma require elementary convex geometry, the third one discrete geometry.

LEMMA 5.1. *For a polytope $P \subset E^d$ with $r(P)/R(P) > \varepsilon_0 > 0$ and for $\varrho > 0$*

$$V(P + \varrho B^d) = V(P)\left(1 + \mu^{-1}\varrho \sum_{i=1}^{k} f_i(P) + O(\mu^{-2})\right),$$

where $O(\mu^{-2})$ depends on ε_0.

PROOF. The Steiner formula (cf [**12**]) gives

$$V(P + \varrho B^d) = \sum_{j=0}^{d} V_j(P)\kappa_{d-j}\varrho^{d-j}$$

With $2V_{d-1}(P) = F(P)$ and $F(P)/(V(P))^{(d-1)/d} = \sum_{i=1}^{k} f_i(P)$ follows

$$V(P + \varrho B^d) = V(P)\left[1 + \mu^{-1}\varrho \sum_{i=1}^{k} f_i(P) + \sum_{j=0}^{d-2}(V_j(P)/V(P))\kappa_{d-j}\varrho^{d-j}\right].$$

With $rB^d \subset P \subset RB^d$ follows $V_j(P)/V(P) \le c_j(d)(R/r)^j r^{j-d}$, $j = 0, \dots, d-2$, with some $c_j(d) > 0$.

So $\sum_{j=0}^{d-2}(V_j(P)/V(P))\kappa_{d-j}\varrho^{d-j}r^{j-d} \le r^{-2}\sum_{j=0}^{d-2} c_j(d)\kappa_{d-j}\varrho^{d-j}\varepsilon_0^{-j}r^{j-(d-2)} = O(\mu^{-2})$.

□

LEMMA 5.2. *Let P and P' be the polytopes in Definition 3.1 and let $r(P)/R(P) > \varepsilon_0 > 0$. Then $V(P) - V(P') = O(\mu^{d-2})$, where $O(\mu^{d-2})$ depends on M, U and ε_0.*

PROOF. For each $(d-2)$–face e_i of P let $Z(r_i)$ be the cylinder with axis e_i and radius $r_i > 0$. We choose the r_i such that just $P \backslash P' \subset \bigcup_{i=1}^{k} (Z(r_i) \cap P)$.

The r_i depend only on L and U, and k is fixed. So

$$V\left(\bigcup_{i=1}^{k}(Z(r_i)\cap P)\right) < \sum_{i=1}^{k} V(Z(r_i)) = \kappa_{d-1}\sum_{i=1}^{k} r_i^{d-1}\cdot V_{d-2}(e_i) \le C\sum_{i=1}^{k} V_{d-2}(e_i) = $$
$$= O(\mu^{d-2}).$$

□

For the last lemma we need a generalization of lattice polytopes.

DEFINITION 5.3.

a) Let $L \subset E^d$ be a lattice and L^* its polar lattice, let $u^i \in L^*, i = 1, \ldots, k$ and $c_i \in \mathbb{R}$, $i = 1 \ldots k$. Then

$$P = \{x \in E^d \mid \langle x, u^i\rangle \le c_i, i = 1, \ldots, k\}$$

is called a lattice normal polytope.

b) If $c_i \in \mathbb{Z}$, $i = 1, \ldots, k$, then P is called a lattice facet polytope.

REMARKS. P might be degenerate or empty. If $P \cap L \neq \emptyset$, we tacitly assume that $0 \in P$. So we only consider P with $0 \in P$. Any lattice polytope (i.e. the convex hull of lattice points) is a lattice facet polytope and hence a lattice normal polytope.

LEMMA 5.4. *Let $L \subset E^d$ be a lattice and P a lattice normal polytope with $U = \{u^1, \ldots, u^k\} \subset L^*$ and $r(P)/R(P) > \varepsilon_0$. Then*

$$\textit{card}\,(L\cap P) = V(P)(\det L)^{-1}\left[1 + \mu^{-1}\sum_{i=1}^{k} f_i(P)(\|u^i\|)^{-1}(\frac{1}{2} - \varepsilon_i) + O(\mu^{-2})\right],$$

where $\varepsilon_i \in [0,1]$ and $\varepsilon_i/\|u^i\|$ is the distance between $f_i(P)$ and the next inner $(d-1)$-sublattice of L parallel to f_i. Further $O(\mu^{-2})$ depends on U and ε_0.

COR. 1 Lemma 5.4 also holds if L is replaced by a grid L_j.

PROOF. (of Lemma 5.4)

Let P' be a lattice normal polytope with $r(P')/R(P') > \varepsilon_0 > 0$, without restriction $0 \in P'$ and $P \subset P'$ the next inner lattice facet polytope, i.e. with same facet normals u^i and distance

$$\varepsilon_i/\|u^i\|, \varepsilon_i \in [0,1)$$

from the corresponding facet F_i' of P'. Then

$$\begin{aligned} V(P) &= V(P') - \sum_{i=1}^{k} F_i'(P)\varepsilon_i/\|u^i\| + O(\mu^{d-2}) = \\ &= V(P')\left[1 - \mu^{-1}\sum_{i=1}^{k} f_i'(P)\varepsilon_i/\|u^i\| + O(\mu^{-2})\right], \end{aligned} \tag{5.1}$$

where $O(\mu^{-2})$ depends only on ε_0.

From now on we only consider the lattice facet polytope P with its facets F_i. For each $x \in L$ let D_x denote its Dirichlet–Voronoi cell and

$$c(x) = \text{card}\,\{i|((\text{int}\,D_x)\cap(bd\,P))\cap F_i \neq \emptyset,\ F_i \text{ facet of } P\}.$$

This gives a classification of all $x \in L$ into those with

1) $c(x) = 0$, 2) $c(x) = 1$, and 3) $c(x) \geq 2$.

We now consider the equation

$$\text{card}\,(P \cap L) = V(P)/\det L + \frac{1}{2}\sum_{1=1}^{k} F_i(P)/\Delta_i + O(\mu^{d-2}). \tag{5.2}$$

Here Δ_i denotes the $(d-1)$–determinant of the $(d-1)$–sublattice $L_i^{d-1} = \operatorname{aff} F_i \cap L$.

1) For the $x \in L$ with $c(x) = 0$ either holds $x \notin P$ or $x \in P$. In the first case x and its D_x contribute 0 to each side of (5.2), and in the second case 1 to each side.
2) For the second class, i.e. those $x \in L$ with $c(x) = 1$, we need some geometry of $L_i^{d-1} = \operatorname{aff} F_i \cap L$. Each L_i^{d-1} induces a tiling in $\operatorname{aff} F_i$, where the tiles are the $(d-1)$–dimensional DV–cells with respect to the lattice points of L_i^{d-1}. These DV–cells D_{ij}, $j = 1, \ldots, s_i$ have $(d-1)$–volume Δ_i. If $D_{ij} \subset F_i$, we write D_{ij}^+, otherwise D_{ij}^-. For each D_{ij}^+ we consider all D_x with $D_x \cap D_{ij}^+ \neq \emptyset$. Clearly for all these D_x is $c(x) > 0$. If for all these D_x is $c(x) = 1$, we write D_{ij}^{++}, otherwise D_{ij}^{+-}. We observe that the D_{ij}^- and D_{ij}^{+-} lie 'near' the boundary of F_i.

   Now we consider a D_{ij}^{++} and all D_x with $D_x \cap D_{ij}^{++} \neq \emptyset$. For all these D_x holds $c(x) = 1$. It is easy to see that these D_x contribute in (5.2) with the same value on the left and on the right:

   If $x \in \operatorname{int} P$, then there is exactly one $x' \notin P$, such that the volumes of their corresponding $D_x \cap P$ add up to $\det L$, hence $V(P)/\det L$ to 1.

   Further there is exactly one $x \in D_{ij}^{++}$, and this counts 1 on the left of (5.2) and 1/2 in the first and in the second summand on the right of (5.2).
3) All $x \in L$ with $c(x) \geq 2$ or which correspond to a D_{ij}^- or D_{ij}^{+-}, lie in a neighborhood of the $(d-2)$–skeleton $\operatorname{skel}_{d-2} P$ of P, i.e. in appropriate cylinders along the $(d-2)$–faces of P. The volume of this set is

$$O(V_{d-2}(\operatorname{skel}_{d-2} P)) = O(\mu^{d-2}).$$

Hence (5.2) is proved, and we rewrite (5.2) as follows:

$$\text{card}\,(P \cap L) = V(P)(\det L)^{-1}\left[1 + \mu^{-1}\frac{1}{2}\sum_{i=1}^{k} f_i(P)/\|u^i\| + O(\mu^{-2})\right]. \tag{5.3}$$

From (5.1) and (5.3) and $f_i(P') = f_i(P) + O(\mu^{-1})$ and $\text{card}\,(P \cap L) = \text{card}\,(P' \cap L)$ the lemma follows, if we replace again P' by P. □

PROOF. (Proof of Theorem 4.1) Under the assumptions of Theorem 4.1 let P be an (M, U)–polytope. Then P is a lattice normal polytope with respect to the grids L_j, $j = 1, \ldots, n$. For each facet F_i of P let j_i such that $F_i \cap L_{j_i} \neq \emptyset$. Then we apply Lemma 5.4 and Cor. 1 and get

$$\begin{aligned} L_j(P) &= \text{card}\,(P \cap L_j) = \\ &= V(P)(\det L)^{-1}\left[1 + \mu^{-1}\sum_{i=1}^{k} f_i(P)\left(\frac{1}{2} - \{\varepsilon_{ij_i} - \varepsilon_{ij}\}\right)/\|u^i\| + O(\mu^{-2})\right]. \end{aligned}$$

Now we get from Definitions 2.1, 3.1, 3.2 and Lemmas 5.1, 5.2:

$$\begin{aligned}\delta_M(\mathfrak{V}, P, \varrho) &= \left(\sum_{j=1}^{n} v_j\, L_j(P)\right) (V(P+\varrho B^d))^{-1} = \\ &= \delta_M \left[1 + \mu^{-1} \sum_{i=1}^{k} f_i(P) \tau(i, j_i)/\|u^i\| + O(\mu^{-2})\right]. \\ &\left[1 + \mu^{-1} \varrho \sum_{i=1}^{k} f_i(P) + O(\mu^{-2})\right]^{-1} = \\ = \delta_M &\left[1 - \mu^{-1} \sum_{i=1}^{k} f_i(P)(\varrho - \tau(i, j_i)/\|u^i\|) + O(\mu^{-2})\right],\end{aligned}$$

hence (4.1) and (4.2). So Theorem 4.1 is proved. □

PROOF. (Proof of the Proposition (U. Schnell))

If $M = L$ is a lattice, then $\tau_i = \frac{1}{2}$. For the lower bound let $v_j/v = \lambda_j$, $j = 1, \dots, n$. Then $\sum_{j=1}^{n} \lambda_j = 1$ and hence $\sum_{j=1}^{n} \lambda_j^2 \geq n^{-1}$. With $\{\varepsilon_{im} - \varepsilon_{ij}\} + \{\varepsilon_{ij} - \varepsilon_{im}\} \leq 1$ we get

$$\begin{aligned}\sum_{m=1}^{n} \lambda_m \sum_{j=1}^{n} \lambda_j \{\varepsilon_{im} - \varepsilon_{ij}\} &= \sum_{m \neq j} \lambda_j \lambda_m \{\varepsilon_{im} - \varepsilon_{ij}\} \leq \\ &\leq \sum_{j<m} \lambda_j \lambda_m = \frac{1}{2}\left(1 - \sum_{j=1}^{n} \lambda_j^2\right) \leq \frac{1}{2} - \frac{1}{2n}.\end{aligned}$$

So there is an $m \in [1, n]$ with $\sum_{j=1}^{n} \lambda_j \{\varepsilon_{im} - \varepsilon_{ij}\} \leq \frac{1}{2} - \frac{1}{2n}$, i.e. $\tau_i \geq \frac{1}{2n}$. Equality holds for $\lambda_j = 1/n$ and $\varepsilon_{ij} = (j-1)/n$, $j = 1, \dots, n$. □

6. Applications: Wulff–Shapes of Crystals

Before we come to various examples, we emphasize that the method does not work for nonconvex crystals like snow flakes or nonconvex clusters like Fullerenes. For convex clusters or degenerate crystals like whiskers one needs related but different packing methods [**15**]. It is worth mentioning that our method with density deviation is an averaging process, which reflects real crystal growth, whereas individual best packings usually do not, as simple examples show. Of course densest packing cannot model all energetic events, in particular not at the atomic level. But it is more than surprising how far reaching the approach via packing density is. We now come to the applications.

With the previous methods one gets the Wulff–shapes which depend on the prescribed sphere packing and the parameter ϱ. The essential point for application is, how good it models ideal crystals. Clearly here we are only interested in the 3–dimensional case. In [**13, 14**] the lattice case was investigated, which leads easily to the expected Wulff–shapes, e.g. the cube and its truncations if $L = \mathbb{Z}^3$. Further for fcc the octahedron with its truncations (enlarging ϱ enlarges the truncations). Clearly these coincide with 1–atomic crystals, which occur in nature, e.g. metals and noble gases. Of course the general periodic case is much more fruitful. In [**9**] Schnell showed for salt (Na Cl) that its Wulff-shape is the cube and its truncations, for zinc–blend (ZnS) the tetrahedron and its truncations and for the diamond (which is no lattice packing) the octahedron and its truncations. In [**10**] he compared the fcc–lattice with all densest nonlattice packings, as e.g. hcp and proved that fcc has densest Wulff–shape. As in the previous cases this coincides with reality; here with the dominant shape of noble gases (except Helium). It is remarkable that the Lennard–Jones potential in this case does not coincide with reality.

A natural question is, if these methods also work for quasicrystals, i.e. for quasiperiodic packings, which by definition are not periodic. In [**1**] and [**2**] the methods could be applied with a slight modification. For planar quasiperiodic packings it was shown that the Wulff–shapes are 10–gons, 20–gons, etc., i.e. just the observed shapes. Even more remarkable: For 3–dimensional quasicrystals it was shown in [**2**], that the Wulff–shapes are the dodecahedron or the rhombic triacontahedron and their truncations (depending on the underlying 6–dimensional lattice). This again coincides with reality. Of course there may be crystals (besides the above mentioned ones) which cannot be modeled this way, probably some with very complicated structure. But the previous results show the quality of the model.

References

[1] K. Böröczky, Jr., and U. Schnell, *Quasicrystals and the Wulff–Shape*, Discr. Comp. Geom., to appear.

[2] K. Böröczky, Jr., and U. Schnell, *Wulff Shape for Non–Periodic Arrangements*, Letters in Math. Phys., to appear.

[3] J.H. Conway, and N.J.A. Sloane, *Sphere Packings, Lattices and Groups*, Springer, New York, 1988.

[4] A. Dinghas, Über einen geometrischen Satz von Wulff über die Gleichgewichtsform von Kristallen, *Z. Kristallogr.*, **105** (1943) 304–314.

[5] P.M. Gruber, *Mixed Volumes*, Manuscript 1994.

[6] P.M. Gruber, and C.G. Lekkerkerker, Geometry of Numbers, North Holland, Amsterdam, 1987.

[7] M. v. Laue, Der Wulffsche Satz für die Gleichgewichtsform von Kristallen, *Z. Kristallogr.*, **105** (1943) 124–133.

[8] C.A. Rogers, *Packing and Covering*, Cambridge University Press (1964).

[9] U. Schnell, *Periodic Sphere Packings and the Wulff–Shape*, to appear.

[10] , *On the Wulff–Shape of Periodic Sphere Packings*, to appear.

[11] —, *Wulff–Shape and Density Deviation*, manuscript.

[12] J.R. Sangwine–Yager, *Mixed Volumes, Ch. 1.2*, Handbook of Convex Geometry, edit. P.M. Gruber, J.M. Wills, North Holland, Amsterdam, 1993.

[13] J.M. Wills, On Large Lattice Packings of Spheres, *Geometriae Dedicata*, **65** (1997) 117-126.

[14] —, Lattice Packings of Spheres and the Wulff–Shape, *Mathematika*, **86** (1996) 229–236.

[15] —, Parametric Density, Online Packings and Crystal Growth *Rendic. di Palermo*, Ser.II 50 (1997) 413–424.

MATHEMATISCHES INSTITUT, UNIVERSITÄT SIEGEN, D–57068 SIEGEN, GERMANY

E-mail address: `wills@mathematik.uni-siegen.de`

DIMACS Series in Discrete Mathematics
and Theoretical Computer Science
Volume **51**, 2000

On Hypercycles and Hypercircuits in Hypergraphs

Andrew V. Zeigarnik

ABSTRACT. Chemical reaction mechanisms can be represented as (1) chemical equations; (2) bipartite multigraphs; and (3) hypergraphs. Outside of chemistry, the concepts of a hypercircuit and a hypercycle have been found suitable for application in computer science, combinatorial optimization, transportation analysis, and other fields of science. The precise definition of these concepts, however, do not seem suitable for some applications in chemistry; also, they do not preserve the Euler formula which relates the cyclomatic number with the numbers of edges, vertices, and components of a graph. An alternative definition of hypercircuit and hypercycle is suggested that removes the cited drawbacks. Some open problems are mentioned that are of interest within the theory of chemical reaction mechanisms.

1. Introduction

Graph theory traditionally plays an important role in chemistry. Several hundreds (or maybe thousands) of articles and several monographs have been published, which are devoted to applications of graph theory in chemistry. Chemical applications of hypergraph theory so far attracted smaller attention, although graphs are not fully correct models of molecules, because several types of chemical structures cannot be described by graphs (first of all, delocalized bonds, multicentered bonds, electron-deficient structures, and η^n-bonding in organometallic chemistry). Chemical reaction mechanisms are also described better by hypergraphs.

In this article, I consider several problems in graph theory arising from the analysis and classification of chemical reaction mechanisms [**1**].

Chemical reaction mechanisms. Chemical reaction mechanisms (or simply mechanisms) are the sets of chemical equations, which are similar to linear algebraic equations. For example,

$$A + X \rightleftarrows Y$$
$$Y + X \rightarrow Z$$
$$B + Z \rightarrow X + X + P$$
$$C + Y \rightarrow X + T$$

A, B, C, P, T, X, Y, and Z are called *species* (or molecules). Chemical equations listed above are called *steps*. Species are usually categorized as *reactants* (say,

1991 *Mathematics Subject Classification.* Primary 05C65; Secondary 94C15, 05C38.

A, B, and C), target *products* (say, P and T) and intermediates (X, Y, and Z). Usually, it's a chemist's job to define which species are reactants, products, and intermediates, but in most cases some simple heuristic rules may also work [**1**].

In a mechanism, a step can be bidirectional ($\rightleftarrows$) or unidirectional ($\rightarrow$). If a mechanism has bidirectional steps, as in the above example, it can always be rewritten by replacing every bidirectional step with two unidirectional. Returning to the above example, the mechanism can be rewritten as follows (Mechanism 1):

$$\begin{array}{ll} 1. & A + X \rightarrow Y \\ 2. & Y \rightarrow A + X \\ 3. & Y + X \rightarrow Z \\ 4. & B + Z \rightarrow X + X + P \\ 5. & C + Y \rightarrow X + T \end{array}$$

Taking into account this feature, I will consider only mechanisms with "homogeneously" unidirectional steps.

A mechanism is said to be *linear* if each step has at most one intermediate before and at most one intermediate after the arrow. Otherwise, the mechanism is *nonlinear.* The above mechanism is nonlinear because step 3 has two intermediates on the left-hands side and step 4 has two intermediates on the right-hand side, although in the latter case, both of them are X.

A mechanism can be reduced by deleting all reactants and products. Let us consider for simplicity that each step contains at least one intermediate on each side (the ideas presented below can easily be extended to the case when some reduced steps have "empty sides" by adding a fictitious $\emptyset$-intermediate). Thus, the reduced form of the above mechanism looks like Mechanism 2 below:

$$\begin{array}{ll} 1. & X \rightarrow Y \\ 2. & Y \rightarrow X \\ 3. & Y + X \rightarrow Z \\ 4. & Z \rightarrow X + X \\ 5. & Y \rightarrow X \end{array}$$

Digraphs of mechanisms. Consider a reduced form of a linear mechanism and a digraph in which each a vertex represents an intermediate, and an arc begins at the vertex X and ends at the vertex Y if a reduced form of a mechanism contains a step $X \rightarrow Y$. Such a digraph (call it **D**) may have multiple arcs, and the arcs joining two vertices do not necessarily have the same direction. Self-loops, however, cannot be present since I will not consider reactions like $X \rightarrow X$. (These exotic chemical steps may exist, but they are rare.) Clearly, digraphs cannot represent nonlinear mechanisms.

Bipartite digraphs of mechanisms. To overcome the problem of representing nonlinear mechanisms, one can use bipartite digraphs (call them **B**-graphs). A vertex set V of a **B**-graph consists of two proper subsets S (for species) and R (for steps) such that $V = S \cup R$ (and $S \cap R = \emptyset$). An arc shows the "flow" of species transformations in chemical reactions and the number (called a *stoichiometric coefficient*) of species appearances on a side of a step. None of the arcs joins two vertices from the same subset. Arcs can be multiple, and if two (or more) arcs join two vertices, then the directions of arcs must be the same. The bipartite graph of the reduced form of a mechanism is presented in Figure 1:

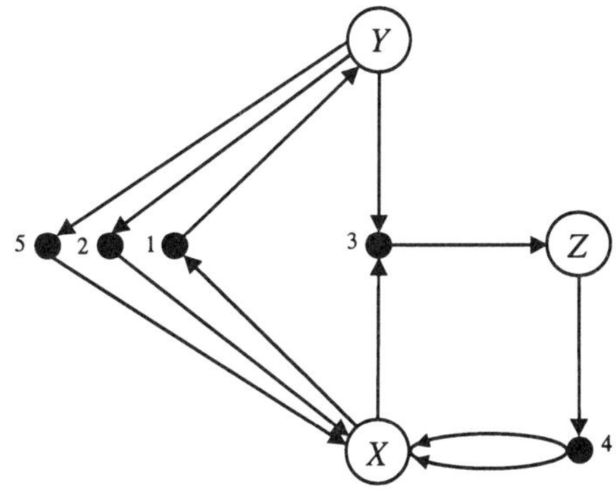

FIGURE 1

Directed hypergraphs of mechanisms. A hypergraph $\mathcal{H}$ is a pair $\mathcal{H}(V,E)$, where V is a (nonempty) set of vertices and E is a set of (nonempty) subsets of V. A bipartite graph can be viewed as a representation of a hypergraph. A directed bipartite graph with multiple arcs can, in turn, be viewed as a representation of a directed hypergraph (call it $\mathcal{D}$) defined as follows. A directed hypergraph is a pair $\mathcal{D}(V,E)$, where V is a (nonempty) set of vertices, and E (a set of hyperarcs) is a set of pairs $E(H,T)$ of families H(heads) and T (tails) consisting of vertices from the set V. Vertices belonging to H or T of a hyperarc may be repeated, but their order is insignificant. A subgraph $\mathcal{D}^*$ of a directed hypergraph $\mathcal{D}$ is a hypergraph with a vertex set $V(\mathcal{D}^*) \subseteq V(\mathcal{D})$ and a hyperarc set $E(\mathcal{D}^*) \subseteq V(\mathcal{D})$.

Directed hypergraphs have found applications in computer science, combinatorial optimization, transportation analysis, and other fields (see, for example, [**2**]). Also, they adequately describe chemical reaction mechanisms and their reduced forms. Thus, the above reduced mechanism is $\mathcal{D}(V,E)$, where $V = \{X,Y,Z\}$, $E = \{e_1,e_2,e_3,e_4,e_5\}$ and $c_1 = (X|Y)$, $c_2 = (Y|X)$, $c_3 = (Y,X|Z)$, $c_4 = (Z|X,X)$, and $e_5 = (Y|X)$. In each arc, vertices belonging to the head and tail stand before and after a separation line, respectively.

The in-degree id(v) (out-degree od(v)) of a vertex v is the number of times v is included in heads (tails) of hyperarcs in a directed hypergraph. For example, in the above directed hypergraph, id$(X) = 1+2+1 = 4$ and od$(X) = 1+1 = 2$. Note that X is included in the tail of e_4 twice.

2. Cyclomatic Number of a Hypergraph

According to the definition given in [**3, 4**], a hypercycle in a hypergraph $\mathcal{H}$ is an alternating sequence of vertices and hyperedges $v_1, e_1, v_2, e_2, \dots, e_q, v_{q+1}, q > 1$, where all vertices and all hyperedges are distinct except for $v_1 = v_{q+1}$ and such that each vertex belongs to its neighboring hyperedges.

Then, it would be logical to assume that a hypercircuit in a directed hypergraph is an alternating sequence of vertices and hyperarcs $v_1, c_1, v_2, c_2, \dots, c_q, v_{q+1}, q > 1$, where all the vertices and all the hyperarcs are distinct except for $v_1 = v_{q+1}$ and such that for each $i = 1, \dots, q, v_i \in H_{i-1}$ and $v_i \in T_i$ [**2**]. However, the above definition of a hypercycle leads to the formula for the cyclomatic number $\gamma(\mathcal{H})$ of a hypergraph H, which is substantially different from the Euler formula for graphs [5]:

$$\gamma(\mathcal{H}) = \sum_{e\in E(\mathcal{H})} |e| - |V(\mathcal{H})| - |E(\mathbf{T}(\mathcal{H}))| \tag{2.1}$$

where $\sum_{e\in E(\mathcal{H})}$ is the sum of cardinalities of edges in $\mathcal{H}$ (instead of the number of edges) and $|E(\mathbf{T}(\mathcal{H}))|$ is the number of edges in the maximal spanning multiforest of the intersection multigraph of $\mathcal{H}$ (instead of the number of components). Also, the sign of the last term in formula (2.1) is different than in the Euler formula. A similar formula is given in [3] as a definition of the cyclomatic number. Therefore, an intriguing question is: *What would be another definition of a hypercycle in a hypergraph that results in the Euler formula identical to that for graphs?* For simplicity, I will consider only one-component hypergraphs and directed hypergraphs. The rest of the article is devoted to the generalizations of directed hypergraphs that follow from chemical applications of hypergraph theory and provide an idea of preserving the form of the Euler formula.

3. Routes

In theory of reaction mechanisms, an important concept is used that defines certain subsets of steps of a mechanism. A subset $M' = \{s_i, s_{i+1}, \dots, s_K\}$ of steps of mechanism M is called a *route* [6] if there exists a set of positive integers $\{\xi_i, \xi_{i+1}, \dots, \xi_K\}$, such that upon multiplying each step s_j by ξ_j and taking the sum of steps, one has an overall chemical equation that does not contain intermediates. If a mechanism is given in a reduced form, then the overall equation is $0 \to 0$.

Suppose a directed hypergraph $\mathcal{D}$ has q hyperarcs and a vector $\mathbf{N}(n_1, n_2, \dots, n_q)$ consists of q entries. The multiplication of the hypergraph by the vector giving a product hypergraph $\mathcal{M} = \mathcal{D} \otimes \mathbf{N}$ is an operation that transforms (i) a hyperarc $e_i(H_i|T_i)$ in $\mathcal{D}$ into the hyperarc $e_i(n_iH_i|n_iT_i)$ in $\mathcal{M}$ if n_i is positive; (ii) a hyperarc $e_i(H_i|T_i)$ in $\mathcal{D}$ into the hyperarc $e_i(m_iT_i|m_iH_i)$ in $\mathcal{M}$, where $m_i = -n_i$ if n_i is negative; and (iii) deletes the hyperarc in $\mathcal{D}$ if $n_i = 0$. By writing nH, I mean that each element of family H is repeated n times in the family nH.

In graph-theoretical terms, a subhypergraph $\mathcal{D}^*$ of $\mathcal{D}$ is a route if one can find a vector of positive integers $\mathbf{N}$ such that $\mathcal{D}^* \otimes \mathbf{N}$ is a directed hypergraph in which for every vertex v_i, $\mathrm{id}(v_i) = \mathrm{od}(v_i)$. A route is *simple* if none of its subgraphs is a route. Simple routes of the directed hypergraph for mechanism 2 can be characterized by vertices and hyperarcs:

	Vertices	Hyperarcs
Route 1	X and Y	$e_1 = (X|Y)$ and $e_2 = (Y|X)$
Route 2	X, Y, and Z	$e_1 = (X|Y)$, $e_3 = (Y, X|Z)$, and $e_4 = (Z|X, X)$
Route 3	X and Y	$e_1 = (X|Y)$ and $e_5 = (Y|X)$

Simple routes of a directed hypergraph can be regarded as hypercircuits. In routes 1 and 3, the out-degree and in-degree of each vertex are equal to unity; therefore, these routes are linear.

CONJECTURE. A directed hypergraph $\mathcal{D}$ has a hypercircuit if for each vertex v of $\mathcal{D}$, $\mathrm{id}(v) = \mathrm{od}(v)$.

CONJECTURE. A directed hypergraph $\mathcal{D}$ is not a hypercircuit if it is disconnected or contains at least one vertex v with $\mathrm{id}(v) < \mathrm{od}(v)(\mathrm{id}(v) > \mathrm{od}(v))$, and for any other vertex w of $\mathcal{D}$ one of the two following conditions is met: $\mathrm{id}(w) \leq \mathrm{od}(w)(\mathrm{id}(w) \geq \mathrm{od}(w))$.

The proof of the second conjecture is probably trivial.

4. Analogy between Graphs Derived from Digraphs and Hypergraphs Derived from Directed Hypergraphs

Let us define a *skeleton graph* of a digraph $\mathbf{D}$ and call it $\mathbf{S}(\mathbf{D})$ as a graph that can be obtained from $\mathbf{D}$ by neglecting the directions of arcs. Each circuit in $\mathbf{D}$ corresponds to a cycle in $\mathbf{S}(\mathbf{D})$, and $\mathbf{S}(\mathbf{D})$ may also contain other cycles for which there are no circuits in $\mathbf{D}$. Each elementary circuit in $\mathbf{D}$ corresponds to an elementary cycle in $\mathbf{S}(\mathbf{D})$.

Cycles in $\mathbf{S}(\mathbf{D})$ possess several properties, which are well studied in graph theory, among which two of the most important are as follows. Cycles in $\mathbf{S}(\mathbf{D})$ induce a *cycle space* of $\mathbf{S}(\mathbf{D})$ over the two-element field $F_2 = \{0, 1\}$ [**7**]. The basis set of cycles consists of γ linearly independent cycles, and γ equals the rank of the cycle $(0, 1)$-matrix, in which rows stand for elementary cycles and columns stand for edges [**7**]. The rank of the cycle space is equal to the cyclomatic number as determined by the Euler formula:

$$\gamma(\mathbf{S}(\mathbf{D})) = \left|E(\mathbf{S}(\mathbf{D}))\right| - \left|V(\mathbf{S}(\mathbf{D}))\right| + q \tag{4.1}$$

where q is the number of components of $\mathbf{S}(\mathbf{D})$.

A finite set of edges of $\mathbf{S}(\mathbf{D})$ and elementary cycles form a matroid because they satisfy the following axioms: (i) no proper subset of edges of a cycle form a cycle; (ii) if an edge $x \in C_1 \cap C_2$, where C_1 and C_2 are elementary cycles of $\mathbf{S}(\mathbf{D})$, then $C_1 \cup C_2 - \{x\}$ contains a cycle.

A *skeleton hypergraph* $\mathcal{S}(\mathcal{D})$ of a directed hypergraph $\mathcal{D}$ is a hypergraph that can be obtained from $\mathcal{D}$ by neglecting the directions of hyperarcs. That is, no distinction is made in $\mathcal{S}(\mathcal{D})$ between former tails and heads, but a hyperedge of $\mathcal{S}(\mathcal{D})$ still consists of two families, which, however, are "equal in rights". Let us call a graph of this sort a *semihypergraph*. The most important distinction between semihypergraphs and hypergraphs as they defined in [**3**]–[**5**] is that hyperedges of "normal" hypergraphs do not consist of two subsets (subfamilies). A subgraph $\Omega(\mathcal{S})$ of a semihypergraph $\mathcal{S}$ is called a hypercycle if, for each hyperedge $E_i(Q_i, R_i)$, where Q_i and R_i are the families of vertices from $V(\mathcal{S})$ (call them *hypervertices*), a positive integer n_i can be found such that the hyperedges of a product hypergraph $\mathcal{S} \otimes \mathbf{N}(n_1, n_2, \ldots, n_K)$ form a sequence of edges

$$(n_1Q_1, n_1R_1)(n_2Q_2, n_2R_2) \cdots (n_KQ_K, n_KR_K)$$

where $n_iQ_i = n_{i-1}R_{i-1}$ for $i = 2, \ldots, K$ and $n_1Q_1 = n_KR_K$, and K is the cardinality $|E(\Omega(\mathcal{S}))|$ of the hyperedge set of a subgraph Ω of the semihypergraph $\mathcal{S}$.

It is easy to prove that hypercycles of a semihypergraph induce a *hypercycle space* of $\mathcal{S}(\mathcal{D})$ over the two-element field $F_2 = \{0, 1\}$. The basis set of hypercycles consists of γ linearly independent hypercycles such that the rank of the cycle space is equal to the cyclomatic number as determined by the Euler formula, which is completely identical to (4.1):

$$\gamma(\mathcal{S}(\mathcal{D})) = \left|E(\mathcal{S}(\mathcal{D}))\right| - \left|V(\mathcal{S}(\mathcal{D}))\right| + \theta \tag{4.2}$$

where θ is the number of components of $\mathcal{S}(\mathcal{D})$ and $|V(\mathcal{S}(\mathcal{D}))|$ is the number of hypervertices in $\mathcal{S}(\mathcal{D})$. Also, the set of edges of $\mathcal{S}(\mathcal{D})$ and its elementary hypercycles form a matroid. Other properties of hypercycles of semihypergraphs probably are similar to the properties of cycles in graphs as well.

"Normal" hypergraphs (as defined in [**3**]–[**5**]) are a generalization of semihypergraphs. If the distinction between subsets Q and R is removed, then we have hypergraphs.

CONJECTURE. Properties of hypercycles in hypergraphs will remain symmetric to the properties of cycles in graphs and semihypergraphs if a cycle in a hypergraph is defined as a subgraph of a hypergraph in which the set of vertices belonging to an edge E_i can be subdivided into two subsets Q_i and $R_i, Q_i \cup R_i = E_i$ and $Q_i \cap R_i = \emptyset$, such that the resulting semihypergraph is a cycle.

Let us consider a hypergraph, which is a simple hypercycle (according to the above definition):

$$\mathcal{D}\{(1\,2\,3), (2\,4), (1\,3\,4)\}$$

Then, only one semihypergraph for $\mathcal{D}$ exists:

$$\mathcal{S}\{(2|1\,3), (1\,3|4), (4|2)\}$$

which is a hypercycle. Therefore, $\gamma(\mathcal{H}) = 1$; $\theta = 1$; the number of hypervertices is equal to three: $(2), (1\,3)$, and (4). The number of hyperedges is also equal to three, and the Euler formula is fulfilled.

Clearly, if each hyperedge consists of two vertices, then (i) the hypergraph reduces to the ordinary graph, (ii) hyperedges are edges, and (iii) hypervertices are vertices since only one subdivision of each edge is possible. In this case, formula (4.2) reduces to formula (4.1).

Probably, paths, walks, and other invariants can be defined for semihypergraphs and hypergraphs by analogy.

5. Immediate Result for Directed Hypergraphs Following from Linear Algebra [1]

The incidence matrix of a directed hypergraph $\mathcal{D}(V, E)$ is a $|V| \times |E|$ matrix $\mathbf{B}$ whose element b_{ij} is positive if vertex v_i belongs to the head of E_j and negative if it belongs to the tail of E_j; in other cases, $b_{ij} = 0$. The absolute value of b_{ij} is equal to the number of times the vertex is involved in E_j. Then, clearly, the maximum number of linearly independent hypercircuits is

$$\gamma(\mathcal{D})_{\max} = |E(\mathcal{D})| - \text{rank}\mathbf{B}.$$

and the directed hypergraph is circuit-free if $|E(\mathcal{D})| = \text{rank}\mathbf{B}$. This is a sufficient but not necessary condition.

6. Open Problems Relevant to Applications

PROBLEM 1. Is there a (reasonably) simple graph-theoretical criterion (rather than an algorithm) that will allow one to define without enumerating routes whether all steps/hyperedges or species/vertices of a mechanism are involved in routes/hypercircuits.

PROBLEM 2. All cycles in graphs differ only in their length. Circuits in directed hypergraphs may also differ in their structures, which can be expressed, for example, in terms of directed bipartite graphs. An interesting problem to solve is: Given the number of species (vertices) define how many types of hypercircuits may exist.

PROBLEM 3. Define the criterion that says if a directed hypergraph contains a hypercircuit.

References

[1] O.N. Temkin, A.V. Zeigarnik, and D. Bonchev, *Chemical Reaction Networks: A Graph-Theoretical Approach*, CRC Press, Boca Raton, FL, 1996.
[2] G. Gallo, G. Longo, S. Pallottino, and S. Nguyen, Directed Hypergraphs and Applications, *Discrete Appl. Math.*, **42** (1993) 177–201.
[3] C. Berge, *Hypergraphs: Combinatorics of Finite Sets*, North Holland, Amsterdam, 1989.
[4] M. Lewin, On Hypergraphs without Significant Cycles, *J. Combinatorial Theory*, **Sec. B. 20** (1976) 80–83.
[5] B.D. Acharya, On the Cyclomatic Number of a Hypergraph, *Discrete Appl. Math.*, **27** (1979) 111–116.
[6] J. Horiuti, and T. Nakamura, On the Theory of Heterogeneous Catalysis, *Adv. Catal.*, **17** (1967) 1–74.
[7] F. Harary, *Graph Theory*, Addison-Wesley, Reading, MA, 1969.

LABORATORY OF CHEMICAL KINETICS AND CATALYSIS, LOMONOSOV ACADEMY OF FINE CHEMICAL TECHNOLOGY, MOSCOW, 117571, RUSSIA
E-mail address: azeigarn@mitht.rssi.ru

DIMACS Series in Discrete Mathematics
and Theoretical Computer Science
Volume 51, 2000

On Maximal Energy Ordering of Acyclic Conjugated Molecules

Fuji Zhang and Huaien Li

ABSTRACT. The energy of a graph is defined as the sum of the absolute values of all the eigenvalues of the graph. This paper focuses on the trees of conjugated hydrocarbons and using 'quasiordering' gives roughly the first four trees in the class in the decreasing order of their energies.

1. Introduction

Chemists have known that the experimental heats from the formation of conjugated hydrocarbons are closely related to the total π-electron energy. And the calculation of the total energy of all π-electrons in a conjugated hydrocarbon can be reduced (within the framework of HMO approximation) [**1**] to that of

$$E(T) = |\lambda_1| + |\lambda_2| + \cdots + |\lambda_n|,$$

where λ_i are the eigenvalues of the corresponding graph. It is also expressed in the Coulson Integral Formula [**2**] as

$$E(T) = \frac{2}{\pi} \int_0^{+\infty} x^{-2} ln[1 + \sum_{k=1}^{h} m(T,k)x^{2k}]dx,$$

where $m(T,k)$ is the number of k-matchings of T, the graph of the conjugated hydrocarbon. The fact that $E(T)$ is a strictly monotonously increasing function of all matching numbers $m(T,k)$, $k = 0, 1, \cdots, [\frac{n}{2}]$, provides us with a way of comparing the energies of trees. Thus the quasiordering is introduced: if for two graphs G_1 and $G_2, m(G_1,k) \preceq m(G_2,k)$ holds for all $k \geq 0$, we say G_1 is *m-smaller* than G_2 written as $G_1 \preceq G_2$ or $G_2 \succeq G_1$. G_1 and G_2 are *m-equivalent*, written as $G_1 \sim G_2$, if $G_1 \preceq G_2$ and $G_2 \preceq G_1$. If $G_1 \preceq G_2$ but they are not m-equivalent, then $G_1 \prec G_2$. If neither $G_1 \preceq G_2$ nor $G_2 \preceq G_1$, then G_1 and G_2 are said to be *m-incomparable*. By the monotonity of $E(T)$, if $T_1 \preceq T_2$ for two trees T_1 and T_2, then $E(T_1) \leq E(T_2)$, and $E(T_1) < E(T_2)$ if $T_1 \prec T_2$. For the study of quasiordering a number of results have been reached [**3**]–[**8**]. A typical one, [**4**] which we frequently use, is that if we denote by $G \cup H$ the graph whose components are G and H, then

1991 *Mathematics Subject Classification.* 05C90.

$$P_l \succeq P_2 \cup P_{l-2} \succeq P_4 \cup P_{l-4} \succeq \cdots \succeq P_{2k} \cup P_{l-2k} \succeq P_{2k+1} \cup P_{l-2k-1} \succeq$$
$$\succeq P_{2k-1} \cup P_{l-2k+1} \succeq \cdots \succeq P_3 \cup P_{l-3} \succeq P_1 \cup P_{l-1},$$

where P_i is the path with i vertices and $l = 4k + r$, $0 \leq r \leq 3$.

For the case of maximal energy, Ivan Gutman and Li Qiao have proved that the tree of the largest energy is the path P_n. Let ${}^lT_b^r$ denote the T shaped graph shown in the following figure (Fig. 2).

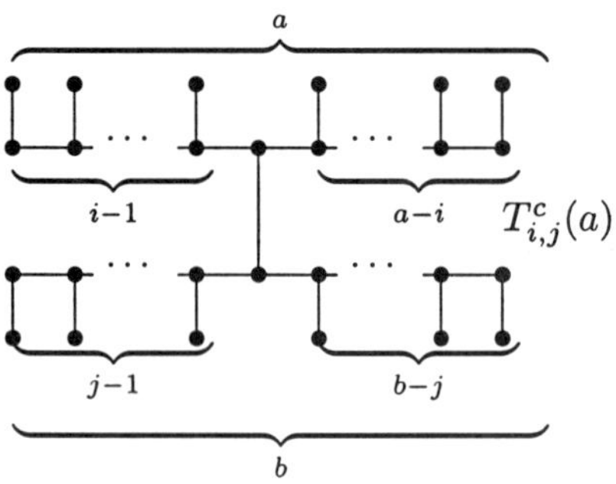

FIGURE 1. $T_{i,j}^c(a)$, where $a + b = c = \frac{n}{2} + 1$.

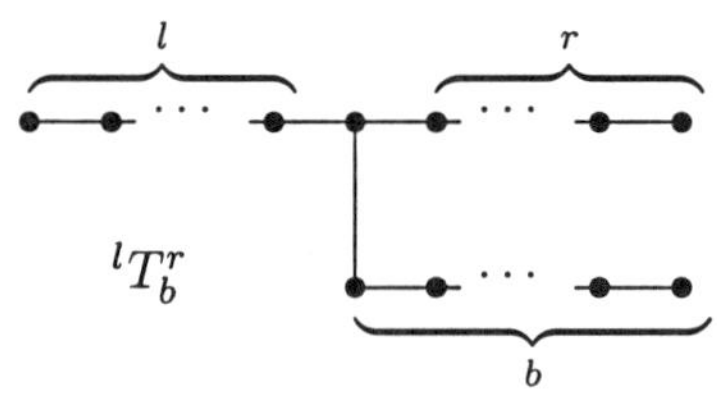

FIGURE 2. ${}^lT_b^r$, where $l + r + b = n - 1$.

I. Gutman proved that ${}^2T_{n-5}^2$ is the graph of the second largest energy in the class of trees with a perfect matching. Fuji Zhang and Lai Zaikang pointed out that Hasse graph can be constructed through quasiordering. However this graph is not a lattice nor planar. In addition, for the case of the smallest energy, the present authors have given the first and second smallest trees and shown that the third one is among two trees which quasiordering fails to compare [**9**]. And curiously Huaien Li has given roughly the first $n/2$ trees of n vertices in the increasing order of their energies in the class of trees with a perfect matching whose degrees do not exceed three [**10**], the first $n/2$ trees are of the form shown in Fig. 1. And the sequence is still open-ended. The difficulties of such energy comparison lies in that we are now deciding the precise positions of trees in the energy ordering in a certain class while before only totally-ordered subsets are discussed. The quasiordering, as shown above, doesn't seem so effective. To solve the problem completely, we should seek other methods. Denote by Φ_n the class of trees with n vertices which have a perfect matching.

2. Some Totally-Ordered Sets Relative to Maximal Energies in Φ_n

We divide Φ_n into two parts:the first class, denoted by Π_n, includes all the T shaped trees ${}^lT_b^r$ mentioned above with P_n (a special case of ${}^lT_b^r$, when $l = 0$) counted in for convenience, where $l + r + b + 1 = n$; all the rest are included in the second one, denoted by Γ_n. Since ${}^lT_b^r$ has a perfect matching, one of l, r and b, say b, must be odd. Hence l and r are all even. Assume that $l \leq r$. We have that $0 \leq l \leq 2[n - b - 1\ 4] = t_b$. To begin with, we list some totally-ordered sets. Let ${}^lT_H^r$ be the graph shown in Fig. 3.

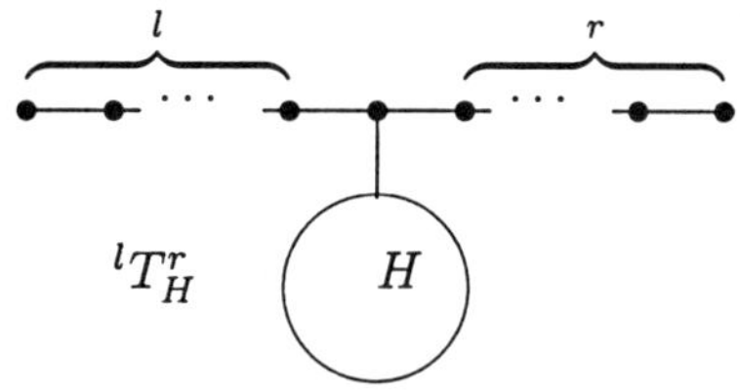

FIGURE 3. ${}^lT^r_H$, where$H$ is an arbitrary graph.

THEOREM 2.1. ${}^0T^s_H \succ {}^2T^{s-2}_H \succ \cdots \succ {}^3T^{s-3}_H \succ {}^1T^{s-1}_H$,
where the series is composed of two parts:the former part is of the form ${}^iT^{s-i}_H$ with even i no greater than $[s/2]$ and the latter part with odd i no greater than $[s/2]$.

PROOF.

$$m({}^iT^{s-i}_H, k) = m(P_s \cup H, k) + m(P_i \cup P_{s-i} \cup H - v, k-1).$$

From the result of I. Gutman in Section 1, we are done. In fact, the quasiorderings are actually strict. See [**10**]. □

Taking $H = P_b$ in Theorem 2.1, we get

COROLLARY 2.2.

$${}^2T^{n-b-3}_b \succ {}^4T^{n-b-5}_b \succ \cdots \succ {}^{t_b}T^{n-b-t_b-1}_b, \quad 1 \leq b \leq n-5.$$

COROLLARY 2.3.

$${}^lT^{n-l-2}_1 \succ {}^lT^{n-l-4}_3 \succ \cdots \succ {}^lT^2_{n-l-3}, 2 \leq l \leq n-4.$$

From the two corollaries, we have immediately

THEOREM 2.4.

$${}^2T^2_{n-5} \succ {}^2T^4_{n-7} \succ {}^2T^6_{n-9} \succ T \in \Pi_n,$$

where $T \neq {}^2T^2_{n-5}, {}^2T^4_{n-7}, {}^2T^6_{n-9}$.

PROOF. The first two quasiorderings are straightforward from Corollary 2.3. For the third one, let $T = {}^l T^r_b$. If $l > 2$, then from Corollary 2.3 $T \preceq {}^{8-l}T^l_{n-9}$. And by Corollary 2.2,

$$T = {}^l T^r_b \preceq {}^{8-l}T^l_{n-9} \preceq {}^2T^6_{n-9}.$$

Hence ${}^2T^6_{n-9} \succ T$, since ${}^2T^6_{n-9} \neq T$.
If $l = 2$, then by Corollary 2.3,

$$T = {}^2 T^r_b \prec {}^2T^6_{n-9}.$$

□

3. Cases in Γ_n $(n > 10)$

By the two corollaries and Theorem 2.4, we have got the first three trees in the decreasing order in Π_n, that is, ${}^2T^2_{n-5} \succ {}^2T^4_{n-7} \succ {}^2T^6_{n-9}$. Now we compare them with the trees in $\Gamma_n = \Phi_n - \Pi_n$. We first give a result.
Suppose G has a vertex v of degree $d(v) \geq 3$ and branches out into two paths P_a and P_b at v. Then moving one of the paths along the other to its pendant vertex, the resulting graph G' has greater energy (see Fig. 4).

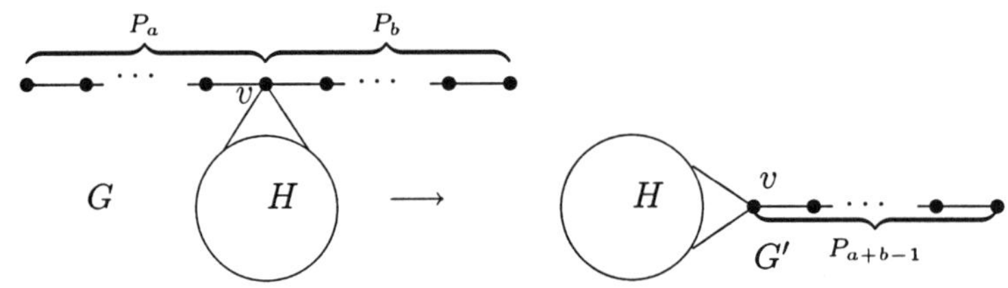

FIGURE 4. Transform G into G'.

That is

LEMMA 3.1. $G' \succ G$.

PROOF. We count the differences of the numbers of their k-matchings.

$m(G, k) = m\Big($ 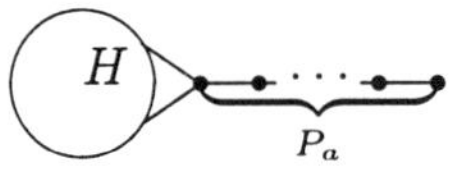 $\cup P_{b-1}, k\Big) + m(H \cup P_{a-1} \cup P_{b-2}, k-1)$

$m(G', k) = m\Big($ 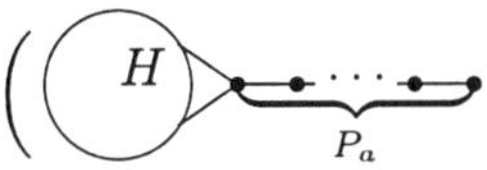 $\cup P_{b-1}, k\Big)$

$+m\Big($ 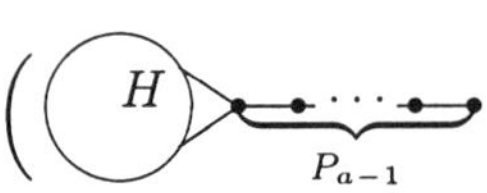 $\cup P_{b-2}, k-1\Big)$

Thus

$m(G', k) - m(G, k)$

$= m\Big($ 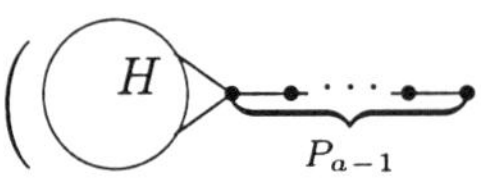 $\cup P_{b-2}, k-1\Big) - m(H \cup P_{a-1} \cup P_{b-2}, k-1)$

$\geq 0.$

And when $k = 2$, $m(G', k) - m(G, k) > 0$. □

Let $T \in \Gamma_n$. By the definition of Γ_n, it either has a vertex v with degree > 3, or has two vertices u and v with degrees ≥ 3.
We can use the above Lemma to transform T into the shape of T_1 or T_2 shown in Fig. 5. By the above Lemma, we have either $T \preceq T_1$ or $T \preceq T_2$. We note that the transformation in the Lemma guarantees that the new graph has a perfect matching. In fact, if wv is an edge of the perfect matching, then when $w \in H$, both a and b are odd. Hence the fact that $a+b-1$ is odd implies G' has a perfect matching.Similarly when $w \notin H$, a and b must be an odd and an even, and $a+b-1$ is even. Thus G' has a perfect matching. We denote $T_n - T_1$ if $a_1 = a_2 = a_3 = a_4 = 2$

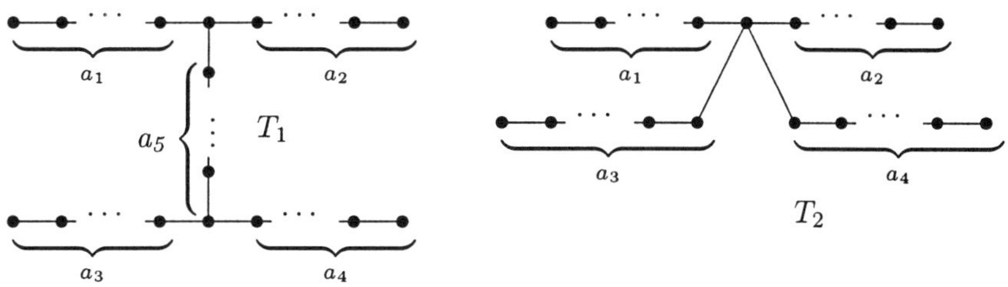

FIGURE 5. T_1 and T_2.

First suppose $T \preceq T_2$. (In Fig. 5, we suppose $a_i \geq 1, i = 1, 2, 3, 4, 5$).

THEOREM 3.2. *$T_2 \prec I_n$ or $T_2 \prec^2 T_{n-9}^6$.*

PROOF. Since T_2 has a perfect matching, one and only one of a_1, a_2, a_3, a_4, say a_1, is odd.
Case 1. One of $a_2, a_3, a_4 \geq 4$, Let $a_2 \geq 4$. By Corollary 2.3 and the Lemma,

$$T_2 \prec {}^{a_3}T_{a_1}^{a_2+a_4} \preceq {}^2T_{n-a_2-a_4-3}^{a_2+a_4} \preceq {}^2T_{n-9}^6.$$

Case 2. $a_2 = a_3 = a_4 = 2$. Then we have

$$\begin{aligned} & m(I_n, k) - m(T_2, k) \\ = \; & [m({}^2T_{n-7}^2 \cup P_2, k) + m({}^2T_{n-10}^2 \cup P_2, k-1)] \\ & -[m({}^2T_{n-7}^2 \cup P_2, k) + m(P_2 \cup P_2 \cup P_{n-7}, k-1)] \\ = \; & [m(P_2 \cup P_2 \cup P_{n-7}, k-1) + m(P_2 \cup P_{n-10}, k-2)] \\ & -m(P_2 \cup P_2 \cup P_{n-7}, k-1) \\ = \; & m(P_2 \cup P_{n-10}, k-2) \geq 0 \end{aligned}$$

when $k = 2$, $m(I_n, k) - m(T_2, k) > 0$, which implies $T_2 \prec I_n$. □

Now let $T \preceq T_1$.

THEOREM 3.3. *$T_1 \prec I_n$ or $T_1 \prec^2 T_{n-9}^6$, or $T_1 = I_n$.*

PROOF. Case 1. One of $a_i (i = 1, 2, 3, 4)$, say a_1, is odd. Then by the Lemma, $T_1 \prec {}^{a_2}T_{a_1}^{a_3+a_4+a_5}$. Since T_1 has a perfect matching, both a_2 and $a_3 + a_4 + a_5$ are even. And $a_3 + a_4 + a_5 \geq 3$ implies $a_3 + a_4 + a_5 \geq 4$. By the same reason, $a_2 \geq 2$. If $a_3 + a_4 + a_5 = 4$, then one of a_3 and a_4 must be 1. Say $a_3 = 1$. Hence $a_1 + a_2 + a_4 + a_5 = n - 2 > 10 - 2 = 8$. Consequently

$$T_1 \prec {}^{a_4}T_1^{a_2+a_1+a_5} \preceq {}^2T_{n-9}^6.$$

Case 2. a_1, a_2, a_3, a_4 are all even and one of $a_1 + a_2$ and $a_3 + a_4$ is no less than 8, let $a_1 + a_2 \geq 8$. By the Lemma and Theorem 2.4,

$$T_1 \prec {}^{a_1}T_{n-a_1-a_2-1}^{a_2} \preceq {}^2T_{n-9}^6.$$

Case 3. $a_1 = a_2 = a_3 = 2$, $a_4 = 4$. Then

$$\begin{aligned} & m(I_n, k) - m(T_1, k) \\ &\quad = [m(^2T^2_{n-7} \cup P_2, k) + m(^2T^2_{n-10} \cup P_2, k-1)] \\ &\qquad -[m(^2T^4_{n-9} \cup P_2, k) + m(^2T^4_{n-12} \cup P_2, k-1)] \\ &\quad \geq m(^2T^2_{n-10} \cup P_2, k-1) - m(^2T^4_{n-12} \cup P_2, k-1) \\ &\quad \geq 0. \text{(Theorem 2.4)} \end{aligned}$$

From Theorem 2.4, there must be a k such that the above inequation is sharp. Thus $T_1 \prec I_n$.

Case 4. $a_2 = a_3 = 2$, $a_1 = a_4 = 4$ ($n \geq 14$). In this case, we write the T_1 in Case 3 as T_1'. Then

$$\begin{aligned} & m(T_1', k) - m(T_1, k) \\ &\quad = [m(^2T^2_{n-7} \cup P_2, k) + m(^2T^2_{n-12} \cup P_4, k-1)] \\ &\qquad -[m(^2T^4_{n-9} \cup P_2, k) + m(^2T^4_{n-14} \cup P_4, k-1)] \\ &\quad \geq m(^2T^2_{n-12} \cup P_4, k-1) - m(^2T^4_{n-14} \cup P_4, k-1) \\ &\quad \geq 0. \text{(Theorem 2.4)} \end{aligned}$$

And from Theorem 2.4, there must be a k such that the above inequation is sharp. Thus $T_1 \prec T_1' \prec I_n$. □

THEOREM 3.4. $I_n \prec {}^2T^4_{n-7}$.

PROOF.

$$\begin{aligned} & m(^2T^4_{n-7}, k) \\ &\quad = m(P_2 \cup P_{n-2}, k) + m(P_4 \cup P_{n-7}, k-1) \\ &\quad = [m(P_2 \cup P_2 \cup P_{n-4}, k) + m(P_2 \cup P_{n-5}, k-1) \\ &\qquad + m(P_2 \cup P_2 \cup P_{n-7}, k-1) + m(P_{n-7}, k-2)] \\ &\quad = m(P_2 \cup P_2 \cup P_{n-4}, k) + [m(P_2 \cup P_2 \cup P_{n-7}, k-1) \\ &\qquad + m(P_2 \cup P_{n-8}, k-2)] \\ &\qquad + m(P_2 \cup P_2 \cup P_{n-7}, k-1) + m(P_{n-7}, k-2) \\ &\quad = m(P_2 \cup P_2 \cup P_{n-4}, k) + m(P_2 \cup P_2 \cup P_{n-7}, k-1) \\ &\qquad + m(P_2 \cup P_2 \cup P_{n-10}, k-2) + m(P_2 \cup P_{n-11}, k-3) \\ &\qquad + m(P_2 \cup P_2 \cup P_{n-7}, k-1) + m(P_{n-7}, k-2), \\ &\quad\ \ m(I_n, k) \\ &\quad = m(P_2 \cup {}^2T^2_{n-7}, k) + m(P_2 \cup {}^2T^2_{n-10}, k-1) \\ &\quad = [m(P_2 \cup P_2 \cup P_{n-4}, k) + m(P_2 \cup P_2 \cup P_{n-7}, k-1)] \\ &\qquad + [m(P_2 \cup P_2 \cup P_{n-7}, k-1) + m(P_2 \cup P_2 \cup P_{n-10}, k-2)]. \end{aligned}$$

Thus

$$m(^2T^4_{n-7}, k) - m(I_n, k) = m(P_2 \cup P_{n-11}, k-3) + m(P_{n-7}, k-2) \geq 0.$$

When $k = 2$, $m(^2T^4_{n-7}, k) - m(I_n, k) > 0$, hence $I_n \prec {}^2T^4_{n-7}$. □

By the above Theorem 2.4, 3.2, 3.3, 3.4, we get the trees with the first, second and third largest energies are P_n, ${}^2T^2_{n-5}$, ${}^2T^4_{n-7}$. And the tree with the fourth largest energy is either I_n or ${}^2T^6_{n-9}$. However

THEOREM 3.5. *I_n and ${}^2T^6_{n-9}$ are incomparable ($n > 14$).*

PROOF. The difference of k-matchings of the two graphs is

$$\begin{aligned}
& m({}^2T^6_{n-9}, k) - m(I_n, k) \\
&\quad = m(P_2 \cup P_{n-2}, k) - m(P_6 \cup P_{n-9}, k-1) \\
&\qquad -[m(P_2 \cup {}^2T^2_{n-7}, k) + m(P_2 \cup {}^2T^2_{n-10}, k-1)] \\
&\quad = [m(P_2 \cup P_2 \cup P_{n-4}, k) + m(P_2 \cup P_{n-5}, k-1)] \\
&\qquad +m(P_6 \cup P_{n-9}, k-1) \\
&\qquad -[m(P_2 \cup P_2 \cup P_{n-4}, k) + m(P_2 \cup P_2 \cup P_{n-7}, k-1)] \\
&\qquad -[m(P_2 \cup P_2 \cup P_{n-7}, k-1) + m(P_2 \cup P_2 \cup P_{n-10}, k-2)] \\
&\quad = [m(P_2 \cup P_2 \cup P_{n-7}, k-1) + m(P_2 \cup P_{n-8}, k-2)] \\
&\qquad +[m(P_2 \cup P_4 \cup P_{n-9}, k-1) + m(P_3 \cup P_{n-9}, k-2)] \\
&\qquad -2m(P_2 \cup P_2 \cup P_{n-7}, k-1) - m(P_2 \cup P_2 \cup P_{n-10}, k-2) \\
&\quad = [m(P_2 \cup P_2 \cup P_{n-10}, k-2) + m(P_2 \cup P_{n-11}, k-3)] \\
&\qquad +m(P_2 \cup P_4 \cup P_{n-9}, k-1) + m(P_3 \cup P_{n-9}, k-2) \\
&\qquad -m(P_2 \cup P_2 \cup P_{n-7}, k-1) - m(P_2 \cup P_2 \cup P_{n-10}, k-2) \\
&\quad = m(P_2 \cup P_{n-11}, k-3) \\
&\qquad +[m(P_2 \cup P_{n-6}, k-1) + m(P_2 \cup P_2 \cup P_{n-11}, k-3)] \\
&\qquad +[m(P_{n-7}, k-2) + m(P_{n-11}, k-4)] \\
&\qquad -[m(P_2 \cup P_{n-6}, k-1) + m(P_2 \cup P_{n-9}, k-3)] \\
&\quad = m(P_2 \cup P_{n-11}, k-3) + m(P_2 \cup P_2 \cup P_{n-11}, k-3) \\
&\qquad +m(P_{n-7}, k-2) + m(P_{n-11}, k-4) \\
&\qquad -[m(P_2 \cup m(P_2 \cup P_{n-11}, k-3) + m(P_2 \cup P_{n-12}, k-4)] \\
&\quad = m(P_2 \cup P_{n-11}, k-3) + m(P_{n-7}, k-2) + m(P_{n-11}, k-4) \\
&\qquad -[m(P_{n-11}, k-4) + m(P_{n-14}, k-6)] \\
&\quad = m(P_2 \cup P_{n-11}, k-3) + m(P_{n-7}, k-2) - m(P_{n-14}, k-6).
\end{aligned}$$

When $k = 3$, the above expression > 0. When $k = \frac{n}{2} - 1$, $m(P_2 \cup P_{n-11}, k-3) = 0$, $m(P_{n-7}, k-2) = 0$, $m(P_{n-14}, k-6) = 1$, hence the above expression < 0. That is to say, I_n and ${}^2T^6_{n-9}$ are incomparable. □

References

[1] Ivan Gutman and Oskar E. Polansky, Mathematical Concepts in Organic Chemistry, (Springer, Berlin, 1986), pp. 135–142.

[2] Ivan Gutman, Acyclic Conjugated Molecules, Trees and their Energies, *J. Mathematical Chemistry*, **1** (1987) 123–143.

[3] Ivan Gutman, Acyclic Systems with Extremal Hückel π-Electron Energy, *Theoret. Chim. Acta.* (Berl.), **45** (1977) 79–87.

[4] Ivan Gutman and Fuji Zhang, On the Ordering of Graphs with Respect to their Matching Numbers, *Discrete Applied Mathematics*, (North-Holland) **15** (1986) 25–33.

[5] Ivan Gutman and Fuji Zhang, On the Quasiordering of Bipartite Graphs, Publications de L'institut Mathématique, Nouvelle Série, tome 40 (54) (1986) 11–15.

[6] Fuji Zhang and Zaikang Lai, The Totally Ordered Subset in a Sort of Hasse Graphs of Trees, *J. Xinjiang Univ.*, (1983) 12–20.

[7] Fuji Zhang and Zaikang Lai, Three Theorems of Comparison of Trees by their Energy, *Science Exploration*, **3** (1983) 35–36.

[8] Qiao Li and Keqin Feng, Some Results about the Spectral Properties of Graphs, *J. Univ. of Sciences and Technology of China*, (1979) 53–56.

[9] Fuji Zhang and Huaien Li, On Acyclic Conjugated Molecules with Minimal Energies, *Discrete Applied Mathematics*, 92 (1999) 71–84.

[10] Huaien Li, On Minimal Energy Ordering of Acyclic Conjugated Molecules, J Mathematical Chemistry, 25 (1999) 145–169.

DEPARTMENT OF MATHEMATICS, XIAMEN UNIVERSITY, XIAMEN, FUJIAN 361005, P.R.CHINA
E-mail address: fjzhang@jingxian.xmu.edu.cn

DEPARTMENT OF MATHEMATICS, ZHENGZHOU UNIVERSITY, ZHENGZHOU, HENAN 450052, P.R. CHINA
E-mail address: huaienli1@263.net